# 21st Surface Preparation and Cleaning Conference (SPCC 2019)



Portland, Oregon, USA
2 - 3 April 2019

ISBN: 978-1-5108-8502-8

**Printed from e-media with permission by:**

Curran Associates, Inc.
57 Morehouse Lane
Red Hook, NY 12571

**Some format issues inherent in the e-media version may also appear in this print version.**



Printed by Curran Associates, Inc. (2019)



Surface Preparation and Cleaning Conference (SPCC)
c/o Mark Thirsk
Linx Consulting-PO Box 384
Mendon, MA 01756-0384
USA

Phone: 617-273-8837

mthirsk@linx-consulting.com

**Additional copies of this publication are available from:**

Curran Associates, Inc.
57 Morehouse Lane
Red Hook, NY 12571 USA
Phone: 845-758-0400
Fax: 845-758-2633
Email: curran@proceedings.com
Web: www.proceedings.com

# TABLE OF CONTENTS

## SESSION 1



## SESSION 2



## SESSION 4

## SESSION 5



## POSTER SESSION AND NETWORKING RECEPTION

imec

# COBALT PRE-METALLIZATION CLEAN AND FUNCTIONAL WATER RINSE IN BEOL INTERCONNECT

E. KESTERS*[1], Q.T. LE[1], M. VAN DER VEEN[1], Y. AKANISHI[2], H. IINO[3], A. MIZUTANI[4], AND F. HOLSTEYNS[1]

[1*] Imec, Kapeldreef 75, 3001 Heverlee, Belgium – kesterse@imec.be – +32 16 288609
[2] SCREEN Semiconductor Solutions Co
[3] Kurita Ltd.
[4] Fujifilm Electronic Materials (Europe) N.V.

SPCC 2019, Portland, April 1-3rd, 2019

PUBLIC

## OUTLINE

- BEOL pre-metallization cleaning
  - Introduction
  - Galvanic corrosion study
    - Background
    - Different combinations of metals
      - Effect of HF pre-treatment: role of dissolved oxygen and oxygen ambient atmosphere
  - Co pre-metallization clean and functional water rinse
    - Co corrosion prevention
    - 22 nm HP using Co fill at M1in BEOL: electrical, TEM (EDS)
    - Alternative contact Co metal in MOL: FIB-SEM
- Summary and outlook

imec 2

## BEOL PRE-METALLIZATION CLEAN: INTRODUCTION

- Dielectric etch
- HF pre-treatment → 0.05% HF, various [O], different oxygen atmosphere
  → effect on galvanic corrosion (Co/Cu; Cu/Ru and TiN/Co) and on wafer discharge
- UPW rinse
- Formulated chemistry clean → PER and TiN HM removal & optimization for Co metal clean
- UPW rinse → The use of functional water during rinse to prevent metal corrosion
  → slightly alkaline rinse for better control on material loss
- Solvent rinse
- Dry → Rinse and dry optimization for pattern collapse prevention

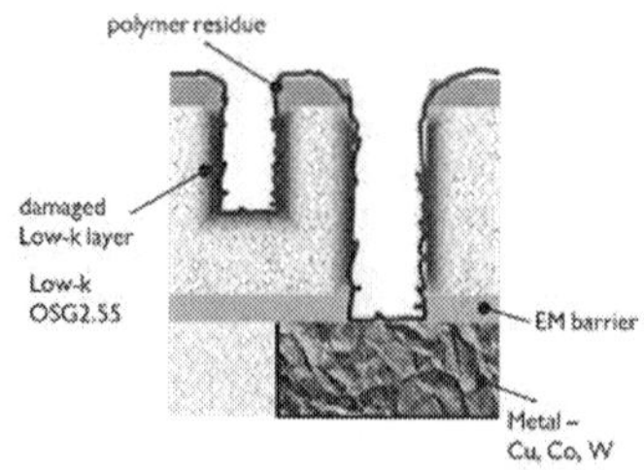


imec

## MATERIAL LOSS PREVENTION DURING WET CLEAN: EFFECT OF HF PRE-TREATMENT ON GALVANIC CORROSION

# BACKGROUND

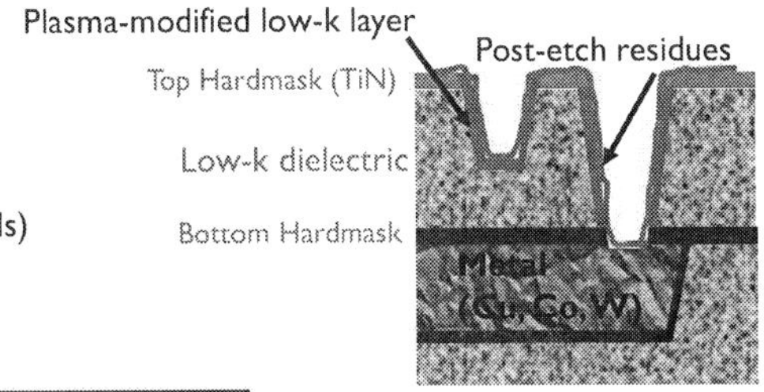


## Pre-metallization clean

- Need for cleaning with minimum etching of exposed metals
- Exposed metal interface (e.g. metal fill and liner/barrier metals) ➔ concern of galvanic corrosion during wet cleaning when bimetal is exposed → causes unwanted metal loss
- Introduction of new interconnect material with barrier/liner

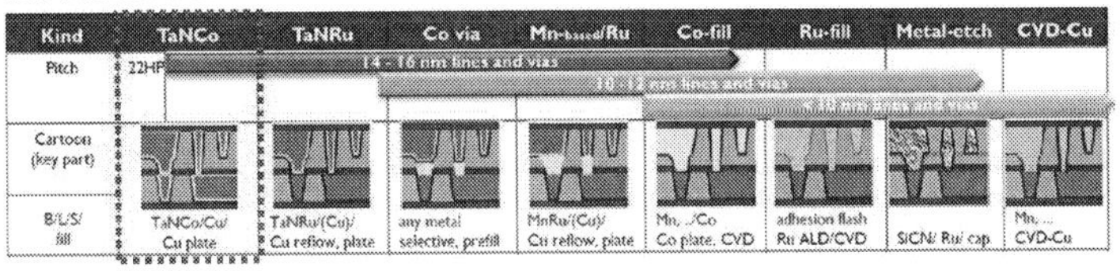


## Objectives

- Investigation & prevention of galvanic corrosion occurrence during wet clean
- Study corrosion characteristics between different metal types
  - Impact of oxygen atmosphere and dissolved oxygen (DO) is investigated in dHF cleaning for different combinations of metals

imec

5

# BACKGROUND

## Galvanic corrosion:

- Large difference of electrode potentials of the metals in contact can cause galvanic corrosion

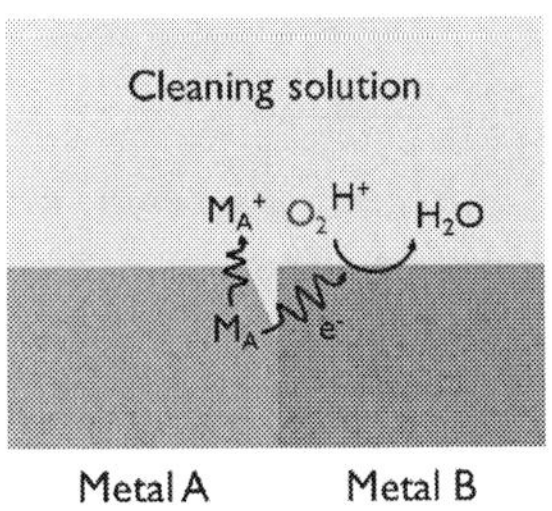


$M_A$ (Electronegative) : Metal dissolution (oxidation)

$M_A$ ➔ $M_A^+ + e^-$

$M_B$ (Electropositive) : Oxygen reduction

$O_2 + 4H^+ + 4e^-$ ➔ $2H_2O$

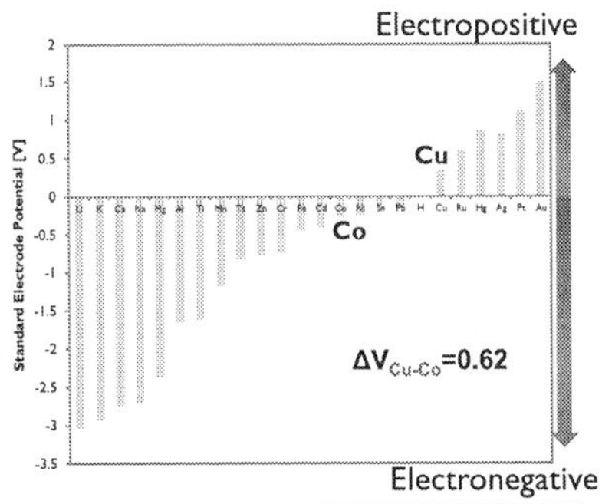


*electron supplied by the "Metal A dissolution reaction", is consumed for the $O_2$ reduction at the Metal B interface*
*→ By eliminating dissolved $O_2$ (DO) from cleaning solution, these reaction is estimated to be suppressed*

imec

6

## GALVANIC CORROSION STUDY: DIFFERENT METAL COMBINATIONS

- Test structure
  - Damascene test vehicle: 45nm HP metal hanging trenches with different barrier/liner

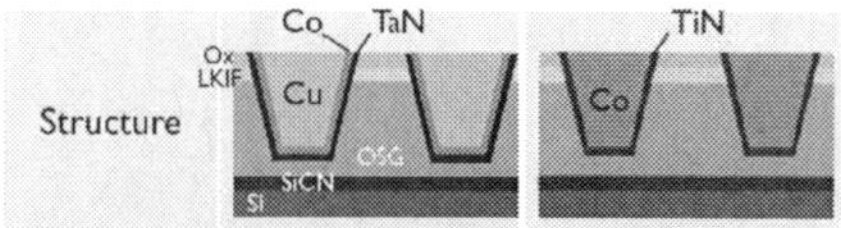


| | Cu/TaN stack | TiN stack |
|---|---|---|
| Fill | Cu | Co |
| Liner | Co 5nm | TiN 3nm |
| Barrier | TaN 5nm | |

**Objectives and experiments**
- Study the impact of dissolved oxygen (DO) on galvanic corrosion at Co/Cu interface
- HF (1:1000, RT) on Screen Single Wafer Cleaning Tool (SU-3200)
  - Variables: dissolved oxygen concentration (30-3000 ppb) and cleaning time (60-180s)

- Test Flow

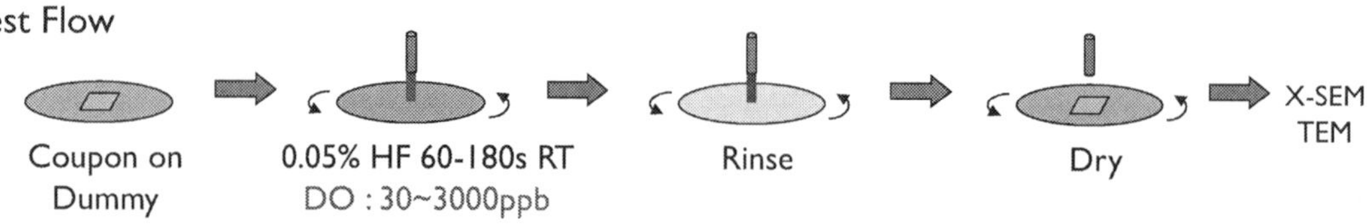


*Ref: Y. Akanishi et al., Solid State Phenomena, 1012-0394, Vol. 282, p.256-262*

imec 7

## EFFECT OF DISSOLVED OXYGEN (DO) ON Co/Cu CORROSION

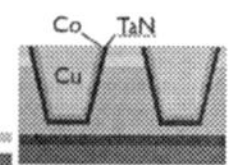


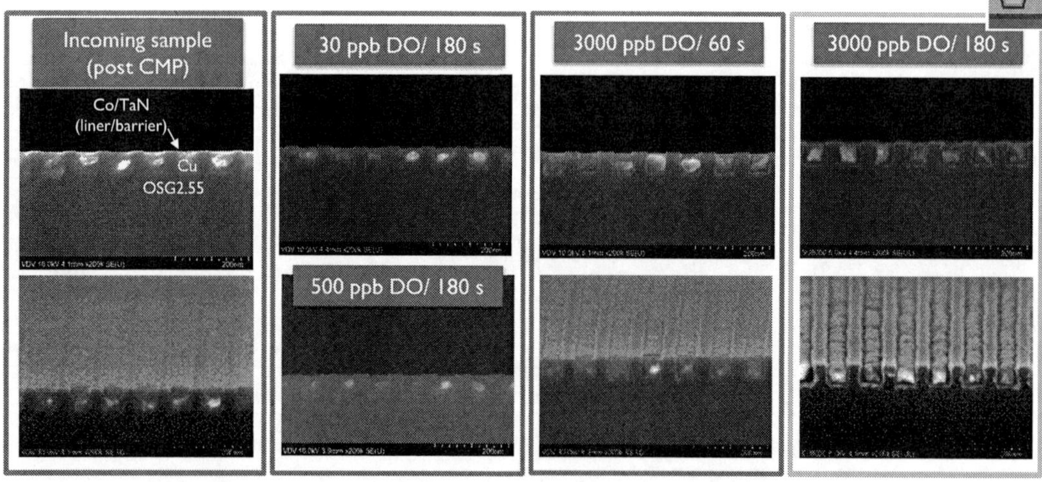


- 30 and 500 ppb of DO: no visible galvanic corrosion observed
- 3000 ppb of DO: clear corrosion occurred at the Cu/Co interface

imec 8

## EFFECT OF DO CONCENTRATION ON Co/Cu CORROSION: TEM RESULTS

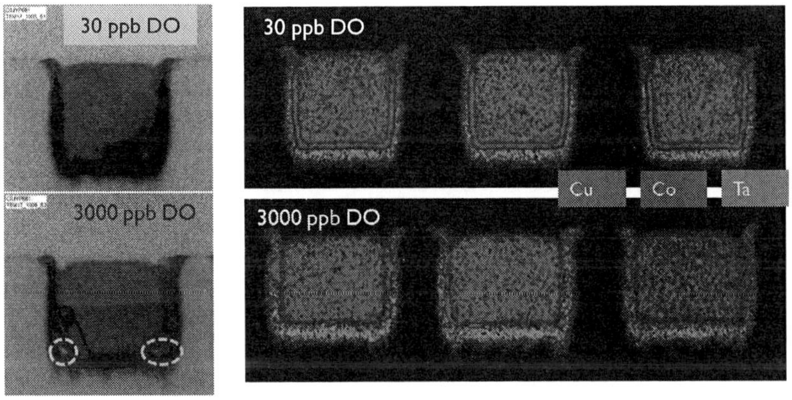


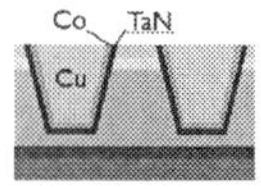


- Galvanic corrosion of Co was observed with higher DO condition, even in very small trenches (~5nm)
  - Chemical etching and galvanic corrosion (both enhanced by dissolved oxygen)
  - By controlling DO in HF, galvanic corrosion of Co can be suppressed
- Cu is also slightly recessed with higher DO condition
  - Chemical etching continuous after Co is lost by galvanic corrosion

imec

9

## CORROSION BEHAVIOR OF TaN/Co/Cu

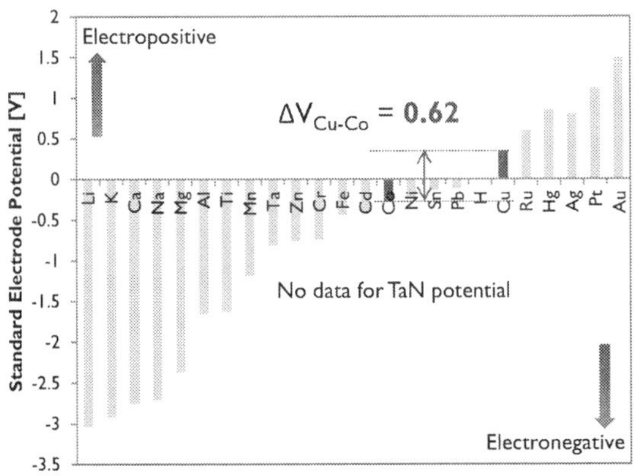


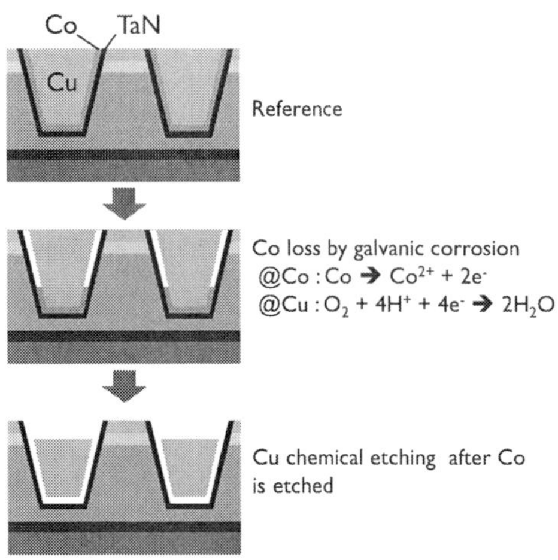


imec

10

## EFFECT OF DISSOLVED OXYGEN (DO) ON TiN/Co CORROSION

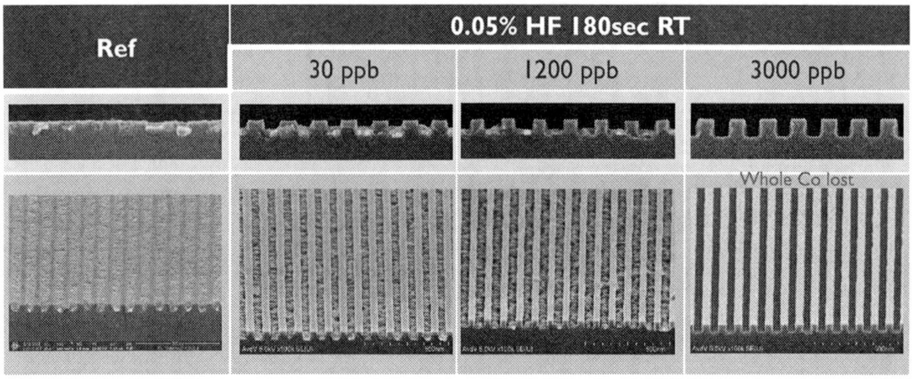


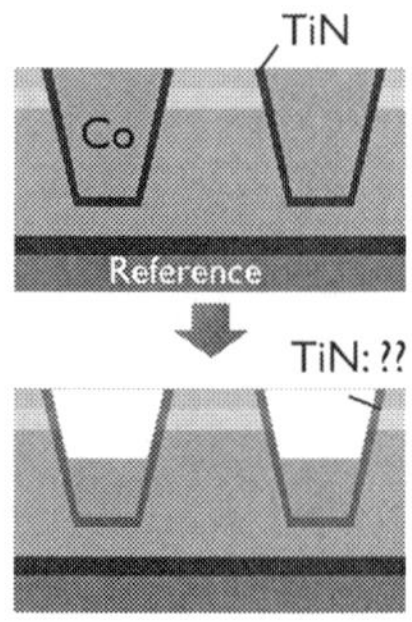


- Incoming surface has already some roughness induced by CMP
- Co recess amount increases as DO increases, Co is completely removed at 3000 ppb
  - Mainly due to chemical etching of HF solution
- Not clear from x-sem if TiN is removed or not

imec 11

## OXYGEN AMBIENT ATMOSPHERE IMPACT ON GALVANIC CORROSION

- Test flow

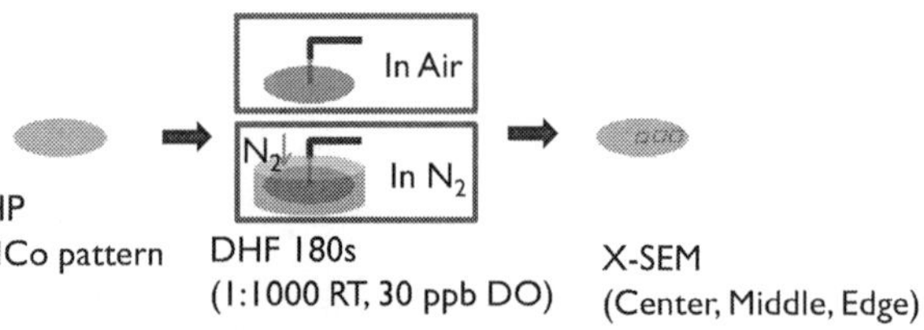


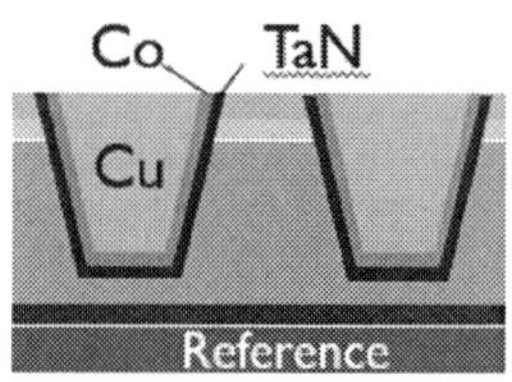


- Cross section

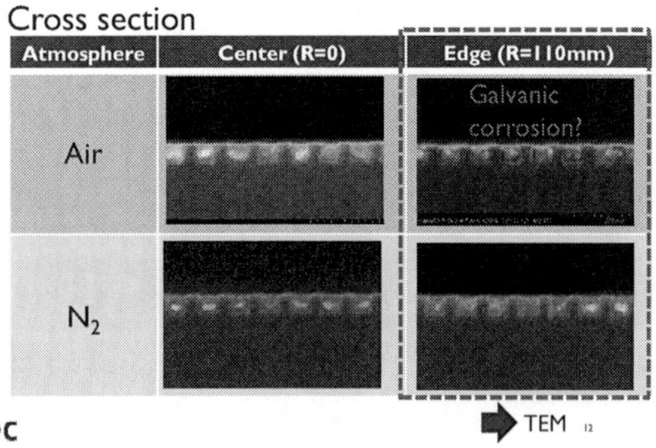


- Co liner (5 nm) dissolved at wafer edge after HF clean in Air atmosphere

Dissolved oxygen concentration & oxygen ambient atmosphere control is needed for galvanic corrosion control

imec 12

## OXYGEN AMBIENT ATMOSPHERE IMPACT (TEM/EDS MAP)

**AIR**

- Void at bottom corner (& partially on sidewall)
- Co signal negligible
- Ta layer diffusion
- Oxide on Cu surface (native oxide)

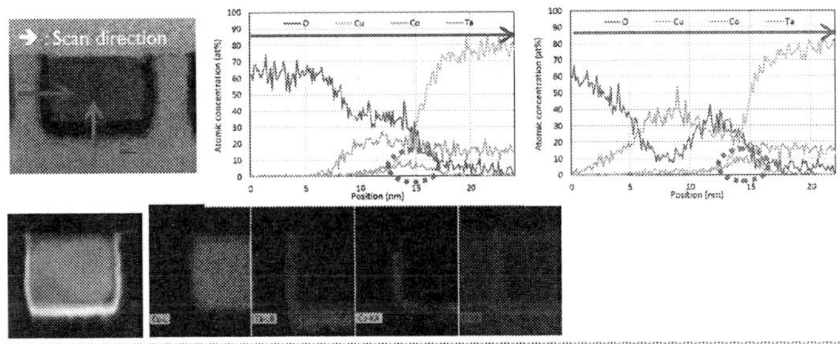


**$N_2$**

- No void or gap between Cu and barrier/liner
- Co liner remains uniformly
- Oxide on Cu surface (native oxide)

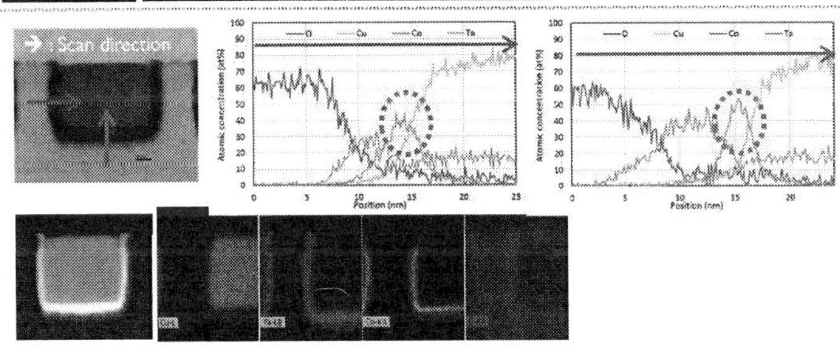


imec

13

## Co PRE-METALLIZATION CLEAN AND FUNCTIONAL WATER RINSE

## Co CORROSION PREVENTION

- For the post-etch residue removal (PERR)
  - FFEM formulated chemistry required to suppress any metal corrosion
- Approach for Co Loss reduction in rinsing water
  - Conventional rinse water ($CO_2W$ / DIW) causes Co corrosion
  - Two approaches assessed to reduce the Co loss during rinse
    - Co loss can be suppressed by pH adjust functional water rinse:
      - $NH_4OH$ water (pH control)
    - Further Co loss reduction is achieved by $H_2O_2$ addition which forms passivation layer on Co surface:
      - $NH_4OH + H_2O_2$ water (Oxidation Reduction Potential - ORP control)

Pourbaix diagram of Co

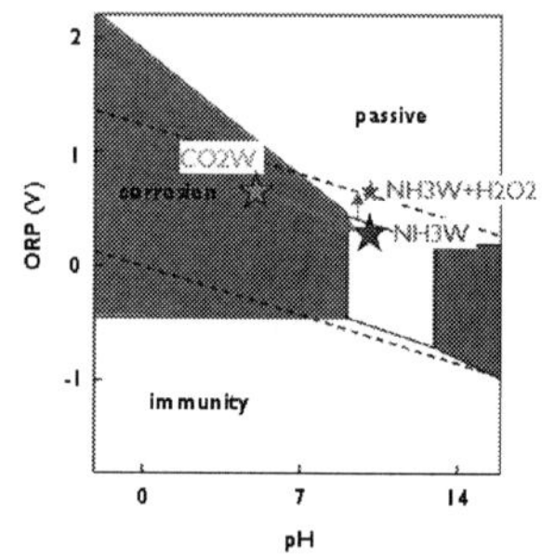


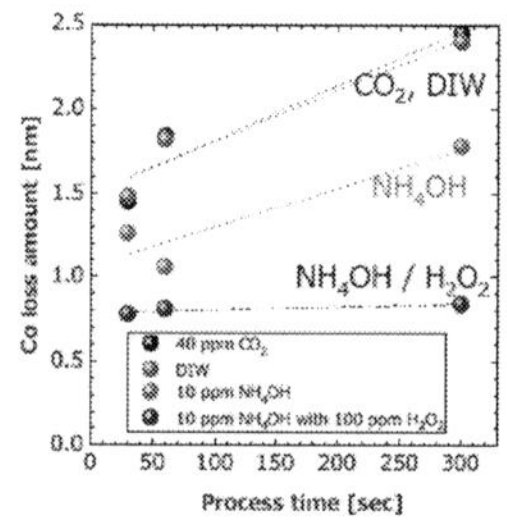


*H. Iino et al., Solid State Phenomena 282, 268-272 (2018)*
*H. Iino et al., SPCC 2019*

imec

## Co PRE-METALLIZATION CLEAN

- Test Structure: 22 nm HP structures
  - M2 Cu & V1 Cu on M1 Co
  - Via resistance (Kelvin via, 25 dies) measured after via clean split performed with different rinse conditions
- Test flow

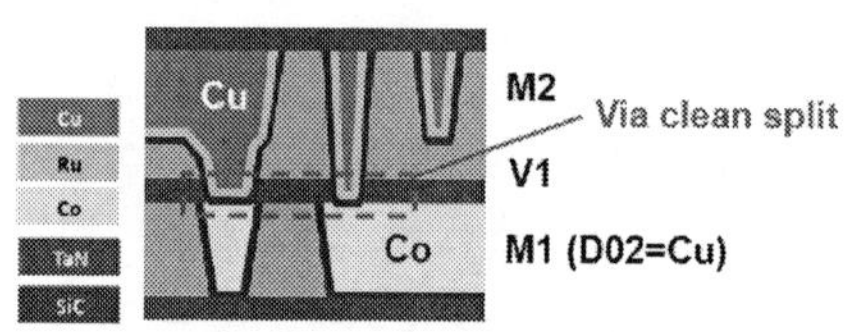


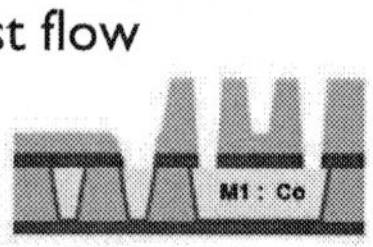


Clean split → V1/M2 metallization →

→ Via resistance measurements and TEM/EDS analysis

| Purpose M2 | Via clean |
|---|---|
| Cu-on-Cu | POR for Cu |
| Co clean test #1 | FFEM PERR chem. + 180s $CO_2W$ rinse (reference) |
| Co clean test #2 | FFEM PERR chem. + 180s $dNH_4OH$ rinse |
| Co clean test #3 | FFEM PERR chem. + 180s $dNH_4OH/H_2O_2$ rinse |
| Co clean test #4 | 20s 0.05%HF (30ppb DO) + PERR chem. + 180s $CO_2W$ rinse |

imec

16

## Co PRE-METALLIZATION CLEAN

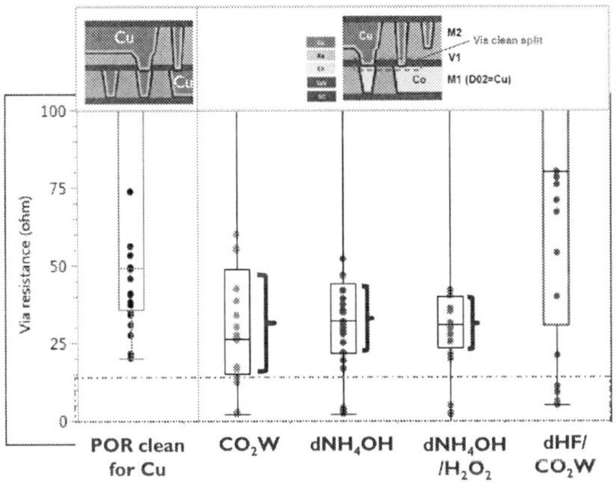


- Via resistance variation reduction confirmed by introducing functional water rinse
  - Variation: ($dNH_4OH/H_2O_2$) < ($dNH_4OH$) < ($CO_2W$)
- Low resistance: caused by shorts (not related to cleaning)

imec

17

## TEM-EDS COMPARISON

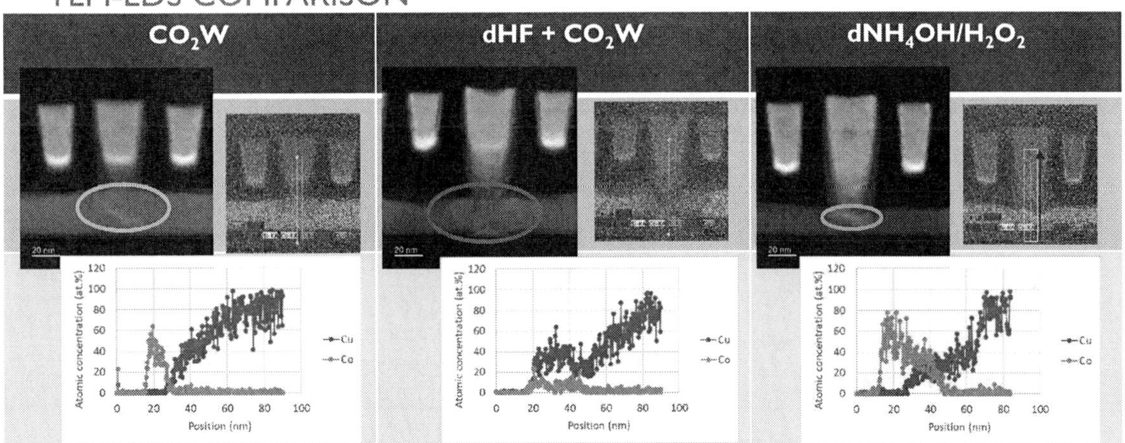


- $CO_2W$ rinse: during via wet clean, the Co trench of M1 is being partially attacked
- dHF +$CO_2W$ rinse: during via wet clean, the entire Co of M1 under the via is removed
  - lateral etch along the Co line is observed and, the etched Co is replaced by Cu during metallization process
- $dNH_4OH/H_2O_2$ rinse: Minor Co etch was observed, confirmation of electrical data

imec

18

## ADDITIONAL RESULT ON Co CLEAN FOR MOL

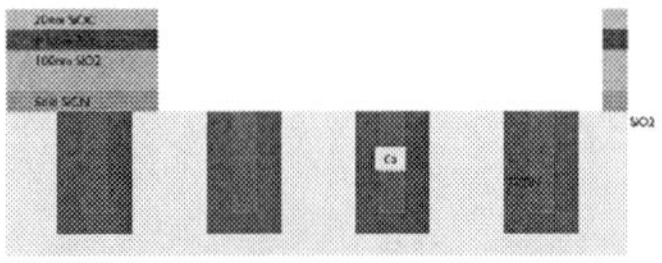

| | Co clean |
|---|---|
| D11 | 10s 0.05%HF + 2 min FFEM PERR chem. + 30s $CO_2W$ |
| D12 | 10s 0.05%HF + 2 min FFEM PERR chem. + 30s $dNH_4OH/H_2O_2$ rinse |
| D13 | 2 min FFEM PERR chem. + 30s $dNH_4OH/H_2O_2$ rinse |
| D14 | 20s 0.05%HF + 2 min FFEM PERR chem. + 30s $dNH_4OH/H_2O_2$ rinse |

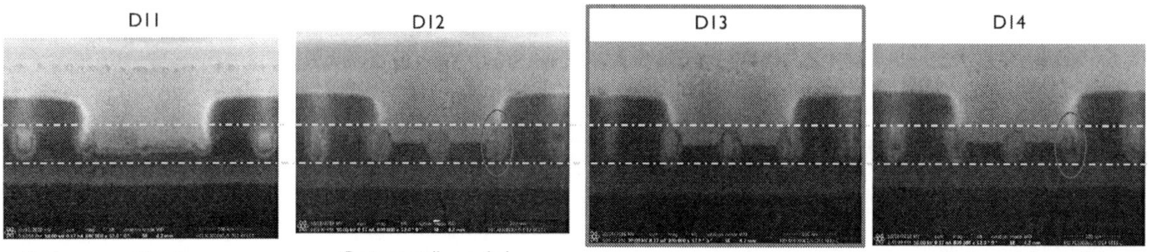


Co is removed by clean — Co is partially eroded — No Co attack observed — Co is partially eroded

imec



## SUMMARY

- Galvanic corrosion behavior of metal in contact with different materials
  - Oxygen reduction in dHF and reduced oxygen ambient atmosphere are mandatory for a good control of galvanic corrosion

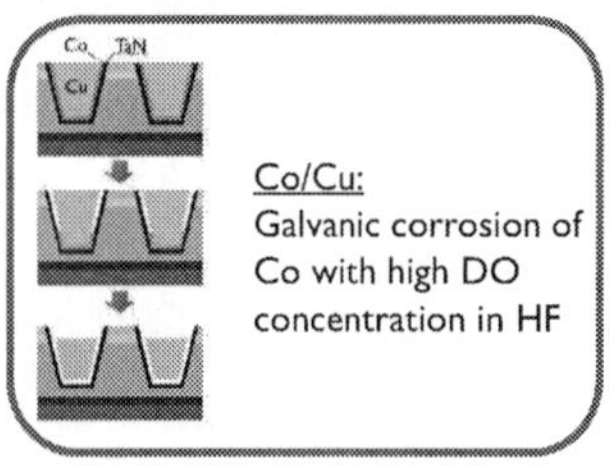


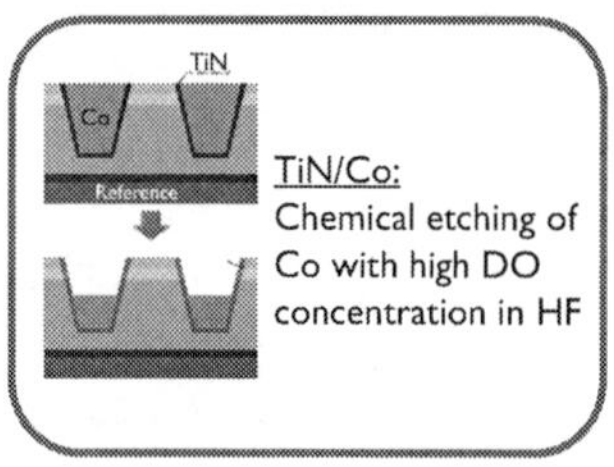


- Co Pre-metallization clean and functional water rinse
  - FFEM PERR chem in combination with functional water rinse: Co clean achieved together with TiN HM removal and without removing ALD TiN liner
  - Less Co wet etching and reduction of via resistance variation by using functional water rinse (Kurita unit)

imec

imec

embracing a better life

imec PUBLIC

imec KURITA

# Functional water solutions to enable wet cleaning process for leading edge semiconductor device manufacturing


*KURITA WATER INDUSTRIES LTD.*

○Hideaki Iino, Yuichi Ogawa, Takeo Fukui hideaki.iino@kurita.co.jp, +81-80-6884-0245

*imec vzw*

Quoc Toan Le, Els Kesters , Yusuke Onìki, Frank Holsteyns




1

## 1. Functional water opportunity for new materials

imec KURITA

**Lanthanoids, alkali earth metals: La, Mg, Ba, ...**

- Dipole for logic multi-Vt RMG and DRAM peri high thermal stability gate stack
- → water soluble

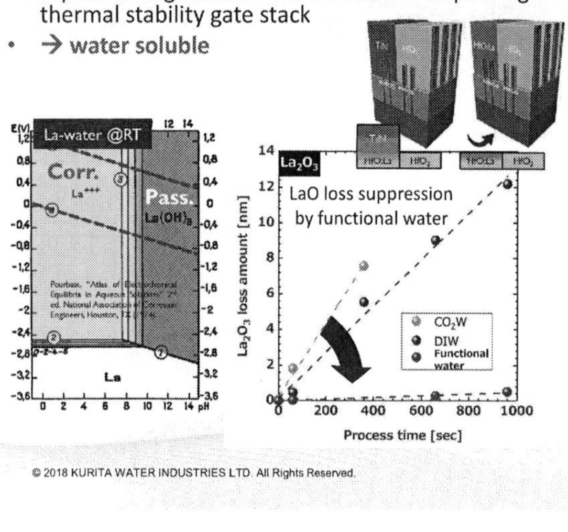


**3d transition metals: Co, Cu, Ni, ...**

- New materials for contact and interconnect
- → **easy to corrode**

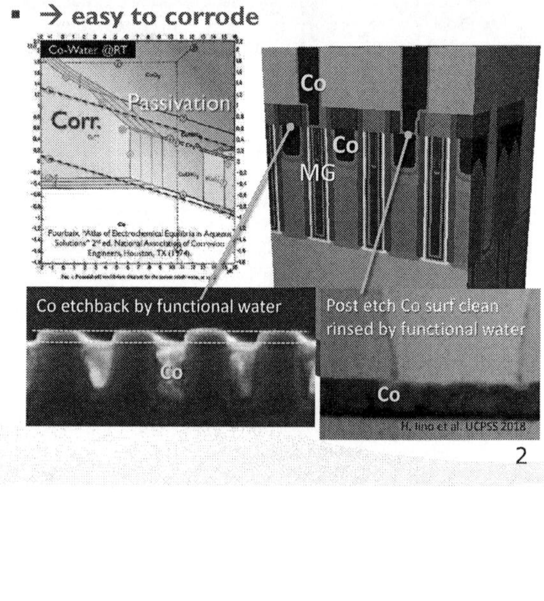




2

## Agenda

imec KURITA

1. Introduction
2. $La_2O_3$ dissolution and its suppression
3. Cobalt corrosion and passivation
4. Atomic scale etching of Cobalt
5. Summary

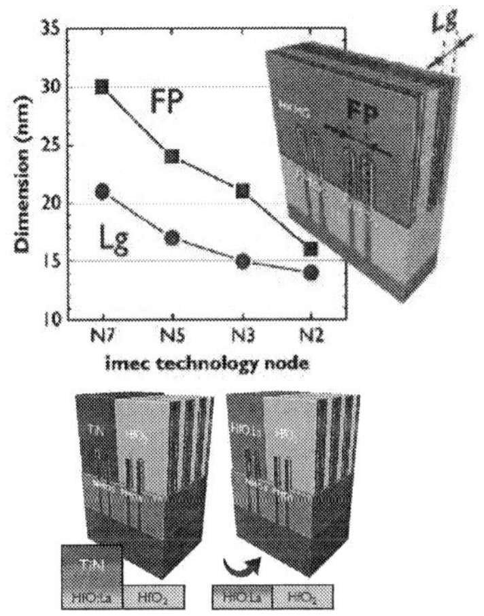


Y.Oniki (imec), SEMICON KOREA 2019





## $La_2O_3$ dissolution and its suppression

imec KURITA

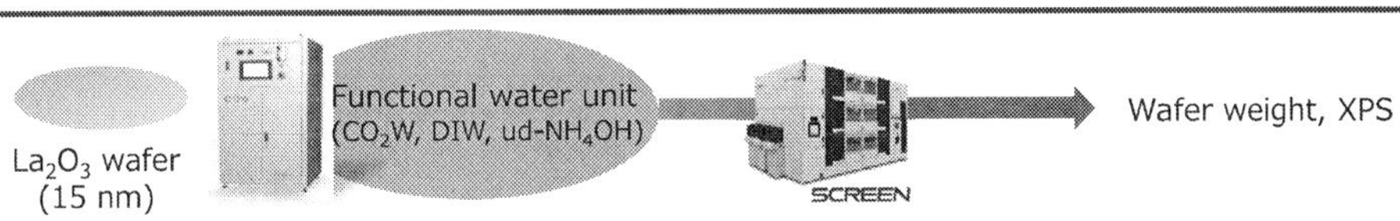

## $La_2O_3$ dissolution and its suppression

imec KURITA

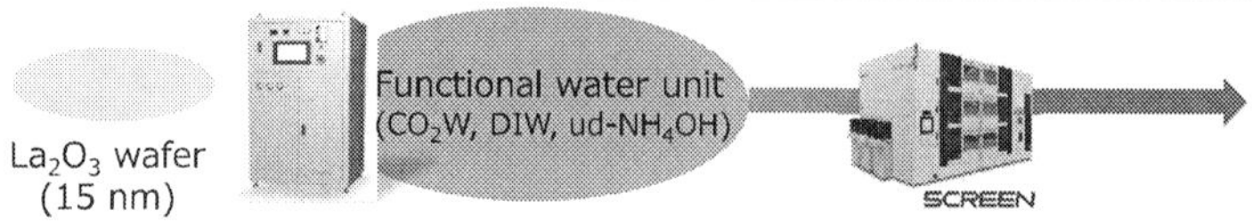


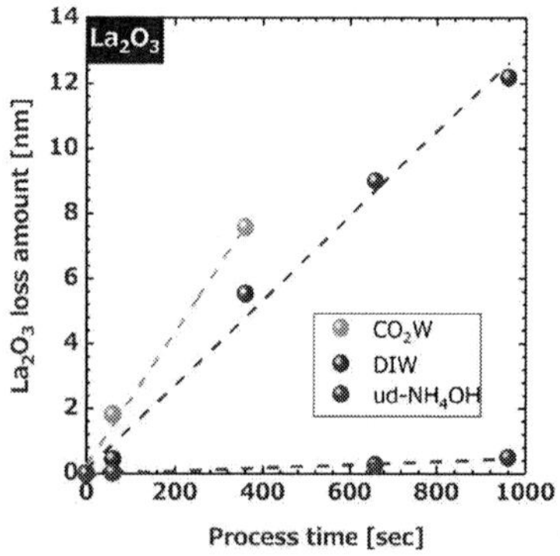


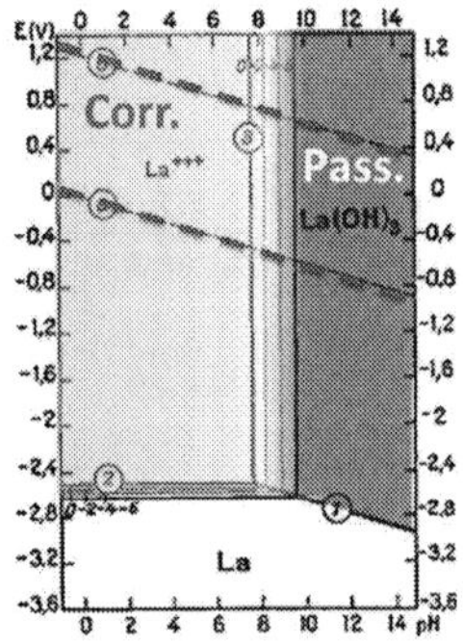


$La_2O_3 + 6H^+ \rightarrow 2La_3^+ + 3H_2O$ ($CO_2W$, DIW)

$La_2O_3 + 3H_2O \rightarrow 2La(OH)_3$ (ud-$NH_4OH$)



- $La_2O_3$ dissolution; $La_2O_3$ E/R < 1 Å/min. @ud-$NH_4OH$

5

## $La_2O_3$ dissolution and its suppression

imec KURITA

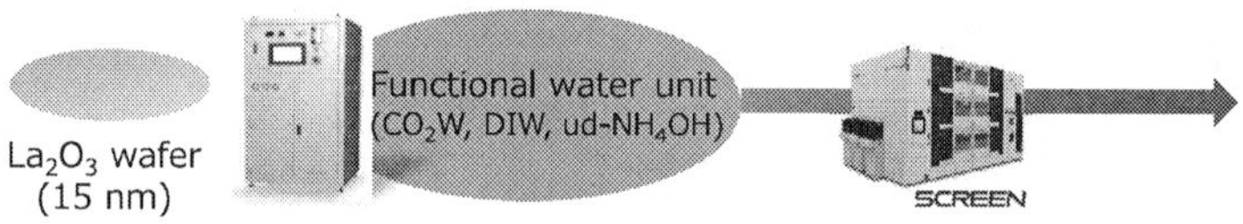


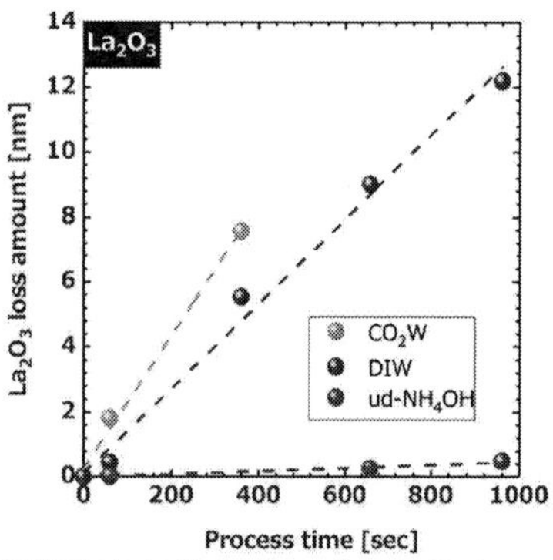


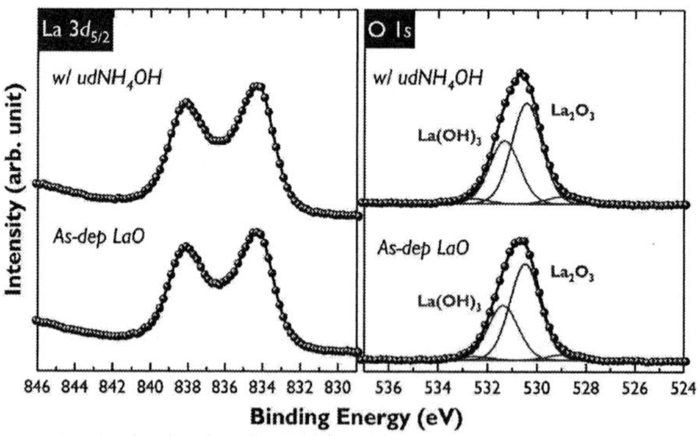




- Passivation layer effectively suppresses $La_2O_3$ dissolution

## $La_2O_3$ dissolution and its suppression

imec KURITA

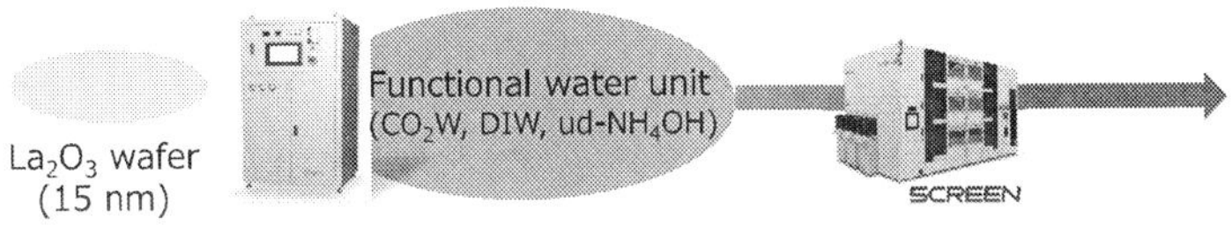


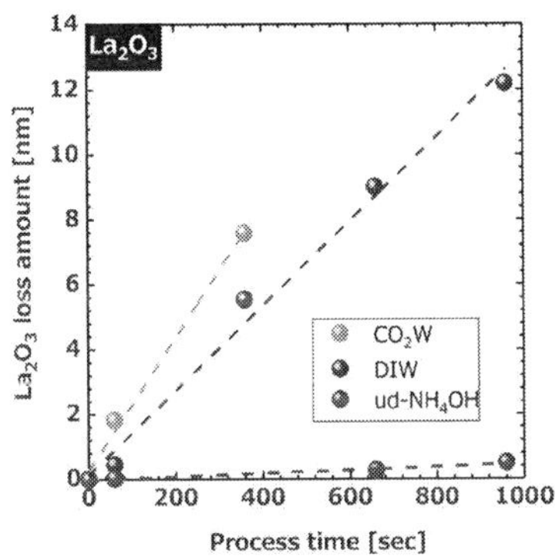


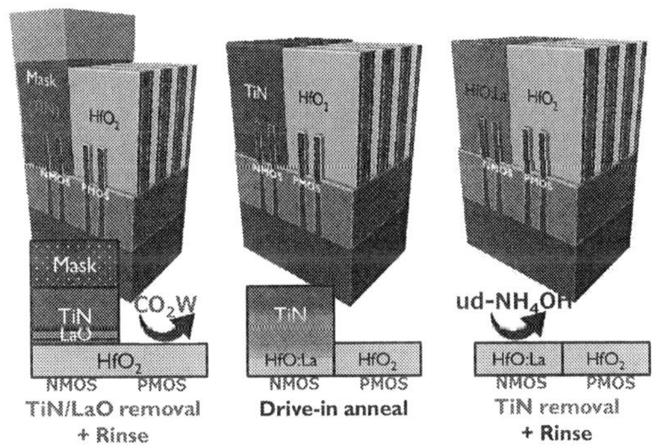


- Regarding the RMG $La_2O_3$ patterning

PMOS : after TiN removal rinse by $CO_2W$
NMOS : after TiN removal rinse by ud-$NH_4OH$



7

## Agenda

imec KURITA

1. Introduction
2. $La_2O_3$ dissolution and its suppression
3. Cobalt corrosion and passivation
4. Atomic scale etching of Cobalt
5. Summary

| Layer | N10 | N7 | N5 |
|---|---|---|---|
| LI | W | Co | Co |
| Vint | Cu | Co | Co |
| M0 | Cu | Co | Co |
| V0 | Cu | Cu | Co |
| M1 | Cu | Cu | Co |
| V1 | Cu | Cu | Cu/Co |
| M2 | Cu | Cu | Cu |

Co compatible cleans required:
1. Contact Co PCMP
2. Co etchback (for SAGC)
3. **Post via etch clean (SD)**
4. Post via etch clean (gate)
5. M0 Co PCMP
6. Post V0 etch clean
7. M1 Co PCMP
8. Post V1 etch clean

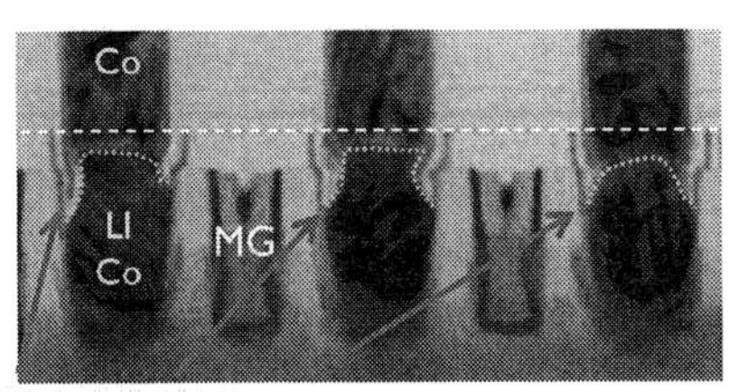


Co corrosion seen

Z. Tokei (imec), IEDM 2017



8

## Cobalt corrosion and passivation

imec KURITA

Process flow: Native oxide removal → Pre measurement → Rinse → Wafer weight

0.05% HF, 20sec.    Co loss test, 30 – 300 sec.

$N_2$ ambient (DO < 30 ppb)    $N_2$ ambient (DO < 30 ppb)

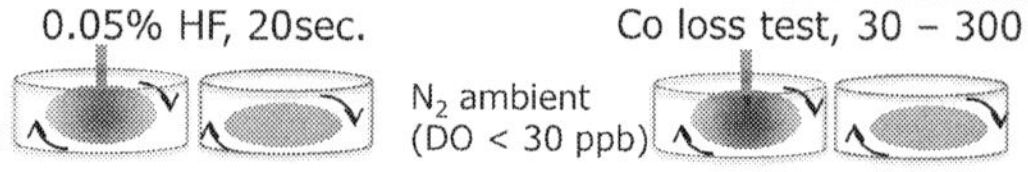



9

## Cobalt corrosion and passivation

imec KURITA

Process flow: Native oxide removal → Pre measurement → Rinse → Wafer weight

0.05% HF, 20sec.    Co loss test, 30 – 300 sec.

$N_2$ ambient (DO < 30 ppb)    $N_2$ ambient (DO < 30 ppb)

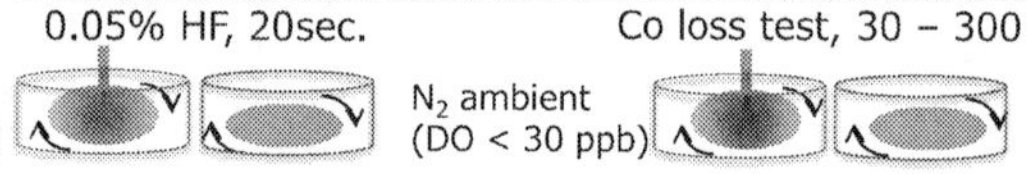

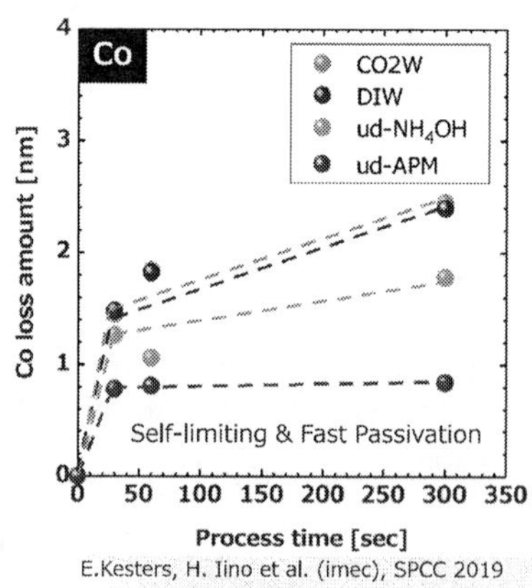


E.Kesters, H. Iino et al. (imec), SPCC 2019

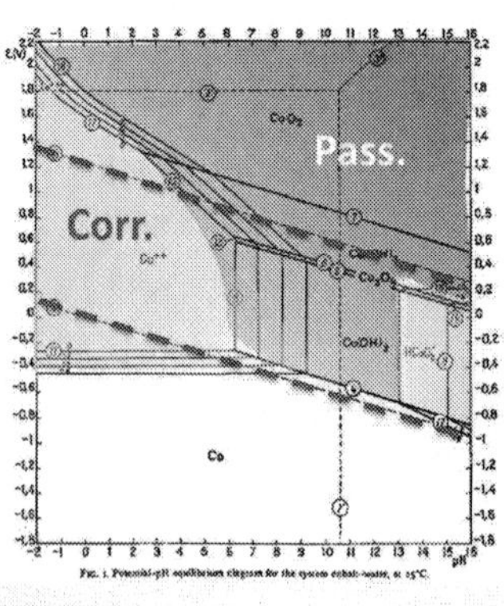


$Co + 2H^+ \rightarrow Co^{2+} + H_2$

$Co + 2H_2O \rightarrow HCoO_2^- + 3H^+ + 2e^-$

$Co + 2H_2O \rightarrow Co(OH)_2 + 2H^+ + 2e^-$

$Co + H_2O \rightarrow CoO + 2H^+ + 2e^-$

$CoO + H_2O \rightarrow HCoO_2^- + H^+$

$\mathbf{Co + H_2O_2 \rightarrow Co(OH)_2}$

$\mathbf{Co(OH)_2 \rightarrow CoO + H_2O}$

Q.T.Le et al. (imec), UCPSS 2018

© 2

10

## Cobalt corrosion and passivation

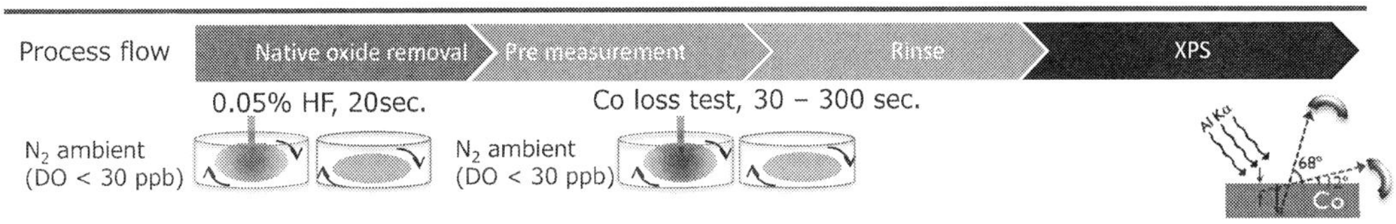


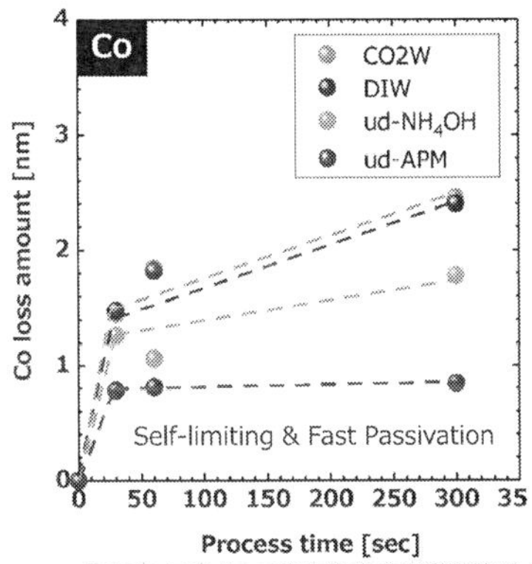


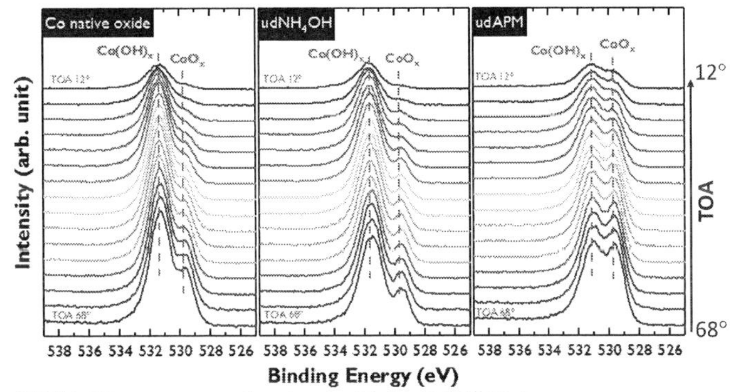


E.Kesters, H. Iino et al. (imec), SPCC 2019

11

## Cobalt corrosion and passivation

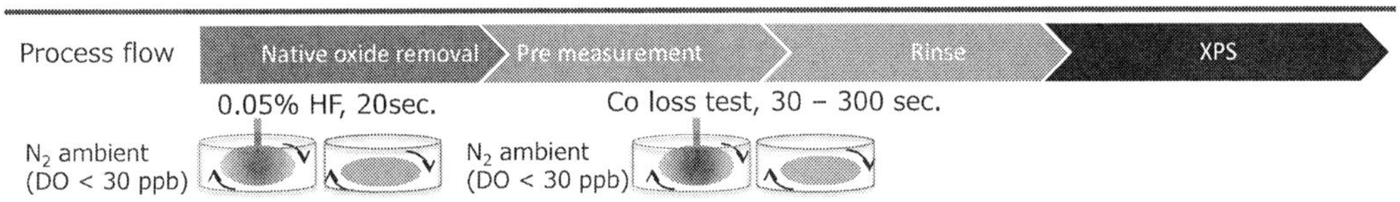


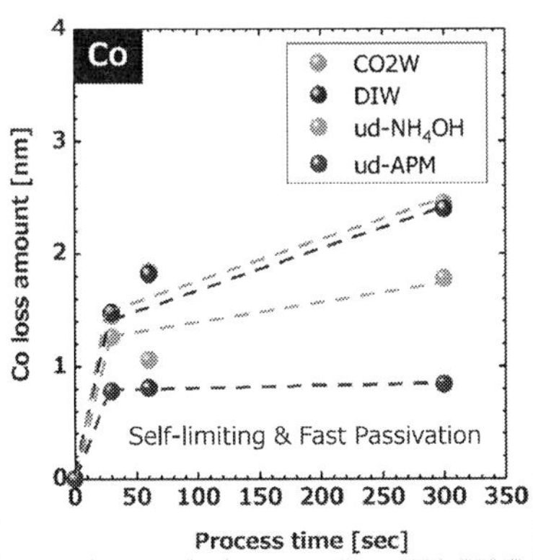


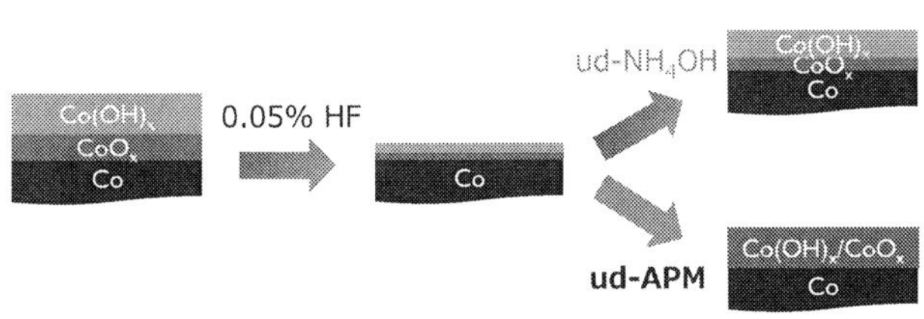


- Co surface is passivated by a thin $Co(OH)_x$ and $CoO_x$ layer

E.Kesters, H. Iino et al. (imec), SPCC 2019

12

## Cobalt corrosion and passivation

imec KURITA

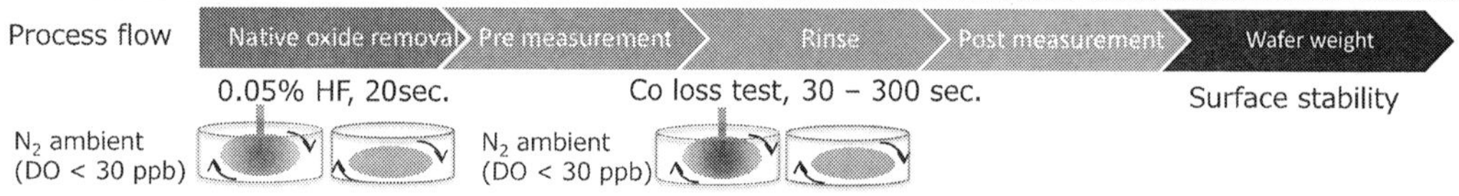


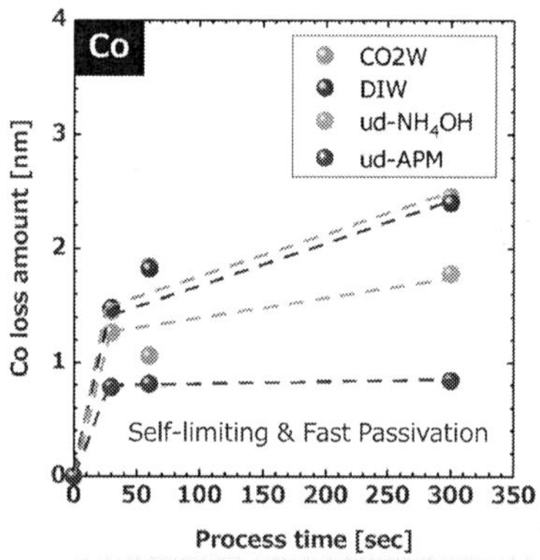


E.Kesters, H. Iino et al. (imec), SPCC 2019

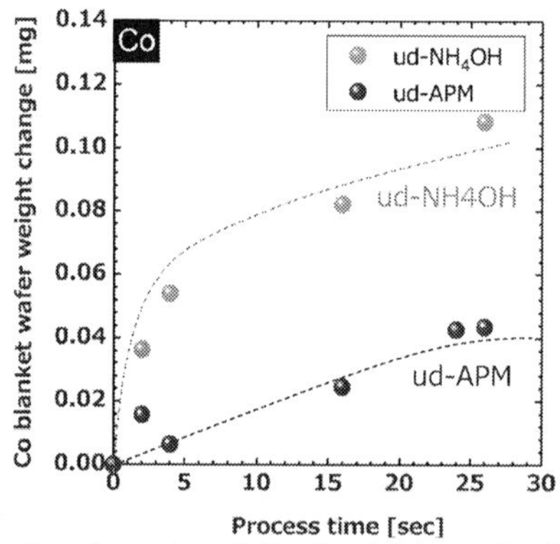


- Co passivation layer is relatively stable under the atmosphere

## Agenda

imec KURITA

1. Introduction
2. $La_2O_3$ dissolution and its suppression
3. Cobalt corrosion and passivation
4. Atomic scale etching of Cobalt
5. Summary

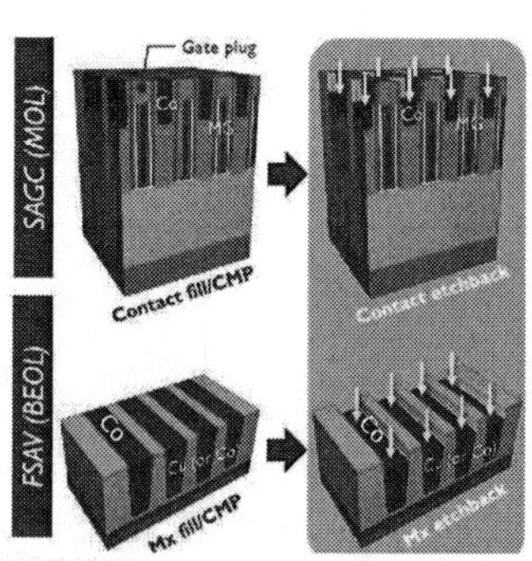




Y.Oniki (imec), SEMICON KOREA 2019 14

## Atomic scale etching of Cobalt

imec KURITA

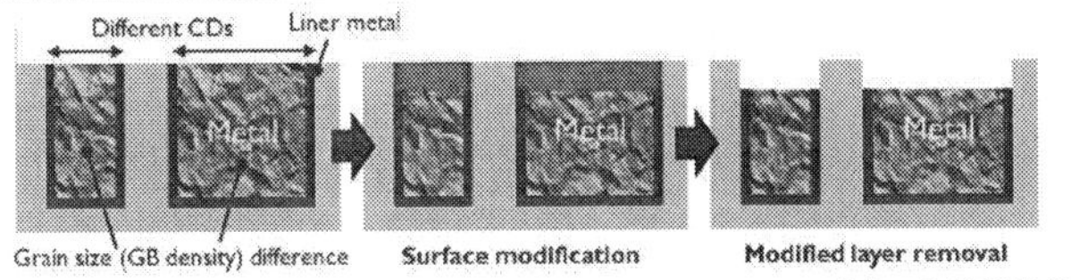


A.Pacco et al. (imec), SPCC 2019

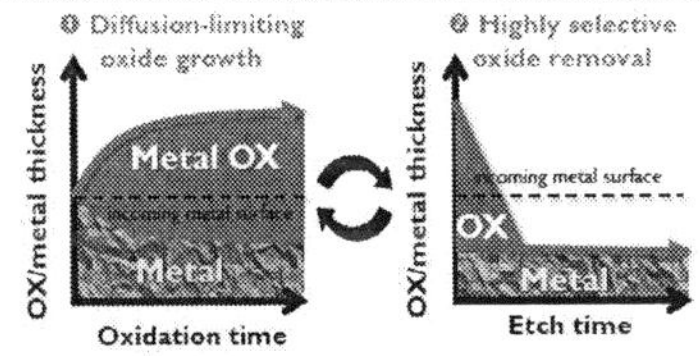


EA determined by formed OX (consumed metal) thickness and cycle#



15

## Atomic scale etching of Cobalt

imec KURITA

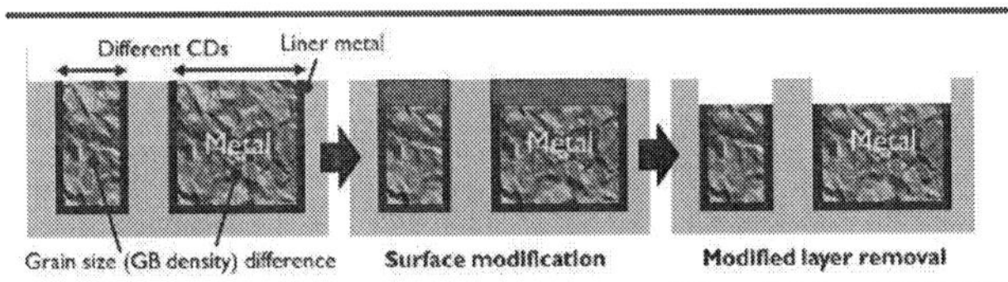


A.Pacco et al. (imec), SPCC 2019

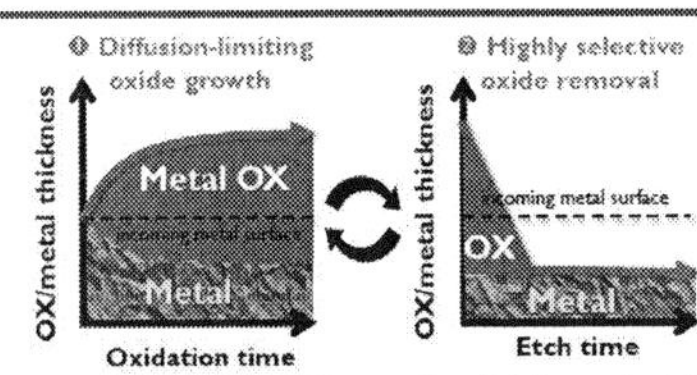


EA determined by formed OX (consumed metal) thickness and cycle#

Y.Oniki et al. (imec), UCPSS 2018

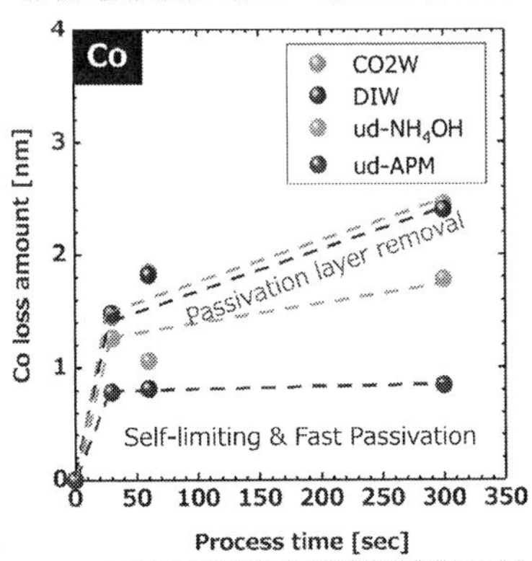


E.Kesters, H. Iino et al. (imec), SPCC 2019

- **General digital wet etching**

- **Fully functional water digital etching**

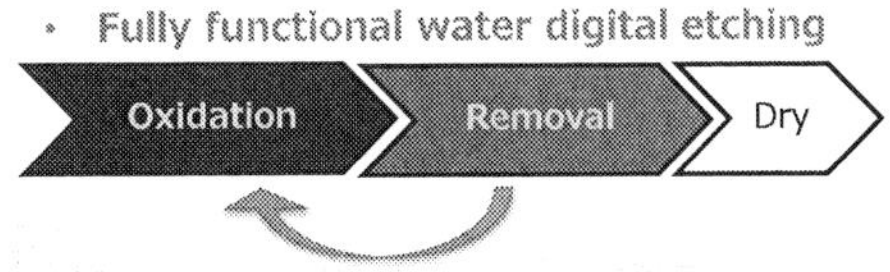


- No rinsing step between each etching step

→ Improve the throughput and the cost of ownership!

16

## Atomic scale etching of Cobalt

imec KURITA

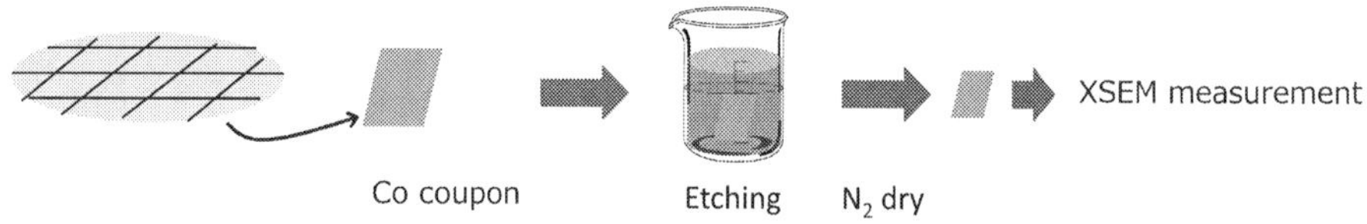


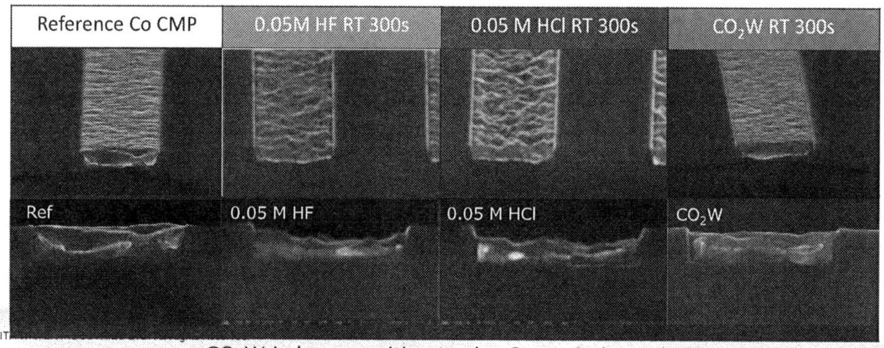




- $CO_2W$ is less sensitive to the Co grain boundaries

17

## Atomic scale etching of Cobalt

imec KURITA

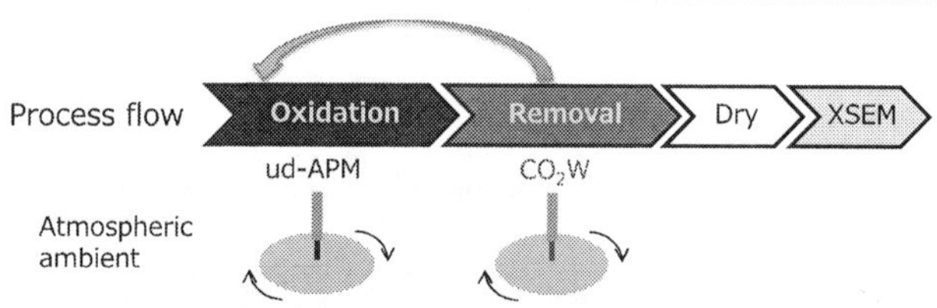


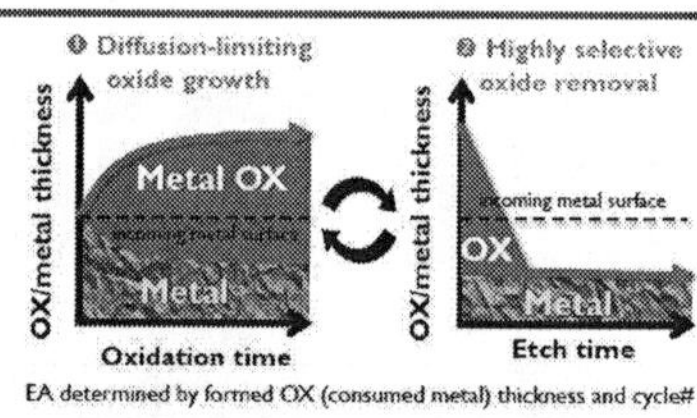


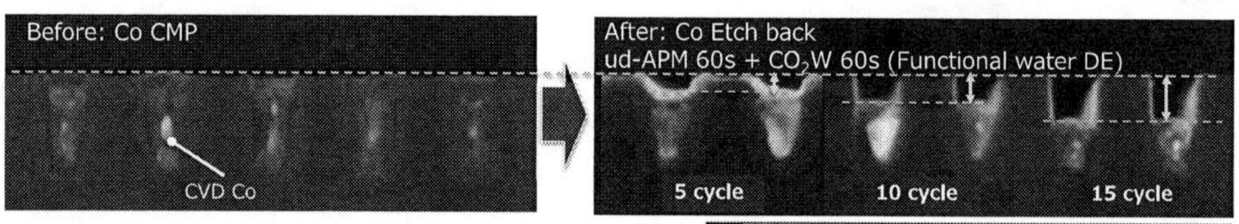


- Atomic scale etching by Fully functional water
- Co ER directly proportional to the etch cycle#

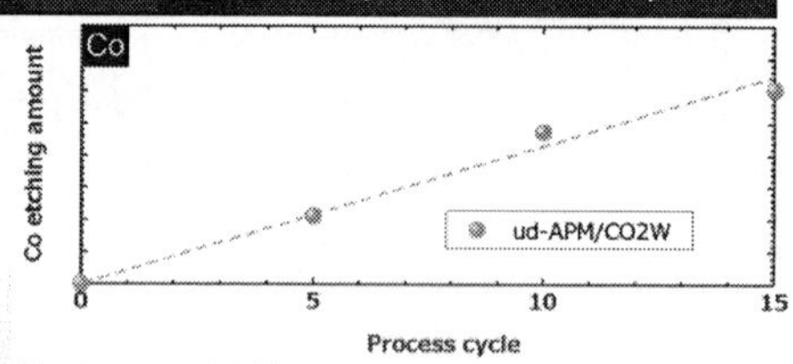




18

## Atomic scale etching of Cobalt

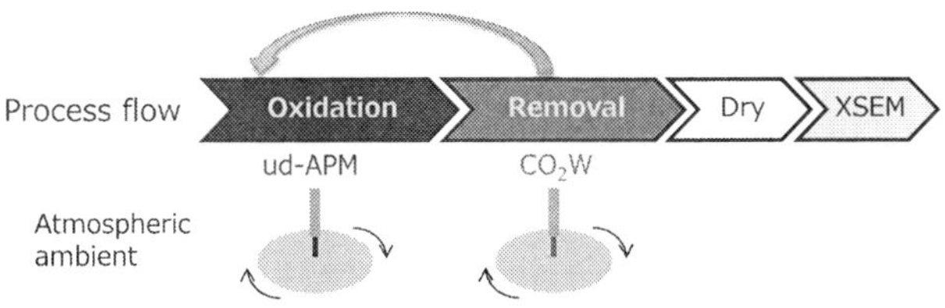


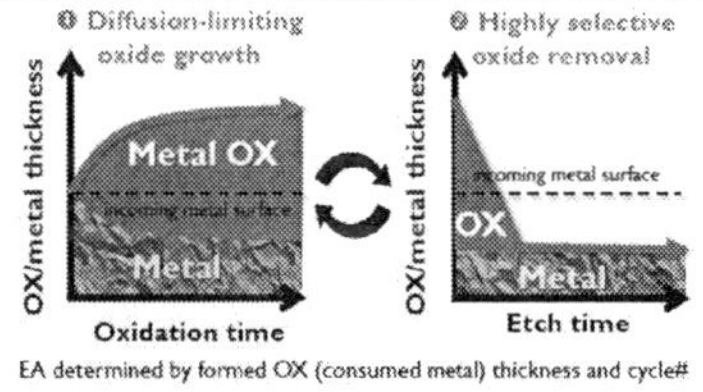


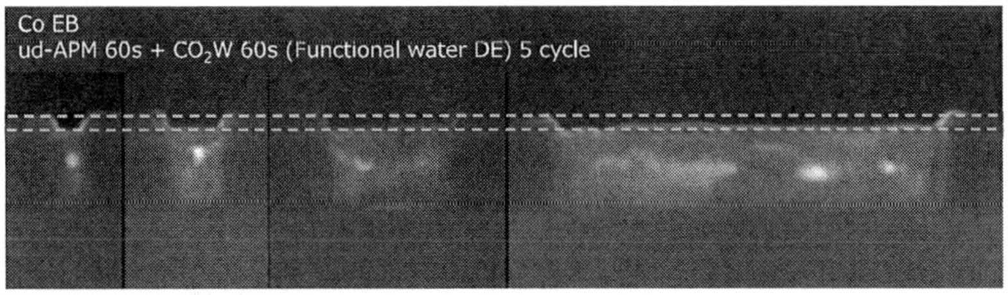


- Fully functional water atomic layer etching is less sensitive to the Co grain boundaries/sizes.
- Controlled local uniformity and negligible pattern loading seen.




## Summary

➢ <u>HKMG patterning</u>
In case HKMG patterning required to suppress $La_2O_3$, ud-$NH_4OH$ rinsing could preserve the $La_2O_3$ layer.
Meantime, removal $La_2O_3$, $CO_2W$ rinsing is preferred.

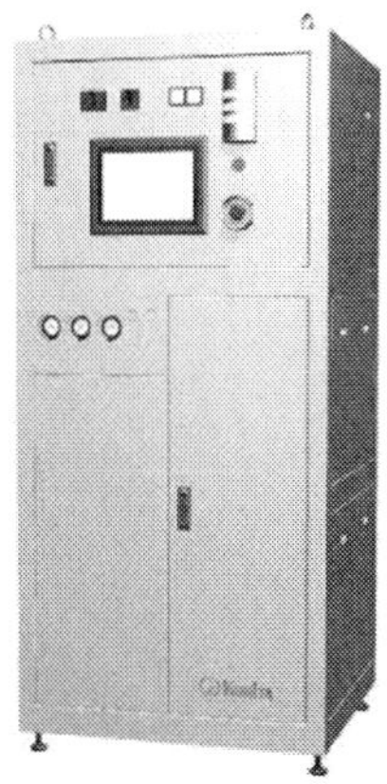

➢ <u>Co surface cleaning in the MOL and BEOL</u>
ud-APM successfully passivates the Co surface, resulting in negligible Co loss during the wafer rinsing step.

➢ <u>Controlled Co recess for SAGC and FSAV</u>
Fully functional water digital etching could demonstrate atomic scale etching of Co.

KURITA

Thank you for your attention

# CONTROLLED COBALT RECESS FOR ADVANCED INTERCONNECT METALLIZATION.

**Antoine Pacco**[a*], Y. Akanishi[b], Q.T. Le[a], E. Kesters[a], G. Murdoch[a], F. Holsteyns[a]

[A] IMEC VZW, KAPELDREEF 75, 3001 LEUVEN, BELGIUM
[B]SCREEN, SEMICONDUCTOR SOLUTIONS CO., LTD., 480-1, TAKAMIYA-CHO, HIKONE, SHIGA 522-0292, JAPAN
*EMAIL: ANTOINE.PACCO@IMEC.BE

PUBLIC

## OVERVIEW

- copper → cobalt transition
- challenges for metal recess & specifically for cobalt recess
  - specifications for wet recess
- approach & results: Co recess by digital wet etch
  - effect of different oxidizing solutions
  - effect of oxide dissolution step & rinse step times
  - effect of the ambient
- wet recess of cobalt for FSAV
  - Electrical testing
  - Morphological results
- conclusions

imec 2

## WHEN & WHY **COBALT** OFFERS AN ADVANTAGE OVER **COPPER** AS AN INTERCONNECT METAL

1. The electron mean free path of cobalt is considerably lower than copper, reducing surface scattering in small trenches → lower line resistivities compared to copper.
2. A thinner combined barrier-nucleation layer can be used for cobalt → lower cross-sectional resistance.
3. Cobalt can be annealed at reasonable thermal budgets thereby enabling grain growth which reduces resistivity.

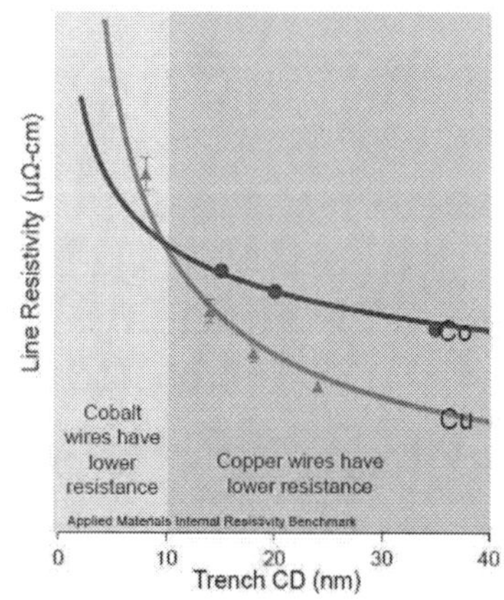


*Ref.: Applied Materials Internal Resistivity Benchmark*

imec

## WHEN & WHY **COBALT** OFFERS AN ADVANTAGE OVER **COPPER** AS AN INTERCONNECT METAL

1. The electron mean free path of cobalt is considerably lower than copper, reducing surface scattering in small trenches → lower line resistivities compared to copper.
2. A thinner combined barrier-nucleation layer can be used for cobalt → lower cross-sectional resistance.
3. Cobalt can be annealed at reasonable thermal budgets thereby enabling grain growth which reduces resistivity.

*will cobalt climb up the interconnect stack?*

| Layer | N14/10 MMP44-52 | N10/7 MMP36-40 | N7/5 MMP28-32 | N5/3 MMP20-24 | N3/2 MMP14-18 |
|---|---|---|---|---|---|
| Mx/Vx-1 | Cu | Cu | Cu | Cu | Cu |
| M3/V2 | Cu | Cu | Cu | Cu or Co or Mo or Ru | Cu or Co or Mo or Ru |
| M2/V1 | Cu | Cu | Cu or Co | Cu or Co or Mo or Ru | Cu or Co or Mo or Ru |
| M1/V0 | Cu | Cu or Co | Cu or Co | Cu or Co or Mo or Ru | Cu or Co or Mo or Ru |
| M0 | Cu | Cu or Co | Cu or Co | Cu or Co or Mo or Ru | Cu or Co or Mo or Ru |
| MOL | W | W or Co | Co | Co or Mo or Ru | Co or Mo or Ru |
| BPR | - | - | - | - | W or Mo or Ru |

*taking place....*

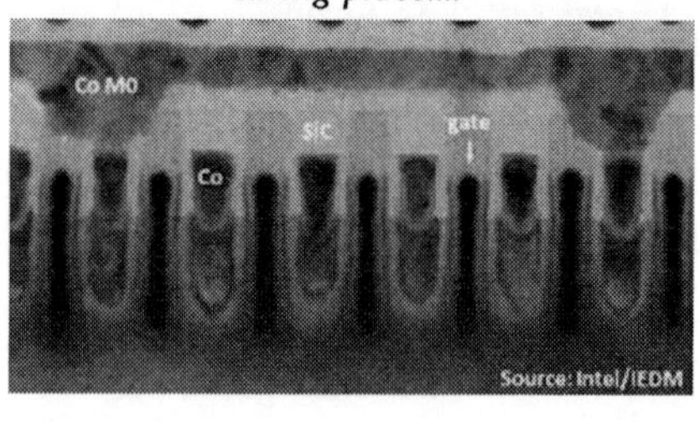


imec

## COBALT RECESS FOR FSAV (FULLY SELF ALIGNED VIA)

### M1 recess

- Target = 10nm metal recess

→ Topography for SAV

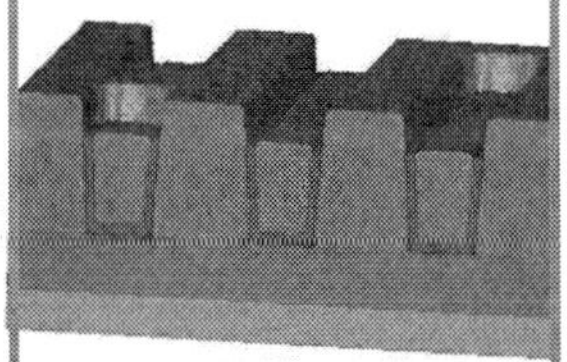

### M2V1 Etching

- Via confined in y-direction by M2 HM
- Via-first etch lands selectively on SiCN barrier
- Topography is maintained by selective removal of SiCN

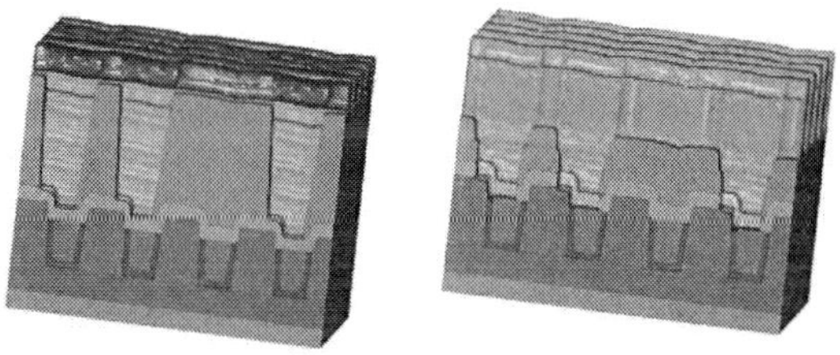

### M2V1 Fill

- FSAV is formed
- Min-distance M1-V1 is maintained by topography

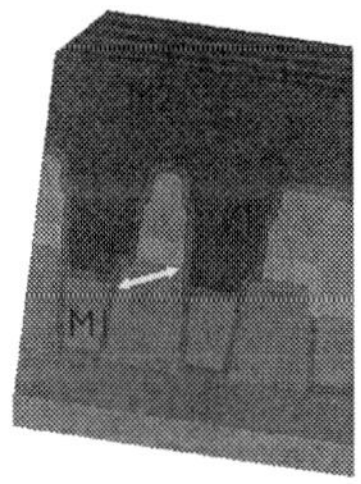


imec

5

COVENTOR

## CHALLENGES FOR METAL RECESS
### NON-IDEAL (NON-UNIFORM RECESS) CASES

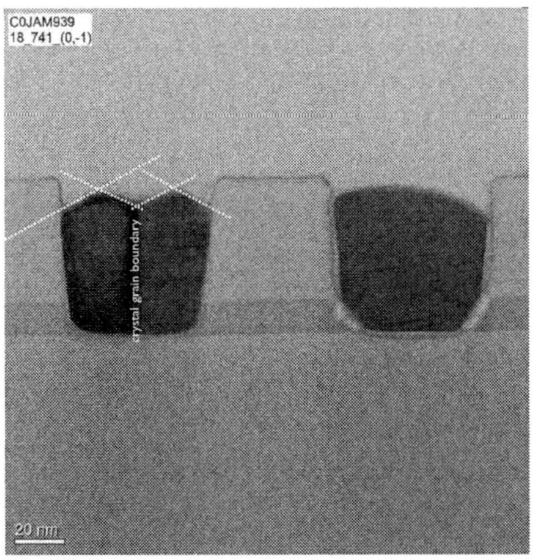


wet metal recess can be affected by:

1. Crystal orientations
   - different crystallographic planes can have different etching rates (anisotropy)
2. Crystal grain boundaries
   - preferential etching can occur at grain boundaries
   - smaller grain size → grain boundary density increases
3. Galvanic corrosion
   - Local increase in oxidation / etch rate due to galvanic contact of the fill metal with the liner/ barrier metals

imec

6

## CHALLENGES FOR METAL RECESS: COBALT

- non-uniform (crystal-grain/boundary dependent) etch
- uncontrolled dissolution of cobalt during slightly acidic / neutral rinse
  → alkaline solutions, $NH_4OH/H_2O_2$ preferred (*,**)
- selective removal of TiN liner/barrier after or simultaneously during Co recess

* *SPCC '18 Y. Ogawa, [Kurita]*
** *SPCC '19 H. Iino, [Kurita]*

## SPECIFICATIONS:

- control of the recess amount within the nanometer range: 10 +/- 1 nm
- small *within-line* roughness ($R_q$ ~ 1nm)
- uniform *within-wafer* recess (center-to-edge NU < 1nm)
- avoid pattern-density-depending etching (CD-independent)

***→ cyclic process (digital etch) = process of choice***

imec

7

## APPROACH FOR CONTROLLED RECESS

"DIGITAL ETCHING": 0-1-0-1-0-1-0-1-....

0 = self-limiting metal-oxide growth on the metal surface
1 = metal-oxide removal selective towards metal surface

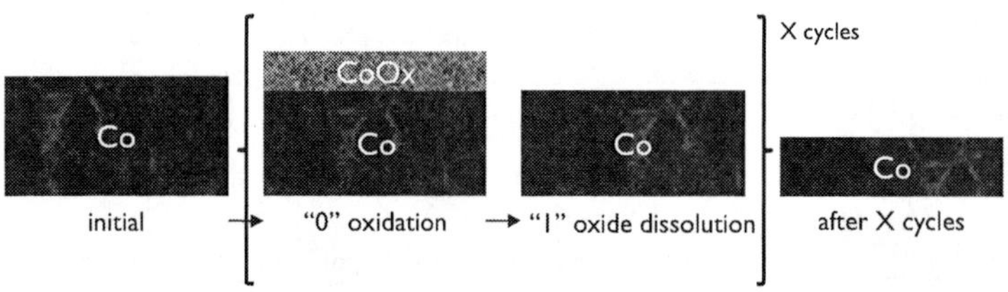


Potential benefits of this approach:

- **recess amount** can be controlled within the nanometer range
- uniform ***within-wafer*** recess (single wafer processing)
- uniform ***within-structures*** recess = non pattern-density-depending etching

imec

8

## EFFECT OF DIFFERENT OXIDIZING SOLUTIONS
### FOR THE SAME OXIDE DISSOLUTION STEP: 10S HF 0.05%

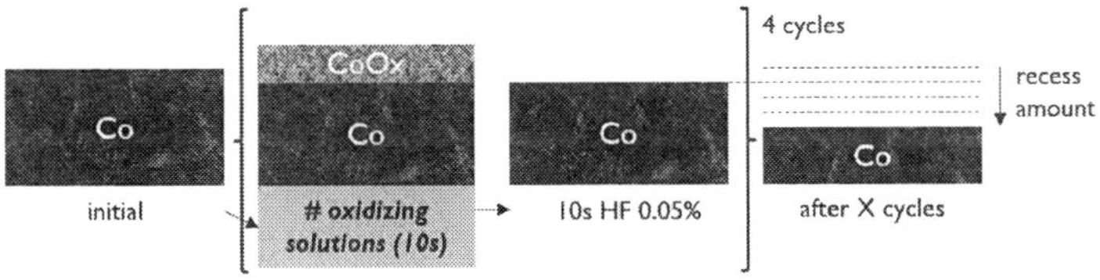


- increasing the peroxide concentration in both slightly acidic $H_2O_2/H_2O$ and alkaline $NH_4OH/H_2O_2/H_2O$ solutions does not significantly affect the total cobalt loss
- removing the oxidizing agent, being it $H_2O_2$ or atmospheric $O_2$ does reduce the total cobalt loss.
- → self-limiting oxide film formation when a sufficiently oxidizing solution is used ($H_2O_2$).

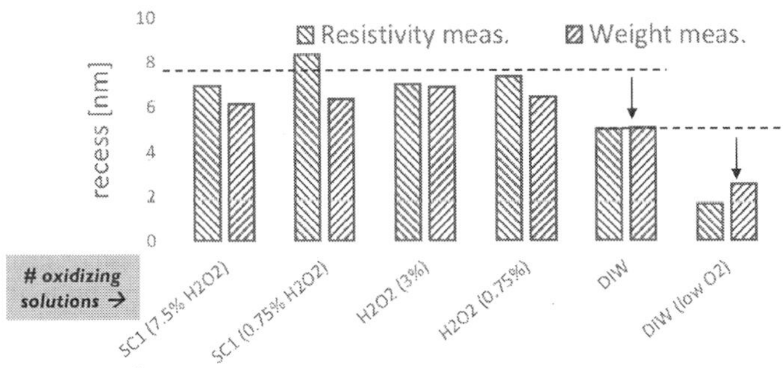


imec

9

## COBALT ROUGHNESS AFTER DIFFERENT OXIDIZING SOLUTIONS
### AFM (ECD COBALT)

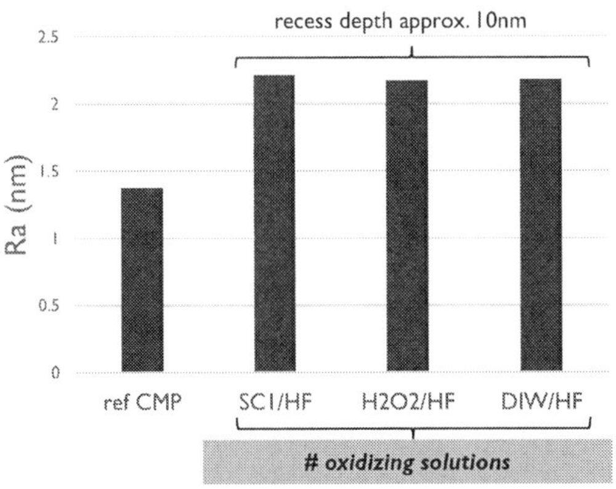


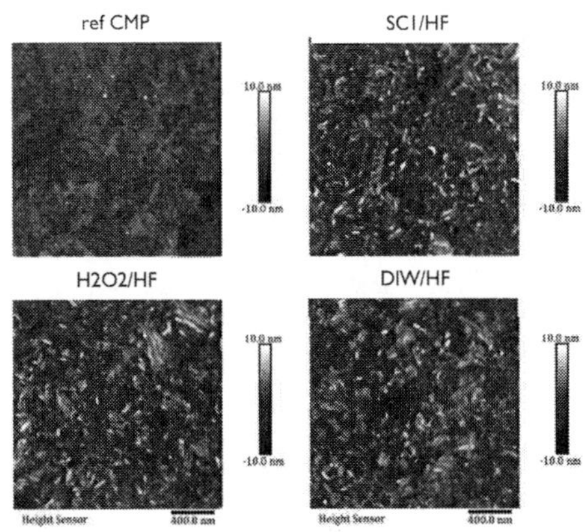


- Slight (<1nm) roughness increase after recess.
- No significant difference between different oxidizing solutions.

imec

10

## IMPACT OF DISSOLVED [$dO_2$] IN DILUTED HF (0.05%)

### EXAGGERATED TIMES & AGGRAVATED CONDITIONS ...

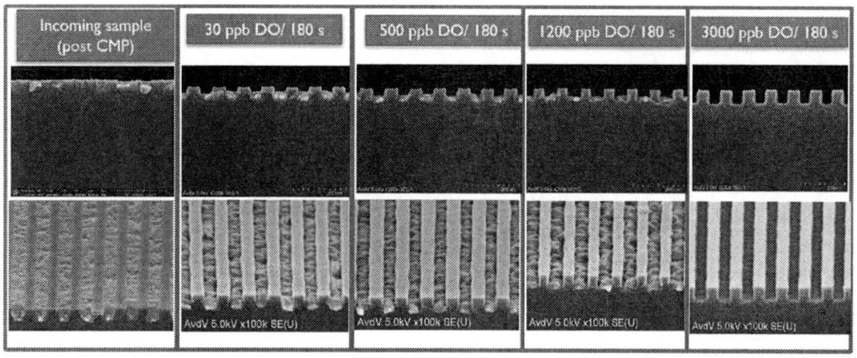


→ non controllable recess

\+ high roughness

## WHAT ABOUT THE IMPACT OF [$O_{2,ATM}$] IN THE PROCESS AMBIENT ?

imec

11

## EFFECT OF OXIDE DISSOLUTION STEP & RINSE STEP TIMES

### STD(NON-CONTROLLED) VS. LOW [$O_2$] AMBIENT

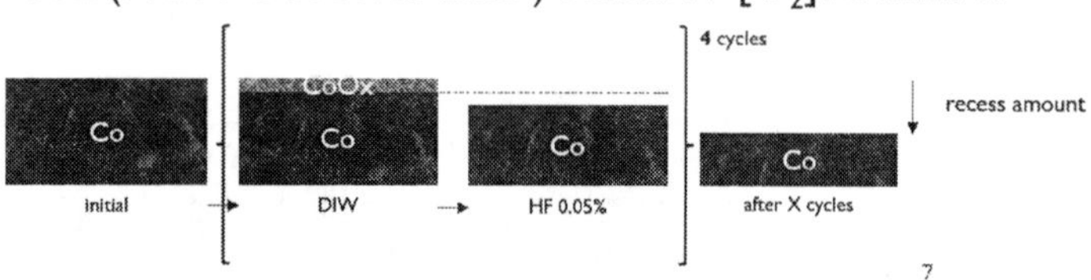


- DIW time has a small impact on cobalt loss.
- Increasing the HF time has a more noticeable impact on the total cobalt loss.
- The parameter that has highest impact is the $O_2$ content in the ambient.

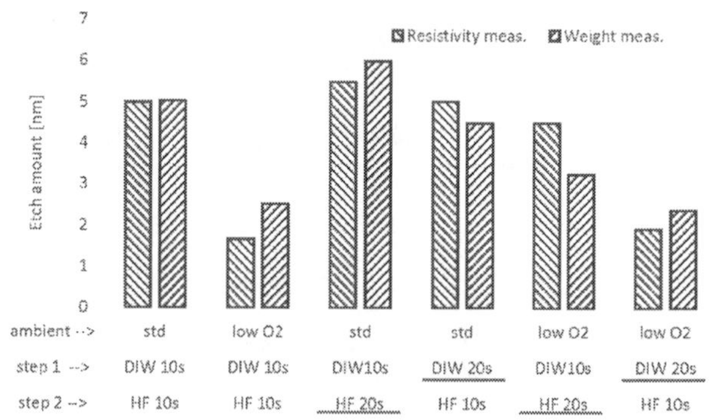


imec

12

# TIN LINER REMOVAL

## TWO APPROACHES: 1-STEP VS. 2-STEPS

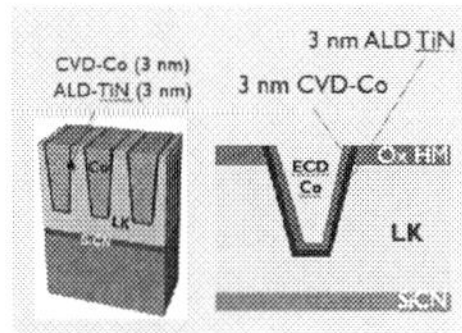


**1-STEP:** combined TiN liner removal & Co recess etch

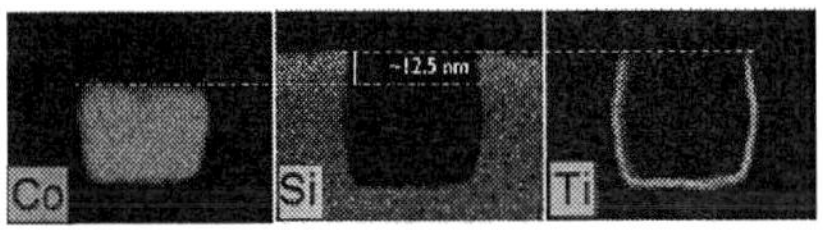


(APM **40°C** /HF) x4

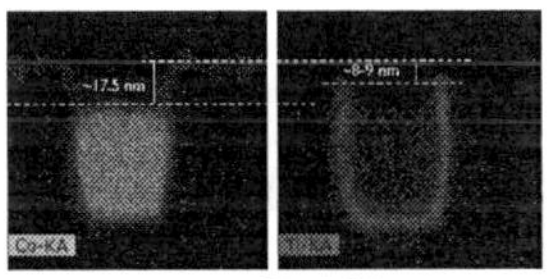


(APM **60°C** /HF) x4

**2-STEPS:** Co recess etch followed by TiN liner removal

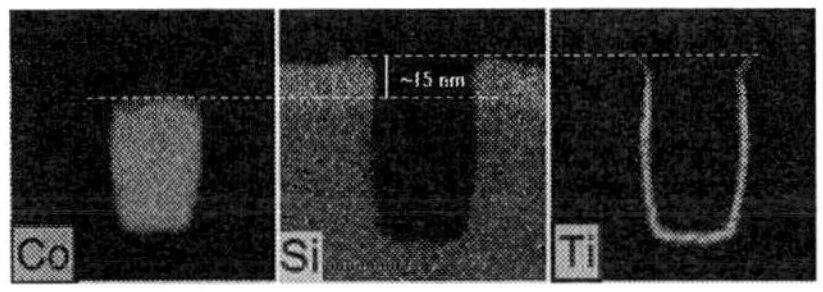


(APM RT /HF) x4 + 30s APM **40°C**

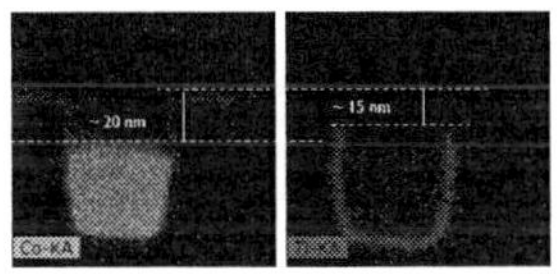


→ TiN removal most effective by using a final 60°C APM

(APM RT /HF) x4 + 40s APM **60°C**

imec

13

# ELECTRICAL TESTING

## TARGET: 10 NM CO RECESS FOR FSAV

- Process: "digital" Co recess etch
  - [APM followed by HF] x times
  - + TiN barrier etch APM 60°C

Line resistance of 10 µm long lines
21 nm → 30 nm CD

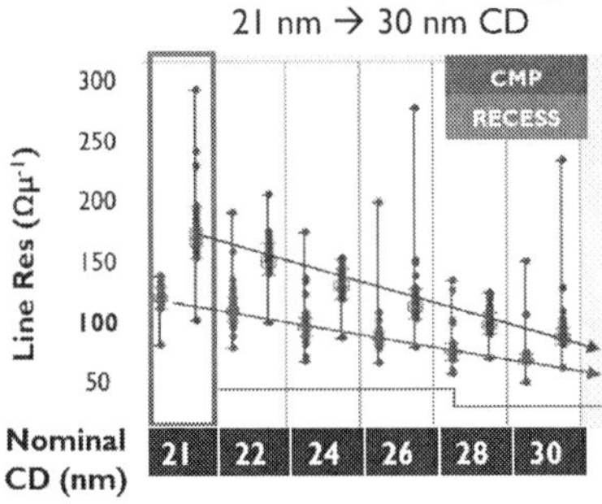


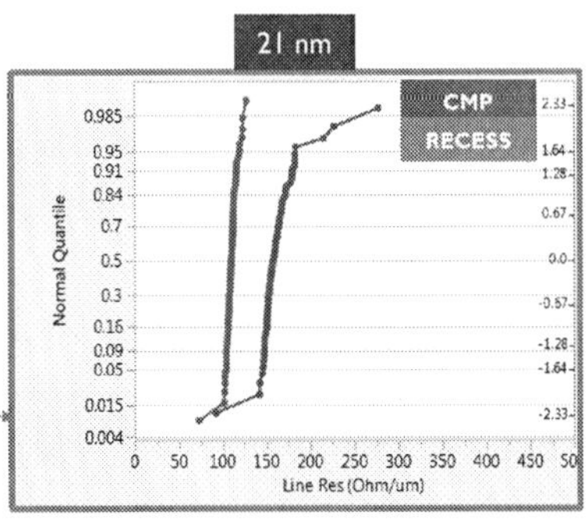


- Uniform distributions
- > 95% yield for smallest CD

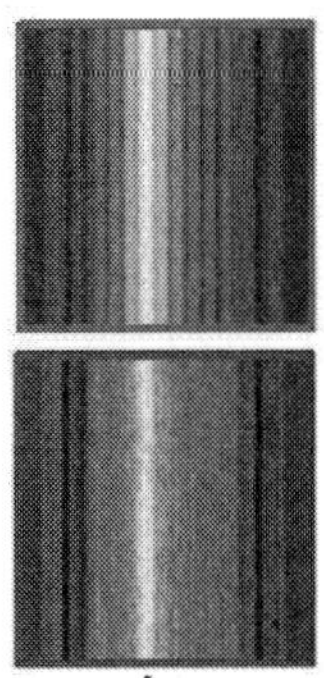


- Electrically active lines

imec

14

## ELECTRICAL TESTING

### TARGET: 10 NM CO RECESS FOR FSAV

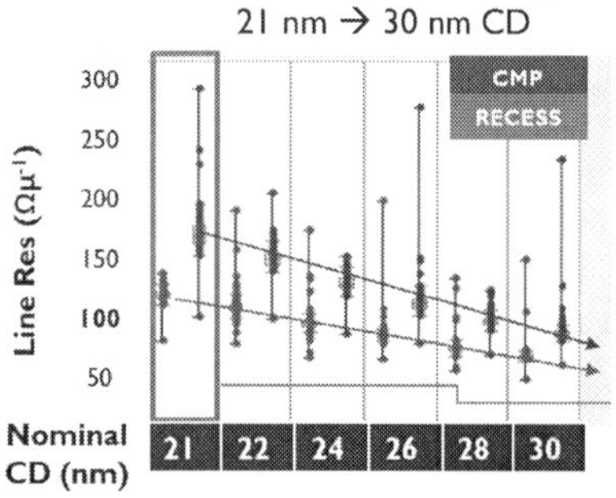


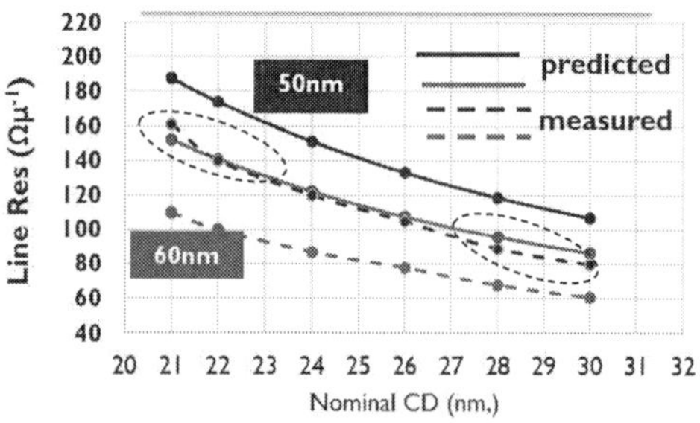


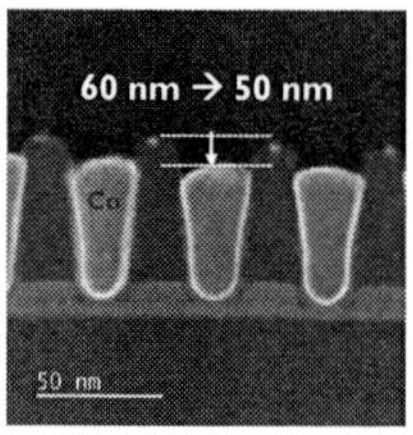


- Data follows the expected trend: line resistance increase for smaller lines due to line X-section reduction (R ~ 1/A)
  → Limited pattern loading

imec



## MORPHOLOGICAL ANALYSIS

### TEM/EDX

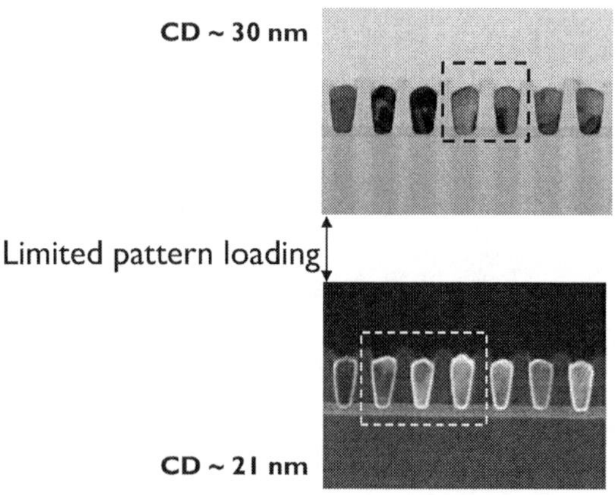


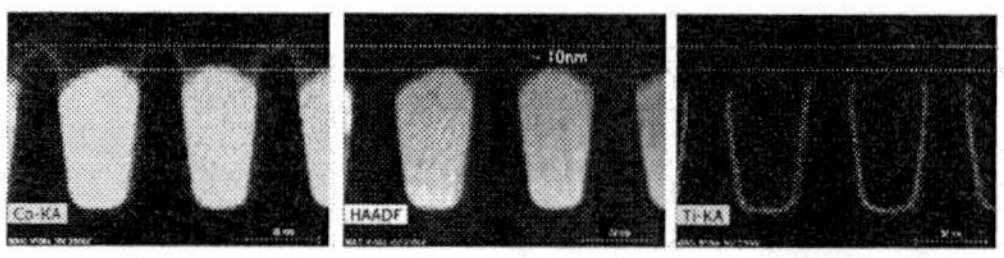


~10 nm Co recess & TiN liner removal

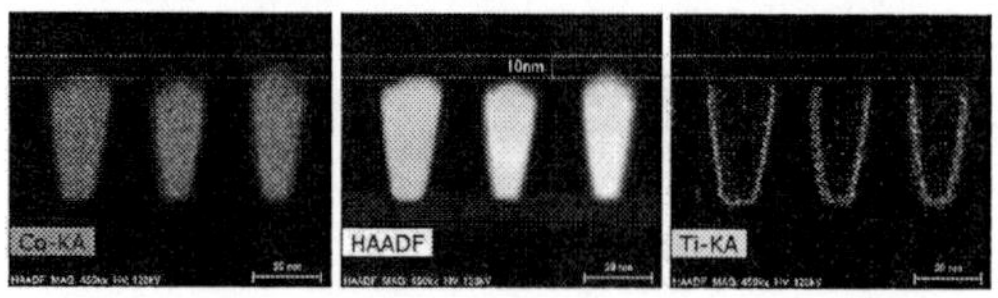


imec

## MORPHOLOGICAL ANALYSIS
### TEM/EDX

EDGE of the wafer ⟷ CENTER of the wafer

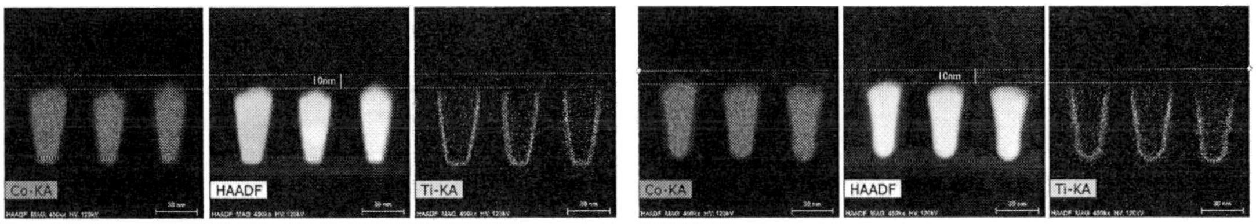


- Good within-wafer uniformity
- TiN removal height ~ Co recess

imec

17

## CONCLUSIONS

Digital etch of cobalt-metal way to go:

1. Promising morphological & electrical results with commodity chemistries
2. Oxidizing step less critical (~ self limiting with $H_2O_2$)
3. Predominant role of the oxide dissolving agent (acid) step & $DO_2$

- The local line roughness was relatively high for the ECD cobalt post-CMP & post-recess etch
- Limited pattern loading observed for CVD cobalt
- Cobalt recess was controlled in the nanometer range (10 +/- 1 nm)
- Good WIW uniformity and yield obtained

imec

18

imec
embracing a better life
imec
PUBLIC

# Development of Metal Free Wet Etching Chemical for Ruthenium Interconnect

TOK (Tokyo Ohka Kogyo Co., Ltd.), Japan
*Takuya Ohashi, Yukihisa Wada, Mai Sugawara, Tomoya Kumagai

IMEC, Belgium
Quoc Toan Le, Els Kesters, Yusuke Oniki, Jens Rip, Frank Holsteyns

*t-ohhashi@tok.co.jp





## OUTLINE

1. Introduction
2. Design of Ru wet etch chemical
3. CVD-Ru Bulk properties
4. Summary

# OUTLINE

1. Introduction
2. Design of Ru wet etch chemical
3. CVD-Ru Bulk properties
4. Summary





# ALTERNATIVE METAL FOR INTERCONNECT

**Ruthenium has been recently considered as a promising candidate to replace copper and cobalt as MOL and BEOL interconnect material.**

- ✓ Low resistivity in the scaled dimensions
- ✓ Better EM

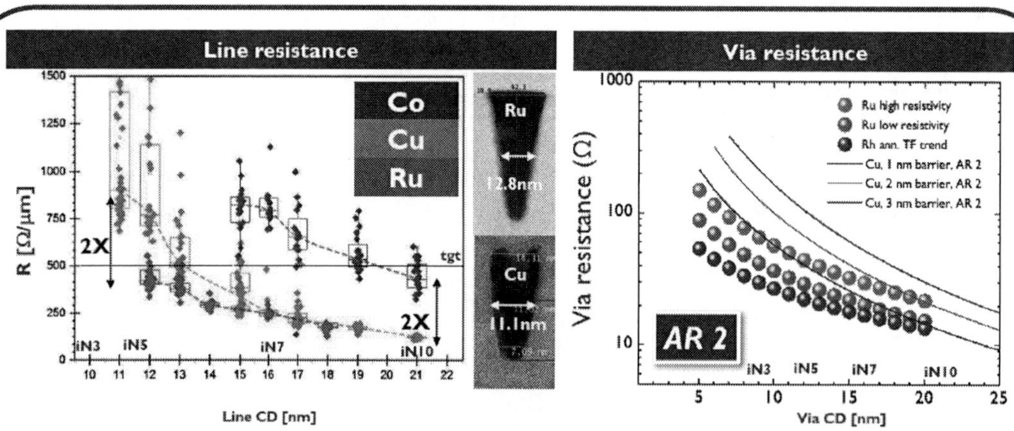


[Marleen H. van der Veen et al, IITC2018, IMEC]

# TARGET RUTHENIUM WET-ETCH APPLICATIONS

## Ru Bevel and Backside clean

**Liner/Metal deposition** → **BV/BS clean** → **CMP**

## Ru recess for FULLY SELF-ALIGNED VIAS

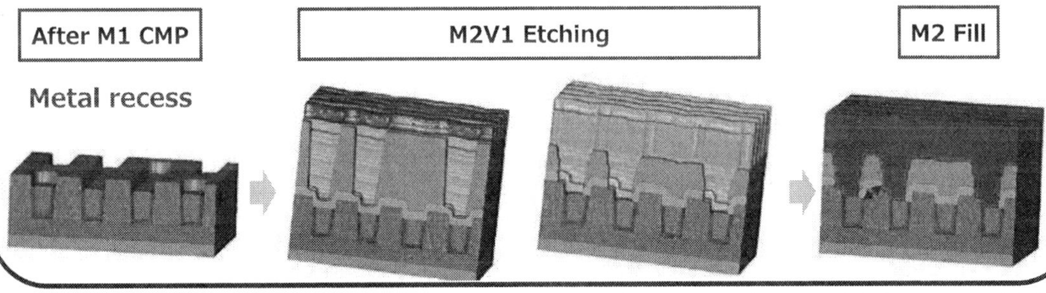


[Efrain Altamirano Sanchez, IITC2018 Short course, IMEC]

tok · imec





# RUTHENIUM IN THE PERIODIC TABLE

## The periodic table

Platinum-Group Metal (Ru, Rh, Pd, Os, Ir, Pt)

| 1 | 2 | 3 | 4 | 5 | 6 | 7 | 8 | 9 | 10 | 11 | 12 | 13 | 14 | 15 | 16 | 17 | 18 |
|---|---|---|---|---|---|---|---|---|---|---|---|---|---|---|---|---|---|
| H | | | | | | | | | | | | | | | | | He |
| Li | Be | | | | | | | | | | | B | C | N | O | F | Ne |
| Na | Mg | | | | | | | | | | | Al | Si | P | S | Cl | Ar |
| K | Ca | Sc | Ti | V | Cr | Mn | Fe | Co | Ni | Cu | Zn | Ga | Ge | As | Se | Br | Kr |
| Rb | Sr | Y | Zr | Nb | Mo | Tc | Ru | Rh | Pd | Ag | Cd | In | Sn | Sb | Te | I | Xe |
| Cs | Ba | La | Hf | Ta | W | Re | Os | Ir | Pt | Au | Hg | Tl | Pb | Bi | Po | At | Rn |
| Fr | Ra | Ac | Rf | Db | Sg | Bh | Hs | Mt | Ds | Rg | Cn | Hn | Fl | Mc | Lv | Ts | Og |
| | | La | Ce | Pr | Nd | Pm | Sm | Eu | Gd | Tb | Dy | Ho | Er | Tm | Yb | Lu | |
| | | Ac | Th | Pa | U | Np | Pu | Am | Cm | Bk | Cf | Es | Fm | Md | No | Lr | |

tok · imec

# CHEMICAL RESISTANCE OF RUTHENIUM

| Commodity chemistry | CVD-Ru* E.R. @R.T. |
|---|---|
| HCl | 0.0nm/min |
| $H_2SO_4$ | 0.0nm/min |
| $HNO_3$ | 0.0nm/min |
| HF | 0.0nm/min |
| $H_2O_2$ | 0.0nm/min |
| SC1 | 0.0nm/min |
| SC2 | 0.0nm/min |
| NH3 | 0.0nm/min |
| KOH | 0.0nm/min |
| TMAH | 0.0nm/min |
| CAN+$HNO_3$** | 1.1nm/min |
| NaOCl 5% | 11.9nm/min |

* As depo CVD-Ru

** 50mM CAN (Ceric Ammonium Nitrate) + $HNO_3$ 1M

Pourbaix diagram of Ru-water system

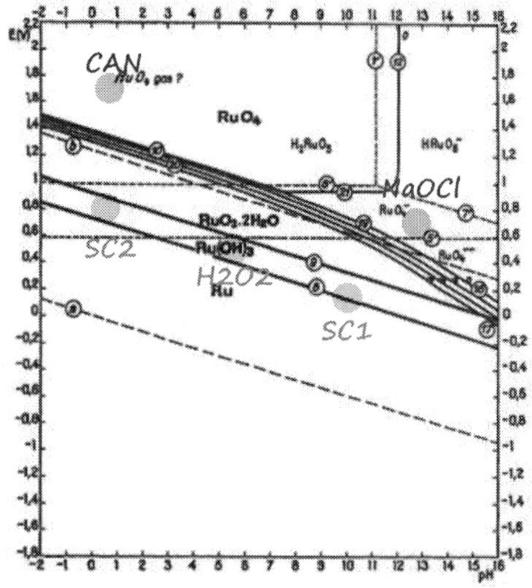


[M. Pourbaix, Atlas of Electrochemical Equilibria in Aqueous Solutions p. 346, National Association of Corrosion Engineers, 1974-644]

**Strong oxidizers (CAN, NaOCl, ...) can etch ruthenium.**

$Ru(s) + 4H_2O \Leftrightarrow RuO_4 + 8H^+ + 8e^-$ $E^0 = 1.032V$

tok imec





# OUTLINE

1. Introduction
2. **Design of Ru wet etch chemical**
3. CVD-Ru Bulk CVD-Ru Bulk properties
4. Summary

tok imec

# DESIGN OF RUTHENIUM WET ETCH CHEMICAL

## TOK Ruthenium Wet Etch chemical is targeting the following requirements.

- **Metal free solution**
  - Metal impurities should be avoided for advanced semiconductor processing.
  - Currently, most of the strong oxidizers include metal elements. (CAN, NaOCl etc.)
- **High selectivity to low-k, SiOx, SiNx, TiN and TaN**
- **Prevent ruthenium tetroxide ($RuO_4$) generation**
  - Volatile and toxic $RuO_4$ should be avoided for EHS reasons.
- **Preservation of surface roughness to be targeted after controlled Ru recess**

tok imec





# DESIGN OF RUTHENIUM WET ETCH CHEMICAL

- **Metal free solution**
  - Metal impurities should be avoided for advanced semiconductor processing.
  - Currently, most of the strong oxidizers include metal elements. (CAN, NaOCl etc.)
- **High selectivity to low-k, SiOx, SiNx, TiN and TaN**
- **Prevent ruthenium tetroxide ($RuO_4$) generation**
  - Volatile and toxic $RuO_4$ should be avoided for EHS reasons.
- **Preservation of surface roughness to be targeted after controlled Ru recess**

| Chemical | Metal | Ru etching | Low-k (2.55) | SiN | TiN | TaN |
|---|---|---|---|---|---|---|
| NaOCl | Yes | Good | - | - | - | - |
| CAN | Yes | Good | - | - | - | - |
| Chemical A | No | Good | <0.1nm/min | <0.1nm/min | <0.1nm/min | <0.1nm/min |
| Chemical B | No | Good | <0.1nm/min | <0.1nm/min | <0.1nm/min | <0.1nm/min |
| Chemical C | No | Good | <0.1nm/min | <0.1nm/min | <0.1nm/min | <0.1nm/min |

tok imec

# DESIGN OF RUTHENIUM WET ETCH CHEMICAL

- **Metal free solution**
  - Metal impurities should be avoided for advanced semiconductor processing.
  - Currently, most of the strong oxidizers include metal elements. (CAN, NaOCl etc.)

- **High selectivity to low-k, SiOx, SiNx, TiN and TaN**

- **Prevent ruthenium tetroxide ($RuO_4$) generation**
  - Volatile and toxic $RuO_4$ should be avoided for EHS reasons.

- **Preservation of surface roughness to be targeted after controlled Ru recess**

tok imec





# RUTHENIUM TETROXIDE

## Mechanism of $RuO_4$ toxic gas generation

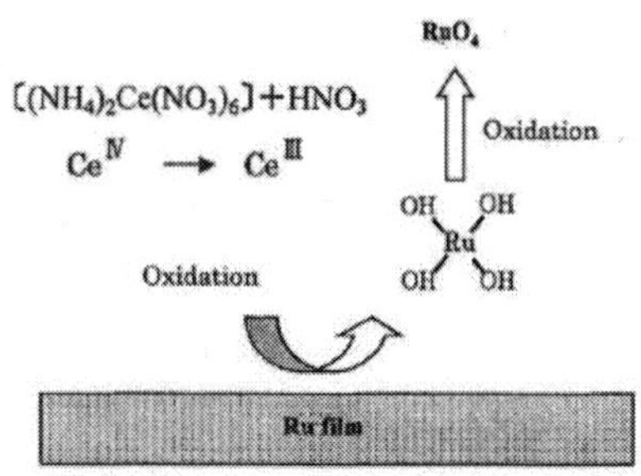


[H. Aoki et. al Extended Abstracts of the 2001 International Conference on Solid State Device and Materials. 2001]

**Ru(s) + *CAN(Ⅳ) ⇒ $RuO_4$↑ + Ce(Ⅲ)↓**

*CAN (Ceric Ammonium Nitrate)

**Ruthenium tetroxide is volatile and toxic.**

**Furthermore, $RuO_4$ is a strong oxidizer.**

## Plastic coloration by $RuO_4$ reaction

**$RuO_4$ and $OsO_4$ are used as staining agent for TEM observation of polymeric materials.**

- ✓ **$OsO_4$ : Oxidize unsaturated type polymer**
- ✓ **$RuO_4$ : Oxidize saturated and unsaturated type polymer**

**Staining reaction by $OsO_4$**

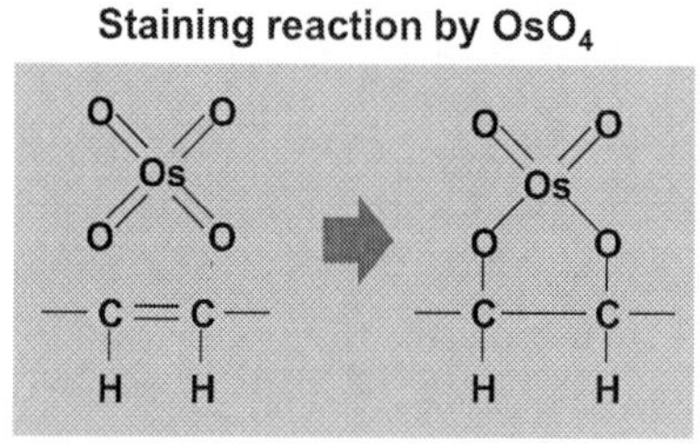


**Staining reaction can be used to detect the generation of $RuO_4$**

# RUTHENIUM TETROXIDE GENERATION TEST

## PLASTIC COLORATION BY RUTHENIUM TETROXIDE GAS

**Bottle A : Reference**

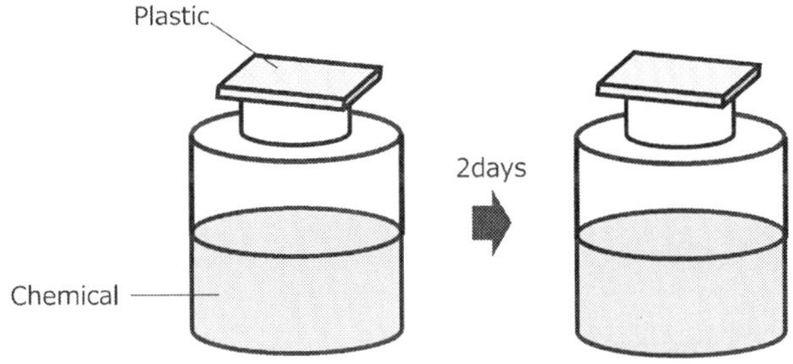


**Bottle B : Ru wafer dipping (Bottle C : Ru Powder dipping)**

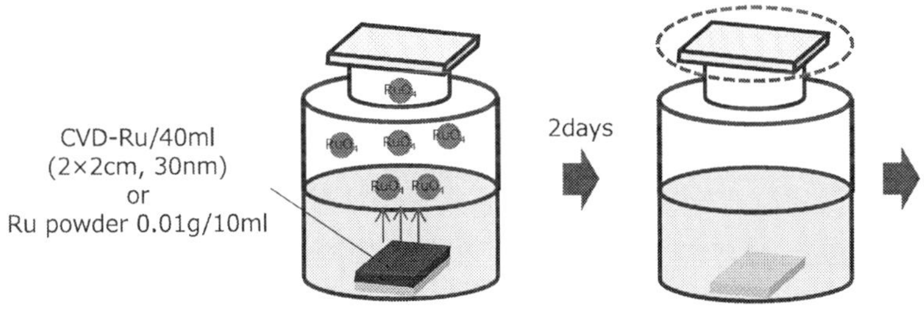


Check discoloration of the plastic on top of the bottle





# RESULT OF PLASTIC COLORATION TEST

## PLASTIC COLORATION BY RUTHENIUM TETROXIDE GAS

| Chemical | Bottle A w/o Ru | Bottle B w/Ru WF | Bottle C w/Ru powder |
|---|---|---|---|
| NaOCl | No change | Change to black | Change to black |
| CAN | No change | Change to black | Change to black |
| Chemical A | No change | Change to black | Change to black |
| Chemical B | No change | No change | No change |
| Chemical C | No change | No change | No change |

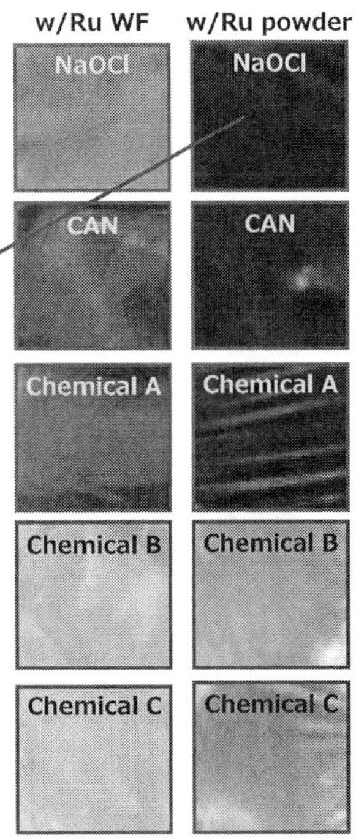


**SEM-EDX of plastic**

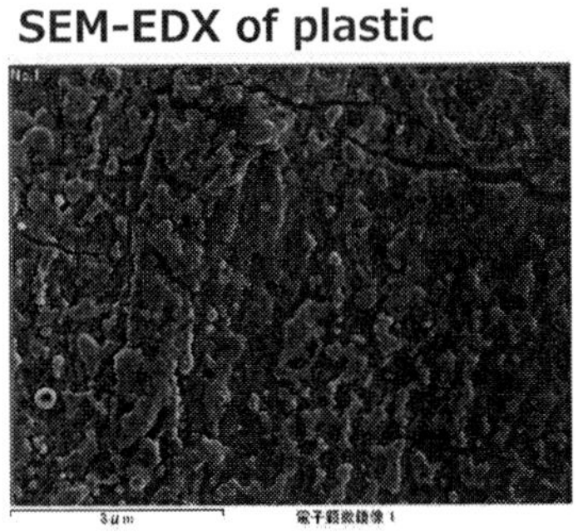

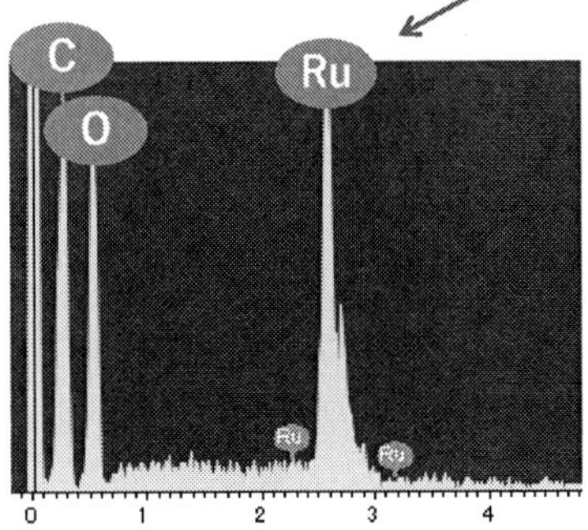


**Chemical B and C can prevent $RuO_4$ gas**

# OUTLINE

1. Introduction
2. Design of Ru wet etch chemical
3. **CVD-Ru Bulk properties**
4. Summary

tok imec





# CVD RUTHENIUM BULK PROPERTIES

IMPACT OF THERMAL BUDGET OF POST DEPOSITION ANNEAL IN $N_2/H_2$ (FGA)

Please see at the conference

tok

# IMPACT ON ETCHING BEHAVIOR

IMPACT OF THERMAL BUDGET OF POST DEPOSITION ANNEAL IN $N_2/H_2$ (FGA)

By NaOCl 5% wet etching @R.T.

(See SPCC2019 Q.T. Le for more detail)

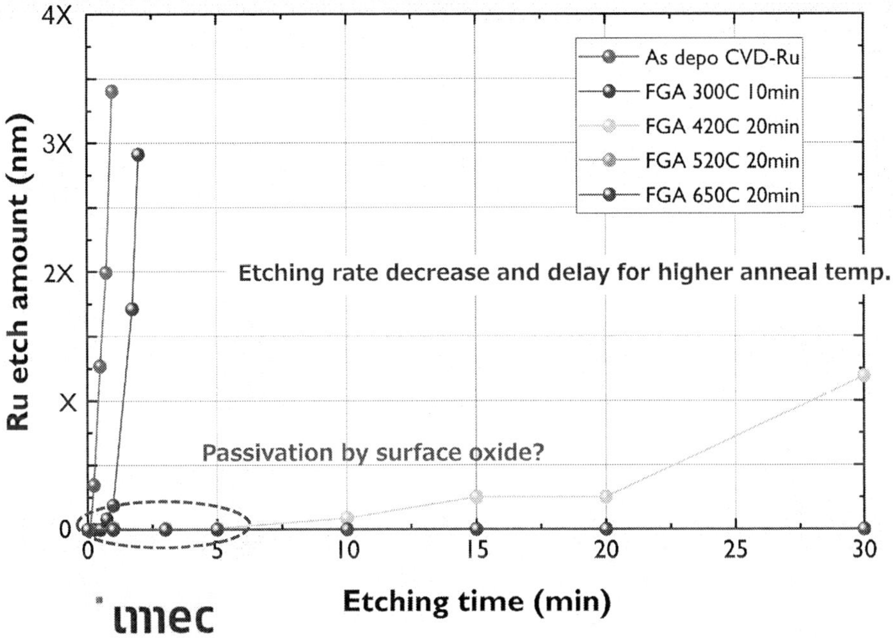






# RUTHENIUM OXIDE GROWTH

IMPACT OF THERMAL BUDGET OF POST DEPOSITION ANNEAL IN $N_2/H_2$ (FGA)

Please see at the conference

tok

# RUTHENIUM WET ETCHING BEHAVIOR

NaOCl 5% @R.T.

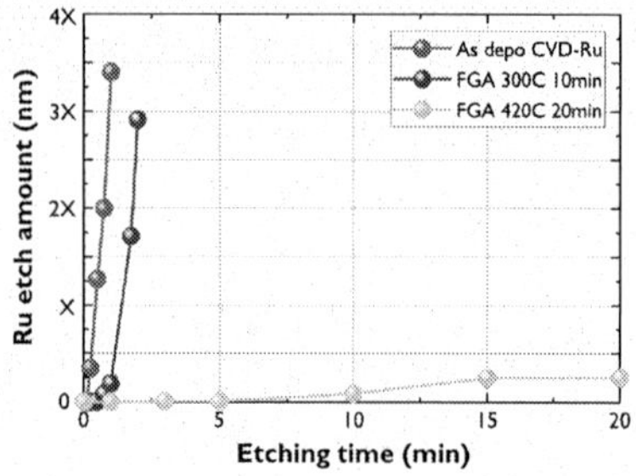


CAN50mM+$HNO_3$ @R.T.

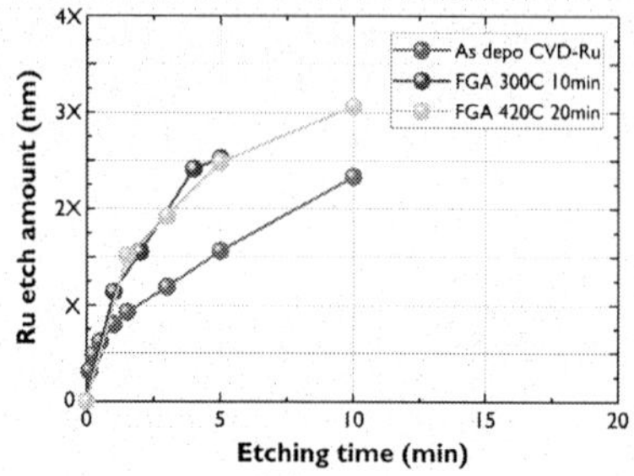


Chemical C @R.T.

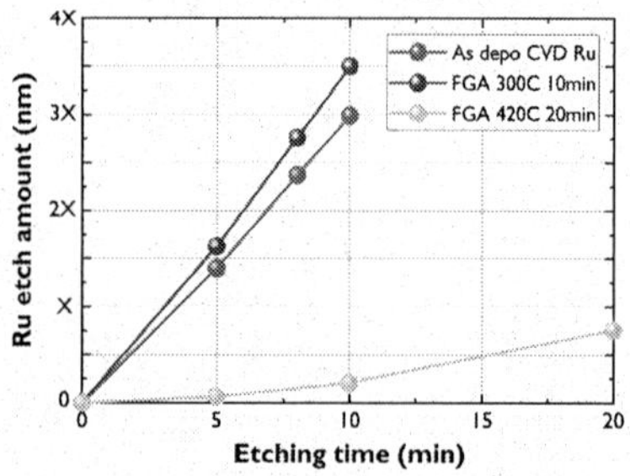


Chemical C @35°C

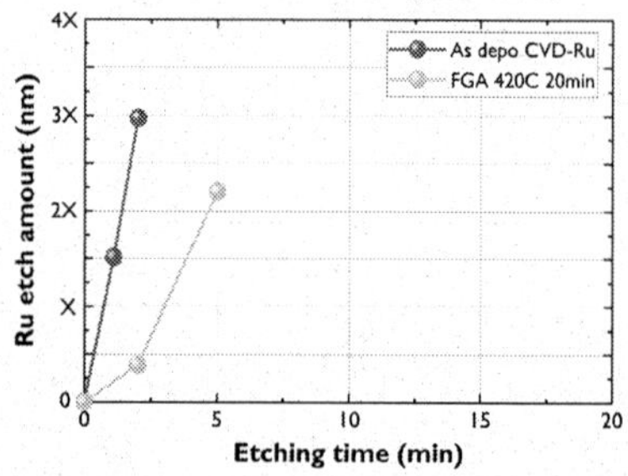


tok imec



**Chemical C @35°C can break through the passivation layer in a short time**



# DESIGN OF RUTHENIUM WET ETCH CHEMICAL

- Metal free solution
  - Metal impurities should be avoided for advanced semiconductor processing.
  - Currently, most of the strong oxidizers include metal elements. (CAN, NaOCl etc.)
- High selectivity to low-k, SiOx, SiNx, TiN and TaN
- Prevent ruthenium tetroxide ($RuO_4$) generation
  - Volatile and toxic $RuO_4$ should be avoided for EHS reasons.
- **Preservation of surface roughness to be targeted after controlled Ru recess**

tok imec

# SURFACE ROUGHNESS

30nm CVD-Ru (FGA 420°C 20min)

AFM 5×5um$^2$

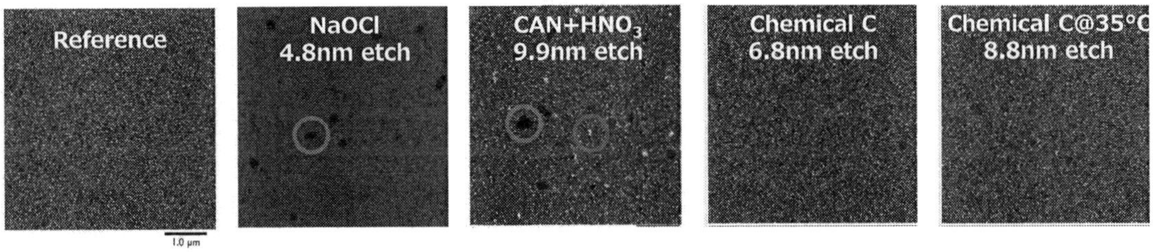


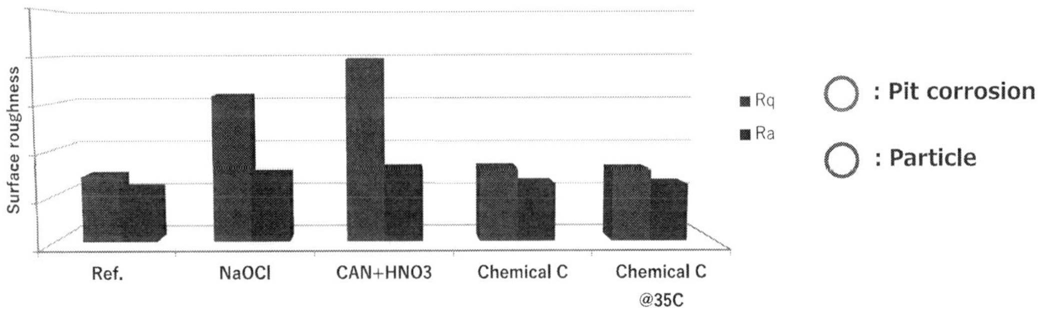


No significant surface roughness change observed with Chemical C

tok imec





# OUTLINE

1. Introduction
2. Design of Ru wet etch chemical
3. CVD-Ru Bulk properties
4. Summary

tok imec

# Summary

- New chemical is developed for Ru BV/BS clean and controlled recess.
- TOK Ru etching chemical targets the following requirements

a. Metal free process (no need for an additional clean step)
b. Eco and tool friendly process (no $RuO_4$ by product)
c. Preservation of surface roughness after controlled Ru recess
d. High selectivity to dielectrics and barrier metals

| Chemical | Metal | $RuO_4$ by product | Ru etching | Roughness | Selectivity* |
|---|---|---|---|---|---|
| NaOCl | Yes | Yes | Good | Non uniform | - |
| CAN | Yes | Yes | Good | Non uniform | - |
| Chemical B | No | Undetected | Good | Good | Good |
| Chemical C | No | Undetected | Good | Good | Good |

*low-k, SiN, TiN and TaN

tok imec





Thank you for your kind attention.

tok imec

# WET ETCHING OF RUTHENIUM: EFFECT OF THERMAL ANNEALING

Q. T. LE, E. KESTERS, H. PHILIPSEN, AND F. HOLSTEYNS
IMEC, LEUVEN, BELGIUM
Email address: QuocToan.Le@imec.be

*SPCC 2019, Portland, April 1-3rd, 2019*

PUBLIC

## OUTLINE

- Introduction to metal recess etch for Fully self-aligned via (FSAV) application
- Etching of "bulk" Ruthenium
  - Effect of annealing on blanket Ru etching
  - XPS and X-SEM characterization
- Etching of Ru liner
  - Effect of annealing on Ru liner etching
  - XPS depth profiling
  - Etching of Ru liner using commodity chemical mixtures
- Summary

## METAL RECESS ETCH FOR FULLY SELF-ALIGNED VIA (FSAV) APPLICATION

Objective: controlled recess etch of metal selectively to other materials in the stack

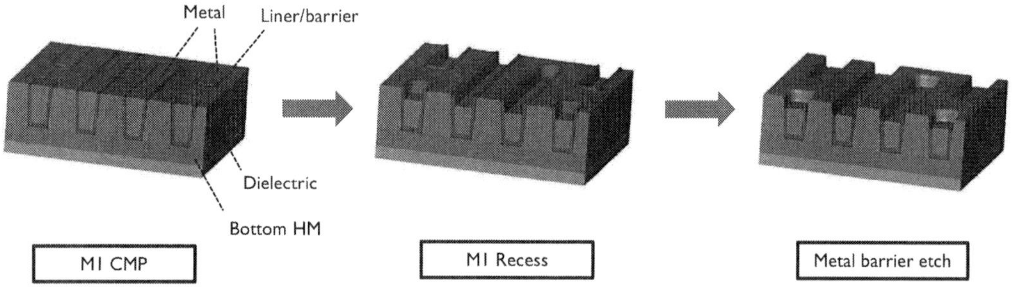


*Example of metals and liner/barrier:*
Metal = Cu; Liner/barrier = Ru/ TaN
Metal = Ru; Barrier = TaN or TiN

Co recess: A. Pacco *et al.*, SPCC 2019

imec — *Q. T. Le et al., SPCC 2019, Portland, April 1-3rd, 2019* — 3

## RUTHENIUM: METAL OF CHOICE FOR BEOL INTERCONNECT

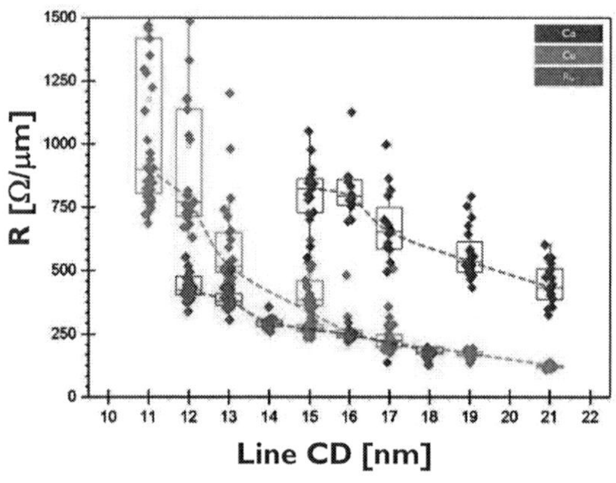


- Ru: Interconnect alternative to Cu
  - Candidate as barrier-less metallization for interconnect
  - Ru vs. Cu: Ru outperforms Cu both for line and via resistance below 12 nm CD
- Ru has been used as a liner layer


- M. van der Veen *et al.*, IITC 2018.
- H. Philipsen *et al.*, Electrochimica Acta (2018), https://doi.org/10.1016/j.electacta.2018.04.093
- H. Philipsen *et al.*, Electrochimica Acta (2019), https://doi.org/10.1016/j.electacta.2019.03.065


imec — *Q. T. Le et al., SPCC 2019, Portland, April 1-3rd, 2019* — 4

## ETCHING OF "BULK" RUTHENIUM

## ETCHING OF Ru USING HYPOCHLORITE SOLUTION

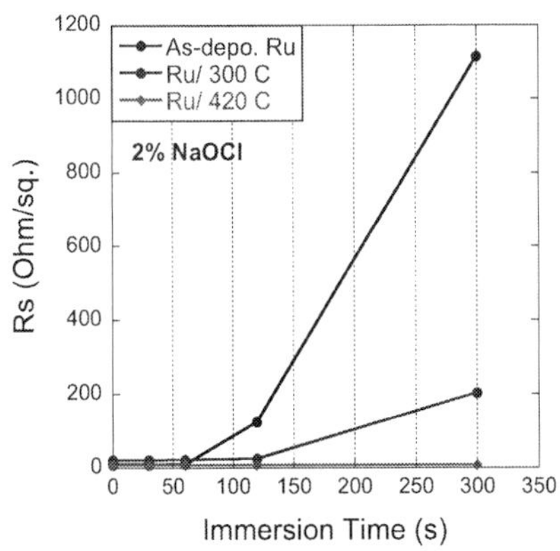


See also T. Ohashi *et al.*, SPCC 2019

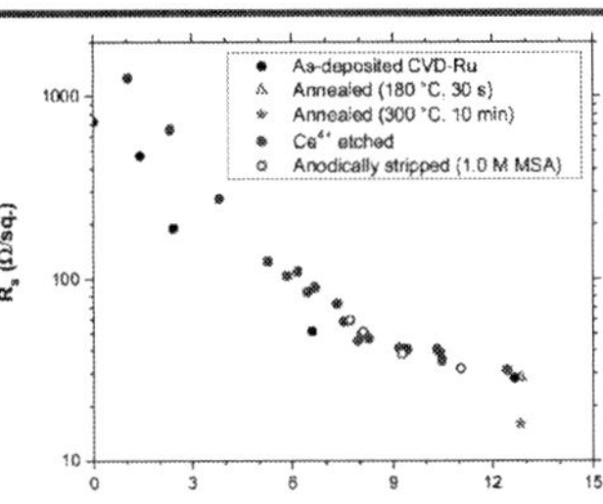


Correlation between Rs and Ru thickness

- H. Philipsen *et al.*, Electrochimica Acta (2019), https://doi.org/10.1016/j.electacta.2019.03.065

- Annealing at elevated temperature (420 °C) significantly affected Ru etch rate
- Annealing made the Ru film more chemical resistant
  - Ru etch rate is inversely proportional to annealing temperature
  - Possible mechanism
    - Bulk change (crystallinity, grain size)
    - Change of surface chemistry

imec

Q. T. Le *et al.*, SPCC 2019, Portland, April 1-3rd, 2019

# EFFECT OF ANNEALING ON Ru ETCH: XPS

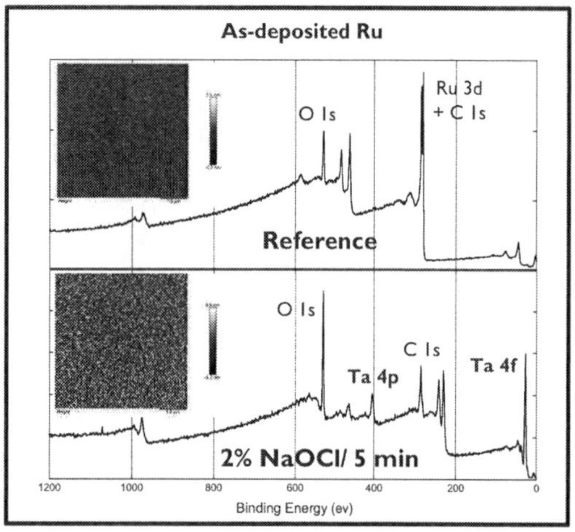


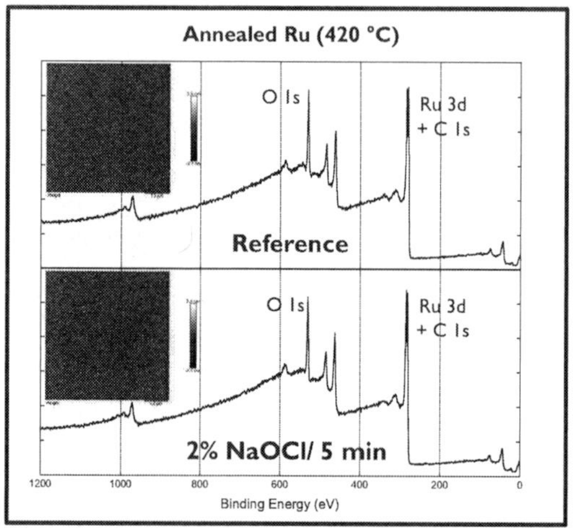


- Ta is detected for the as-depo Ru/ 5 min immersion in NaOCl ➔ Ru was removed
- 420 °C-annealed Ru was not etched in 2% NaOCl/ 5 min

imec Q. T. Le et al., SPCC 2019, Portland, April 1-3rd, 2019 7

# EFFECT OF ANNEALING ON Ru ETCH: XPS

ATOMIC CONCENTRATION (At.%)

| | O | Ru | Ru | Ru | C | N |
|---|---|---|---|---|---|---|
| | | | RuO2 | RuOy | | |
| As-depo. Ru | 24.4 | 10.9 | 5.1 | 4.3 | 54.1 | 1.2 |
| Annealed Ru | 29.9 | 5.7 | 5.7 | 4.7 | 53.3 | 0.6 |
| CVD-Ru + anneal + 5 min NaOCl | 29.4 | 5.6 | 5.6 | 4.5 | 54.4 | 0.6 |

- As-deposited Ru: after immersion in NaOCl solution: only a thin $RuO_2$ layer remained at the surface
- Annealed Ru is significantly oxidized compared to the as-deposited Ru
  - The presence of $RuO_2$ at the surface prevents/ slows down the etching of Ru layer after annealing

**Annealing at elevated temperature (420 °C) significantly affected Ru etch rate**

imec Q. T. Le et al., SPCC 2019, Portland, April 1-3rd, 2019 8

## XPS CHARACTERIZATION – Ru 3d

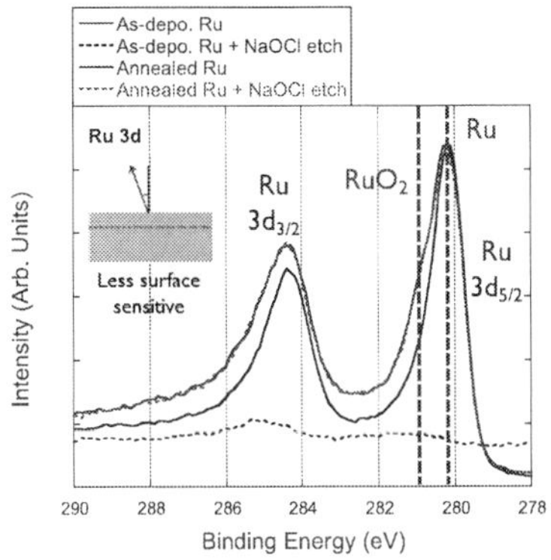


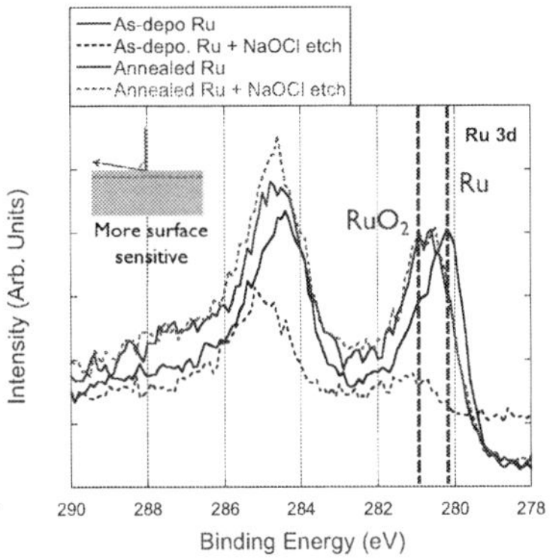


- Main effect of thermal annealing: formation of oxidized Ru at the surface
- The presence of Ru oxide layer is more noticeable when spectra are collected at higher electron take-off angle (more surface sensitive)

XPS database

Ru $3d_{3/2}$ and C 1s    Ru $3d_{5/2}$

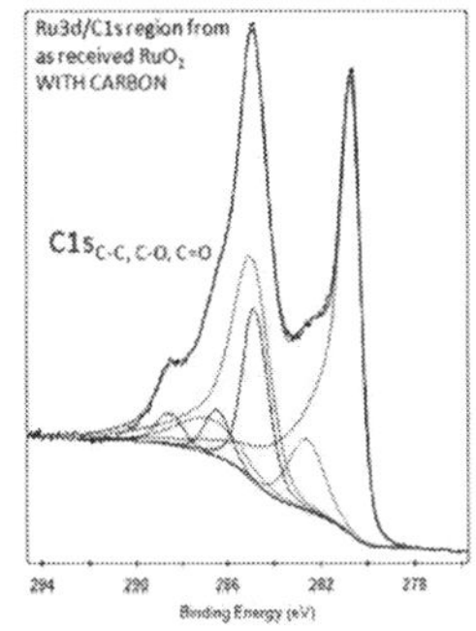


https://xpssimplified.com/elements/ruthenium.php

imec    *Q. T. Le et al., SPCC 2019, Portland, April 1-3rd, 2019*    9

## XPS CHARACTERIZATION – O 1s

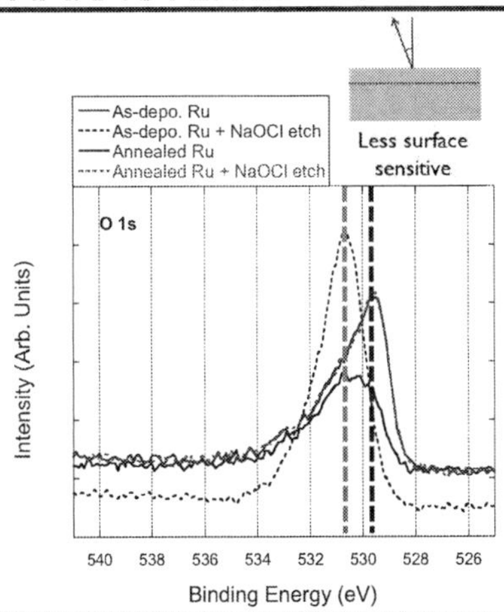


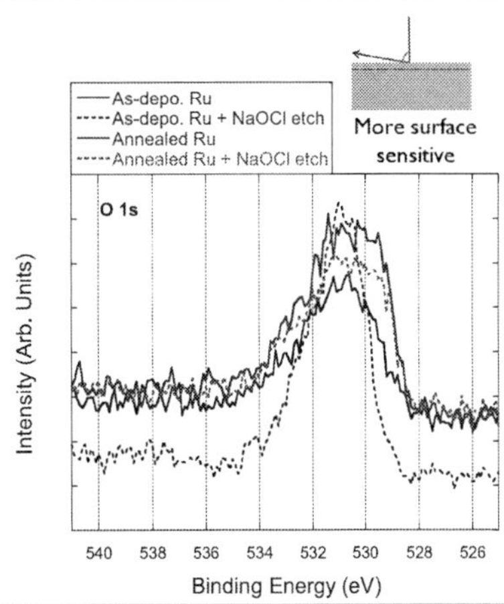


- O 1s intensity increased after annealing at 420 °C
- Significant different spectra for surface and bulk spectra
  - Oxidized Ru is only present at the surface
  - In addition to $RuO_2$, presence of other Ru with higher oxidation states

imec    *Q. T. Le et al., SPCC 2019, Portland, April 1-3rd, 2019*    10

## Ru RECESS ETCH

Blanket annealed Ru (420 °C)

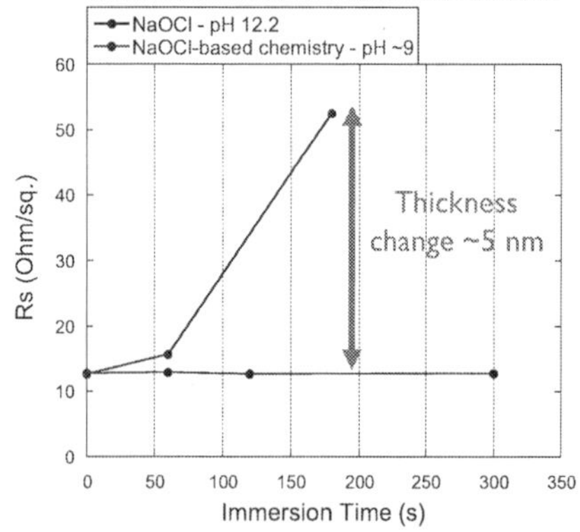


Only focused at pH >7 to avoid formation of $RuO_4$
See for example, T. Oshahi *et al.*, SPCC 2019

Patterned annealed Ru (420 °C)

Incoming structure | pH ~9/ 60 s

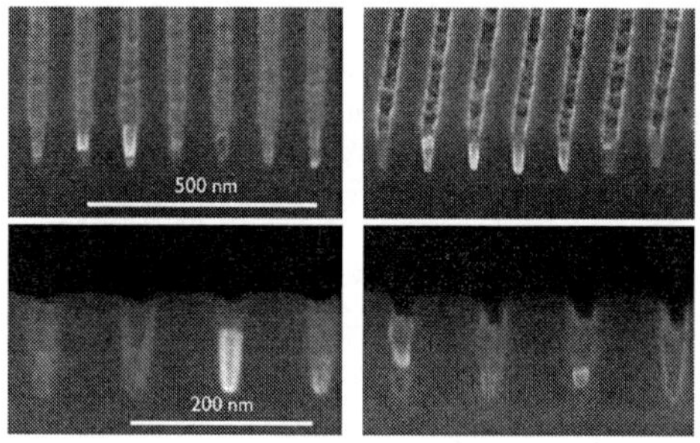


- Ru in trenches was partially etched
- Rougher surface vs. incoming surface
- Non uniform recess

imec *Q. T. Le et al., SPCC 2019, Portland, April 1-3rd, 2019*

## ETCHING OF RUTHENIUM LINER

## Cu RECESS AND Ru LINER ETCH FLOW

SHORT-LOOP STRUCTURE: Cu FILL/ Ru LINER/ TaN BARRIER

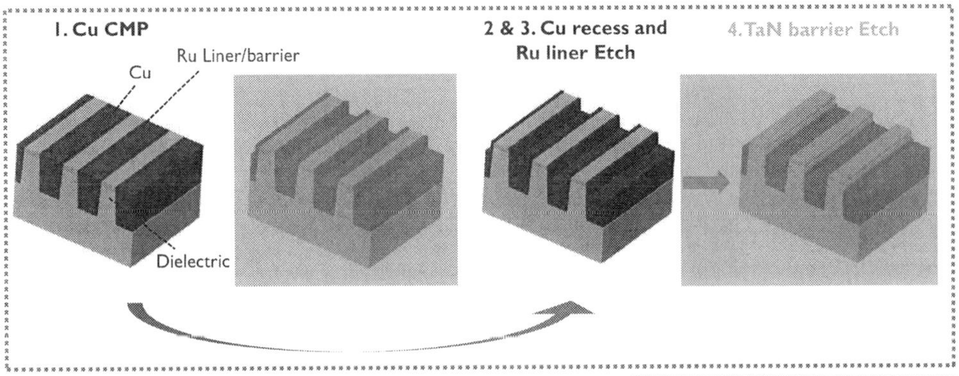


Fully self-align via integration
- G. Murdoch *et al.*, IITC 2017.
- B. D. Briggs *et al.*, IEEE IEDM 2017.



## ETCHING OF Ru LINER IN PATTERNED STRUCTURE

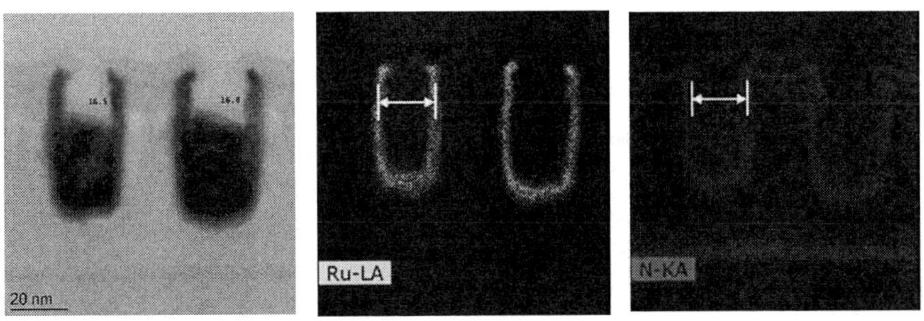


- NaOCl (2-5%) was not efficient for sidewall liner Ru removal (1-2 nm): Ru liner layer was not etched
- Ru profile is very similar to N profile
  - *Possible intermixing of Ru and TaN at the interface. Possibly, formation of $RuN_x$ and/or $RuTa_xN_y$ compound due to annealing.*
    *➔ Stripping onset potential for Ru compound is different vs. metallic Ru*

## EFFECT OF ANNEALING: XPS DEPTH PROFILES

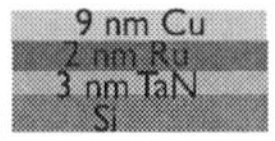


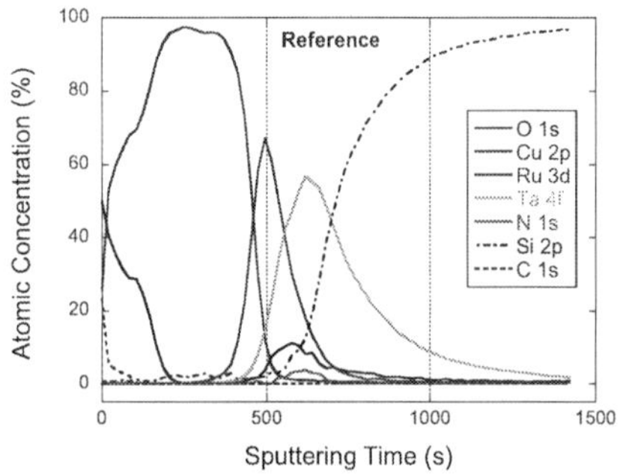


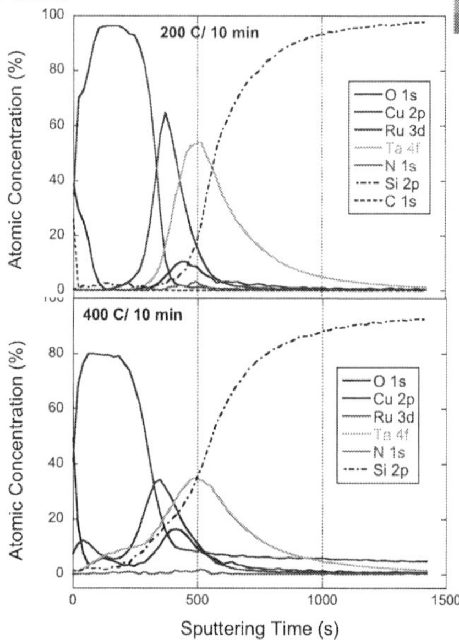


- Annealed at 400 °C resulted in major difference
  - Ru-TaN interface is much broader
  - Formation of an intermixing and/or compound of Ru-Ta could explain the different etching behavior of Ru at the interface

imec *Q. T. Le et al., SPCC 2019, Portland, April 1-3rd, 2019* 15

## SURFACE ROUGHNESS AND RECESS DEPTH EVALUATION

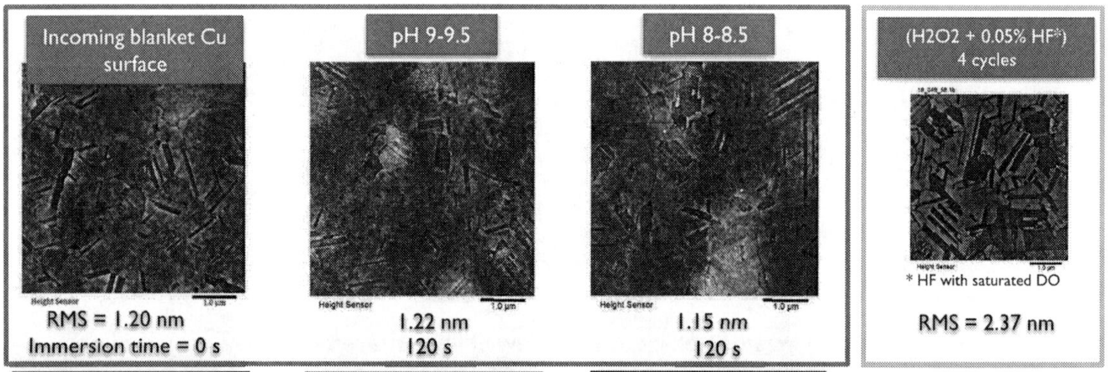


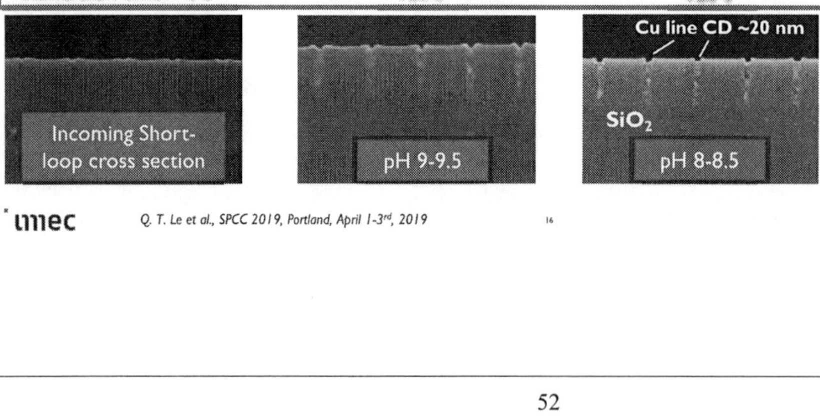


- $H_2O_2$ treatment followed by 0.05% HF etch: high surface roughness
- Promising results for Cond. 4 and Cond. 5

imec *Q. T. Le et al., SPCC 2019, Portland, April 1-3rd, 2019* 16

## SIMULTANEOUS Cu AND Ru LINER ETCH: TEM RESULTS

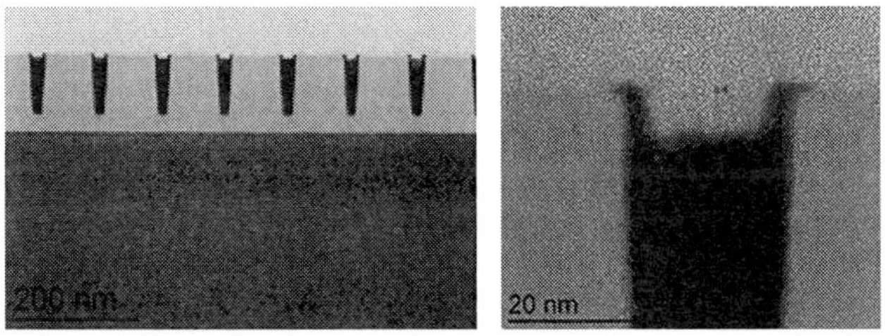


- Cu lines with CD ~20 nm
- Estimation of Cu recess ~7-9 nm
- Simultaneous Cu recess and Ru liner etch demonstrated

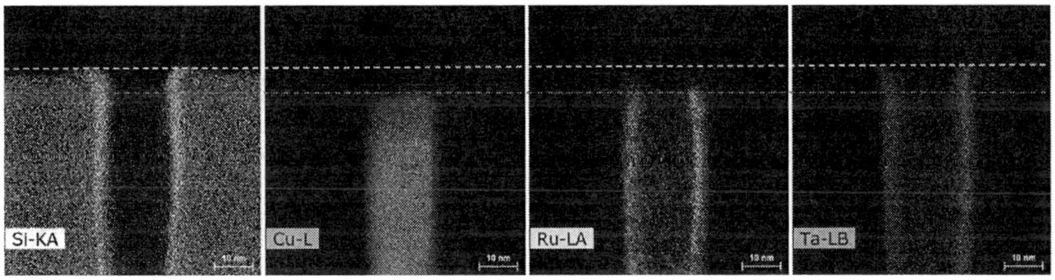




## SUMMARY

## SUMMARY

- Thermal annealing at 420 °C made Ru more resistant to wet etch
- "Bulk" Ru
  - Presence of an oxidized Ru layer at the surface strongly affected the etch rate
  - Annealed Ru can be etched by NaOCl-based chemistry, however it resulted in *rough Ru surfaces*
- Thin layer of Ru (Ru liner)
  - Formation of an intermixing and/or compound at the interface could explain the different etching behavior of Ru → tuning of the etch chemicals, pH, and composition required
- Remark

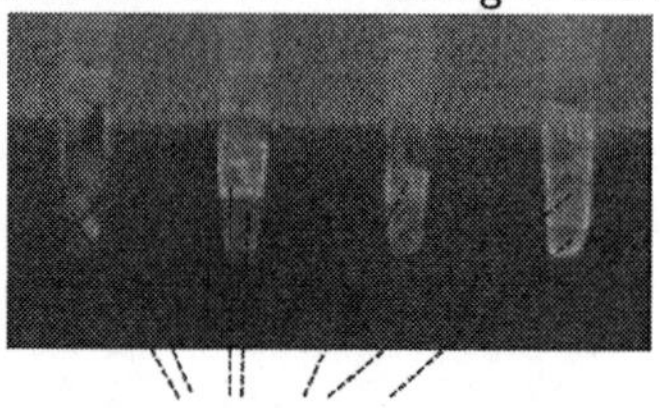

Grains of different sizes (X-SEM)

- Different grain sizes → different grain boundary density
  - Surface roughness after wet etching ↗
  - Etch uniformity ↘

imec Q. T. Le et al., SPCC 2019, Portland, April 1-3rd, 2019 19

## ACKNOWLEDGEMENTS


- **Integration**: Chris Wilson, Gayle Murdoch, Guillaume Boccardi
- **Surface and Interface Processing and CMP**: Antoine Pacco, Nancy Heylen, Katia Devriendt
- **Characterization (XPS, AFM, TEM, RBS)**: Thierry Conard, Ilse Hoflijk, Inge Vaesen, Danielle Vanhaeren, Stefanie Sergeant, Eric Vancoille, Kris Paulussen, Laura Nelissen, Hugo Bender, Johan Meersschaut, Johan Desmet


imec Q. T. Le et al., SPCC 2019, Portland, April 1-3rd, 2019 20

imec

embracing a better life

imec

PUBLIC

# CLEANING CHALLENGES ASSOCIATED WITH SCALING BOOSTERS AND PERFORMANCE ENHANCEMENT FOR ADVANCED LOGIC DEVICES

YUSUKE ONIKI

*ON BEHALF OF THE IMEC SURFACE AND INTERFACE PROCESSING GROUP & LOGIC PROGRAM TEAMS*

*imec, Kapeldreef 75, 3001 Leuven, Belgium, Oniki.Yusuke@imec.be*

*SPCC 2019, 2-3rd Apr. 2019, Portland, OR, US*

PUBLIC

## OUTLINE

- Introduction
  - imec logic scaling roadmap
  - DTCO scaling boosters & performance enhancement knobs / cleaning challenges
1. **Pattern stiction free drying** – ***High aspect ratio***
   - Surface modification & carbon removal
2. **Replacement metal gate** – ***Confined spacing & new materials***
   - Zero-thickness Vt tuning
3. **Interconnect** – ***New materials***
   - MOL/BEOL self-aligned interconnect
4. **Gate-all-around** – ***New device architectures***
   - Nanowire/Nanosheet selective etches & STCO
- Summary

imec 2

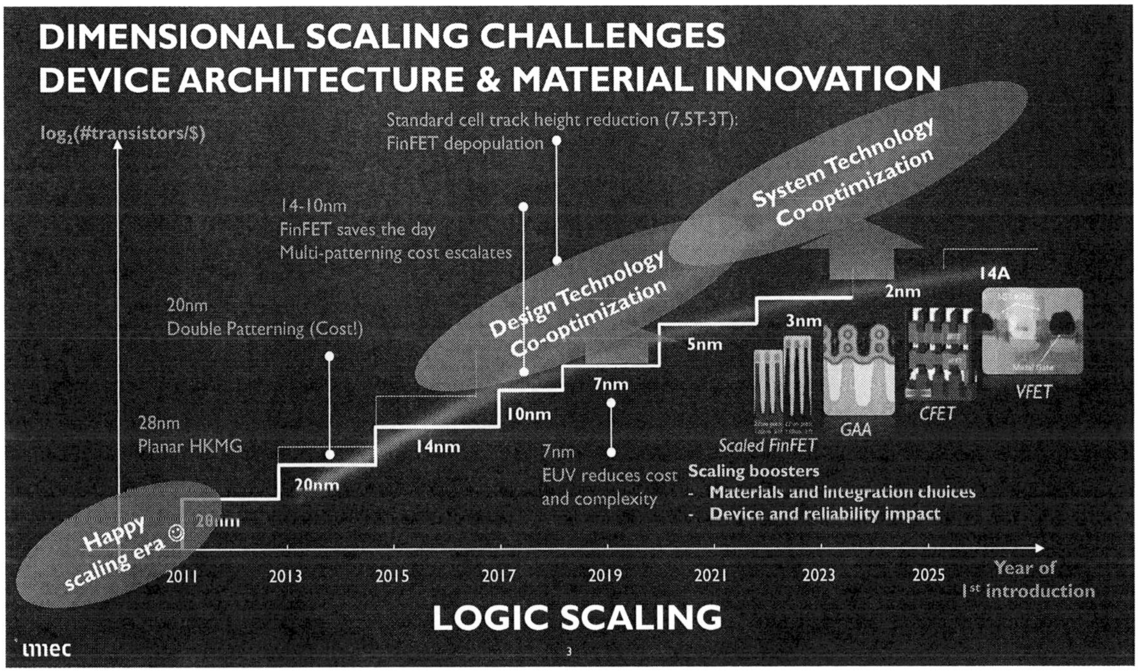

DIMENSIONAL SCALING CHALLENGES
DEVICE ARCHITECTURE & MATERIAL INNOVATION
log2(#transistors/$)
Standard cell track height reduction (7.5T-3T): FinFET depopulation
System Technology Co-optimization
14-10nm
FinFET saves the day
Multi-patterning cost escalates
Design Technology Co-optimization
20nm
Double Patterning (Cost!)
28nm
Planar HKMG
Happy scaling era ☺
28nm
20nm
14nm
10nm
7nm
5nm
3nm
2nm
14A
7nm
EUV reduces cost and complexity
Scaled FinFET
GAA
CFET
VFET
Scaling boosters
- Materials and integration choices
- Device and reliability impact
2011
2013
2015
2017
2019
2021
2023
2025
Year of 1st introduction
LOGIC SCALING
imec
3


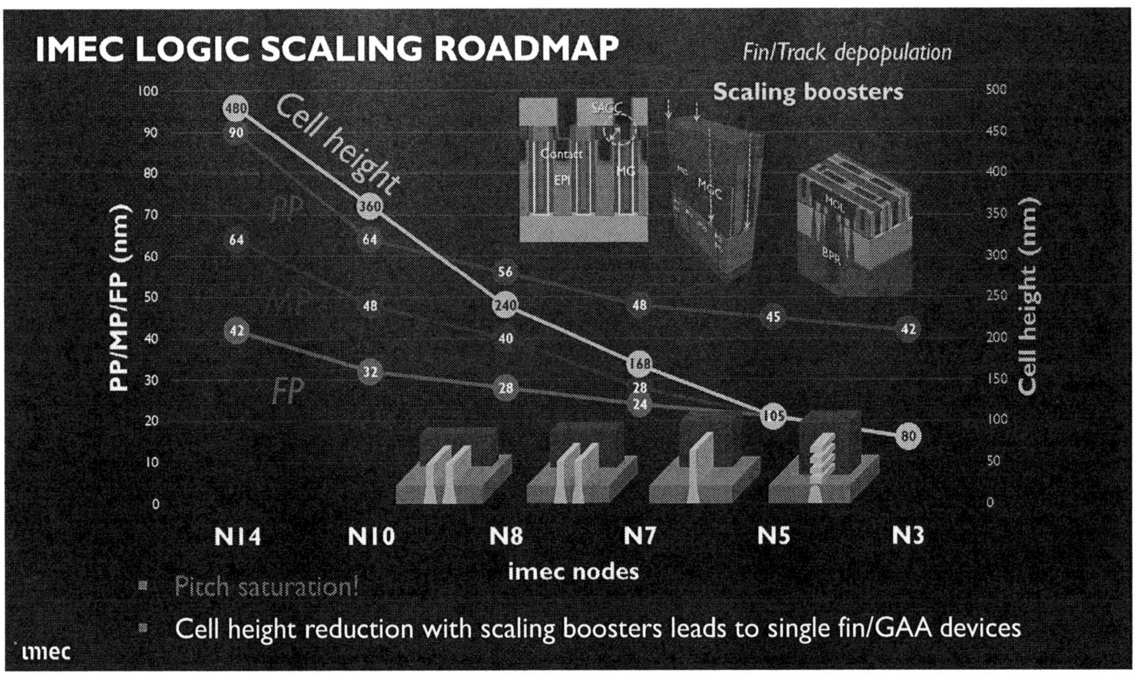

IMEC LOGIC SCALING ROADMAP
Fin/Track depopulation
Scaling boosters
Cell height
PP
MP
FP
PP/MP/FP (nm)
Cell height (nm)
SAGC
Contact
EPI
MG
MOL
BPR
480
360
240
168
105
80
90
64
64
56
48
45
42
48
40
42
32
28
28
24
N14
N10
N8
N7
N5
N3
imec nodes
Pitch saturation!
Cell height reduction with scaling boosters leads to single fin/GAA devices
imec

# DTCO SCALING

## CELL LEVEL – SCALING BOOSTERS

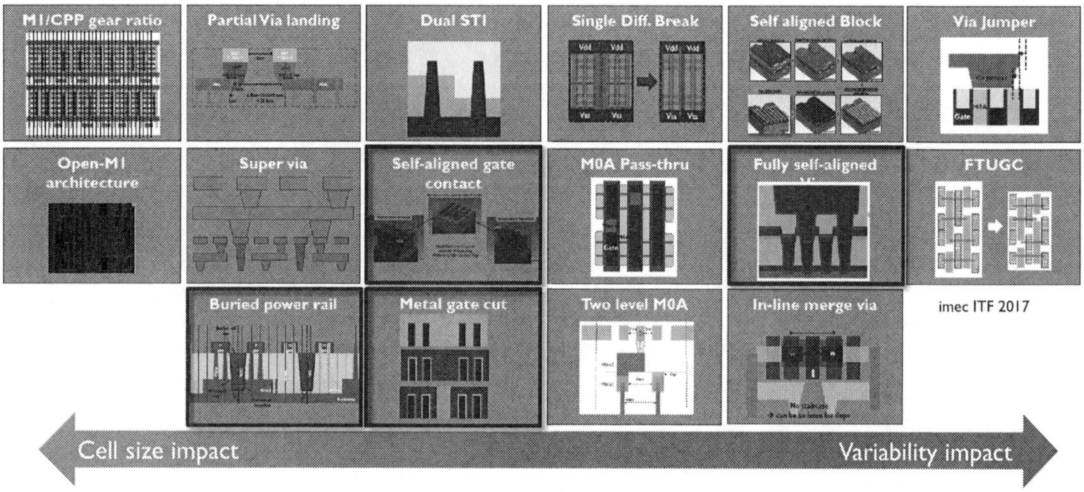


imec

5

# KEY SCALING BOOSTERS

## Buried power rail (BPR)

- Power rail ($V_{DD}/V_{SS}$): BEOL → MOL → FEOL

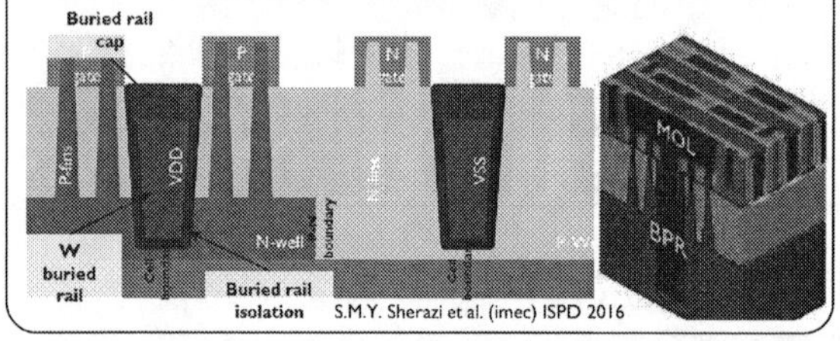


S.M.Y. Sherazi et al. (imec) ISPD 2016

## Metal gate cut (MGC)

- MGC to replace poly cut
- To minimize gate extension

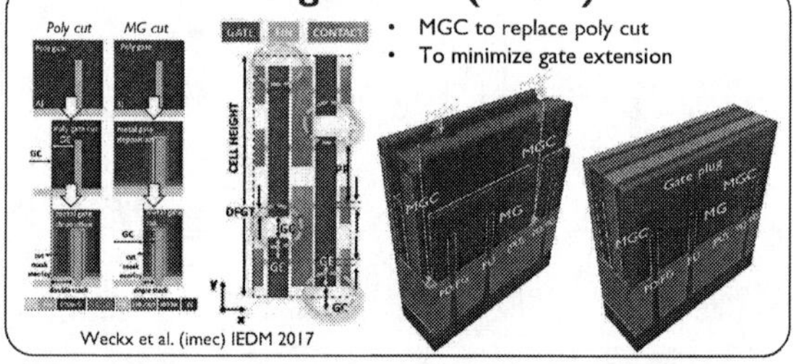


Weckx et al. (imec) IEDM 2017

## Self-aligned gate contact (SAGC)

- Overlay tolerance
- Allowing gate contacting on active

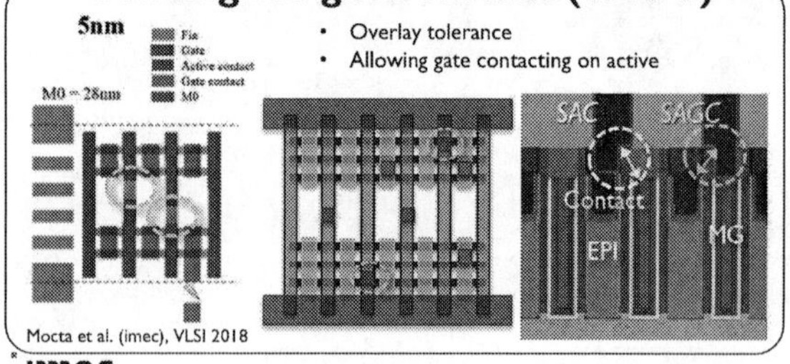


Mocta et al. (imec), VLSI 2018

## Fully self-aligned via (FSAV)

- Overlay tolerance

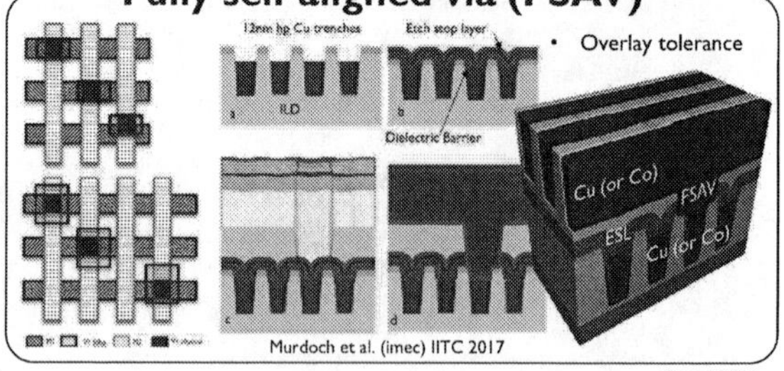


Murdoch et al. (imec) IITC 2017

imec

6

COVENTOR

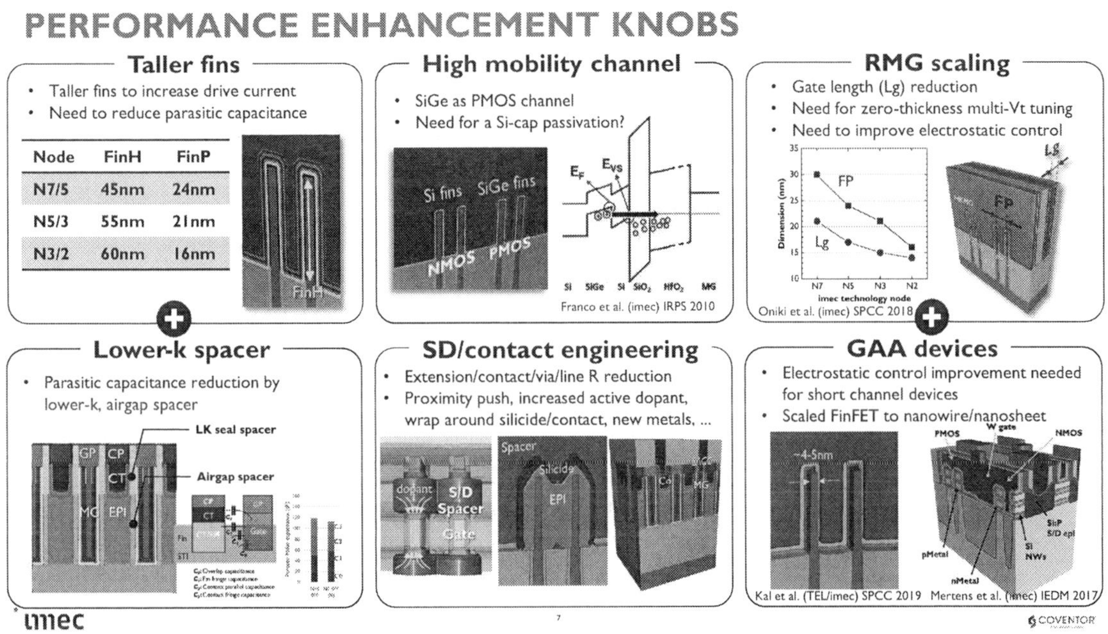
PERFORMANCE ENHANCEMENT KNOBS
Taller fins
• Taller fins to increase drive current
• Need to reduce parasitic capacitance
Node FinH FinP
N7/5 45nm 24nm
N5/3 55nm 21nm
N3/2 60nm 16nm
FinH
High mobility channel
• SiGe as PMOS channel
• Need for a Si-cap passivation?
Si fins SiGe fins
NMOS PMOS
Franco et al. (imec) IRPS 2010
RMG scaling
• Gate length (Lg) reduction
• Need for zero-thickness multi-Vt tuning
• Need to improve electrostatic control
FP
Lg
imec technology node
Oniki et al. (imec) SPCC 2018
Lower-k spacer
• Parasitic capacitance reduction by lower-k, airgap spacer
LK seal spacer
Airgap spacer
GP CP CT MG EPI
SD/contact engineering
• Extension/contact/via/line R reduction
• Proximity push, increased active dopant, wrap around silicide/contact, new metals, ...
dopant S/D Spacer Gate
Spacer Silicide EPI
GAA devices
• Electrostatic control improvement needed for short channel devices
• Scaled FinFET to nanowire/nanosheet
~4-5nm
PMOS W gate NMOS
pMetal nMetal
Kal et al. (TEL/imec) SPCC 2019 Mertens et al. (imec) IEDM 2017
imec
7
COVENTOR

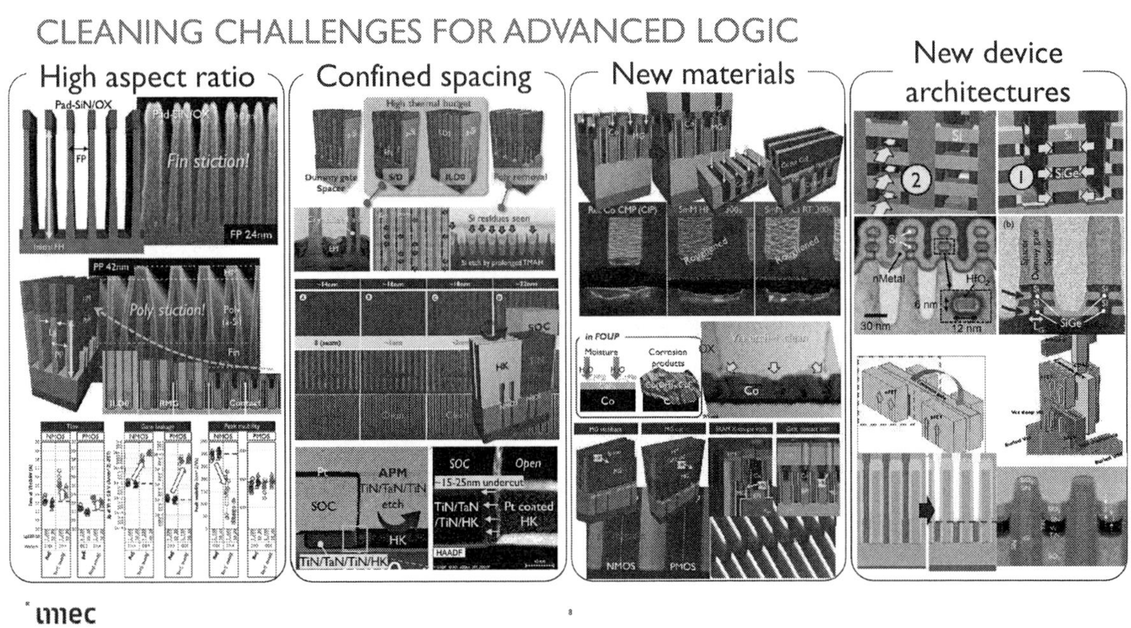
CLEANING CHALLENGES FOR ADVANCED LOGIC
High aspect ratio
Confined spacing
New materials
New device architectures
Pad-SiN/OX
Fin stiction!
FP 24nm
PP 42nm
Poly stiction!
High thermal budget
Si residues seen
SOC
HK
APM TiN/TaN/TiN etch
SOC Open
~15-25nm undercut
TiN/TaN /TiN/HK
Pt coated HK
TiN/TaN/TiN/HK
in FOUP
Co
OX
NMOS PMOS
nMetal
HfO2
SiGe
imec
8

## PATTERN STICTION FREE DRYING

## HIGH ASPECT RATIO PATTERN STICTION ISSUES

### TALL FIN/POLY

**Fin stiction @post STI etch clean**

- Taller fins/deeper STI needed to increase drive current and for isolation → **Fin stiction!**

| Node | Final FH* | FinP |
|---|---|---|
| N7/5 | 45nm | 24nm |
| N5/3 | 55nm | 21nm |
| N3/2 | 60nm | 16nm |

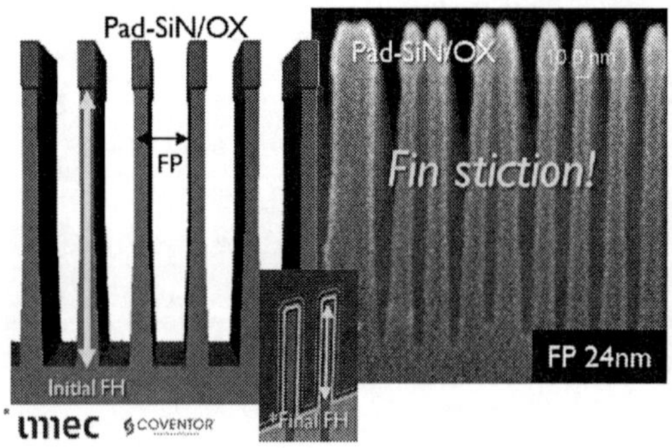


**Poly stiction @post poly etch clean**

- Gate height loss by increased etch/CMP step# in FEOL/MOL → need for a taller dummy poly → **Poly stiction!**

| Node | PP | Lg |
|---|---|---|
| N7/5 | 48nm | 21nm |
| N5/3 | 45nm | 17nm |
| N3/2 | 42nm | 15nm |

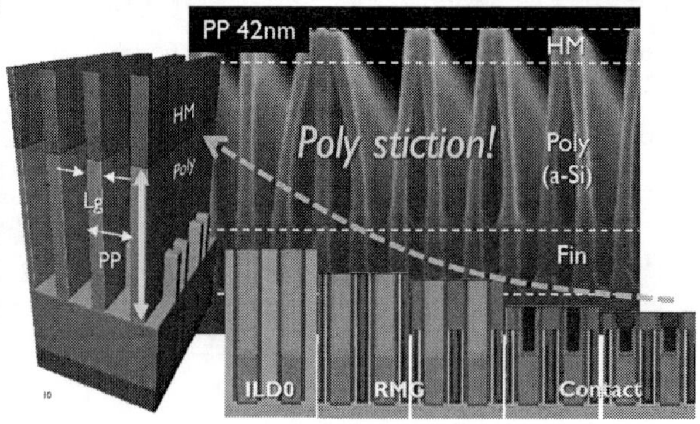

# HIGH ASPECT RATIO STICTION FREE DRYING TECHNIQUES

IPA based → Surf. modification → Sublimation → Supercritical

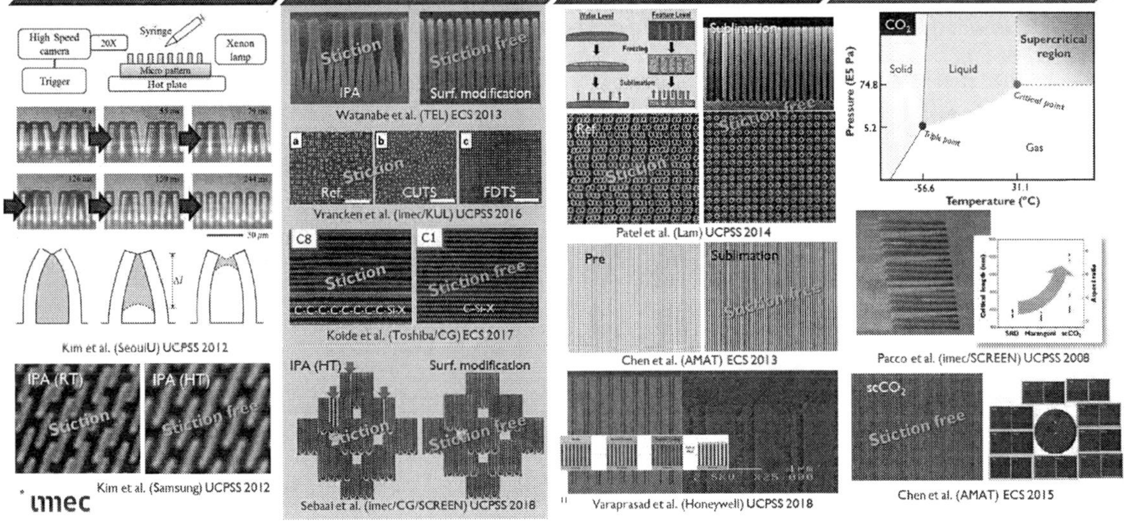


imec

# NEED TO REMOVE CARBON RESIDUES

## NEGATIVE IMPACT ON THE ELECTRICAL PROPERTIES EXPECTED

*Surf. Modification in RMG

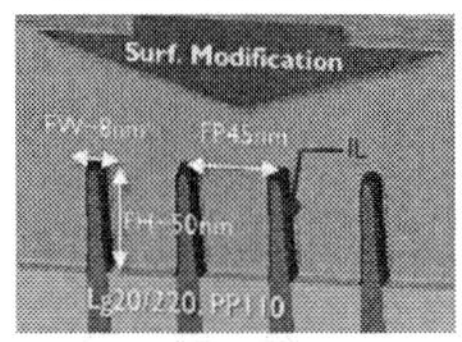


- SRAM: leakage current though STI/Fin interface
  → N/P isolation fail

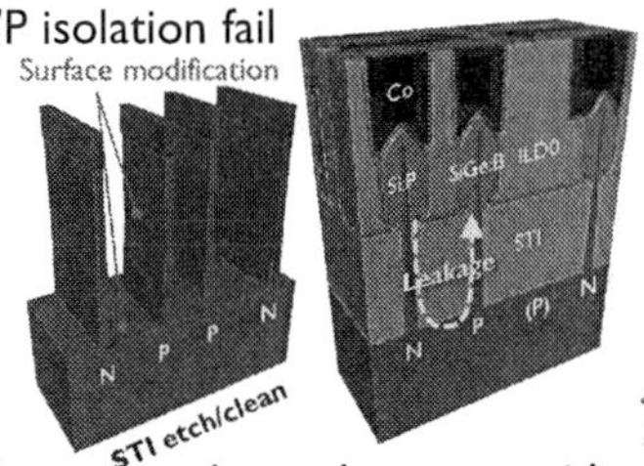


- I/O: dummy oxide used as gate oxide
  → high Dit and Jg expected

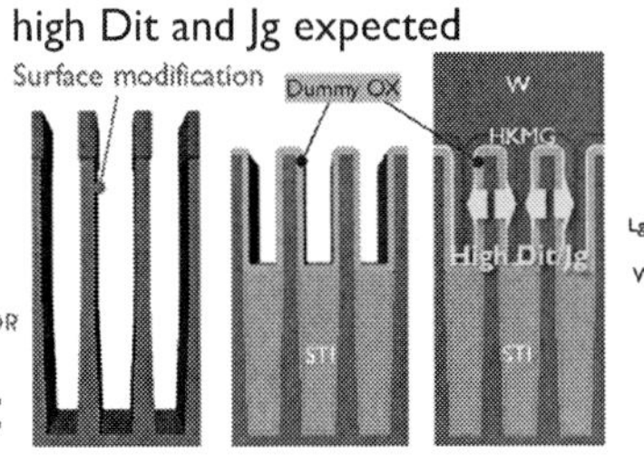


- Electrical properties degraded by surface modification*

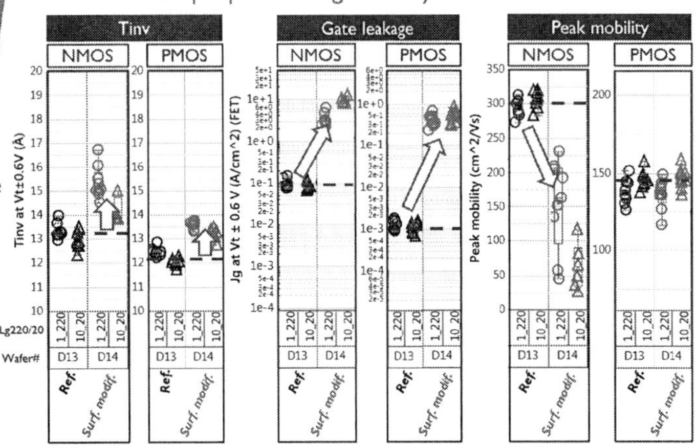


COVENTOR

imec

# CARBON REMOVAL BY UV IRRADIATION
## BLANKET & LINE/SPACE STRUCTURES

VUV λ<200nm
$N_2$ ambient

**Bare Si blanket**

- **WCA:** apparent complete removal with dose ≥4x
- **GATR-FTIR:** disappearance of $Si\text{-}CH_3$ peaks in agreement with WCA / UV strip accompanied by oxide growth

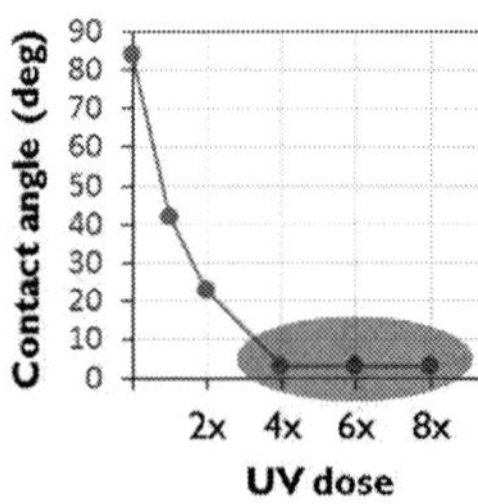


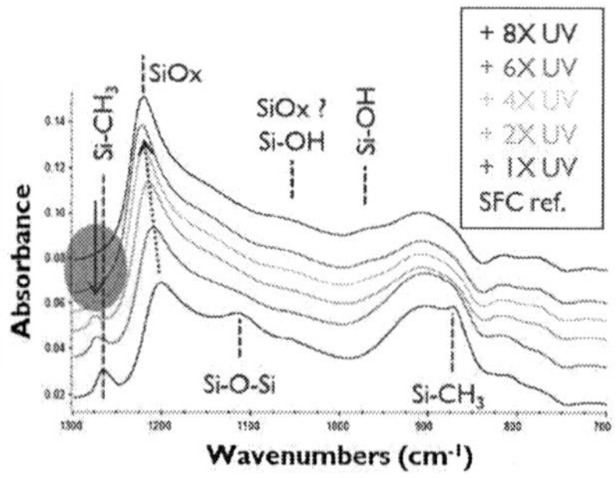


**L/S structure: ~17nm spacing**

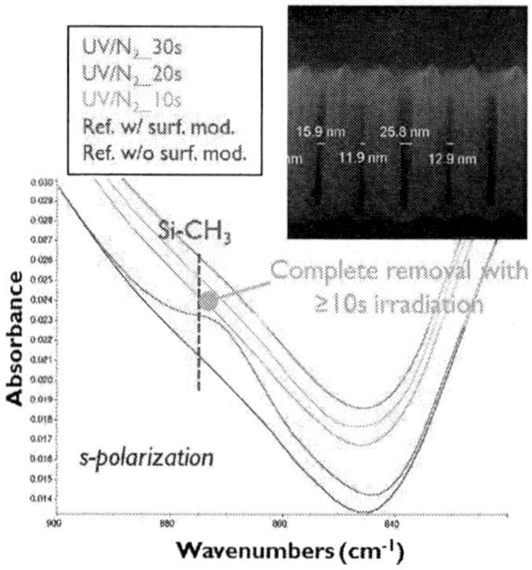


imec

Vereecke et al. (imec/SCREEN) SPCC 2018

13

# SURFACE MODIFICATION/REMOVAL ON FINFET
## @RMG

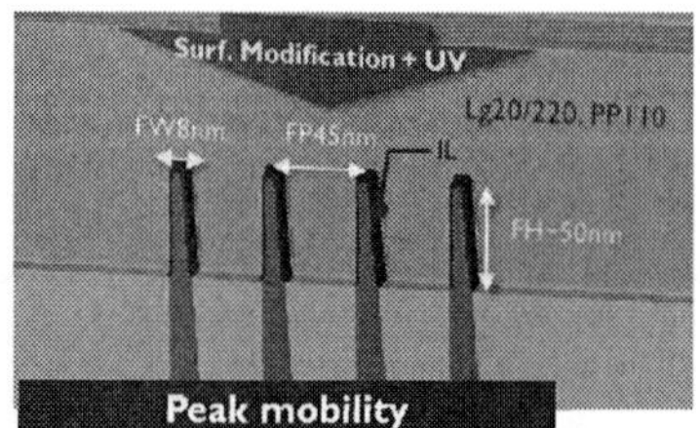


- Electrical properties degraded by surface modification and recovered by UV irradiation

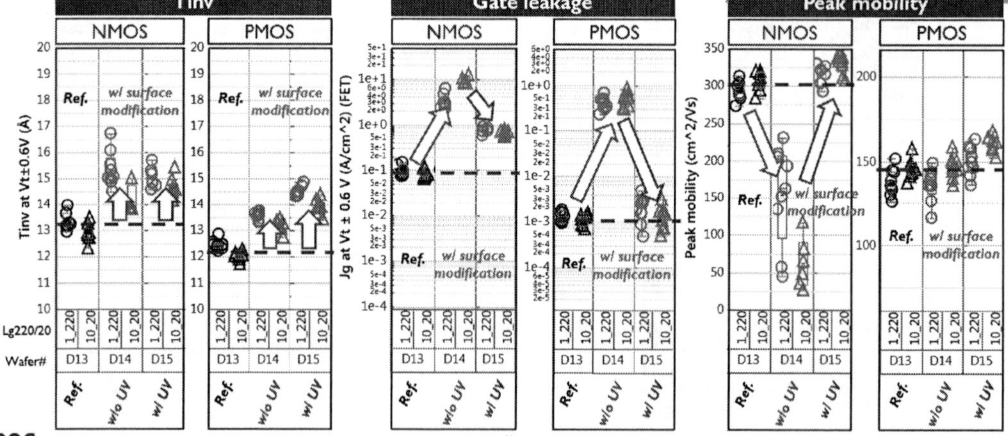


imec

14

## REPLACEMENT METAL GATE

## RMG SCALING

### IMEC ROADMAP

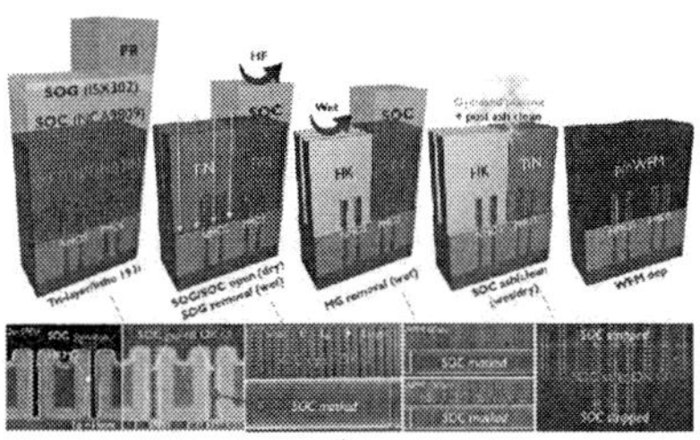


Oniki (imec) SEMICON Korea 2019

- Continuous scaling of fin pitch (FP) & gate length ($L_g$)
  - to increase drive current & to reduce parasitics ($R_{ch}$ & $C_{ch}$)
  - to gain cell area (cell height x CPP)
- Challenge for RMG patterning = confined Fin-Fin/NS-NS/gate open spacing
- Need for MG thickness scaling and **Zero-thickness Vt tuning**

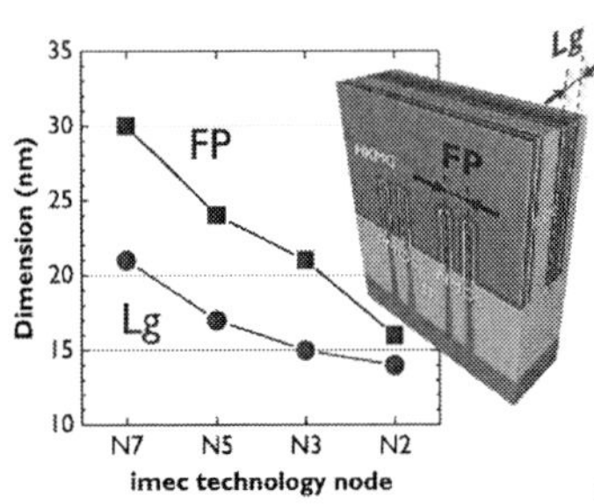


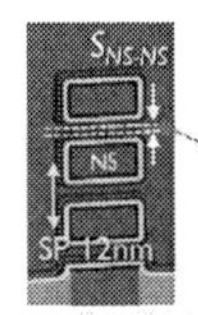


Nanosheet GAA

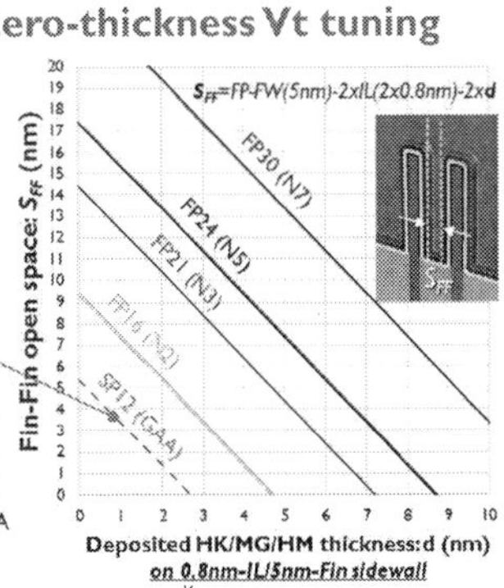


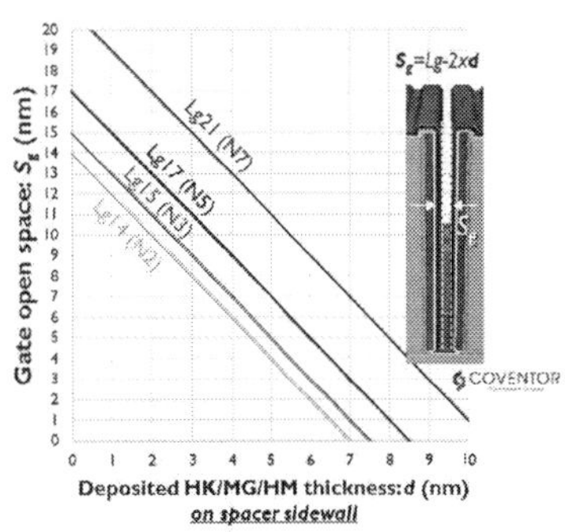


imec

# TOWARDS ZERO-THICKNESS $V_T$ TUNING

## DIPOLE

- nWF shifter: alkaline earth metals (Mg, ...) and lanthanoids (La, ...)
- pWF shifter: Al, ...

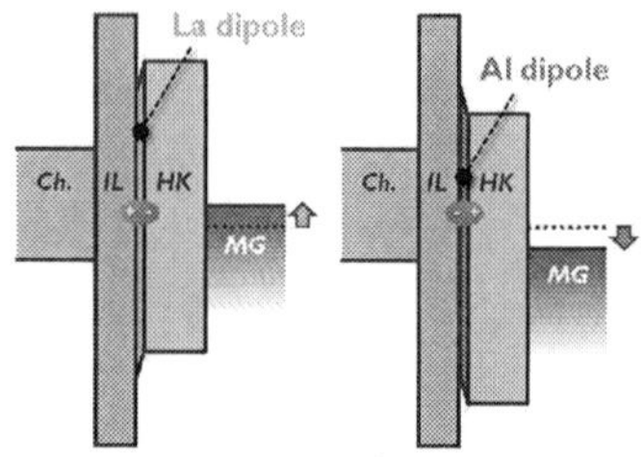


**N32 Gate first planar**

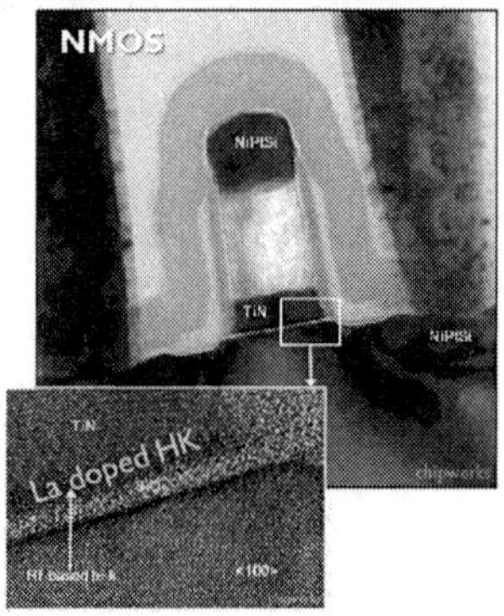


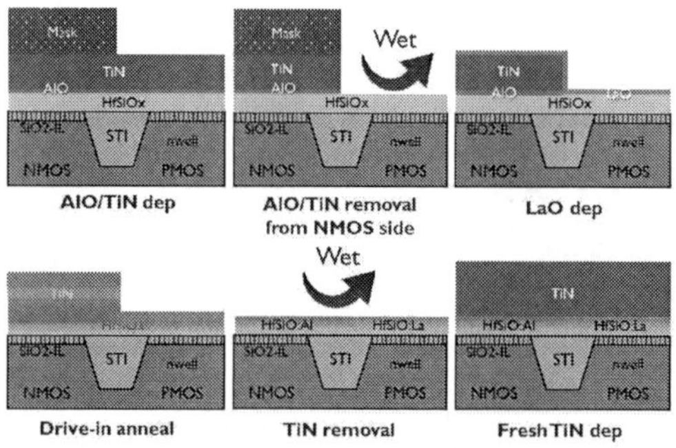


**Diffusion and gate replacement (D&GR)**

Ritzenthaler et al. (imec/Micron/Hynix) IEDM 2014
Dentoni Litta et al. (imec/Micron/Hynix) SSDM 2017

**Dipole to be redeveloped and implemented to:**

**≤N5 Logic gate stack**
- Gate last
- Multi-Vt

**Memory peri gate stack**
- Gate first
- High thermal stability

imec James (Chipworks) ASMC 2012



# CHALLENGES IN DIPOLE PATTERNING

## RISK ASSESSMENT

**Hygroscopic & water soluble**

- Impact of hydr(oxyl)ation / Q-time effect?
- Relevant wet etch/rinse solutions needed

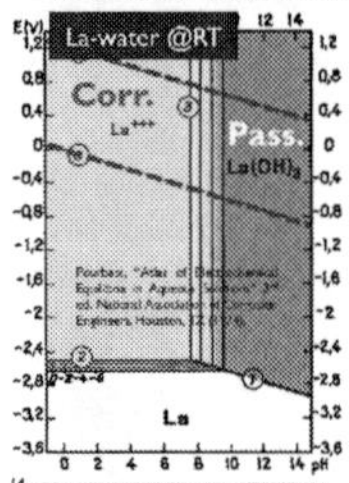


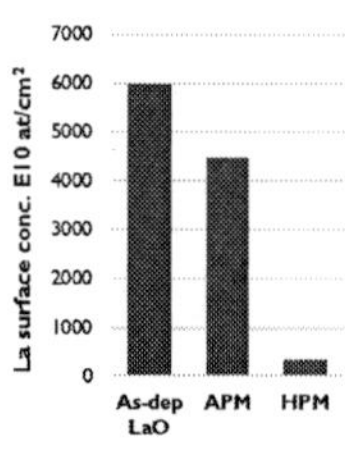


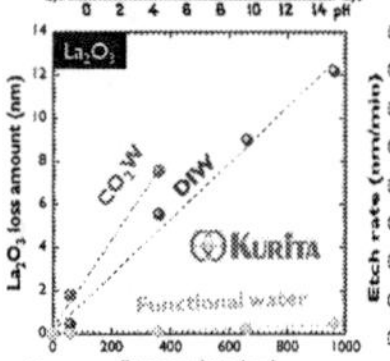


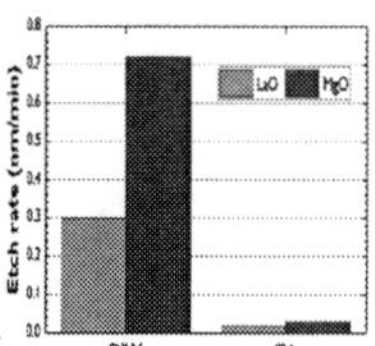


imec Iino et al. (Kurita/imec) SPCC 2019

**Insoluble fluoride residues**

- Insoluble fluoride residues to be generated if HK:Mg, La exposed to fluorine containing chemistries (dry/wet)
- LaFx, MgFx difficult to remove

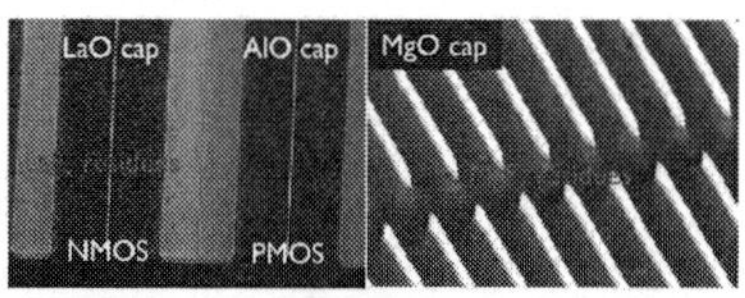


- Concern on PERR: MGEB, MGC, SRAM x-couple (butted contact), gate contact etches

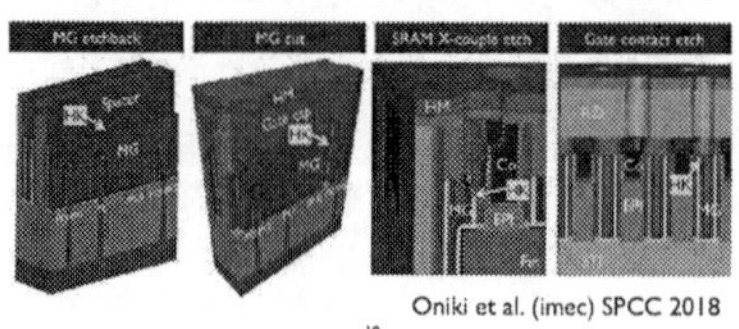


Oniki et al. (imec) SPCC 2018

**Multi-dipole patterning scheme**

*N/P dipole patterning*

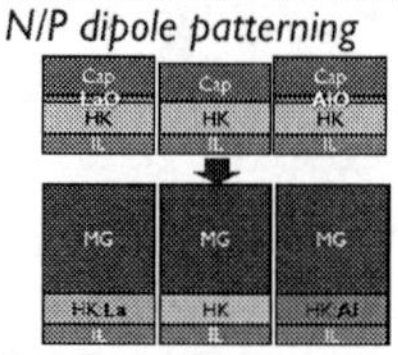


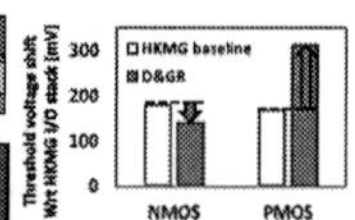


Ritzenthaler et al. (imec/Micron/Hynix) IEDM 2014

*Dipole thickness patterning*

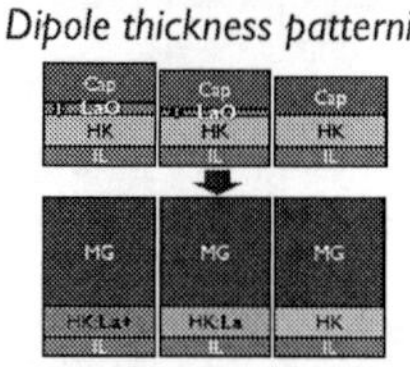


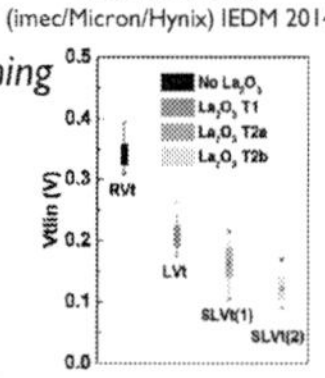


Bao et al. (IBM) IEDM 2018

*Controlling dipole diffusion length*

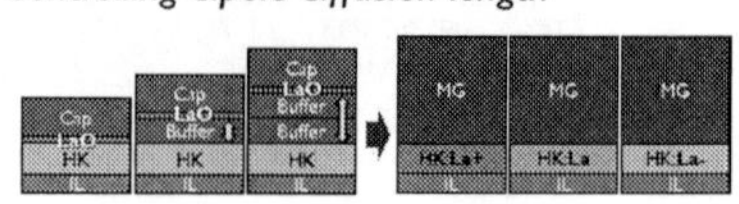

# MULTIPLE DIPOLE PATTERNING

## CONTROLLING LANTHANUM DIFFUSION LENGTH

- eWF tuning by controlling La diffusion length (= TiN buffer thickness) demonstrated

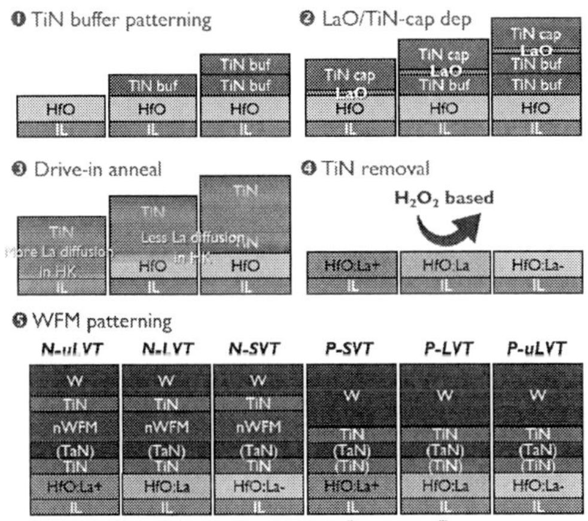


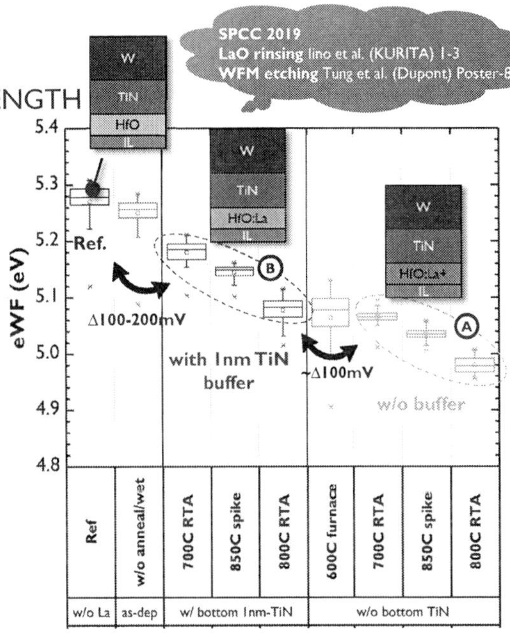


imec 19

# INTERCONNECT

## SELF-ALIGNED INTERCONNECTS

METAL ETCHBACK: DRY OR WET? OR AREA SELECTIVE DEPOSITIONS?

### Self-aligned contact (SAC)

**Metal gate etchback (dry)**

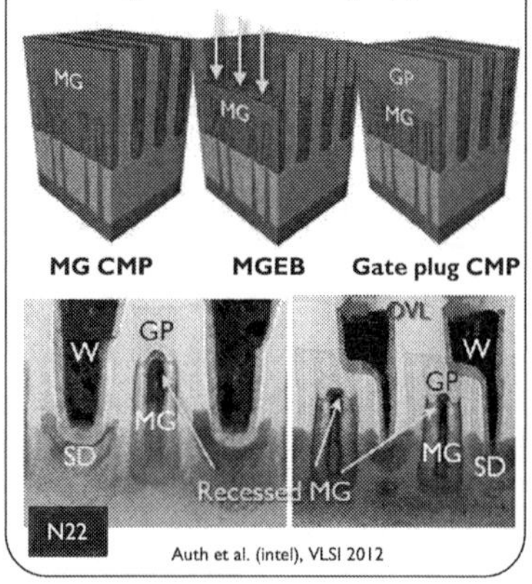


Auth et al. (intel), VLSI 2012

### Self-aligned gate contact (SAGC)

**Contact etchback (dry or wet)**

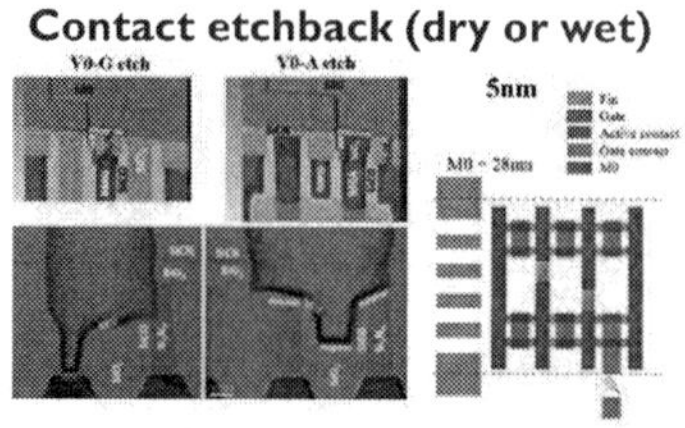


Garcia Bardon et al. (imec/KUL/Qualcomm), VLSI 2016
Mocta et al. (imec), VLSI 2018

Auth et al. (intel), IEDM 2017

### Fully self-aligned via (FSAV)

**Metal line etchback (dry or wet)**

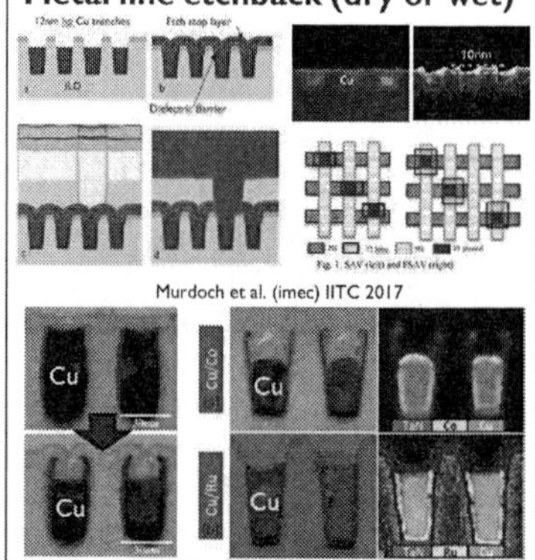


Murdoch et al. (imec) IITC 2017

Iwasaki et al. (SCREEN/imec) ECS 2017

Briggs et al. (IBM/GF/Samsung) IEDM 2017



## CHALLENGES FOR POLYCRYSTALLINE METAL ETCHBACK BY WET

COPPER, COBALT, ...

**Need to understand/control the impact of...**

- Crystal orientations
- Grain sizes/boundaries (GBs)
- Oxidation states
- Liner/barrier metals
- Incoming defects

**Check items:**
❶ Local roughness
❷ Pattern loading
❸ Liner removal
❹ Galvanic/pitting corrosion
❺ Selectivity to dielectrics

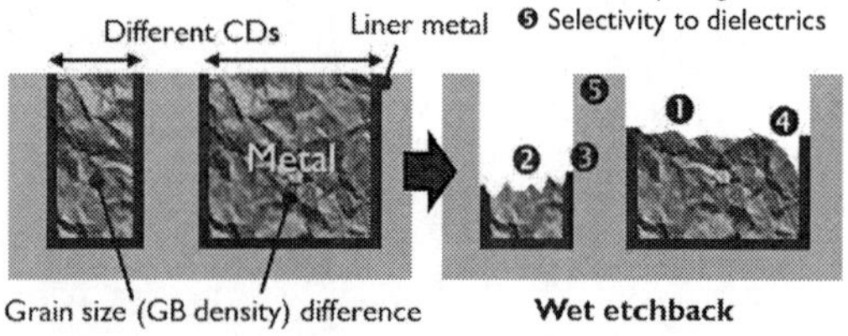


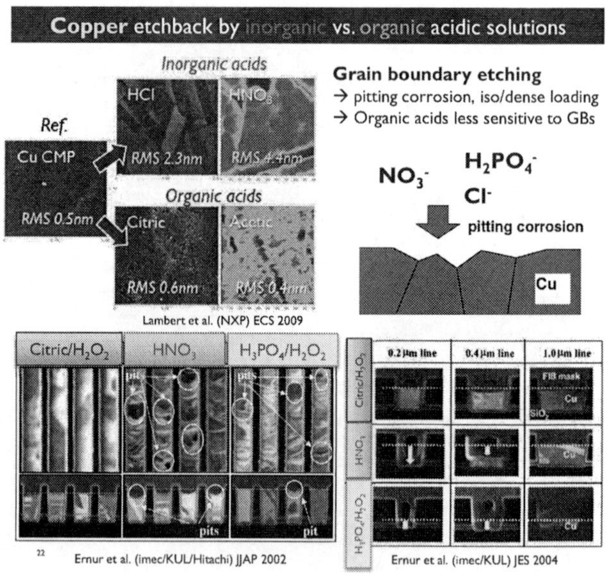


Lambert et al. (NXP) ECS 2009

Ernur et al. (imec/KUL/Hitachi) JJAP 2002

Ernur et al. (imec/KUL) JES 2004

Beker test
DOC not controlled

## COBALT ETCHBACK FEASIBILITY
### INORGANIC VS. ORGANIC ACIDIC SOLUTIONS

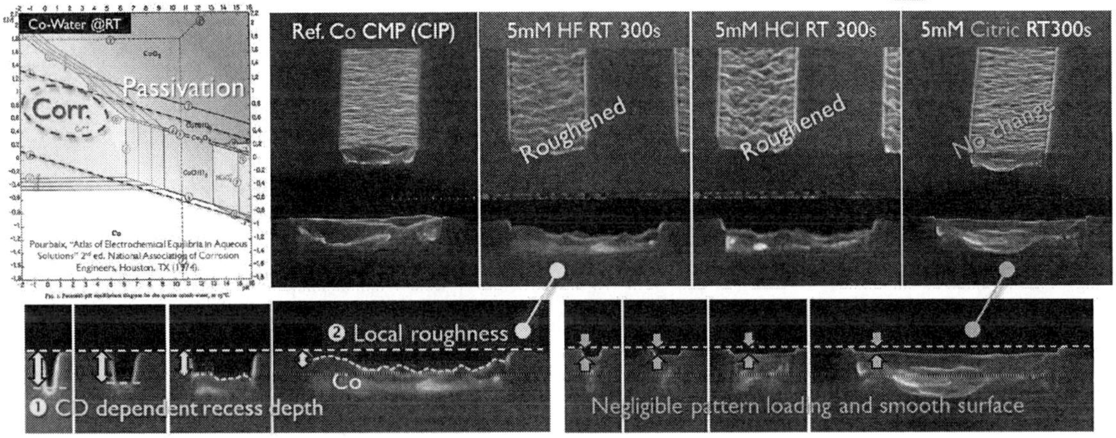


- Organic acid solutions less sensitive to the Co grain boundaries/sizes
- Need to understand the impact of pH, anion size/reactivity, water conc, dissolved oxygen/$H_2O_2$, ...
- Opportunities of formulations

imec

23

## DIGITAL WET ETCH OF POLYCRYSTALLINE METALS
### CONCEPT

- ❶ **Metal surface modification** (self-limiting)
  - **wet**, dry
  - **oxidation**, nitridation, sulfidation, carbonization, silicidation, germanidation, amorphization, ...
- ❷ **Modified layer removal** (highly selective)
  - **wet**, dry
- ❶ + ❷ sequential treatment → **atomic scale etch**
  - Etch amount/rate determined by modified (metal) layer thickness and etch cycle # = process time independence
  - Negligible pattern loading and WiWNU expected

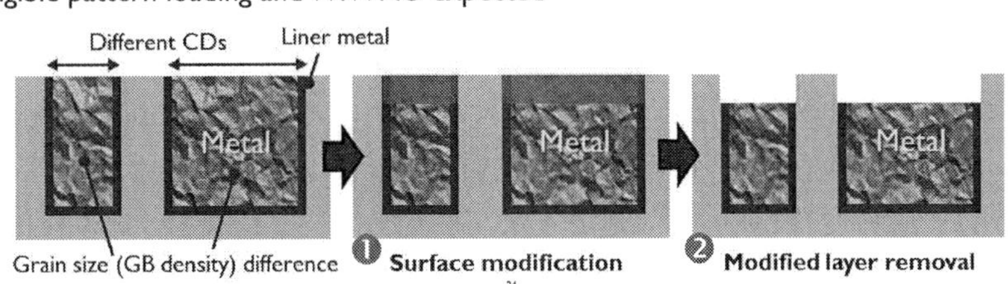


imec

24

Oniki et al. (imec) UCPSS 2018

# DIGITAL WET ETCH OF POLYCRYSTALLINE METALS

## FEASIBILITY

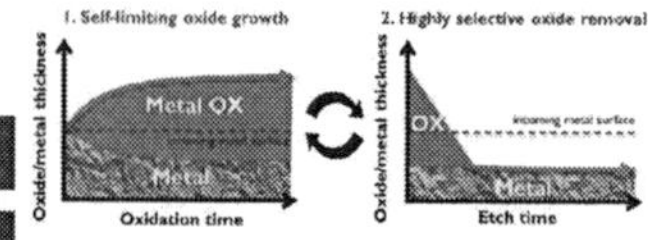


- Uniform metal etchback demonstrated
- No obvious GB/seam attack seen
- Negligible iso/dense loading seen
- Need to improve the throughput of DE
  - Many chambers?
  - One step etch?
  - Alternative dry etch solutions?

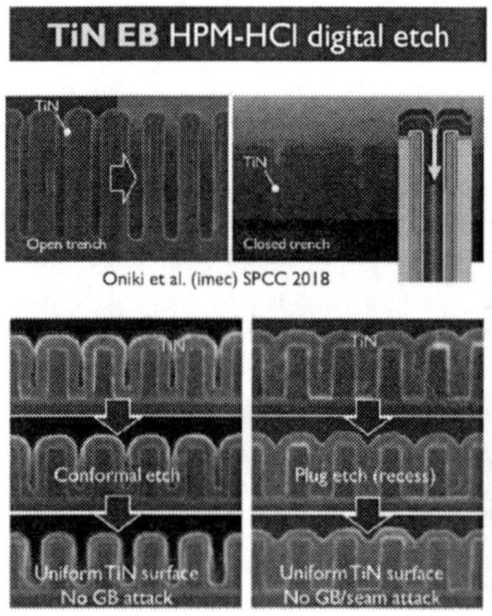


Oniki et al. (imec) SPCC 2018

Oniki et al. (imec) UCPSS 2018

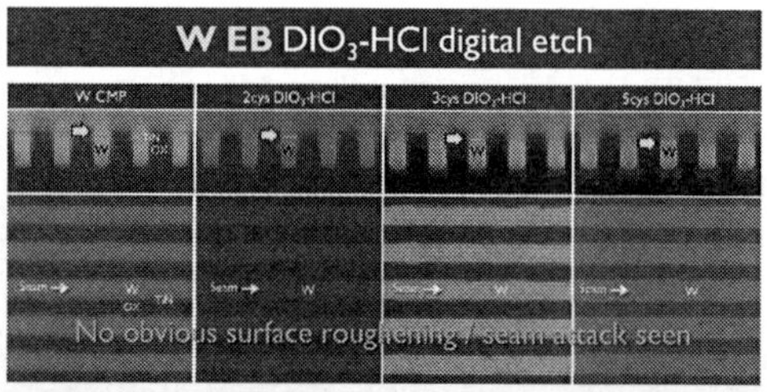


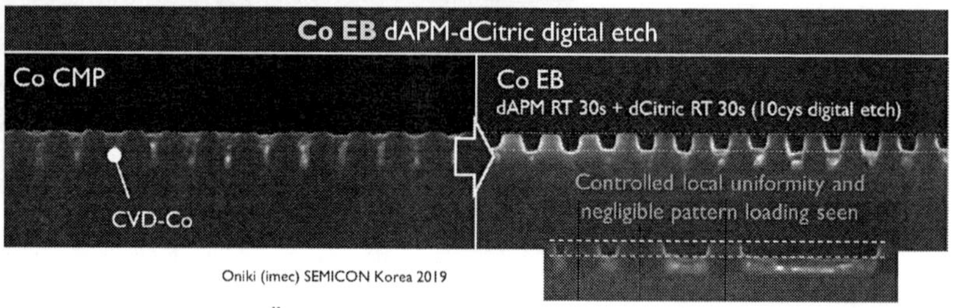


Oniki (imec) SEMICON Korea 2019

imec

25

# COBALT POST ETCH CLEAN ASSESSMENT

## ATMOSPHERIC CORROSION

- Co corrosion inhibition: Post etch treatment(in-situ)/cleaning(ex-situ)/rinsing solutions as well as **Q-time & ambient control** and/or wet strippable etch stop layer needed

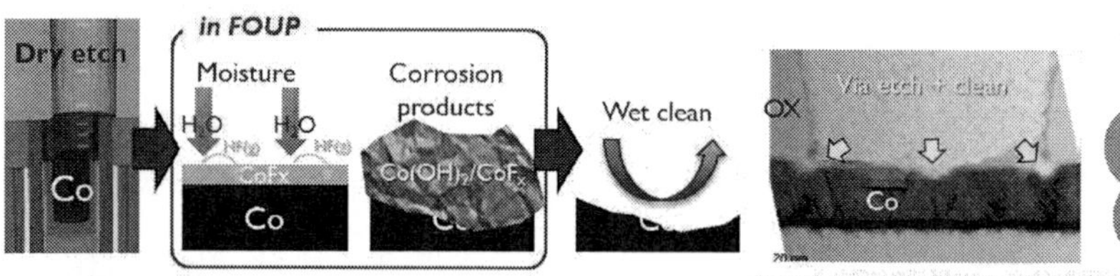


Iino et al. (KURITA/SCREEN/imec) UCPSS 2018

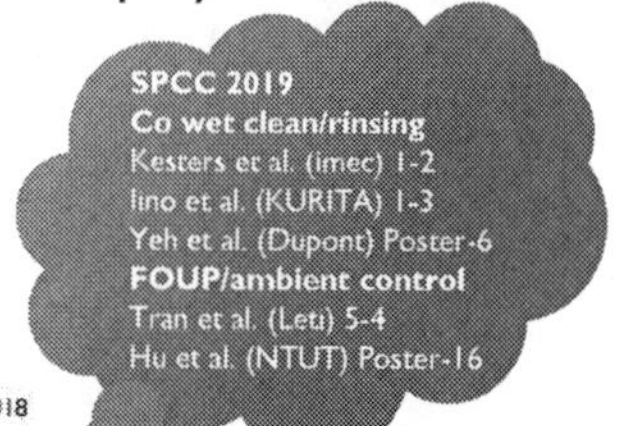


- Co corrosion sensitive to moisture and pollutants

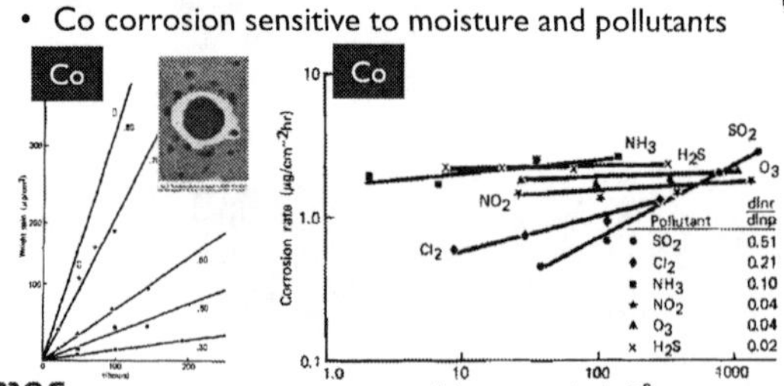


Rice et al. (IBM) JES 1979

- Need to suppress/avoid $CoF_x$ formation

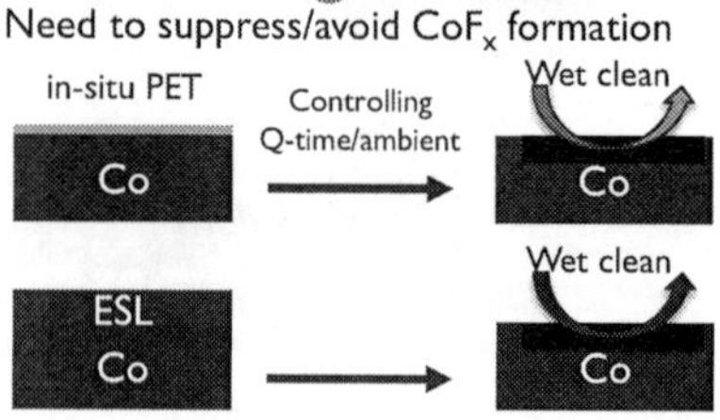


imec

## GATE-ALL-AROUND

## GAA SELECTIVE ETCHES

SILICON GAA

SiGe vs. Si isotropic/selective etches:

① **Inner spacer cavity etch**

② **Wire/sheet release**

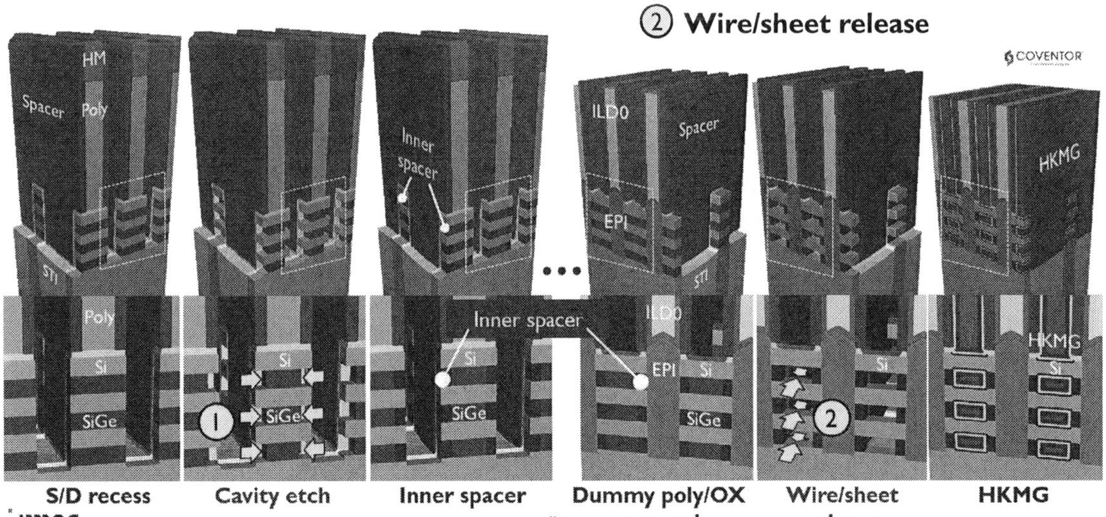


**S/D recess** | **Cavity etch** | **Inner spacer** | **Dummy poly/OX removal** | **Wire/sheet release** | **HKMG**

imec

28

SPCC 2019
SiGe SE Westwood et al. (Avantor) 2-2

# SIGE VS. SI ISOTROPIC/SELECTIVE ETCHES

## DRY OR WET

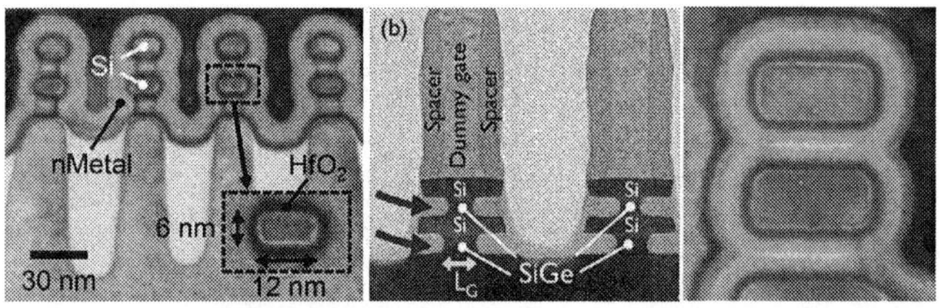


Mertens et al. (imec/AMAT) IEDM 2017

imec ITF 2018

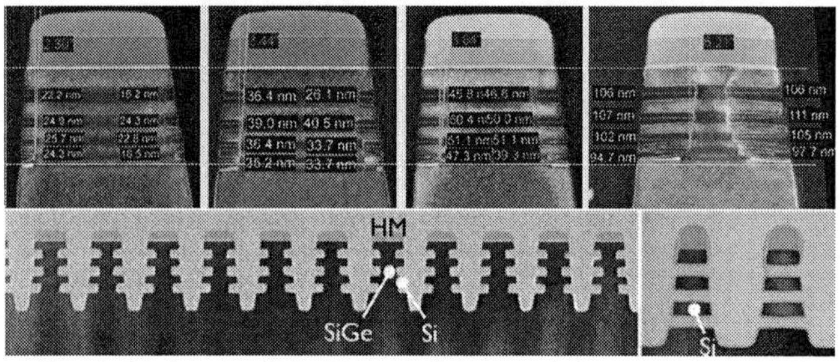


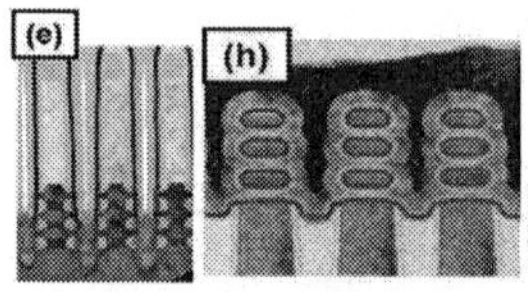


Loubet et al. (IBM/Samsung/GF) VLSI 2017

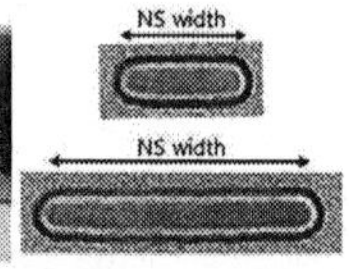


Bae et al. (Samsung) IEDM 2018

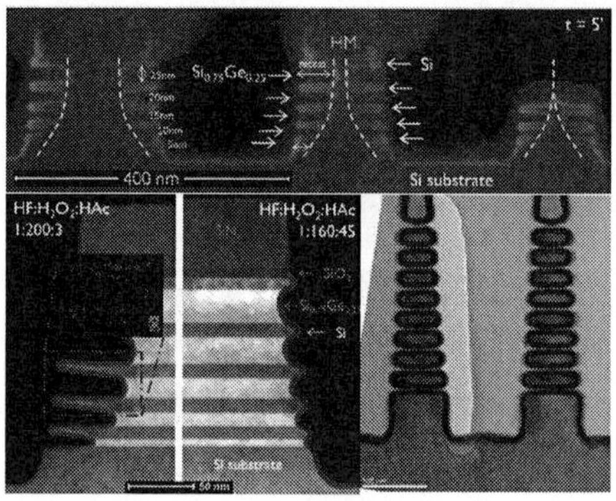


imec

Kal et al. (TEL/imec) SPCC 2018 / SPIE 2019



Komori et al. (SCREEN/imec/Versum) UCPSS 2018

SPCC 2019
CFET & 3D logic Smith (TEL) 4-1

# GAA FAMILY

## Nanowire/sheet

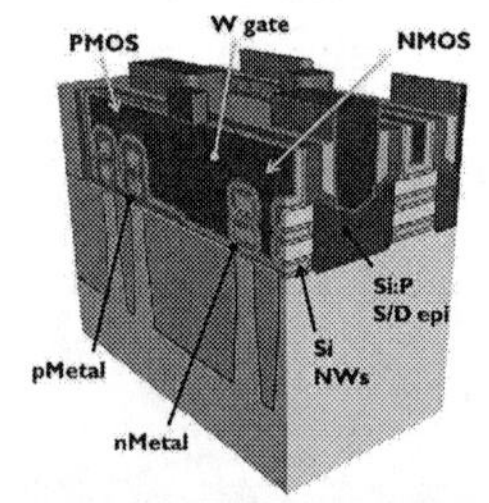


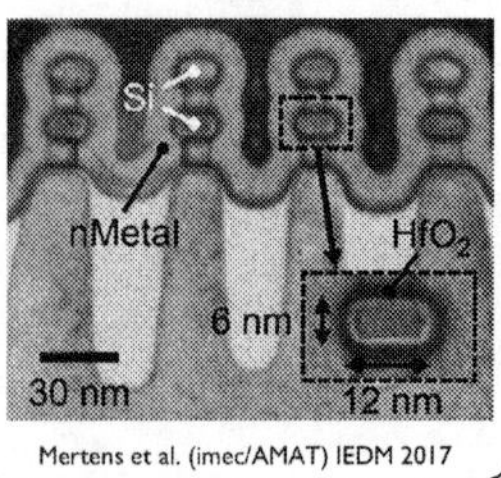


Mertens et al. (imec/AMAT) IEDM 2017

## Fork sheet

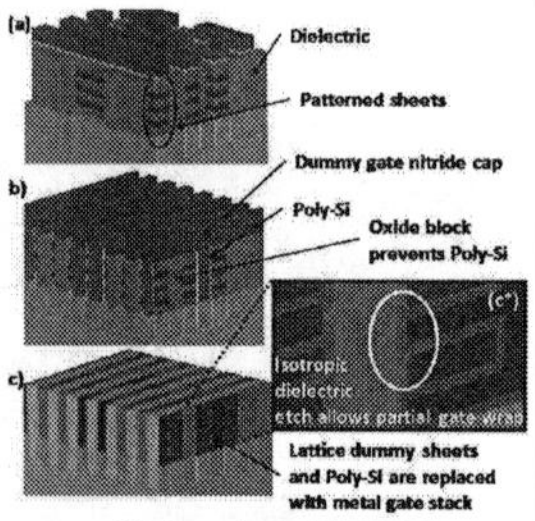


Weckx et al. (imec) IEDM 2017

Put dielectric wall at the NB boundary
→ Ultimate 2D scaling
- To suppress NBB undercut @RMG
- To enable low AR metal gate cut

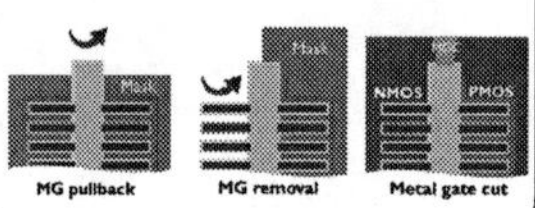


## Complimentary FET (CFET)

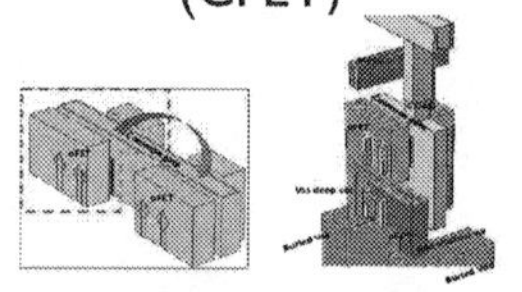

SRAM >40% structural gain

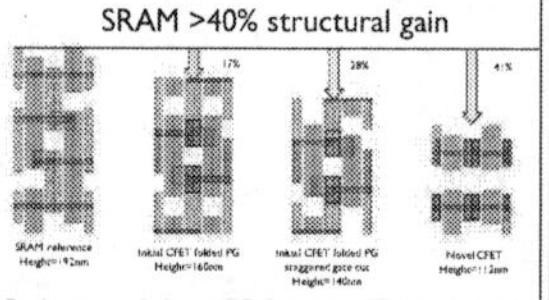

Ryckaert et al. (imec/GF/Coventor/TEL) VLSI 2018

Fill → planarize → etchback → ...

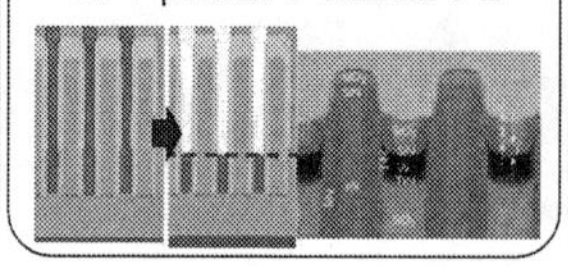

## Vertical FET (VFET)

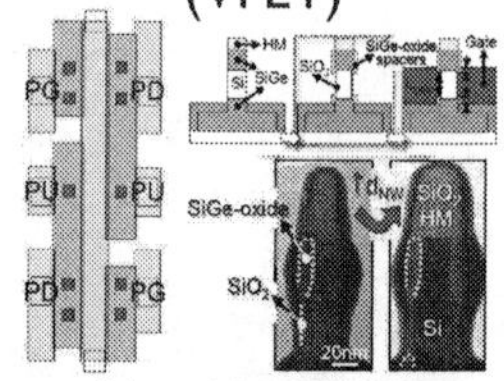


Veloso et al. (imec) SSDM 2018

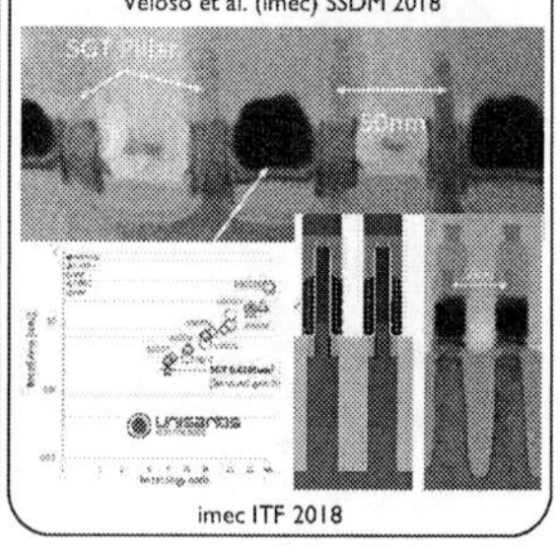


imec ITF 2018

imec

## STCO
## SYSTEM TECHNOLOGY CO-OPTIMIZATION NEEDED

| High performance | High density & Low power | Analog I/O & long channel |
|---|---|---|
| **Nanosheet FET,** Fork sheet FET | **Nanowire FET,** Fork sheet FET, CFET, VFET | **Si FinFET,** Superlattice FinFET |

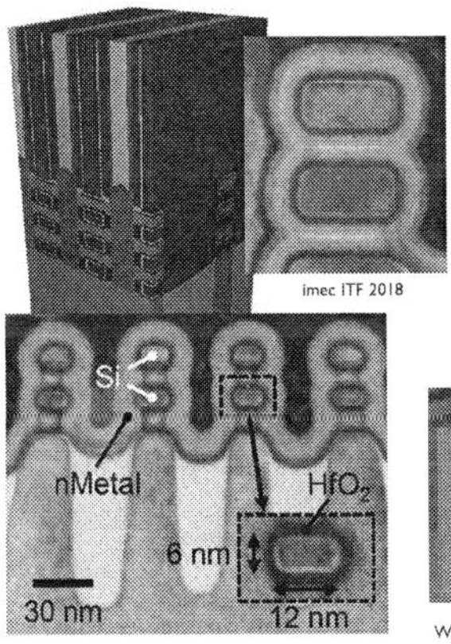

imec ITF 2018

Mertens et al. (imec/AMAT) IEDM 2017

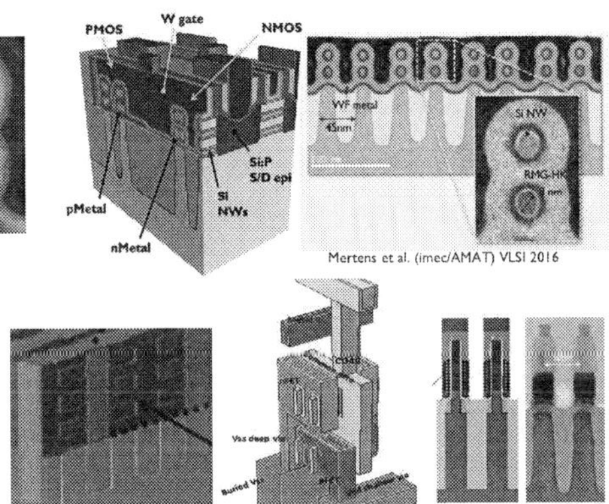

Mertens et al. (imec/AMAT) VLSI 2016

Weckx et al. (imec) IEDM 2017 Ryckaert et al. (imec/GF/Coventor/TEL) VLSI 2018 imec ITF 2018

**GAA for analog I/O and long channel?**

- Concern on wire and sheet collapse/stiction @NW/NS release & wet clean
- Thick gate oxide + MG cannot be fully filled in the wire/wire and sheet/sheet spacing

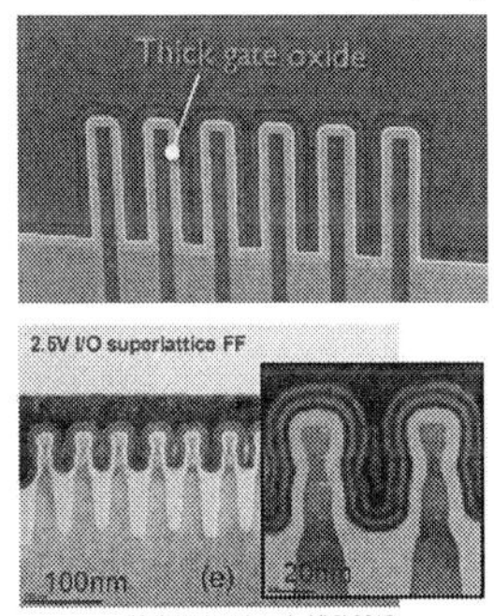

Hellings (imec) et al, VLSI 2018

imec

31

# SUMMARY

## CLEANING CHALLENGES FOR ADVANCED LOGIC

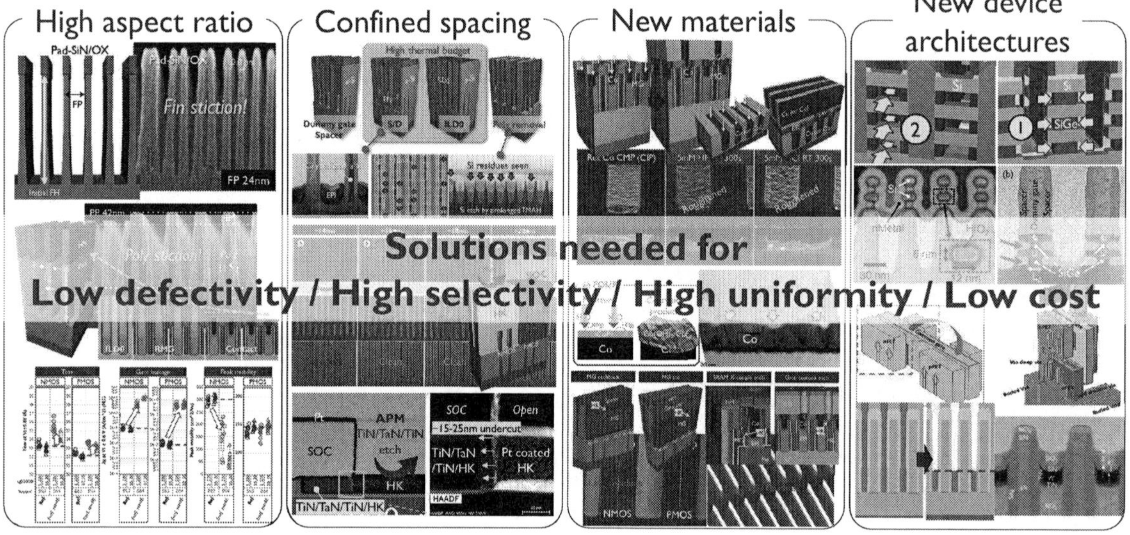


imec

33

## ACKNOWLEDGEMENTS


- imec surface and interface processing group
  - Antoine Pacco, Christophe Lorant, Dennis Van Dorp, Els Kesters, Farid Sebaai, Graniel Abrenica, Guy Vereecke, Jens Rip, Karine Kenis, Kurt Wostyn, Nandi Vrancken, Nina Bazzazian, Quoc Toan Le, Simon Braun, Sophia Arnauts, XiuMei Xu, and Frank Holsteyns
- imec logic program teams
  - Anabela Veloso, BT. Chan, Daire Cott, Elena Capogreco, Eugenio Dentoni Litta, Elie Schapmans, Harold Dekkers, Hans Mertens, Juergen Boemmels, Julien Ryckaert, Katia Devriendt, Lars-Åke Ragnarsson, Min-Soo Kim, Nancy Heylen, Soon Aik Chew, Steven Demuynck, Sujith Subramanian, Sylvain Baudot, Toby Hopf, Yong Kong Siew, Yoshiaki Kikuchi, and Naoto Horiguchi


imec

34

## CORE CMOS PARTNERS

LOGIC / MEMORY IDM & FOUNDRIES

intel, SAMSUNG, tsmc, GLOBALFOUNDRIES, Micron, SK hynix, Western Digital, TOSHIBA, HLMC

FABLESS & FABLITE

HUAWEI, HISILICON, SONY, QUALCOMM, XILINX, socionext, Panasonic, ARM

AND OTHER FABLESS/SYSTEM COMPANIES

EQUIPMENT & MATERIAL SUPPLIERS / OSAT / EDA / JDP PARTNERS

ASML, ASM, APPLIED MATERIALS, Hitachi High-Tech, KLA, Lam RESEARCH, SCREEN, TEL, CYMER, Canon, AIXTRON, NOVA, EVG, ThermoFisher SCIENTIFIC, SPTS, Veeco, SUSS MicroTec, KOKUSAI ELECTRIC, ZEISS, HPSP, Park SYSTEMS, RIBER, TMEIC, TOWA, PVA TePla, Kurita, EBARA, Solmates, BRUKER, NGR, FORMFACTOR, MITSUBISHI CHEMICAL, CENTRAL GLASS, NUFLARE, FRACTILIA, SMIC, Kaneka, JSR Micro, JX Nippon Mining & Metals, VERSUM MATERIALS, LG Chem, Entegris, MERCK, TOPPAN, Inpria, siltronic, Soitec, Nissan Chemical CORPORATION, Shin-Etsu, SUMITOMO CHEMICAL, DONGJIN, tok, ZEON, BASF, FUJIFILM, brewer science, Air Liquide, Amkor Technology, MDLSoft Inc, Panoramic Technology, cadence, Mentor A Siemens Business, SYNOPSYS, global TCAD solutions, COVENTOR

imec

35

# Development of Selective SiGe Etchants

Glenn Westwood and Rebecca Duffy

Avantor, 1013 US Highway 202/206, Bldg. JR-1, Bridgewater, NJ 08807

glenn.westwood@avantorinc.com, rebecca.duffy@avantorinc.com

avantor™



## Introduction

- **Goals**
  - Develop SiGe etchants with high selectivity to Si, $SiO_2$, and $Si_3N_4$.
  - Target etchants for SiGe with <30% Ge.
  - Target Gate-All-Around (GAA) Si applications.
- **Outline of Presentation**
  - Applications
  - Approaches to SiGe/Si Etch Development
  - Etchant Development work – Inhibitor identification, Clouding
  - Performance – Aging effect, Bath life
  - SiGe Thickness Effects – Etch rates and interfacial Ge content.
  - Conclusions and Future work

avantor™

 2

## Applications – Gate All Around (GAA) Transistors

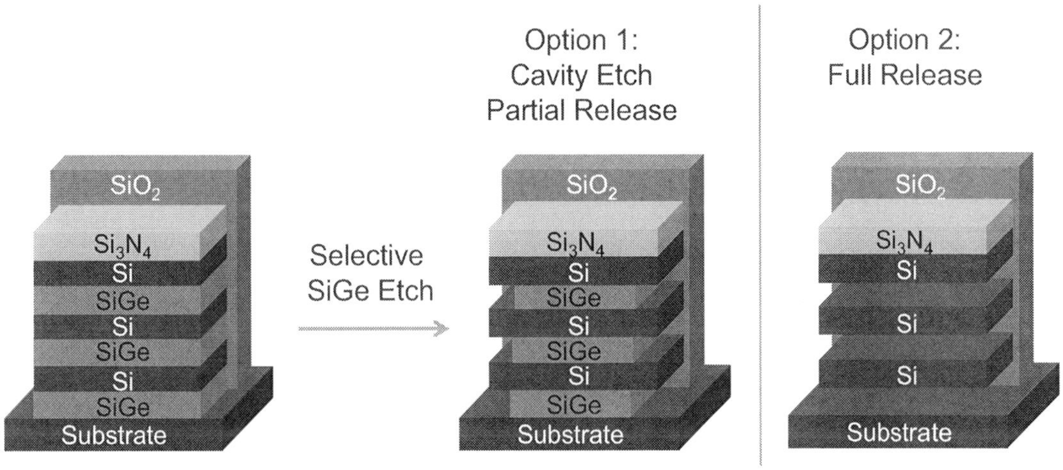


1. Alternating deposition of Si and SiGe layers
2. Partial or full release of the SiGe layers
3. Si nanosheets or nanowires remain




## Approaches to Selectively Etching SiGe (20%) vs. Si

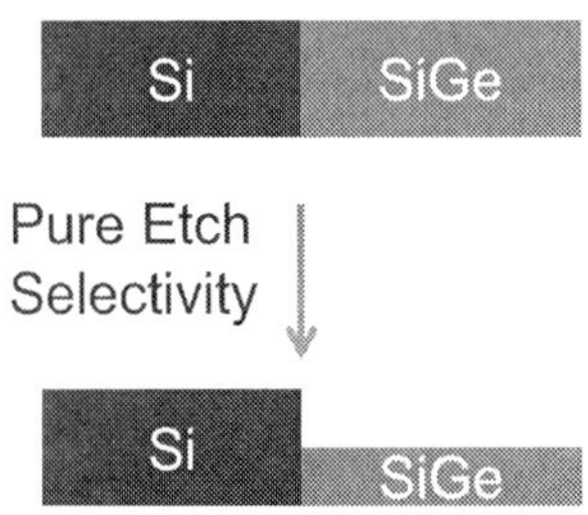


- Effective at Si>>SiGe with TMAH-based chemistries.
- Avantor efforts unsuccessful to date.
- Achieved Si ≈ SiGe with highly alkaline chemistries.

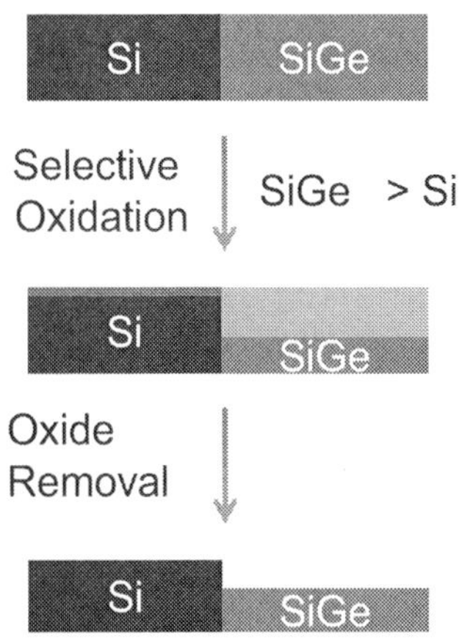


- Limiting step is surface oxidation.
- Proven approach for SiGe>Si etch.
- $SiO_2$ and $Si_3N_4$ compatibility is inconceivable.

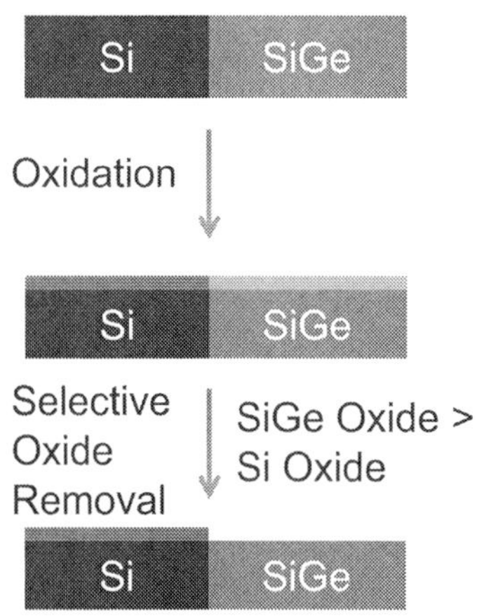


- Limiting step - oxide removal.
- Necessary to achieve $SiO_2$ and $Si_3N_4$ compatibility.

## HF/Acetic Acid/H2O2 Selective SiGe Etchant

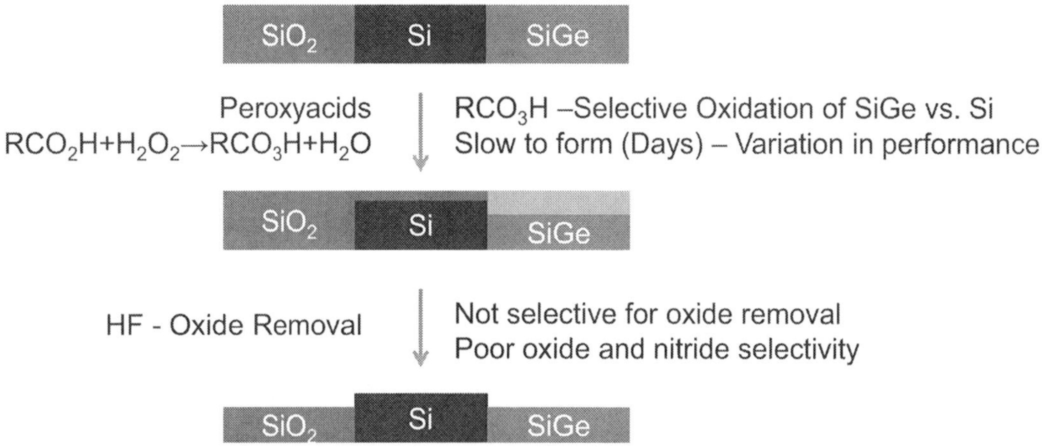


**Very high SiGe etch rates achieved, but reported SiGe:Si Etch selectivity of ~20:1 is still not ideal.**

avantor™

 5

## Avantor Approach to Developing New SiGe Etchant

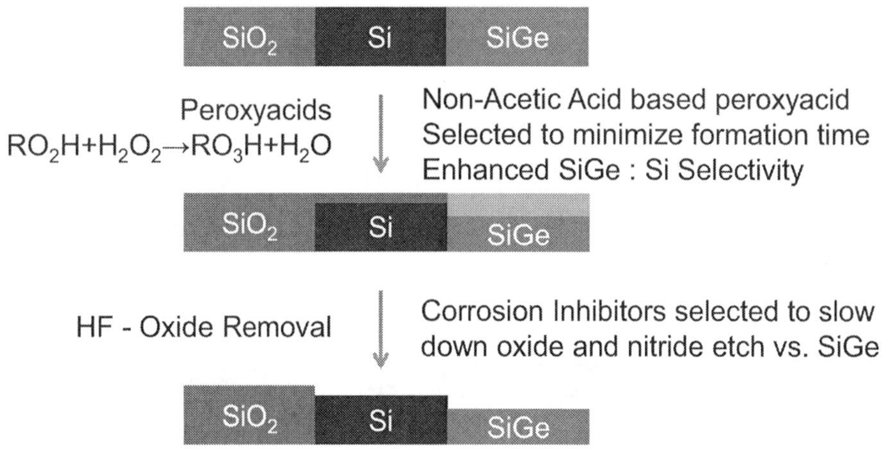


**Formulation Evaluated**

- >95% Water
- Hydrofluoric Acid
- SiGe/Si selectivity enhancer
- Chelator
- Etch Inhibitor Added
- Mixed 5:1 H2O2

avantor™

 6

## $SiO_2$ Etch Inhibitor Screening – First Attempt

**Formulation Evaluated**

- >95% Water
- Hydrofluoric Acid
- SiGe/Si selectivity enhancer
- Chelator
- Etch Inhibitor Added
- Mixed 5:1 $H_2O_2$

**Results**

- No clear winners.
- Many improved pSi selectivity.
- Cationic surfactants (A1-A3) - worse selectivity.
- Picked 4 select additives for further testing.

| Etch Inhibitors | $SiO_2$ Etch Rate (Å/min) | pSi Etch Rate (Å/min) |
|---|---|---|
| None | 6.3 | 13.3 |
| A1 | 130.5 | 20.7 |
| A2 | 17.9 | 61.3 |
| B1 | 11.1 | 3.1 |
| B2 | 4.9 | 1.9 |
| B3 | 6.5 | 1.6 |
| C1 | 5.8 | 4.8 |
| C2 | 5.7 | 3.3 |
| D1 | 5.2 | 1.1 |
| D2 | 6.6 | 1.9 |
| D3 | 10.8 | 1.7 |
| D4 | 2.7 | 1.3 |
| E1 | 11.1 | 1.1 |
| E2 | 6.0 | 1.5 |
| E3 | 10.9 | 1.5 |
| E4 | 10.9 | 3.3 |
| E5 | 26.1 | 0.2 |
| F1 | 11.7 | 2.8 |

avantor™





## $SiO_2$ Etch Inhibitor Screening – Second Attempt

**Formulation Evaluated**

- >95% Water
- Reduced Hydrofluoric Acid
- SiGe/Si selectivity enhancer
- Chelator
- Increased Etch Inhibitor Mixed 5:1 $H_2O_2$

**Results**

- Formulation w/ F1
- Increasing pH dramatically increases SiGe etch rate.
- pSi etch rates also increase with pH.

| Etch Inhibitors | $SiO_2$ Etch Rate (Å/min) | pSi Etch Rate (Å/min) | SiGe (20%) Etch Rate (Å/min.) | SiGe (20%) / pSi Etch Selectivity |
|---|---|---|---|---|
| B1 | 1.1 | 0.6 | 12.1 | 20 |
| D4 | 1.8 | 0.6 | 17.9 | 30 |
| E3 | 1.0 | 0.5 | 12.4 | 25 |
| F1 | 1.5 | 1.3 | 41.6 | 32 |

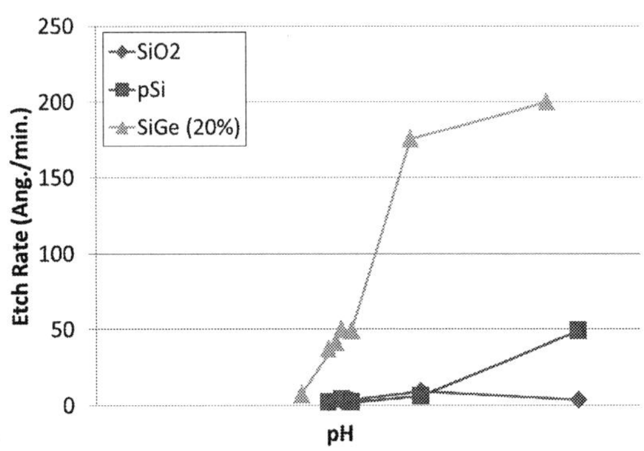


avantor™

## RL-1 and RL-1B – Cloudy Formulations

| Formulation | XM-529-V3 + Etch Inhibitor | $SiO_2$ Etch Rate (Å/min) | $Si_3N_4$ Etch Rate (Å/min.) | pSi Etch Rate (Å/min) | SiGe (20%) Etch Rate (Å/min.) | SiGe (20%) / pSi Etch Selectivity |
|---|---|---|---|---|---|---|
| RL-1* | F1 | 2 | 63 | 2 | 80 | 40 |
| Rl-1B* | F1 + B3 | 2 | 3 | 3 | 58 | 20 |

* Mixed 5:1 $H_2O_2$ at 35 C

- All high selectivity etchants in initial stages of this study were cloudy.
- Cloudiness – precipitation of:

Etch Inhibitor + SiGe/Si Selectivity Enhancer

## Formulations without Clouding

**Prevented clouding by:**

1. Reducing SiGe/Si Selectivity Enhancer – RL-3
2. Modified Etch Inhibitor – RL-4 – Best Etch solution to date.
3. Comparable selectivity, but lower SiGe etch rates.

| Formulation | XM-529-V3 + Etch Inhibitor | $SiO_2$ Etch Rate (Å/min) | $Si_3N_4$ Etch Rate (Å/min.) | pSi Etch Rate (Å/min) | SiGe (20%) Etch Rate (Å/min.) | SiGe (20%) / pSi Etch Selectivity |
|---|---|---|---|---|---|---|
| RL-3* | F1 | 3.4 | 1.8 | 3.2 | 45 | 15 |
| RL-4* | F2 | 0.8 | 1.3 | 1.2 | 34 | 28 |

* Mixed 5:1 $H_2O_2$ at 35 C

# RL-4: Bath Performance Data

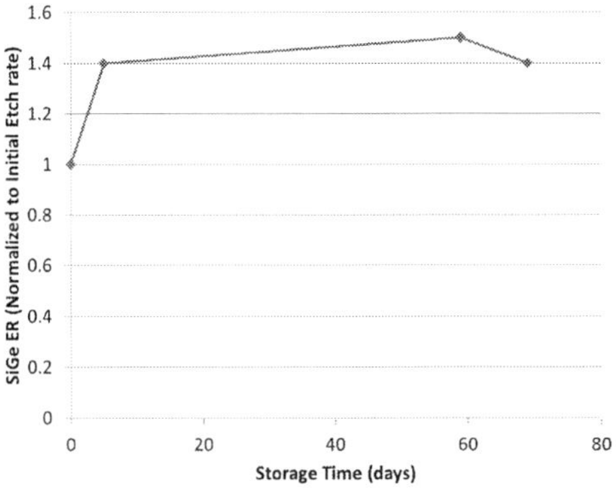


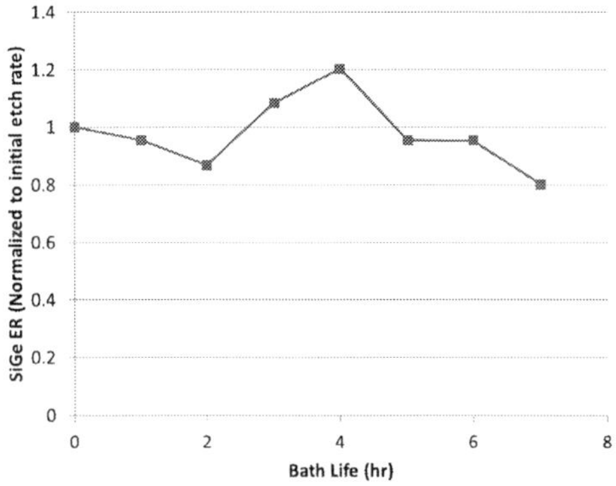


- RL-4 was mixed with $H_2O_2$ (5:1) and tested after days of storage.
- ~50% increase in SiGe etch rates.
- Less increase over time than acetic acid based blends.
- Stable

- 5 RL-4 : 1 $H_2O_2$ at 35 C, 650 rpm.
- RL-4 bath life stable out to 7 hrs.

avantor™





# Etch Rate Dependence on Etch Times

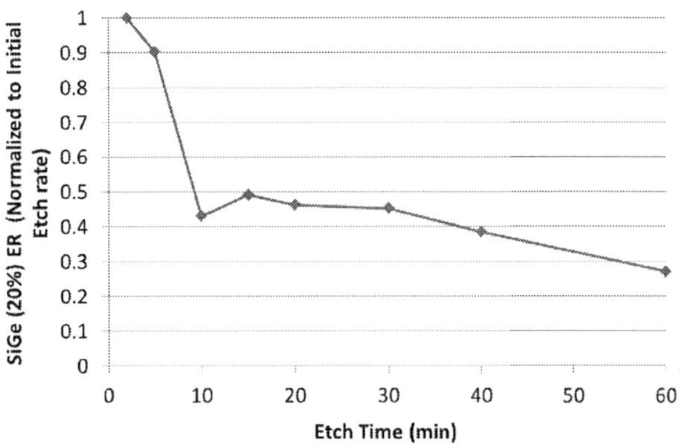


- RL-4 : $H_2O_2$ at 35 C, 650 rpm. Etched SiGe wafers for different etch times.
- The longer the etch time, the lower the etch rate.
- SiGe etch rate does not decrease with bath life, instead the SiGe (20%) etch is slowing because the SiGe thickness is decreasing.

avantor™

## Etch Time Effect (XPS)

| SiGe Etch Depth (Å) | SiGe ER (Normalized) | % Ge |
|---|---|---|
| 0 | 1.00 | 12.9 |
| 156 | 0.90 | 11.3 |
| 519 | 0.46 | 11.0 |
| 917 | 0.38 | 9.2 |

- Longer etch times result in decreased Ge and $GeO_2$ at the interface.
- Etch rates decrease with thinning of SiGe films because of reduced Ge content
- Possible explanations
  - Uneven starting Ge content in SiGe films.
  - Interface becomes depleted in Ge due to the etch Ge>Si.

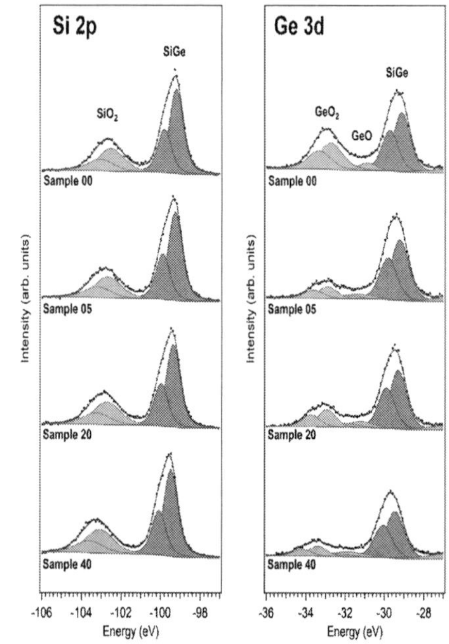


XPS collected by Ryan Thorpe, Rutgers K-Alpha XPS Facility

avantor™

 13

## Additional XPS Results

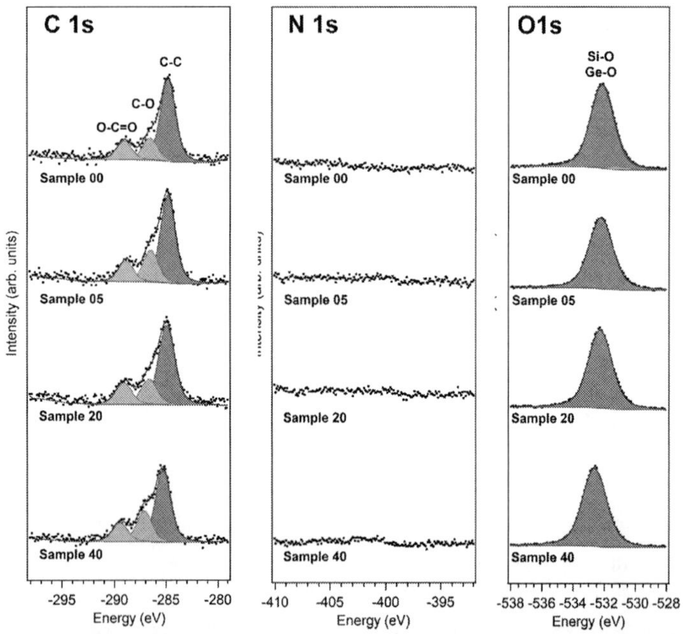


- Carbon – Only adventitious carbon identified
- No nitrogen detected
- No indication that components from chemistry remain on the SiGe

XPS collected by Ryan Thorpe, Rutgers K-Alpha XPS Facility

avantor™

 14

## Conclusions

- **SiGe Etchant Development**
  - Achieving high SiGe:Si selectivity is not difficult.
  - High SiGe:$SiO_2$ selectivity without clouding is a challenge.

- **Best Solution**
  - RL-4
  - Identified $SiO_2$ etch inhibitor that doesn't reduce SiGe etch rates.
  - Etch Rates for partial release not full release.
  - Stable bath life and stable when storing mixed with $H_2O_2$.
  - SiGe etch rate slows with increased SiGe removal due to decrease in Ge at interface.

| Formulation | $SiO_2$ Etch Rate (Å/min) | pSi Etch Rate (Å/min) | SiGe (20%) Etch Rate (Å/min.) | SiGe (20%) / pSi Etch Selectivity |
|---|---|---|---|---|
| RL-4 | 0.8 | 1.2 | 34 | 28 |

avantor™



15

## Next Steps

- Improvements based on cloudy formulations – High SiGe etch rate and selectivities
  - Clear solution separated from cloudiness.

| $SiO_2$ Etch Rate (Å/min) | pSi Etch Rate (Å/min) | SiGe (20%) Etch Rate (Å/min.) | SiGe (20%) / pSi Etch Selectivity |
|---|---|---|---|
| <0.5 | 1.4 | 53 | 38 |

  - Use a cloudy solution

| $SiO_2$ Etch Rate (Å/min) | pSi Etch Rate (Å/min) | SiGe (20%) Etch Rate (Å/min.) | SiGe (20%) / pSi Etch Selectivity |
|---|---|---|---|
| 9 | 6 | 175 | 29 |

- Test Application on devices
- Related GAA etch solutions
  - Si/SiGe, Si/Ge, and SiGe/Ge.
  - III-V release chemistries – InGaAs/GaAs

avantor™



16

# Thermal Atomic Layer Etching (ALE) of Silicon Using an Oxidation and "Conversion-Etch" Mechanism


Steven M. George

Dept. of Chemistry, Univ. of Colorado
Boulder, Colorado 80309
Steven.George@Colorado.edu


# Thermal Atomic Layer Etching (ALE) Using Sequential Surface Reactions

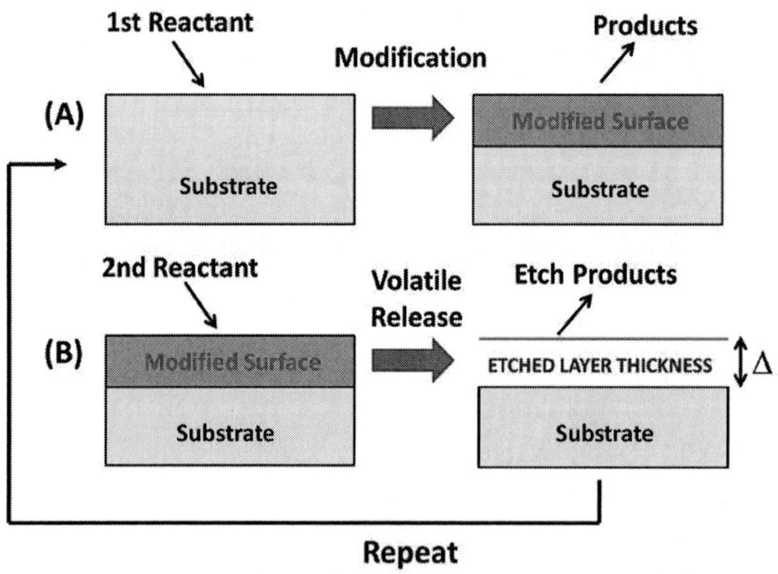


2

# Thermal Si ALE Using an Oxidation and "Conversion-Etch" Mechanism

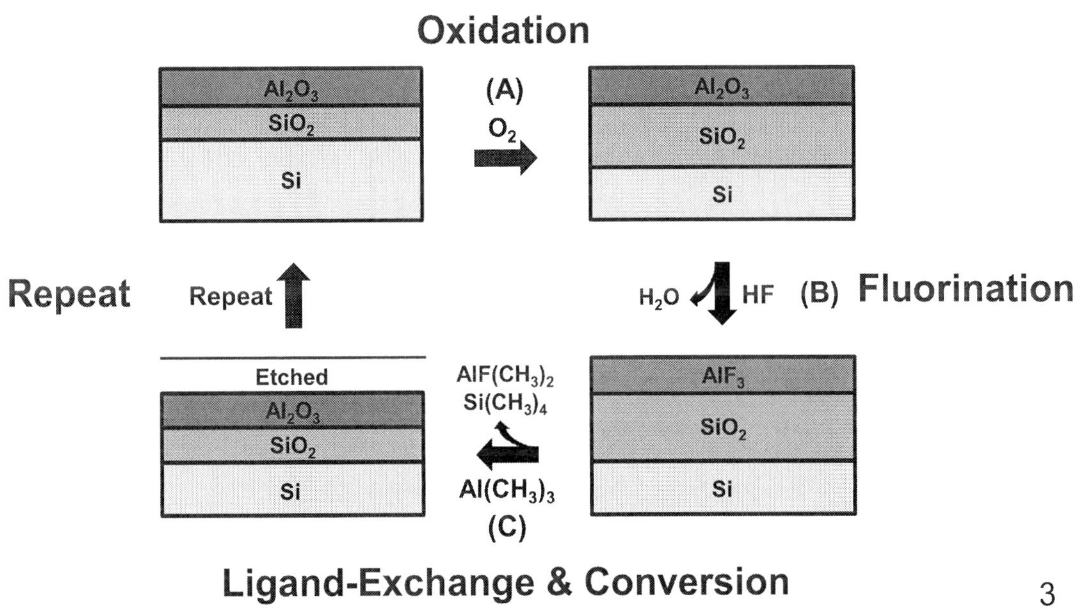


3

# Outline

1. Thermal $Al_2O_3$ ALE Using Fluorination & Ligand-Exchange
2. Thermal $SiO_2$ ALE Using "Conversion Etch"
3. Thermal Si ALE Using Oxidation & "Conversion Etch"

4

# $Al_2O_3$ ALE Using $Al(CH_3)_3$ and HF

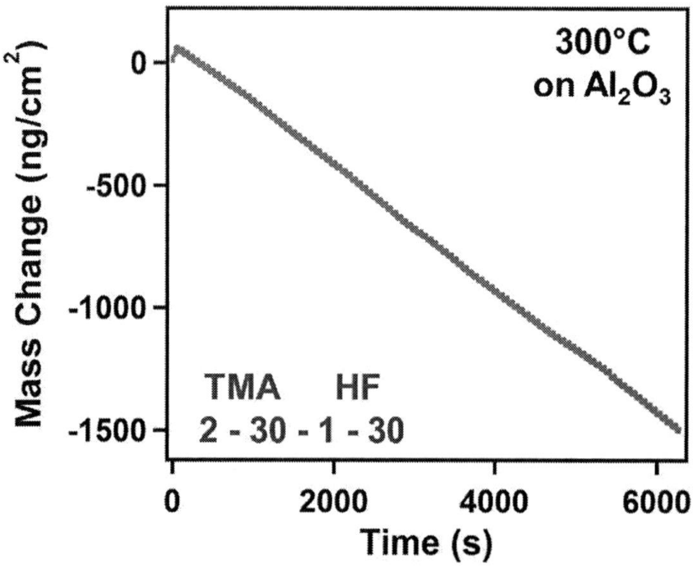


100 ALE Cycles

Mass change per cycle = -15.9 ng/cm$^2$

Etch rate = 0.51 Å/cycle

Y. Lee, J.M. DuMont & S.M. George, *Chem. Mater.* **28**, 2994 (2016). 5

# Mass Loss During $Al(CH_3)_3$ and HF Exposures for 3 Cycles of $Al_2O_3$ ALE

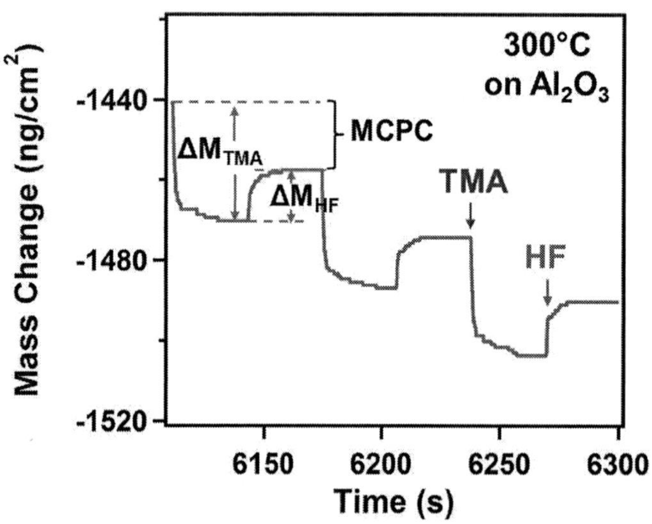


Constant mass changes with each reactant exposure

Y. Lee, J.M. DuMont & S.M. George, *Chem. Mater.* **28**, 2994 (2016). 6

# $Al_2O_3$ ALE via Fluorination & Ligand Exchange

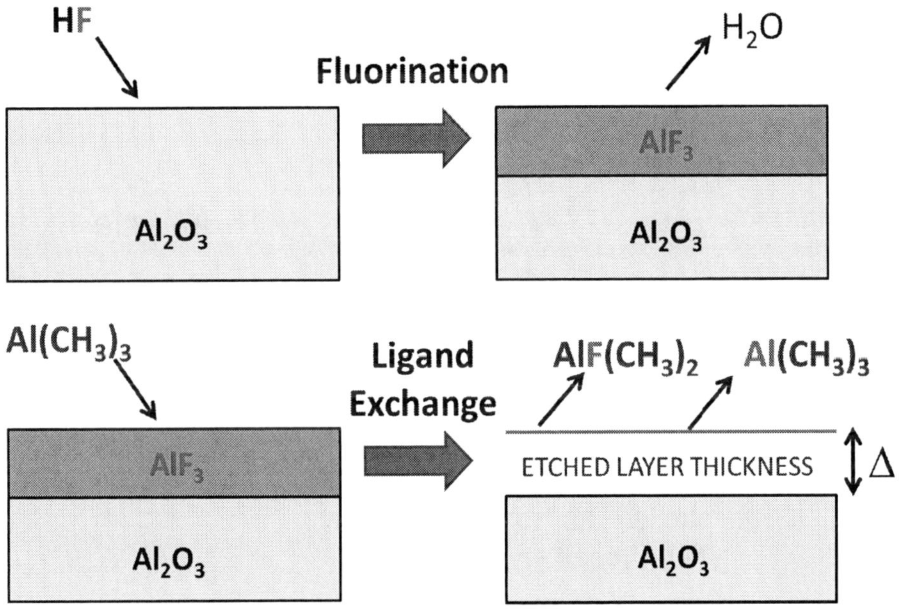


7

# Outline

1. Thermal $Al_2O_3$ ALE Using Fluorination & Ligand-Exchange
2. **Thermal $SiO_2$ ALE Using "Conversion Etch"**
3. Thermal Si ALE Using Oxidation & "Conversion Etch"

8

# $SiO_2$ ALE at 300°C Using $Al(CH_3)_3$ & HF

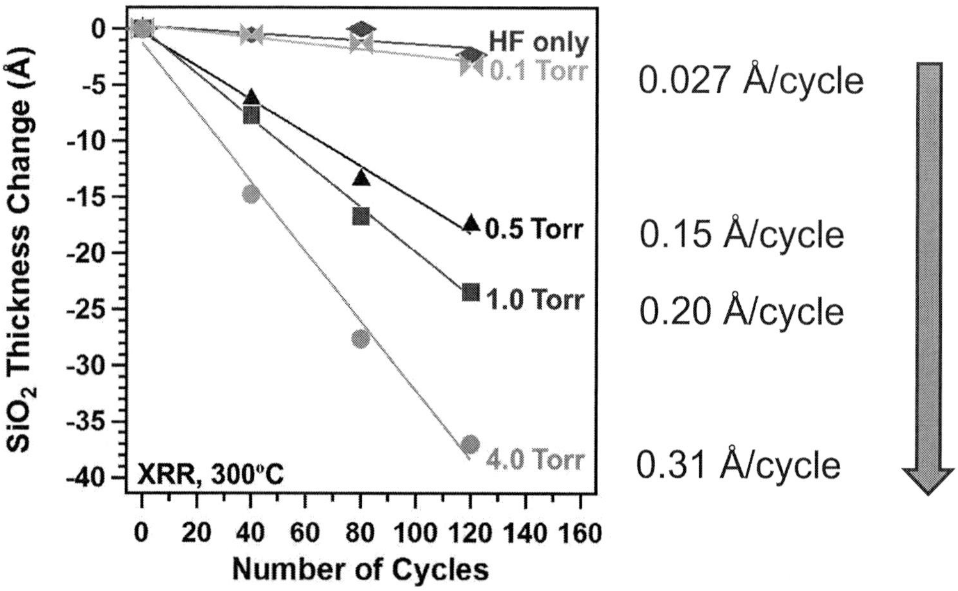


J.W. DuMont et al., *ACS Appl. Mater. Interfaces* **9**, 10296 (2017). 9

# Reactions during "Conversion-Etch"

Conversion of $SiO_2$ to $Al_2O_3$:

$$1.5SiO_2 + 2Al(CH_3)_3 \rightarrow Al_2O_3 + 1.5Si(CH_3)_4$$

Fluorination of $Al_2O_3$:

$$Al_2O_3 + 6HF \rightarrow 2AlF_3 + 3H_2O$$

Etch of $AlF_3$:

$$AlF_3 + 2Al(CH_3)_3 \rightarrow 3AlF(CH_3)_2$$

10

# Pressure-Dependence of TMA Reaction During "Conversion-Etch"

Difference Spectra

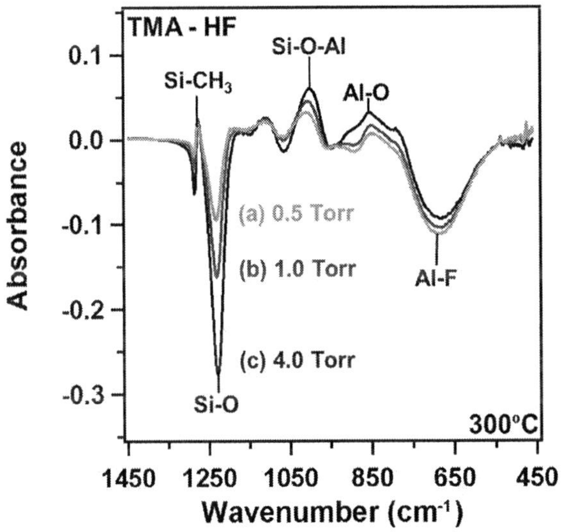


TMA etches $AlF_3$ layer.

$AlF_3 + 2Al(CH_3)_3 \rightarrow 2AlF(CH_3)_2$

TMA converts Si-O species to Al-O species.

$1.5SiO_2 + 2Al(CH_3)_3 \rightarrow Al_2O_3 + 1.5Si(CH_3)_4$

J.W. DuMont et al., *ACS Appl. Mater. Interfaces* **9**, 10296 (2017). 11

# Pressure-Dependence of HF Reaction During "Conversion-Etch"

Difference Spectra

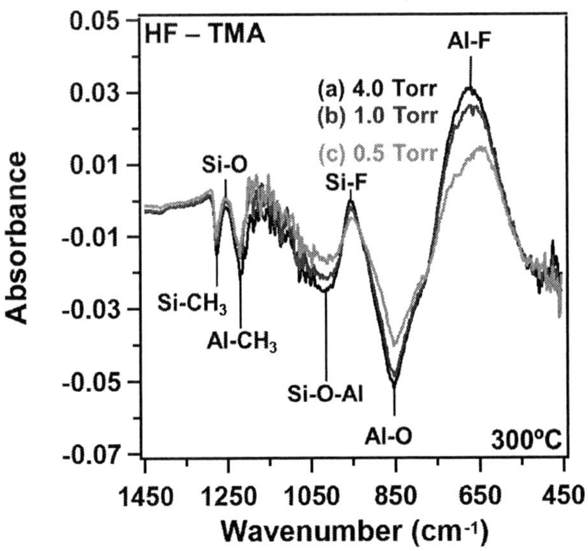


HF Fluorinates Al-O Species

$Al_2O_3 + 6HF \rightarrow 2AlF_3 + 3H_2O$

J.W. DuMont et al., *ACS Appl. Mater. Interfaces* **9**, 10296 (2017). 12

# Mechanism of Thermal $SiO_2$ ALE by "Conversion-Etch"

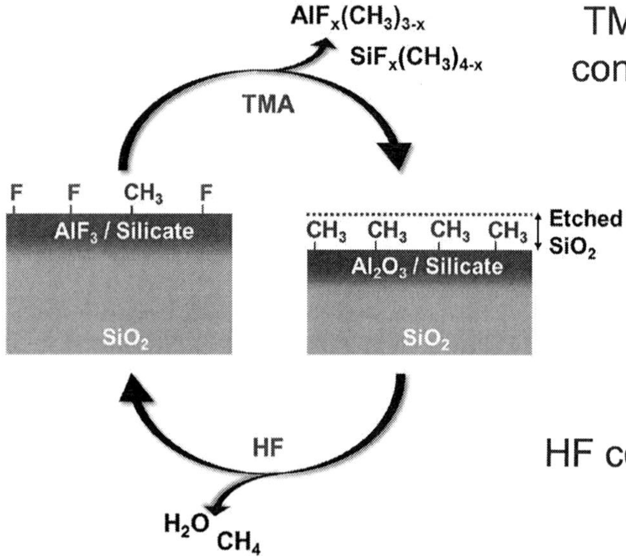


TMA removes $AlF_3$ & converts $SiO_2$ to $Al_2O_3$

HF converts $Al_2O_3$ to $AlF_3$

13

# Outline

1. Thermal $Al_2O_3$ ALE Using Fluorination & Ligand-Exchange
2. Thermal $SiO_2$ ALE Using "Conversion Etch"
3. **Thermal Si ALE Using Oxidation & "Conversion Etch"**

14

# Spectroscopic Ellipsometry Measurement of Both Top $SiO_2$ & Si

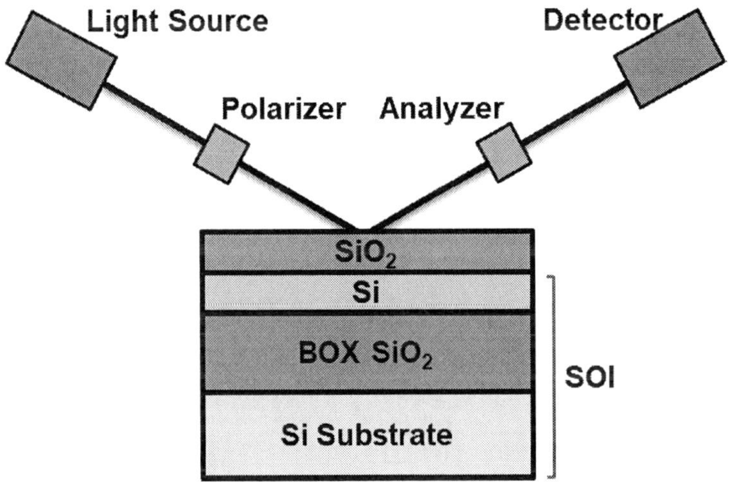


15

# Si ALE Using $O_2$-HF-TMA Sequence

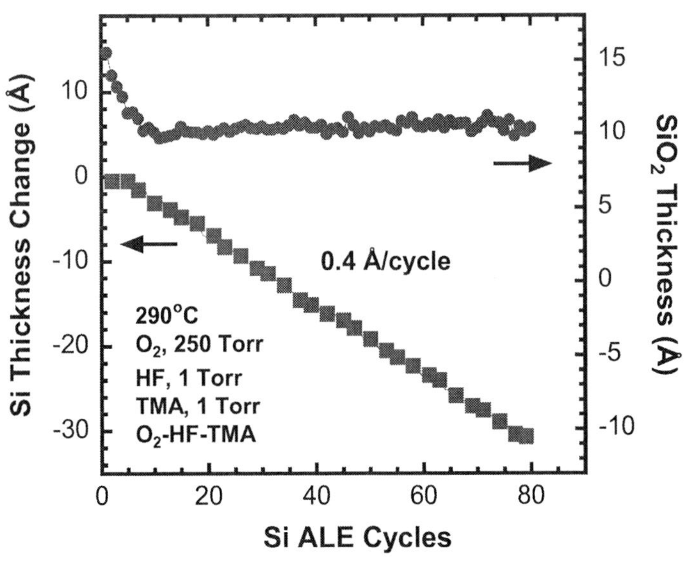


Linear decrease in Si thickness

Si etch rate: 0.4 Å/cycle

Surface oxide remains constant

Surface oxide thickness: ~11 Å

A.I. Abdulagatov & S.M. George, *Chem. Mater.* **30**, 8465 (2018).

16

## Si ALE vs. $O_2$ Pressure

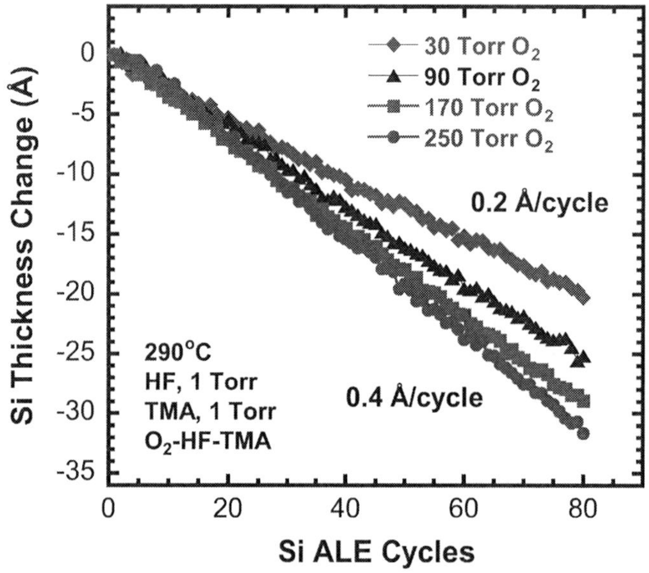


Linear decrease in Si thickness at all $O_2$ pressures

Etch rate increases with $O_2$ pressure

Higher oxidation at higher $O_2$ pressures

A.I. Abdulagatov & S.M. George, *Chem. Mater.* **30**, 8465 (2018). 17

## Si ALE vs. TMA and HF Pressure

TMA Pressure Dependence

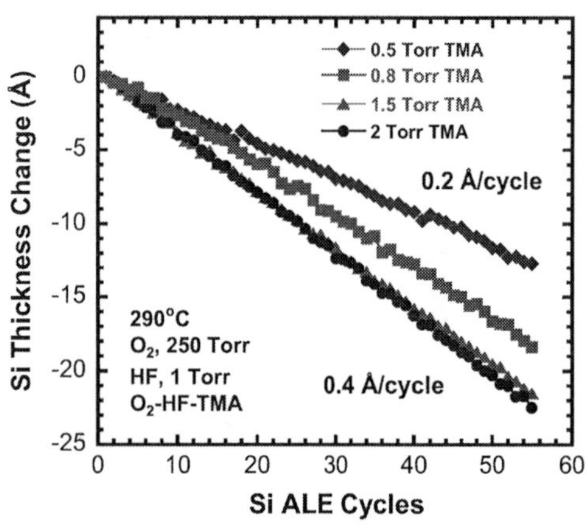


TMA pressure dependence consistent with conversion process

HF Pressure Dependence

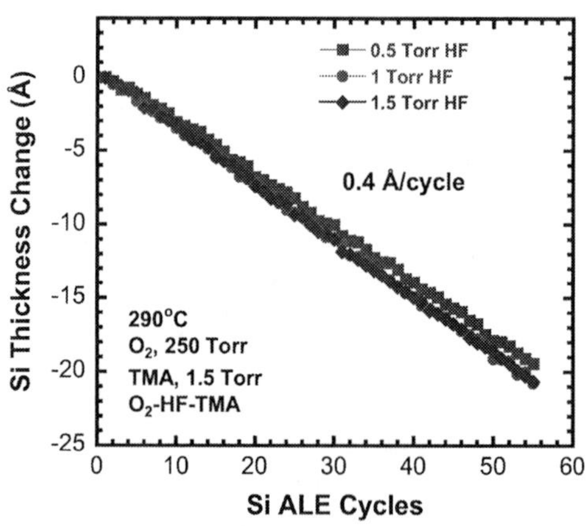


HF pressure independence suggests that HF fluorinates all converted $Al_2O_3$

A.I. Abdulagatov & S.M. George, *Chem. Mater.* **30**, 8465 (2018). 18

# Si ALE of Ultrathin Si Films

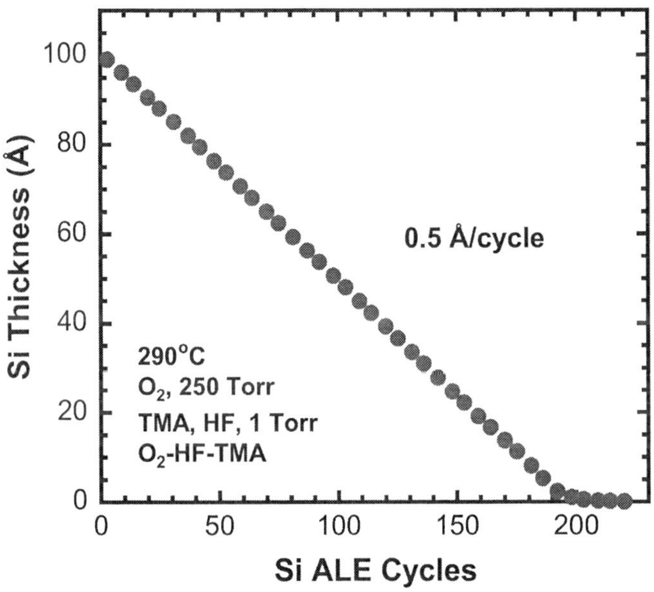


No difficulty etching ultrathin Si films

Linear decrease in film thickness

No effect from silicon quantum confinement

A.I. Abdulagatov & S.M. George, *Chem. Mater.* **30**, 8465 (2018).

19

# Applications for Thermal Si ALE

1. Isotropic etching of three-dimensional structures such as silicon nanowires
2. Surface smoothening to reduce silicon line edge roughness
3. Surface polishing, surface cleaning and removing surface damage

20

# Conclusions – Thermal Si ALE

1. $O_2$ Oxidizes Si to $SiO_2$
2. TMA Converts $SiO_2$ to $Al_2O_3$
3. HF Fluorinates $Al_2O_3$ to $AlF_3$
4. TMA Removes $AlF_3$ by Ligand-Exchange

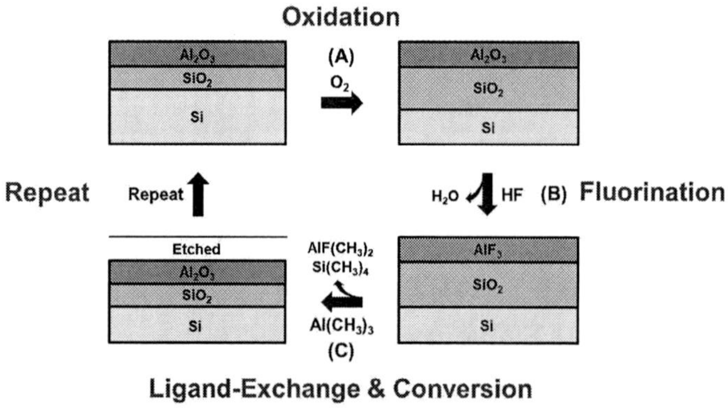


21

# Extra Slides

# QCM Studies in Hot Wall, Viscous Flow Reactor

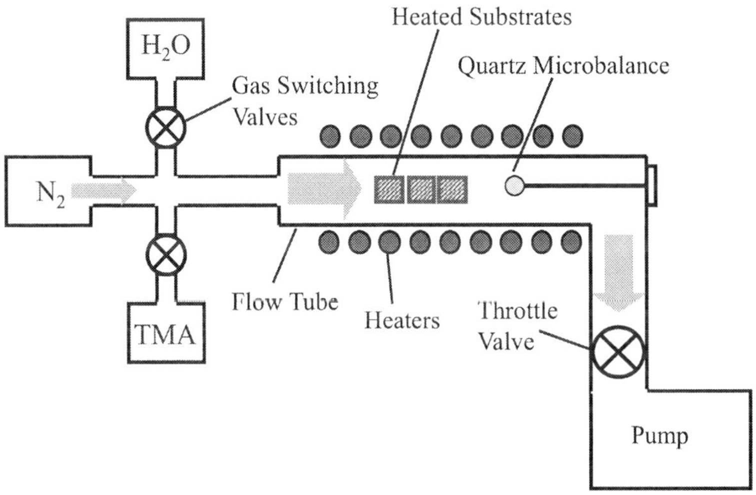


J.W. Elam et al., *Rev. Sci. Instrum.* **73**, 2981 (2002).

# FTIR Spectroscopy Studies of Surface Species & Etched Material

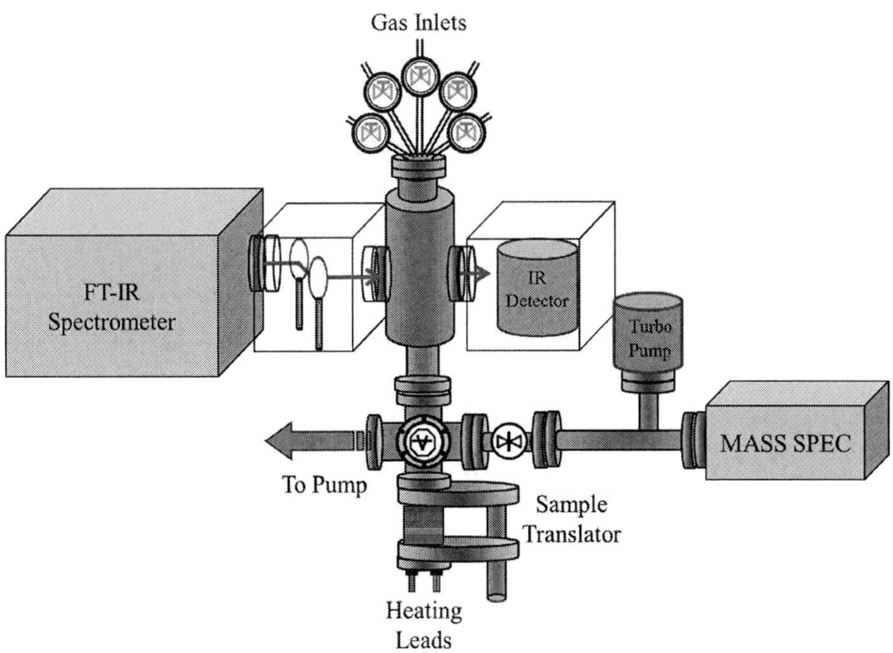

# Film Thickness Measurement with Spectroscopic Ellipsometry

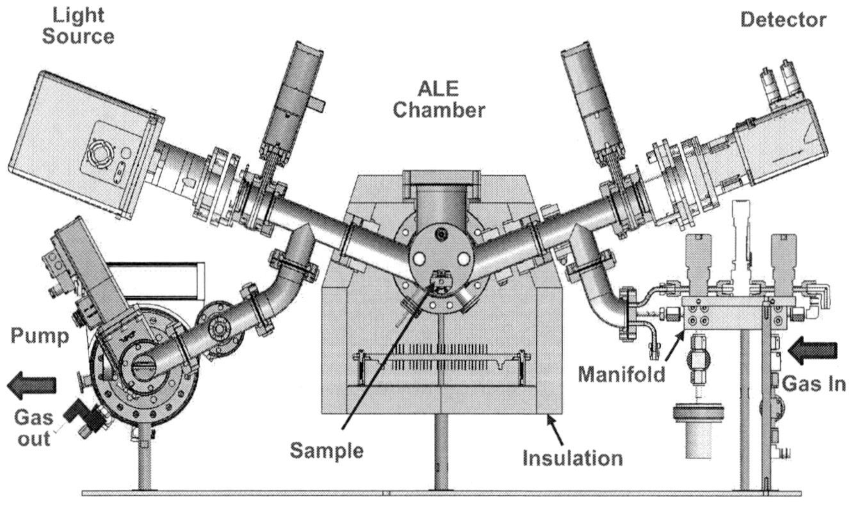

TEL TOKYO ELECTRON · imec

# Si trim applications: benefits and challenges

SPCC, April 2, 2019

S. Kal[1], Y. Oniki[2], C. Alix[1], E. Liu[1], K. Pillai[1], D. Chanemougame[1], F. Holsteyns[2], A. Mosden[1], K. Kumar[1], P. Biolsi[1], T Hurd[1].

[1]TEL Technology Center, America, LLC, USA
[2] Imec, Belgium

Subhadeep.Kal@us.tel.com

## DIMENSIONAL SCALING CHALLENGES
## DEVICE ARCHITECTURE & MATERIAL INNOVATION

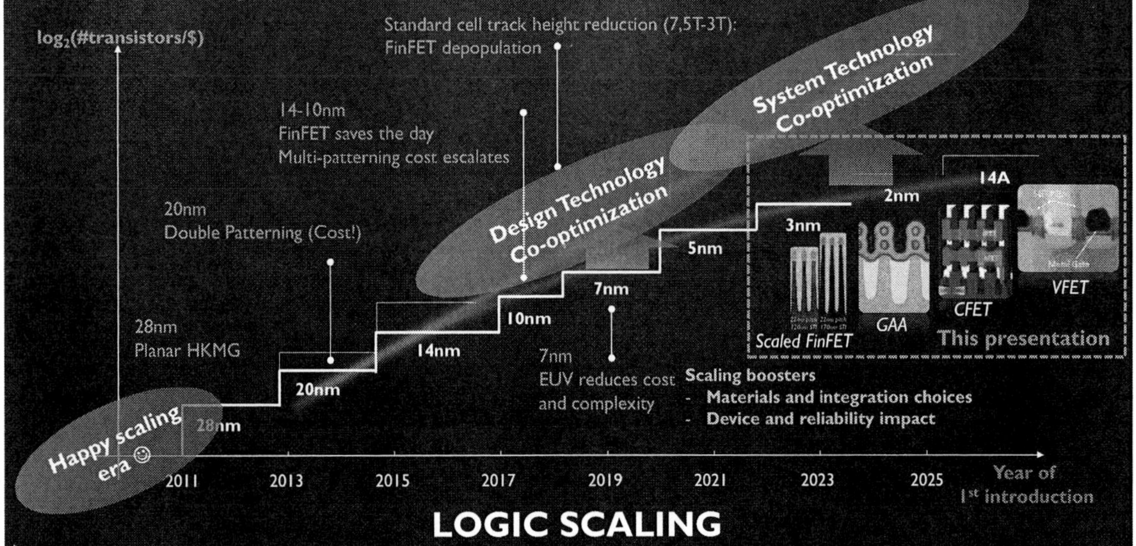


imec 2

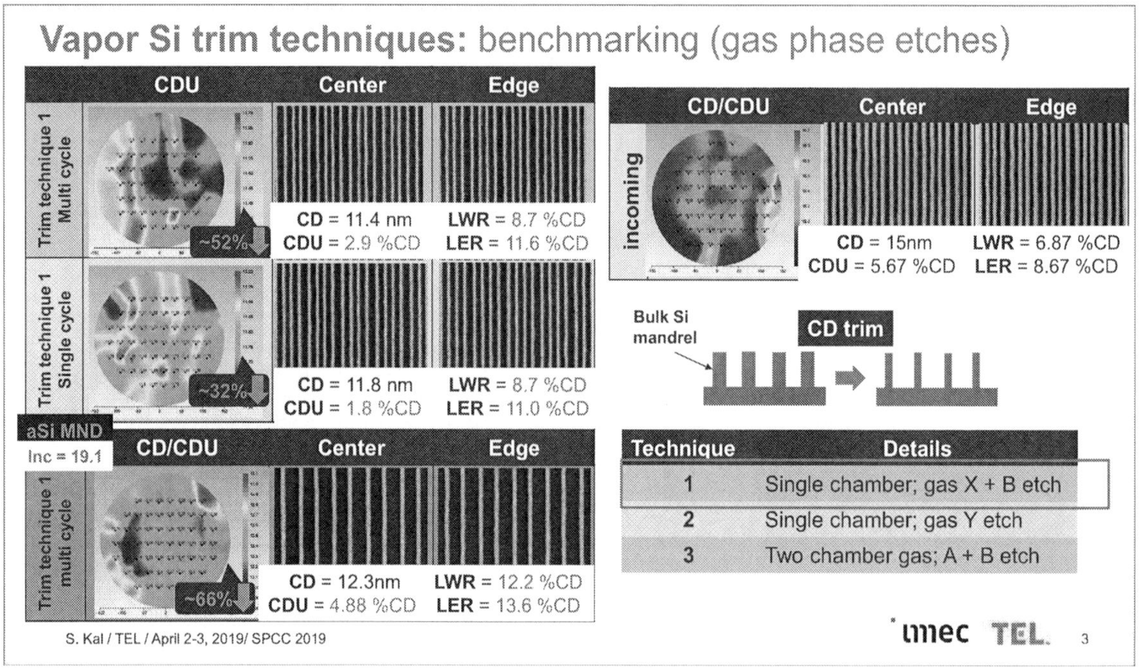
Vapor Si trim techniques: benchmarking (gas phase etches)
CDU
Center
Edge
Trim technique 1 Multi cycle
~52%
CD = 11.4 nm
CDU = 2.9 %CD
LWR = 8.7 %CD
LER = 11.6 %CD
Trim technique 1 Single cycle
~32%
CD = 11.8 nm
CDU = 1.8 %CD
LWR = 8.7 %CD
LER = 11.0 %CD
aSi MND
Inc = 19.1
CD/CDU
Center
Edge
Trim technique 1 multi cycle
~66%
CD = 12.3nm
CDU = 4.88 %CD
LWR = 12.2 %CD
LER = 13.6 %CD
CD/CDU
Center
Edge
incoming
CD = 15nm
CDU = 5.67 %CD
LWR = 6.87 %CD
LER = 8.67 %CD
Bulk Si mandrel
CD trim
Technique
Details
1
Single chamber; gas X + B etch
2
Single chamber; gas Y etch
3
Two chamber gas; A + B etch
S. Kal / TEL / April 2-3, 2019/ SPCC 2019
imec TEL
3

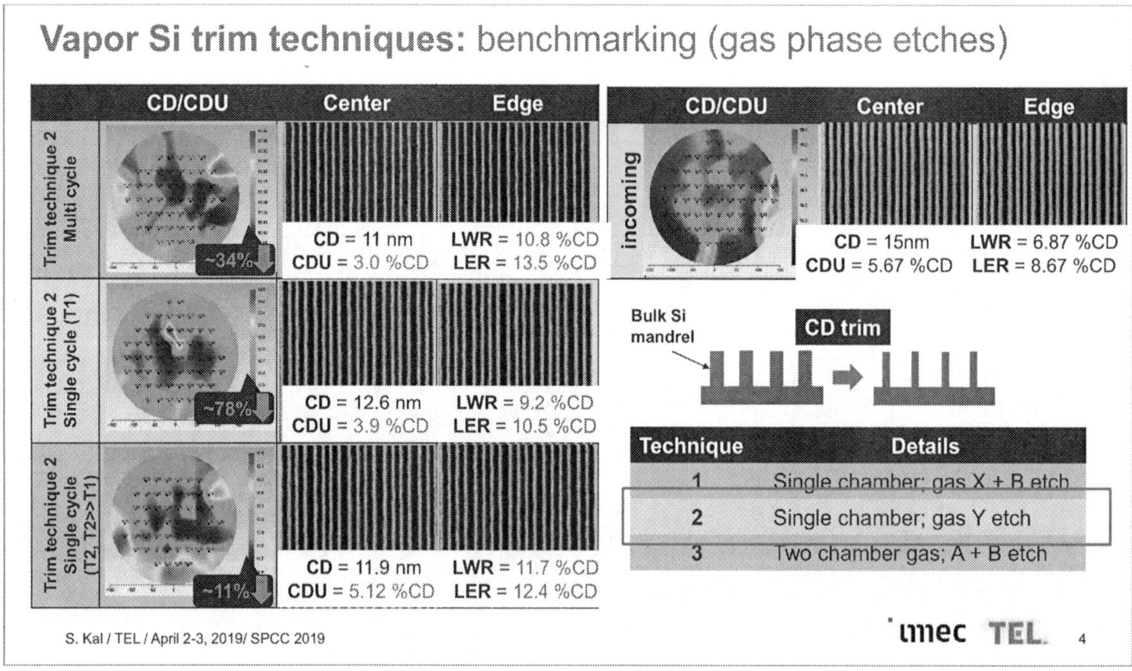
Vapor Si trim techniques: benchmarking (gas phase etches)
CD/CDU
Center
Edge
Trim technique 2 Multi cycle
~34%
CD = 11 nm
CDU = 3.0 %CD
LWR = 10.8 %CD
LER = 13.5 %CD
Trim technique 2 Single cycle (T1)
~78%
CD = 12.6 nm
CDU = 3.9 %CD
LWR = 9.2 %CD
LER = 10.5 %CD
Trim technique 2 Single cycle (T2, T2>>T1)
~11%
CD = 11.9 nm
CDU = 5.12 %CD
LWR = 11.7 %CD
LER = 12.4 %CD
CD/CDU
Center
Edge
incoming
CD = 15nm
CDU = 5.67 %CD
LWR = 6.87 %CD
LER = 8.67 %CD
Bulk Si mandrel
CD trim
Technique
Details
1
Single chamber; gas X + B etch
2
Single chamber; gas Y etch
3
Two chamber gas; A + B etch
S. Kal / TEL / April 2-3, 2019/ SPCC 2019
imec TEL
4

## Vapor Si trim techniques: benchmarking (gas phase etches)

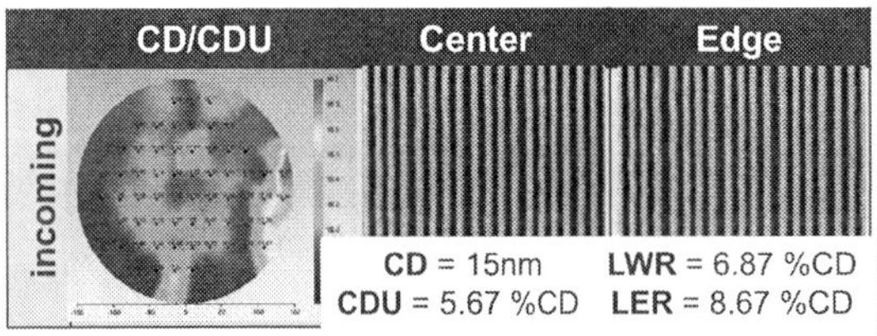


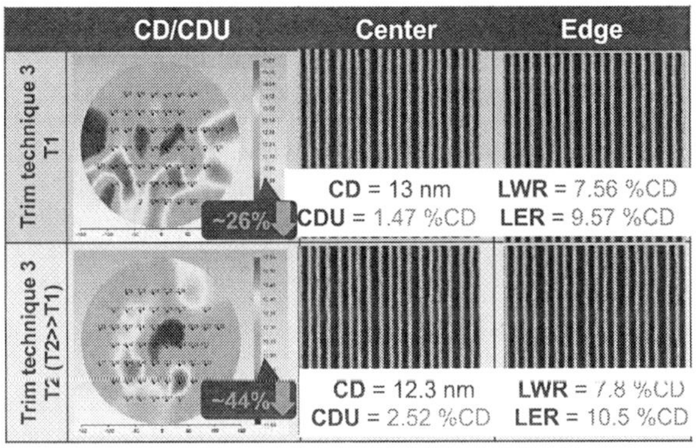


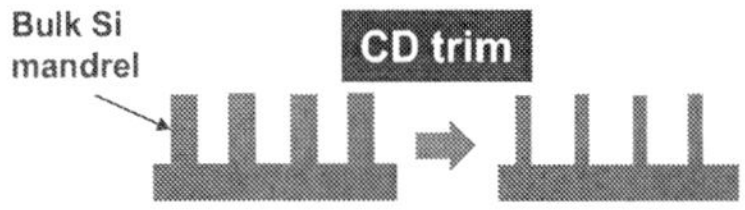


| Technique | Details |
|---|---|
| 1 | Single chamber; gas X + B etch |
| 2 | Single chamber; gas Y etch |
| 3 | Two chamber gas; A + B etch |

S. Kal / TEL / April 2-3, 2019/ SPCC 2019

imec TEL 5

## Vapor Si trim techniques: benchmarking (gas phase etches)

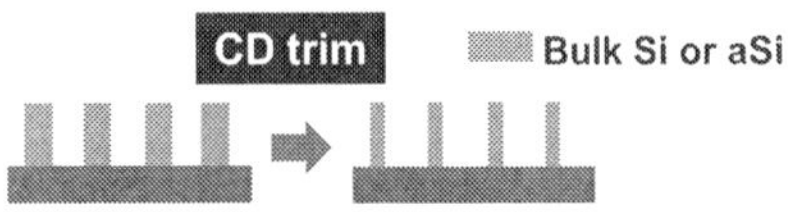


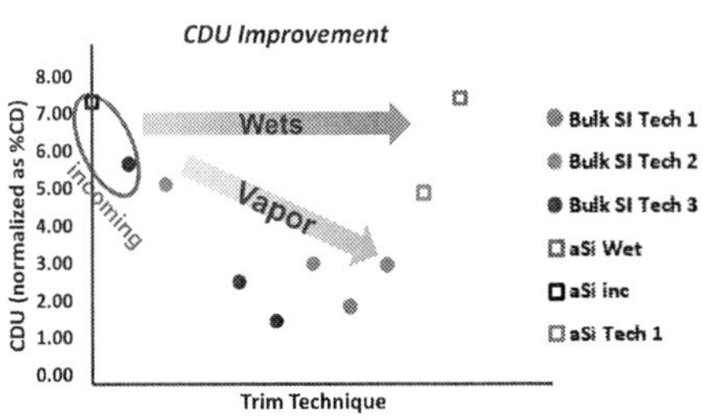


| Technique | Details |
|---|---|
| 1 | Single chamber; gas X + B etch |
| 2 | Single chamber; gas Y etch |
| 3 | Two chamber gas; A + B etch |

**Technique 1 & 3 has the best Si trim CDU improvement**
**Technique 2 however has better post trim surface roughness**
**Wet etch also trim feasibility; maintaining incoming CDU**

S. Kal / TEL / April 2-3, 2019/ SPCC 2019

imec TEL 6

## Vapor Si etch : Roughness Analysis

Technique 1

| Trim Technique | NA (incoming) | 1 | | 3 |
|---|---|---|---|---|
| details | Control | Si etch (process A) | Si etch (process B) | |
| AFM | 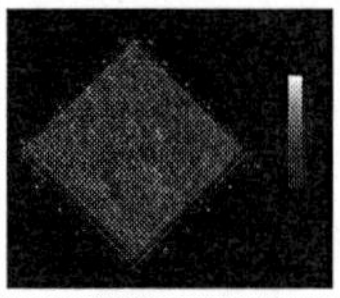 | 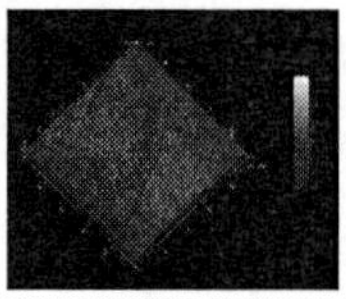 | 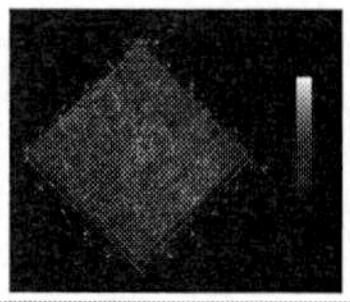 | 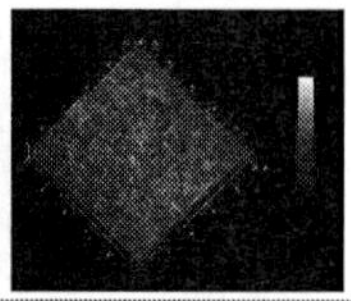 |
| Img. Rq RMS) | 0.112 nm | 0.172 nm | 0.142 nm | 0.104 nm |
| Img. Ra (Avg) | 0.0888 nm | 0.136 nm | 0.110 nm | 0.0821nm |

**Summary:**

- Certas post Si etch roughness *on bare-Si* shows slight increases in roughness (AFM/SEM)
- Roughness can be minimized by recipe tuning
- Major knobs for improving roughness:
  - Temperature
  - Pressure
  - Gas ratio

S. Kal / TEL / April 2-3, 2019/ SPCC 2019

TEL.

## FinFET Extension: Need for fin trim

**NONSELECTIVE SI/SIGE FIN CD TRIM @RMG**

- Fin CD trim @RMG for better electrostatic control w/o sacrificing $R_{ext}$
  - Target FW: ~4-5nm (etch 0.5-1nm/side)
  - Controlled etch Si/SiGe fins at the same time with a similar etch rate
  - Selectivity to: STI, ILD0, dummy OX, spacer, CESL, ...

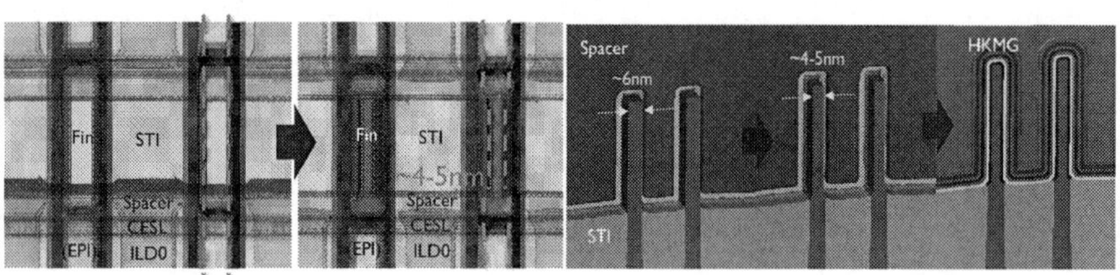


Y. Oniki et al. (imec) SPCC 2018

S. Kal / TEL / April 2-3, 2019/ SPCC 2019

imec TEL. 8

## FinFET Extension: Si and SiGe Fin Trim (simultaneous)

Technique 1

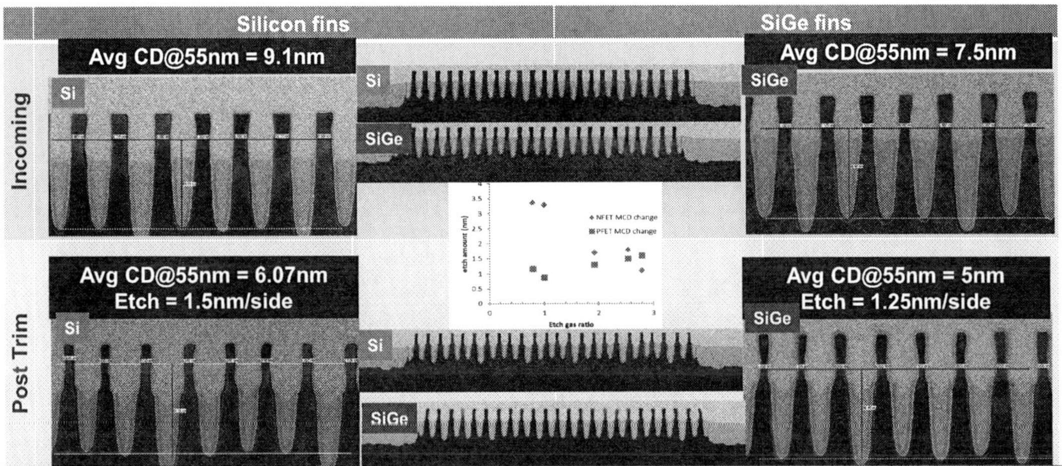


**Technique 1 can trim Si and SiGe fins simultaneously**
**Ability to tune trim rates for preferential trimming**



## FinFET Extension: Si Fin Trim Feasibility

Technique 1

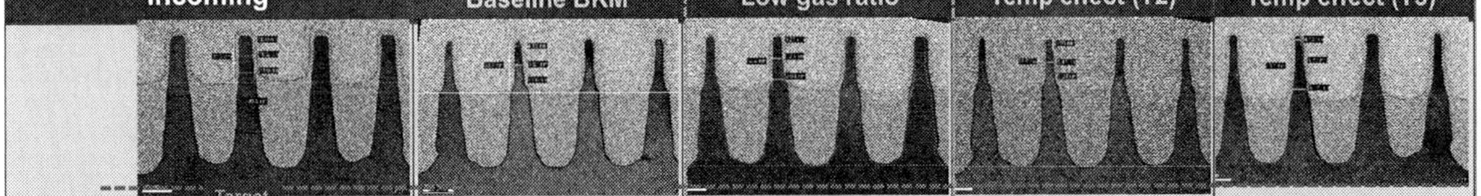

| | Target | Incoming | Baseline BKM | Low gas ratio | Temp effect (T2) | Temp effect (T3) |
|---|---|---|---|---|---|---|
| Fin width (nm) | 5 | 10/ 10/ 11 | 4.2/ 5.8/ 7.1 | 6.2/ 6.2/ 8.6 | 5.9/ 5.4/ 7.2 | 4.1/ 5.6/ 9.1 |
| Fin height (nm) | 31 | 31 | 33 | 34 | 33 | 43 (low temp BT) |
| EA/ DELTA* (nm) | 5/<1 | 0/ 1 | 5.8/4.2/3.9/ 2 | 3.8/3.8/2.4/ 1.4 | 4.1/4.6/3.8/ 2.2 | 5.9/4.4/1.9/ 4 |

*Delta: measure the widest –narrowest fin CD ~ fin width non-uniformity/roughness

Baseline Temp > T2 > T3

**Summary:**

- **Isotropic Si fin trim looks promising**
- Gas ratio tuning in ME step improves fin width variation and fin –fin trim uniformity (~ roughness)

➔ **Post trim FW variation (roughness) is higher than incoming**

- BT step does not introduce field oxide iso-dense loading: fin height maintained

**Technique 1 has ability to improve post trim roughness**
**Trim is very selective to STI oxide (also SiN)**

## VFET: Gate Length Definition

**VERTICAL SI NANOWIRE CHANNEL CD TRIM**

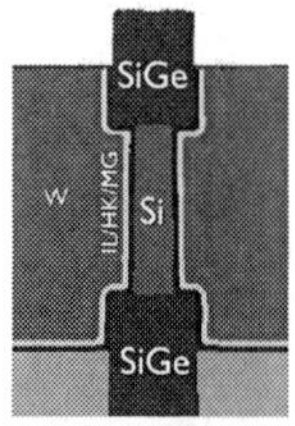


- Si channel trim for self-aligned Lg definition
- Need for a controlled Si etch with selectivity to SiGe S/D

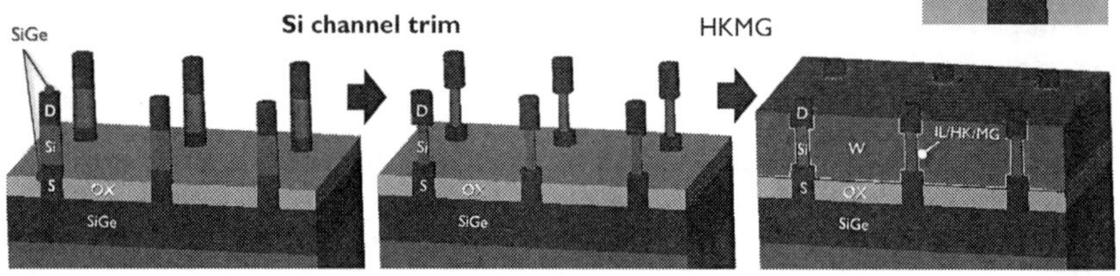




## VFET: Si Channel Trim Feasibility

Technique 1

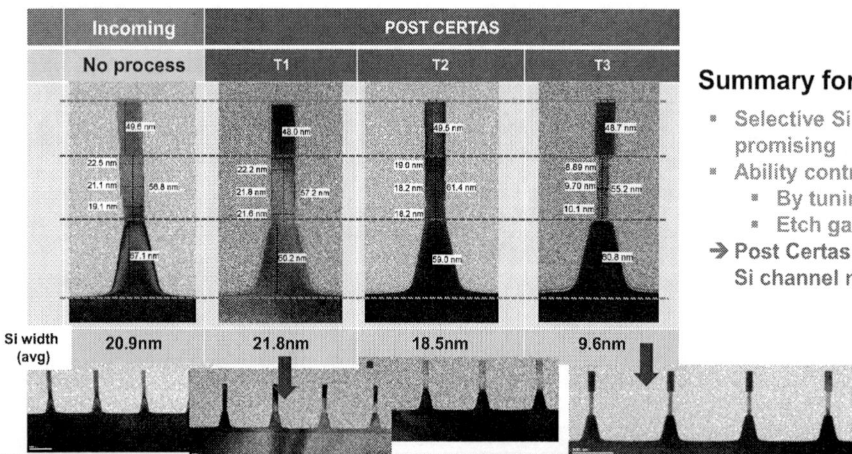


**Summary for Si channel trim:**

- Selective Si trim (wrt SiGe) looks promising
- Ability control the trim amount
  - By tuning etch time
  - Etch gas ratio

→ **Post Certas etch roughness for the Si channel needs improvement**

**Technique 1 shows good Si channel trim selectivity w.r.t SiGe for VFET**
**Facilitate VFET Si gate length definition (self aligned)**

# CFET: Mushroom Gate

## DUMMY POLY CD TRIM PRIOR TO SPACER FORMATION

- Conventional spacer etch needs thicker gate HM >FinH → poly collapse for taller fin devices
- Mushroom gate to enable HM thickness scaling
  - Lg (& spacer width) defined by poly trim amount → need for a controlled a-Si selective etch

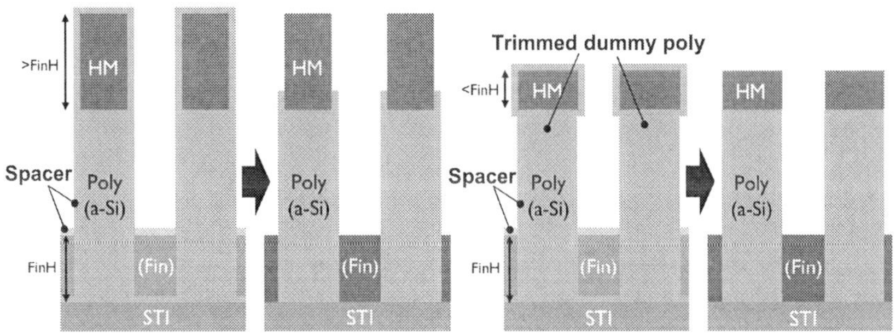




# CFET: Mushroom Gate (Feasibility)

Technique 1

| Incoming (iCT-inc) | T1 | T2 | T3 |
|---|---|---|---|

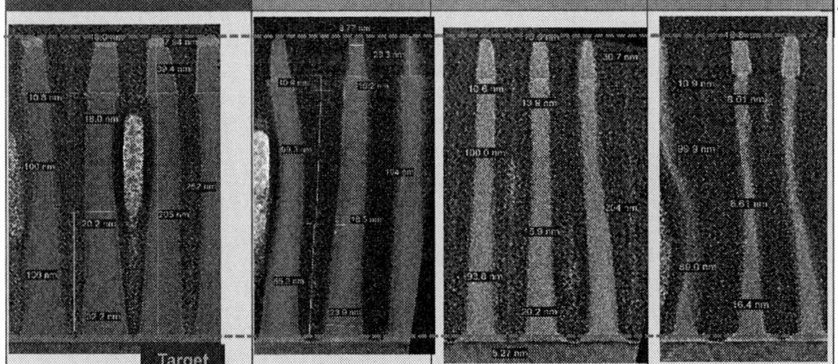

| Width @ | | Target 4nm/side | Incoming (iCT-inc) | T1 | T2 | T3 |
|---|---|---|---|---|---|---|
| | 10nm/EA | | 18/0 | 16.4/0.8 | 13/2.5 | 8/5 |
| | 100nm/EA | | 20/0 | 17.5/1.2 | 15.4/2.3 | 9.5/5.2 |
| | 190nm/EA | | 32/0 | 24.8/3.58 | 20.3/5.8 | 15.2/8.4 |

Summary:

- Isotropic Si fin trim looks promising
- Roughness looks same as incoming

→ Further improvement needed for:

→ Uniformity tuning

- **Incoming gates have slightly wider CD at bottom**
- **4nm trim/side gate bending (may be due to TEM decoration?)**

**Technique 1 shows feasibility for CFET gate trim**

# Conclusion

| Technique | Details | Improvement |
|---|---|---|
| 1 | Single chamber; gas X + B etch | |
| 2 | Single chamber; gas Y etch | CDU |
| 3 | Two chamber gas; A + B etch | |
| Wet | Single chamber | Throughput |

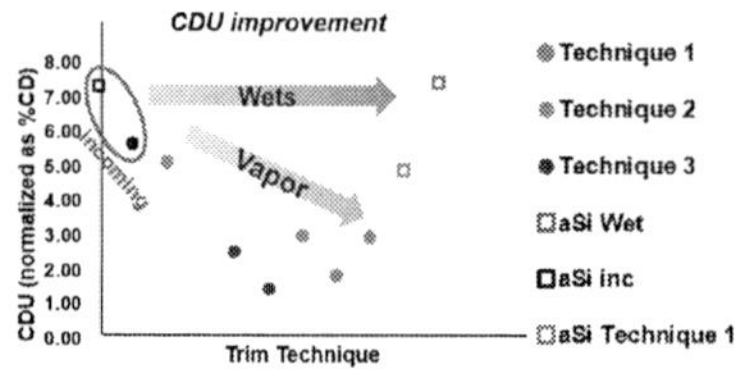


- Established 3 **dry Si trim techniques** for patterning purposes
  - All dry techniques improve CDU post trim
- Demonstrated Si trim feasibility with **wet etch**
- ***Dry plasma free*** etches are advantageous & crucial for FiFET/VFET/CFET integrations applications, due to:
  - High etch selectivity, inherent from the etch mechanism
  - No plasma damage
  - Aspect ratio dependency

S. Kal / TEL / April 2-3, 2019/ SPCC 2019

imec TEL. 15

TEL™

# PHENOMENA AFFECTING THE RATE AND EQUILIBRIUM OF AQUEOUS CHEMICAL REACTIONS IN NANOCHANNELS

GUY VEREECKE[1*], HAROEN DEBRUYN[2], QUINTEN DE KEYSER[2], ABHISHEK DUTTA[2] AND FRANK HOLSTEYNS[1]

[1]IMEC, KAPELDREEF 75, 3001, LEUVEN, BELGIUM
[2]FACULTY OF ENGINEERING TECHNOLOGY CAMPUS GROUP T, K.U. LEUVEN, BELGIUM
*GUY.VEREECKE@IMEC.BE

SURFACE PREPARATION & CLEANING CONFERENCE (SPCC), APRIL 2-3, 2019, PORTLAND, OREGON

PUBLIC

## INTRODUCTION

2

## HF ETCHING OF $SiO_2$ IN FILLED NANOCHANNELS
### KINETICS AFFECTED BY OVERLAP OF ELECTROSTATIC DOUBLE LAYERS (EDL)

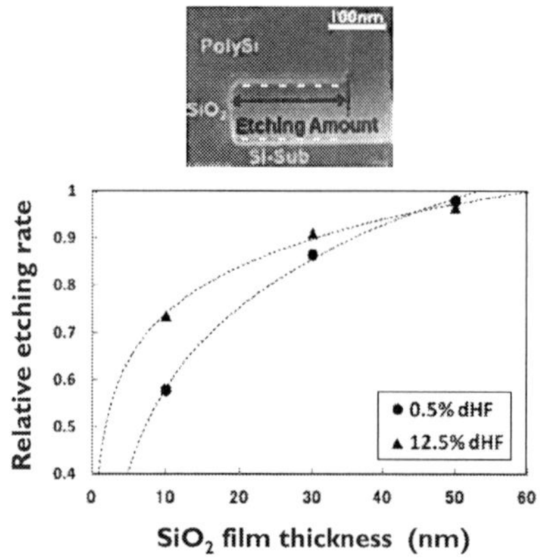


→ Decreased etch rate in narrow channels

- EDL overlap in nanochannel

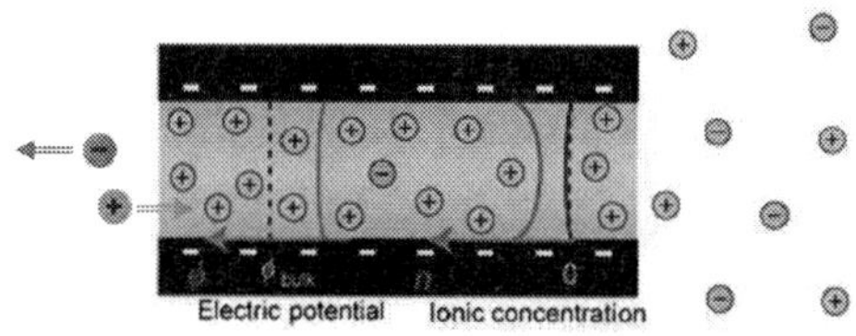


H. Daiguji, *Chem. Soc. Rev.*, 2010, **39**, 901.

- No electroneutral 'bulk' solution
  - Depletion of ions with same charge as surface; enrichment of ...
  - → Acidic pH shift
  - → Decreasing etch rate from increasing depletion of $HF_2^-$

imec A. Okuyama et al., *Solid State Phenom.* 2015, **219**, 115.



## HF ETCHING OF $SiO_2$ IN FILLED NANOCHANNELS
### EVALUATION OF IMPACT OF EDL OVERLAP

- EDL thickness in HF solutions

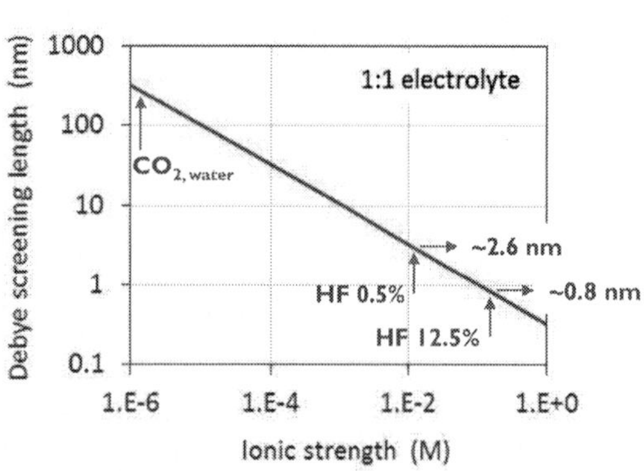


→ No EDL overlap in HF 12.5 %

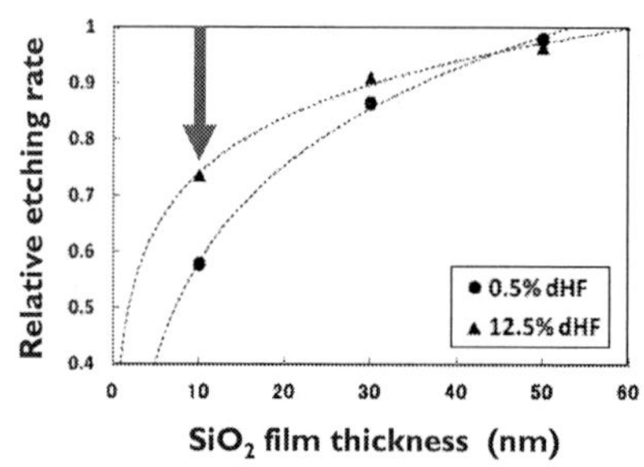


- Etch rate decrease in HF 12.5 % cannot be explained by EDL overlap

→ What else ?

imec

THIS WORK

## PROBING THE CHEMISTRY IN NANO-SPACES USING ATR-FTIR

### ATR CRYSTAL MADE FROM SI WAFER WITH NANO-STRUCTURES[1,2]

Total reflection → $\alpha > 17.1°$ (Si/air)
→ TR at top surface

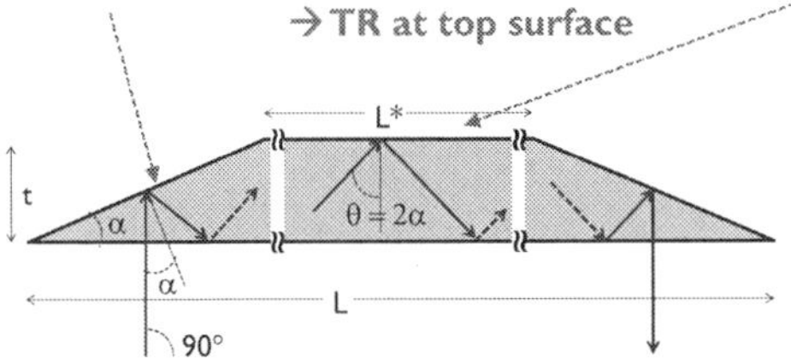


Top surface covered with nanochannels

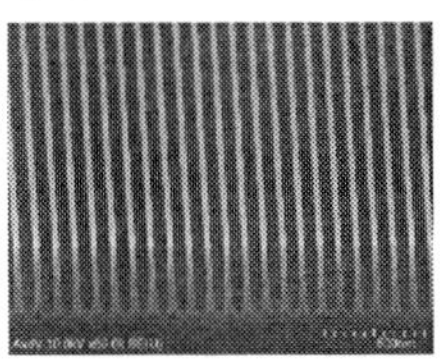


Liquid cell on top of ATR crystal

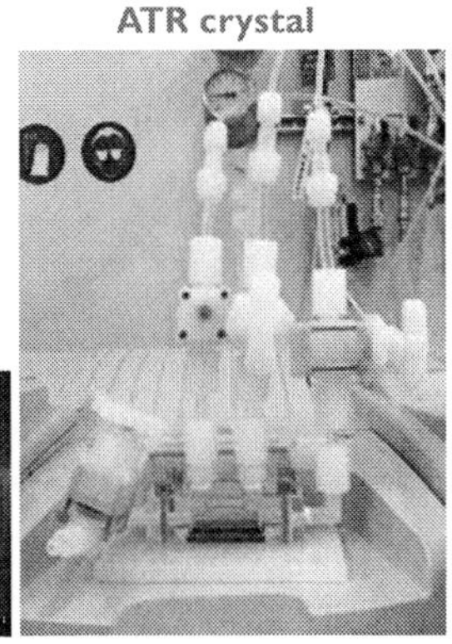

Channel width decreased by $SiO_2$ deposition (PE-ALD)

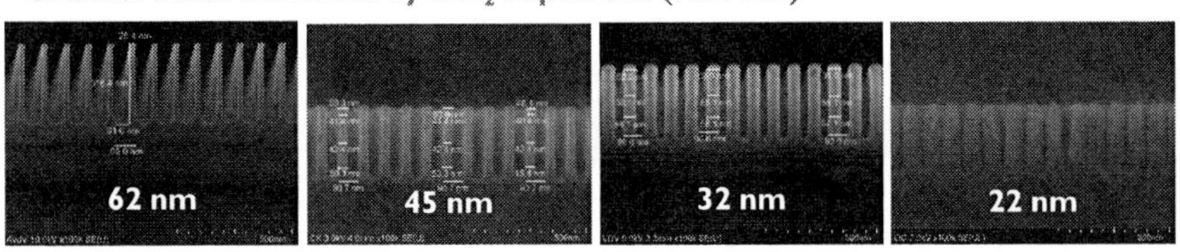


imec

[1]G. Vereecke et al., *ECS J. Solid State Sci. Technol.* **3** (2014) N3095-N3100. [2]N. Vrancken et al., *Langmuir* **33** (2017) 3601–3609.

SPCC 2019

## OUTLINE

1. Measurement of pH in nanochannels
2. Impact of nano-confinement on the kinetics of a model reaction

imec

7

SPCC 2019

## 1. MEASUREMENT OF pH IN NANOCHANNELS

## INDIRECT MEASUREMENT OF pH BY FTIR
### MEASURING THE PROTONATION OF ACIDIC GROUPS

- Citric acid solutions

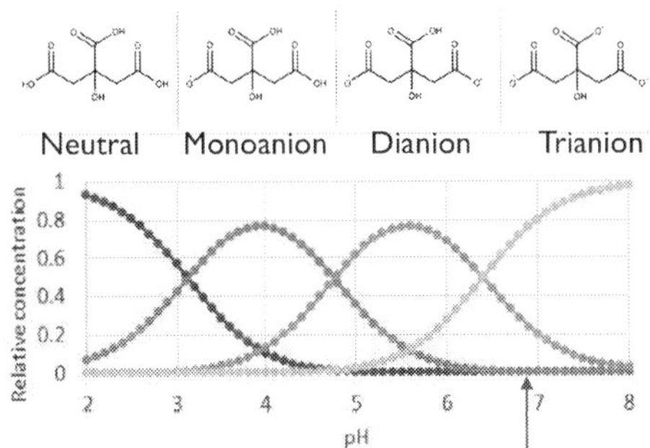


This work:
- 50 mM citric acid (pH = 2.2)
- pH = 6.9 by NaOH addition

- Sodium bicarbonate solutions

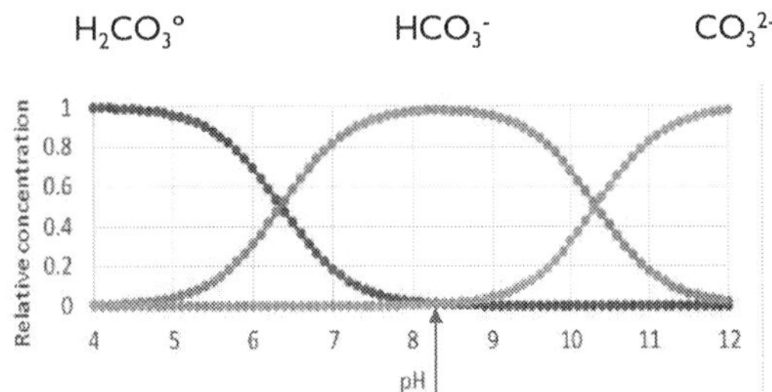


This work:
- 100 mM $NaHCO_3$ (pH = 8.3)
- In $D_2O$

imec 9 SPCC 2019

## FTIR pH CALIBRATION CURVE FOR CITRIC ACID
### pH SET BY HCl OR NaOH ADDITIONS

- ATR-FTIR spectra on planar Si

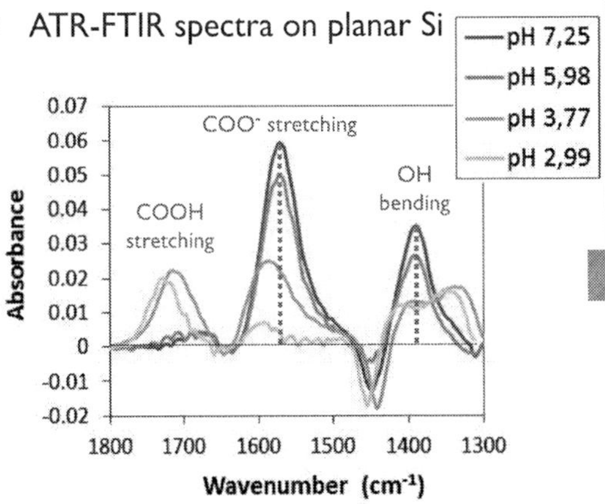


- Calibration curve

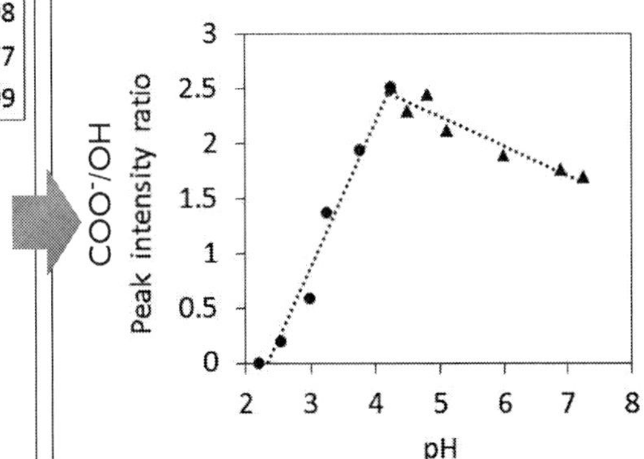


imec 10 SPCC 2019

## MEASURED pH IN NANOCHANNELS
### DISCREPANCY BTW. MEASUREMENTS WITH CITRIC ACID AND NAHCO3

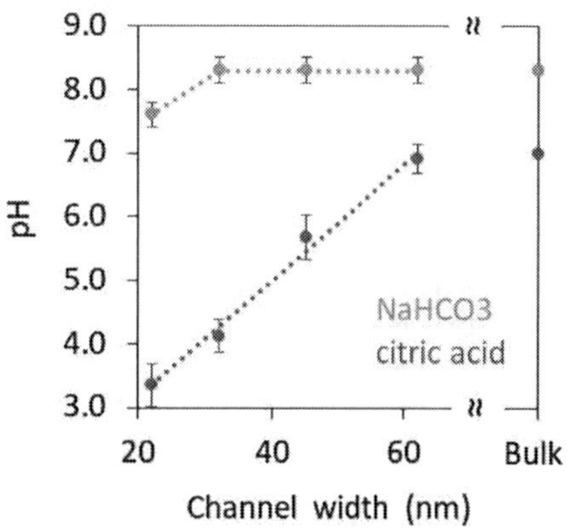


- No large [$H^+$] variation expected at the high ionic strengths used here:

| Probe molecule | Ionic strength (M) | Debye length (nm) |
|---|---|---|
| Citric acid | 0.89 | 0.5 |
| NaHCO3 | 0.1 | 1 |

→ No EDL overlap

- Origin of pH shift with citric acid ?

imec



## CHARACTERIZATION OF WATER H-BOND STRUCTURE BY FTIR
### DECOMPOSITION OF OH STRETCHING BAND

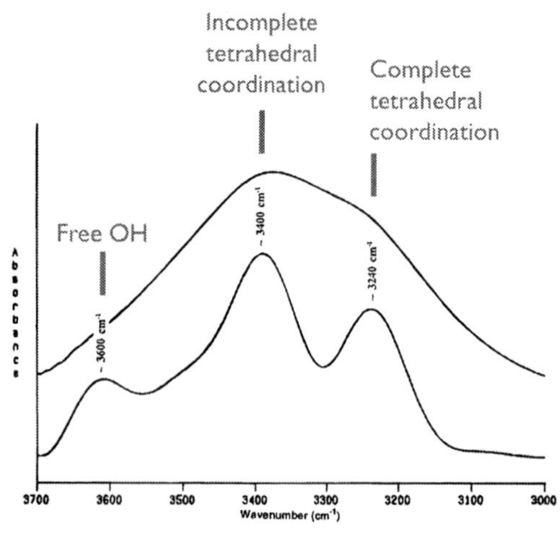


- OH stretching peak of water broadened by H-bonding
- 3 peaks obtained by Fourier self-deconvolution method[1]
  - ~3400 $cm^{-1}$ : Hydrogen bonded water with **in**complete tetrahedral coordination
  - ~3240 $cm^{-1}$ : Hydrogen bonded water with **complete** tetrahedral coordination, sometimes termed "ice-like"
  - ~3600 $cm^{-1}$ : non Hydrogen-bonded or weakly H-bonded "free OH" stretch

imec

[1]M.R. Yalamanchili et al., *Langmuir*, 1996, **12**, 4176.

## WATER STRUCTURING IN NANOCHANNELS
### DIFFERENCE SPECTRA NANOCHANNEL – PLANAR SAMPLE

- Citric acid solutions

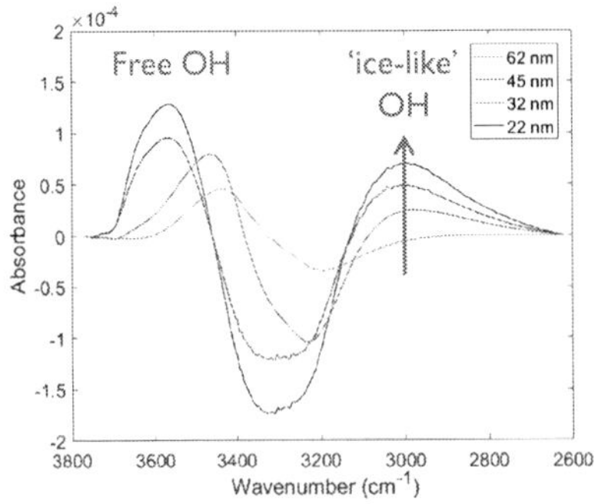


- Sodium bicarbonate solutions

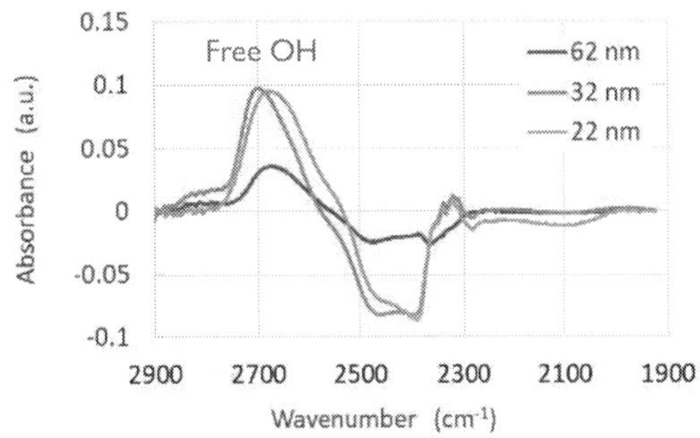


- Good correlation btw. formation of 'ice-like' water and pH shift



## POSSIBLE EXPLANATION FOR pH SHIFT IN NANOCHANNELS

- Lower dielectric constant of aqueous solutions in nanochannels from water structuring[1]

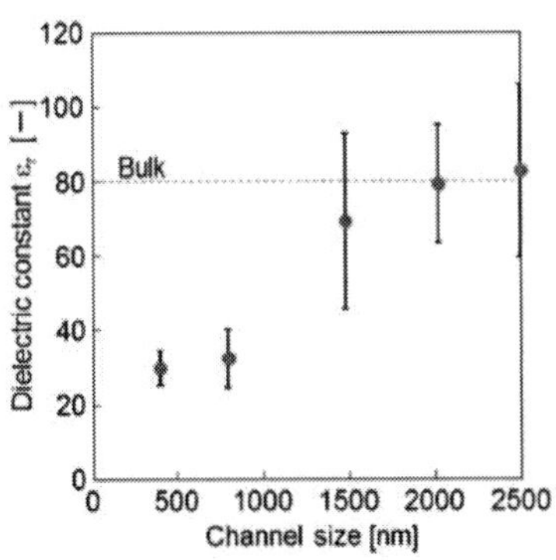


$H_2O$ behaving as a solvent with lower polarity

→ Decreased screening of ions from acid dissociation

→ Increase of the acid p$K_a$'s

→ Depressed ionization of citric acid

→ Shifted equilibrium, but no real pH shift

[1]K. Morikawa et al., *Anal. Chem.*, 2015, **87**, 1475.

# 2. IMPACT OF NANO-CONFINEMENT ON THE KINETICS OF A MODEL REACTION

## MODEL REACTION
### STRAIN PROMOTED ALKYNE - AZIDE CYCLOADDITION (SPAAC)

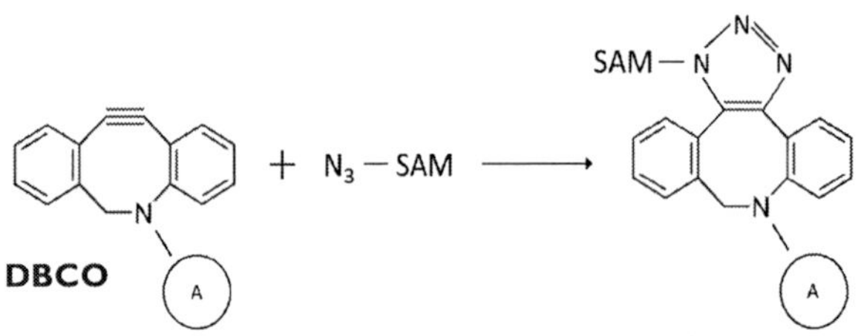


A = $PEG_3$-OH

- Click Reaction
- Used in bio-sensor applications to link DNA strands to surfaces
- 2nd order reaction kinetics w/o side reactions

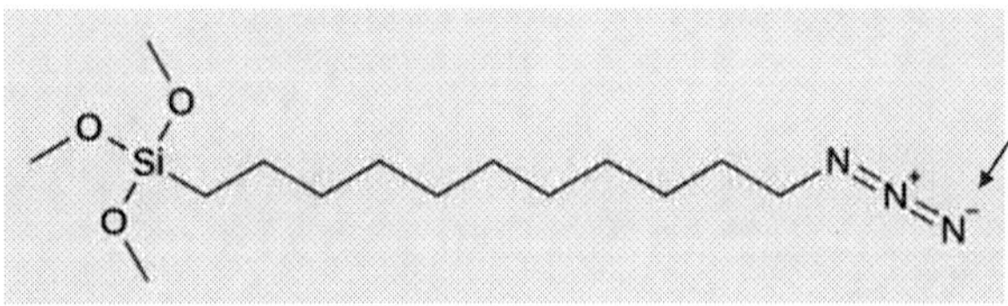


Negative surface charge

- DBCO = dibenzyl-cyclo-octyl
- PEG = poly-ethylene-glycol

imec 16 SPCC 2019

## KINETIC STUDY
### DETERMINATION OF ORDER IN AZIDE CONCENTRATION

- Monitoring the disappearance of the azide peak at 2100 $cm^{-1}$

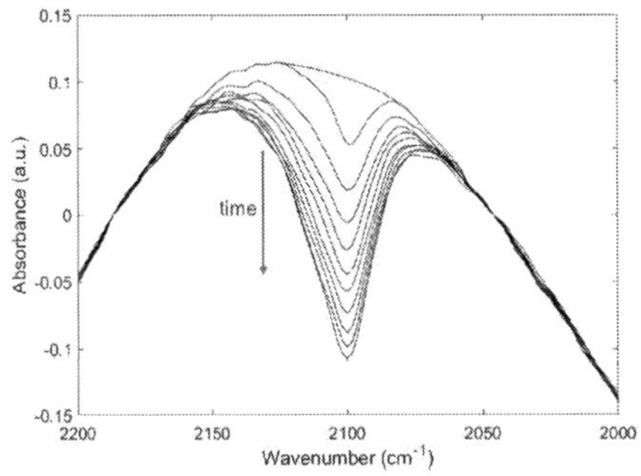


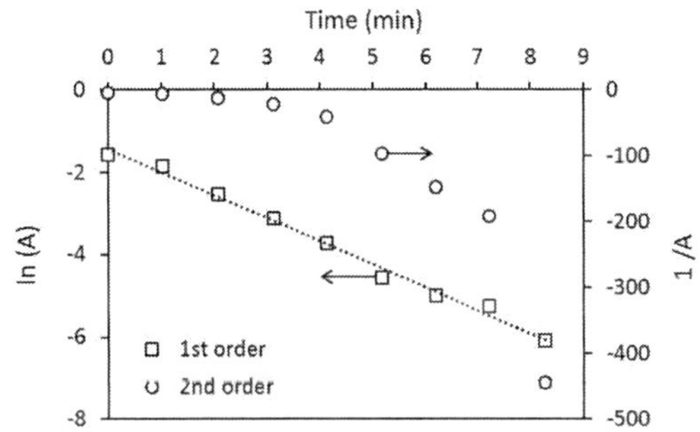


→ Pseudo first order confirmed



## SLOWER KINETICS IN NANOCHANNELS VS. PLANAR
### 2ND ORDER REACTION CONSTANTS

| Pattern | $k_2$ ($M^{-1}$ $s^{-1}$)* | $k_{planar}$ / $k_{channel}$ |
|---|---|---|
| Planar crystal | 285 ± 27 | 1 |
| 64 nm channel | 66 ± 2 | 4.3 ± 0.4 |
| 45 nm channel | 54 ± 1 | 5.3 ± 0.5 |
| 32 nm channel | 41 ± 0.5 | 7.0 ± 0.7 |

* $k_2 = \frac{k'}{[DBCO]_0}$ ; average values from 2-3 tests

- Lower reaction rates in nanochannels, decreasing along with the channel width
- EDL overlap in all conditions

| Ionic strength (M)* | Debye length (nm) |
|---|---|
| 2.2E-6 | 288 |

*water saturated with $CO_2$ gas

→ Nearly constant negative potential in channels in all conditions

→ No impact on concentration of neutral reactant

→ What else ?

## WATER STRUCTURING IN NANOCHANNELS
### SPECTRA FROM KINETIC TESTS

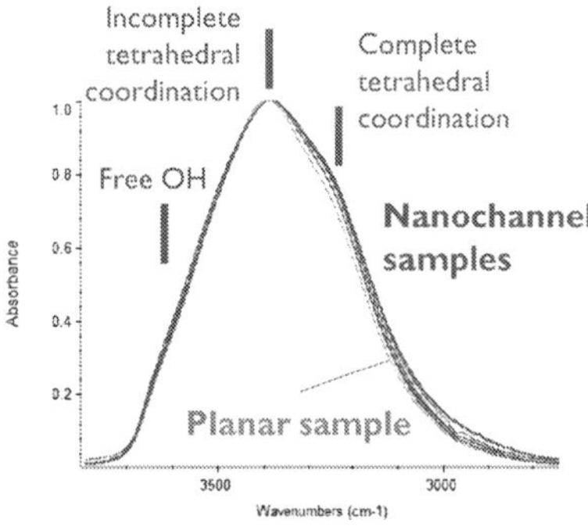


Normalized spectra after ATR correction and baseline subtraction

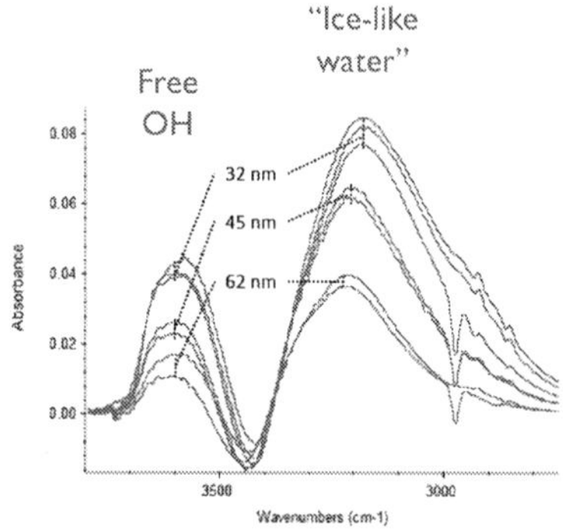


Difference spectra
Nanochannel - planar sample

- Water structuring in nanochannels, increasing as channel width decreases
- Solvent effect on SPAAC reaction ?



## SOLVENT EFFECT ON SPAAC REACTIONS
### REACTION RATE CONSTANTS IN DIFFERENT SOLVENT SYSTEMS[1]

| Entry | Azide structure | 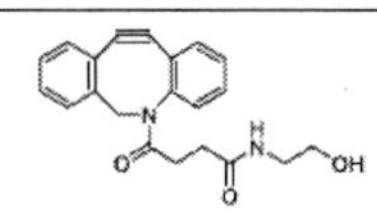 | | |
|---|---|---|---|---|
| | | ε = 69 | 33 | 15 |
| 1 | 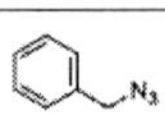 | 1.9 | 0.42 | 0.36 |
| 2 | 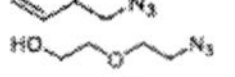 | 0.90 | 0.33 | 0.21[a] |
| 3 | 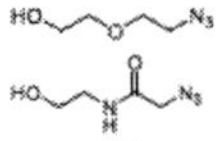 | 0.82 | 0.29 | 0.20[a] |
| 4 | 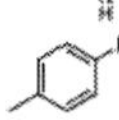 | 0.11 | 0.11 | 0.08 |
| 5 | 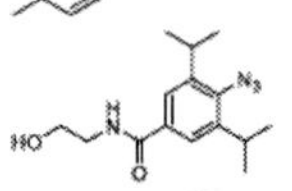 | ~4[b,c] | 2.3 | 0.77[a] |
| 6 | 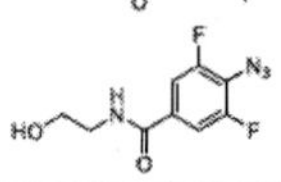 | 0.95 | 0.18 | 0.11[a] |

- Solvents used in study:

| Solvent system | $\varepsilon^{+}$ |
|---|---|
| $D_2O:CD_3CN$ (3:1) | 69 |
| MeOD | 33 |
| $H_2O$:THF (1:9) | 15 |

[+] calculated using eq.(3) in Wang (2001)

- SPAAC proceeds faster in more polar aqueous solvent systems

→ reaction rate decrease in nanochannels likely from water structuring

[1]J. Dommerholt et al., *Top. Curr. Chem.*, 2016, **374**, 16.

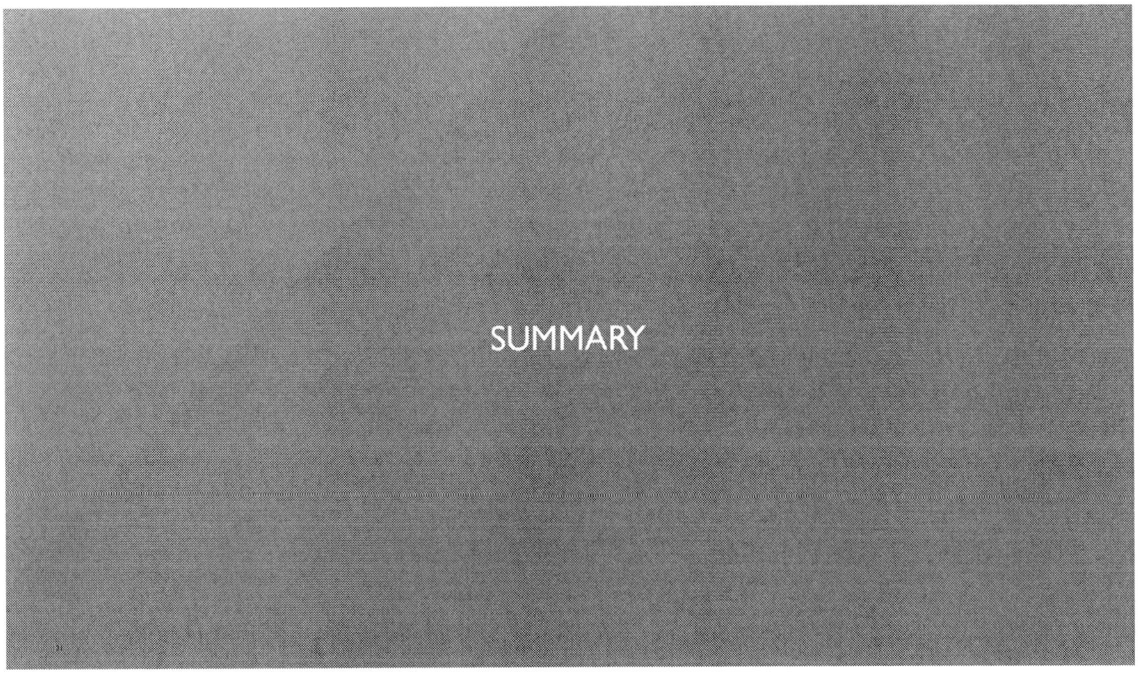


## SUMMARY

- Not all results obtained in this study could be explained by EDL overlap
  - pH shift measured using citric acid
  - SPAAC reaction rate decrease
- Proposed explanation: decrease of $H_2O$ permittivity by $H_2O$ structuring
  - Citric acid dissociation depressed, no real pH shift
  - SPAAC reaction slowed down in less polar solvent
- Water in nanochannels may behave as a solvent of lower polarity, depending on dissolved species – molecules & ions

imec
embracing a better life
imec
CONFIDENTIAL

imec

# IMPROVING CLEAN PERFORMANCE FOR ADVANCED PACKAGING

CHRISTOPHE LORANT, SIMON BRAUN, SAMUEL SUHARD, FRANK HOLSTEYNS

IMEC, Kapeldreef 75, 3001 Heverlee, Belgium
+32 016 28 32 95
Christophe.lorant@imec.be

PUBLIC

## OUTLINE

1. 3D interconnect technology landscape
2. Hybrid collective die to wafer bonding
3. Cleaning of dies on carrier
4. Towards wet process optimization

imec

2

## THE 3D INTERCONNECT TECHNOLOGY LANDSCAPE

| 3D-SIP | | 3D-SIC | 3D-SOC | | 3D-IC |
|---|---|---|---|---|---|
| Package stacking | Muliti-die Packaging<br>Interposer "2.5D"<br>Embedded Die | Die Stacking<br>µbump | Wafer-to-Wafer bonding | Wafer-to-Wafer Sequential Processing | Transistor Stacking |

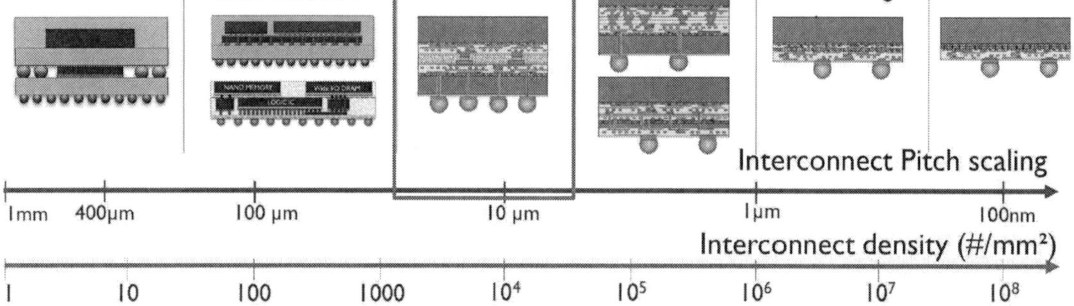


Eric Beyne, "3D System Integration Technologies", 2006 International Symposium on VLSI Technology, Systems, and Applications, 2006, pp.1-9

Eric Beyne, "The 3-D Interconnect Technology Landscape", IEEE Design & Test, 2016, Vol. 33 NR 3, pp. 8 – 20

imec

3

## HYBRID BONDING
### CU-CU DIRECT BONDING

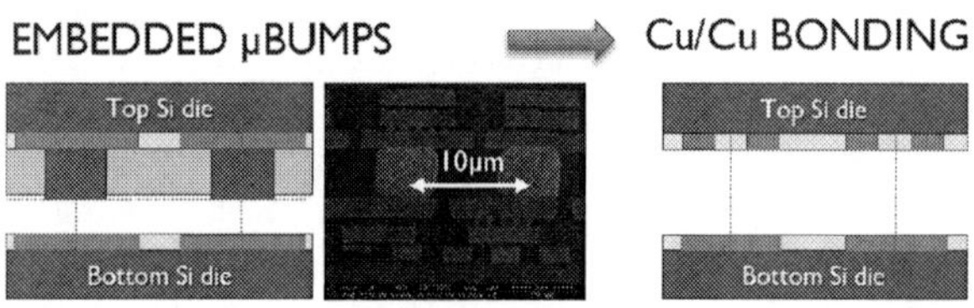


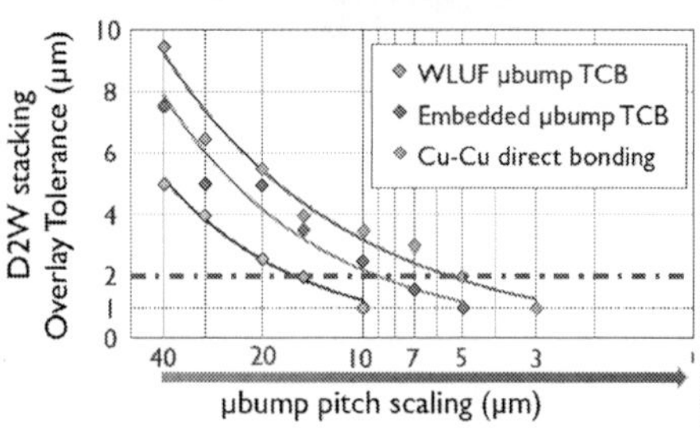


Bonding mechanism
- Dielectric bonding
- Metal-metal interdiffusion

Requires:
- Smooth surfaces Ra <1nm
- Flat surface : slope < 1nm/µm
- Oxide-free metal surface
- Activated dielectric surface

- Vertical tolerance ~ 0
- No tolerance to particles
- Ultra-clean die handling

- Clean Dicing
- Die-to-Carrier bonding
- Wafer-level cleaning
- Carrier-to-Wafer Die transfer

Eric Beyne et al., "Scalable, sub 2µm pitch, Cu/SiCN to Cu/SiCN hybrid wafer-to-wafer bonding technology", 2017 IEEE International Electron Devices Meeting (IEDM), 2017, pp. 32.4.1 - 32.4.4

imec

4

# HYBRID COLLECTIVE DIE TO WAFER BONDING

## PROCESS FLOW

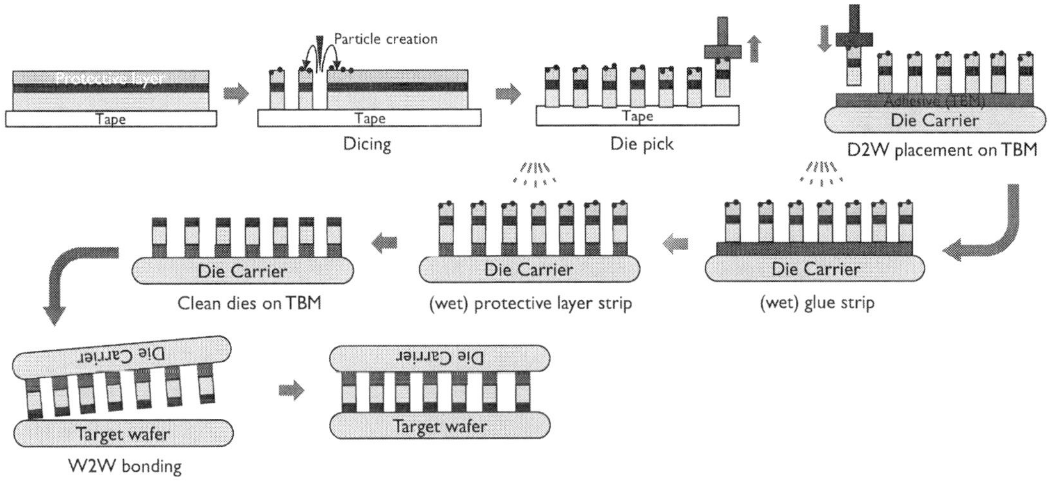


imec

5

# PROTECTIVE LAYER FOR DEFECT REDUCTION

## MOTIVATION

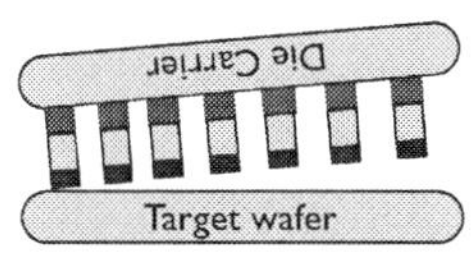


SAM after bonding

**Without** protective layer    **With** protective layer

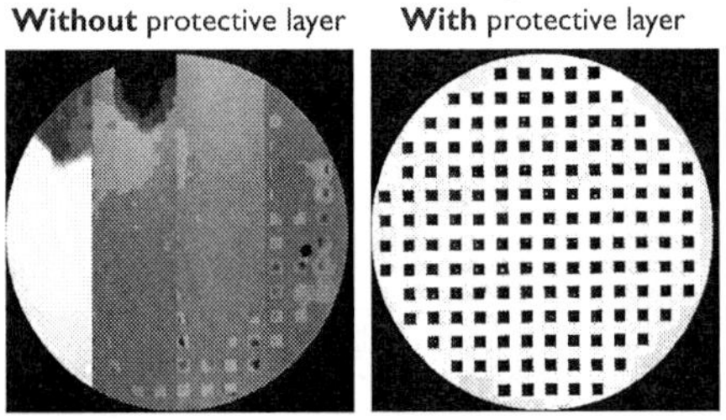

After population    After protective layer strip    After bonding

- Defects can be generated either during dicing or die handling
  - Si particles generation
  - Contact defects during pick & place
- Particulate contamination cause large voids after bonding when they are not removed

➢ A protective layer is a key enabler to achieve a low defect die surface and successful bonding

imec

6

# PROTECTIVE LAYER FOR DEFECTIVITY REDUCTION
## PROCESS REQUIREMENTS

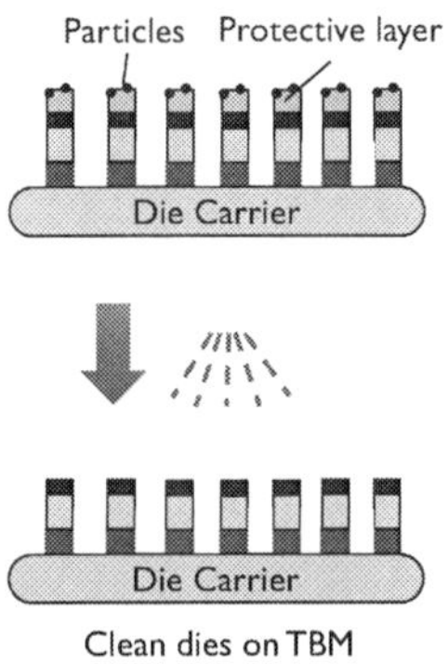


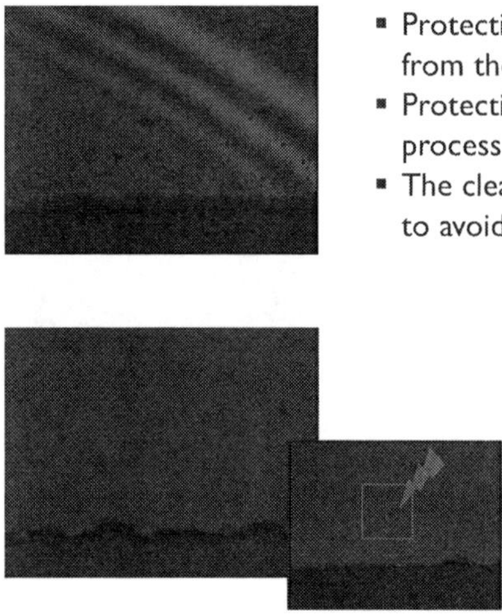

- Protective layer and particles must be removed from the die to enable successful bonding
- Protective layer is removed in a wet strip process compatible with TBM
- The clean needs to remove sub 0.1 µm particles to avoid bonding voids

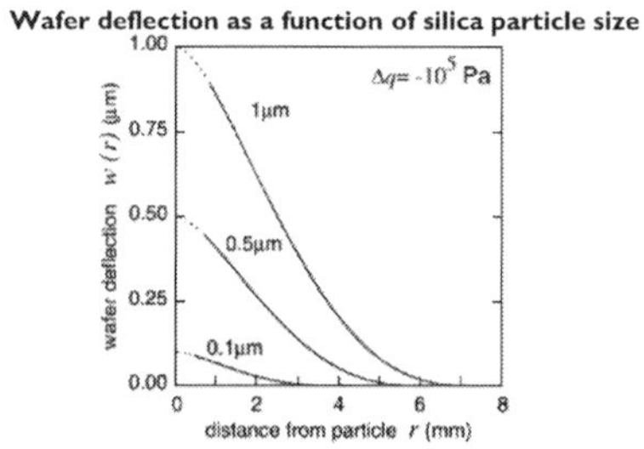


Bearda T. et al., JJAP 44 (10), 7409, 2005

7

# CLEANING OF DIES ON A CARRIER
## TEST MATERIAL

- 30 µm adhesive layer (TBM)
- 14 µm photoresist as protective layer
- 7x7 mm dies, 3 mm spacing
- Photoresist is stripped in a single wafer wet tool

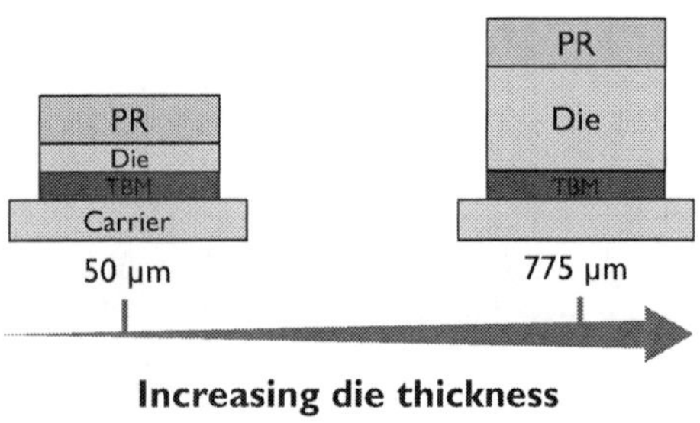


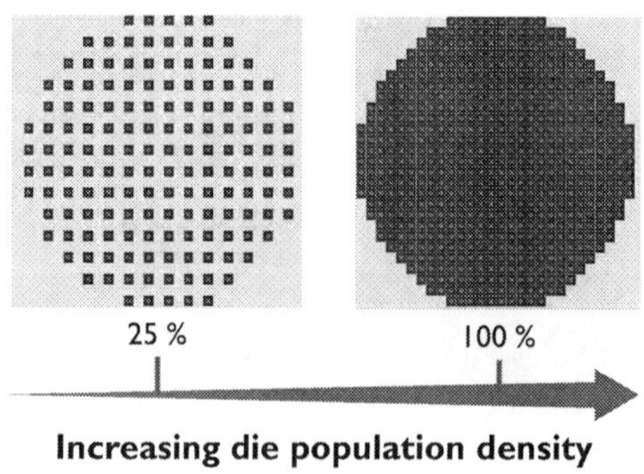


8

# CLEANING OF DIES ON A CARRIER
## PROCESS UNIFORMITY AFTER PROTECTIVE LAYER STRIP

Standard TMAH/DMSO based resist strip on a TEL Cellesta+

**50 μm dies (25% pop.)**

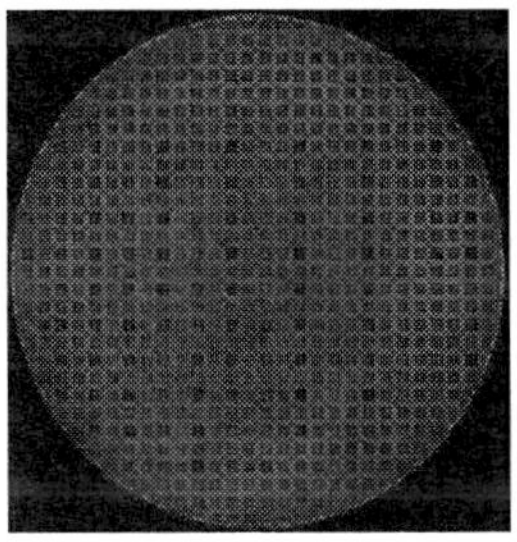

- No PR residues visible

**775 μm dies (25% pop.)**

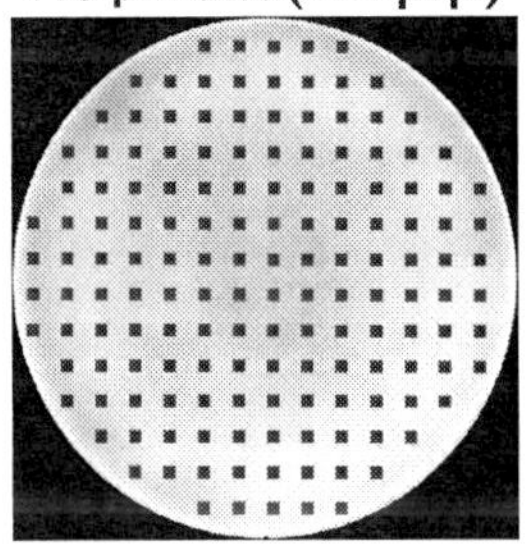

- No PR residues visible

**775 μm dies (100% pop.)**

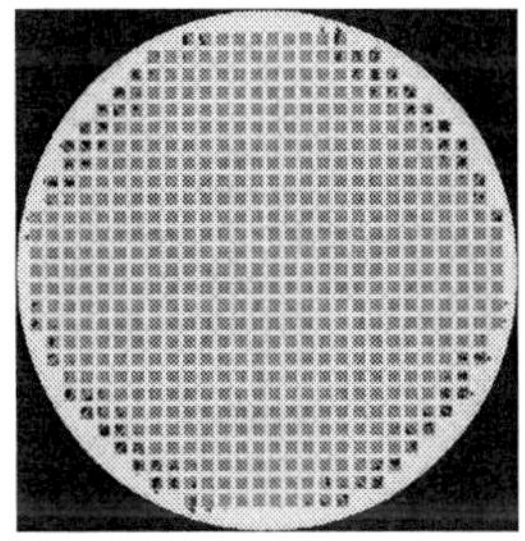

- X-like pattern is clean
- PR residues in 4 "corners"

(V) **Thinned dies**: Good cleaning performance on thinned dies, residues free surface

(X) **Thick dies**: strong process non-uniformity observed and increases with population density

imec 9

# CLEANING OF DIES ON A CARRIER
## FLOW PATTERN DURING CENTRAL DISPENSE ON FULLY POPULATED WAFERS

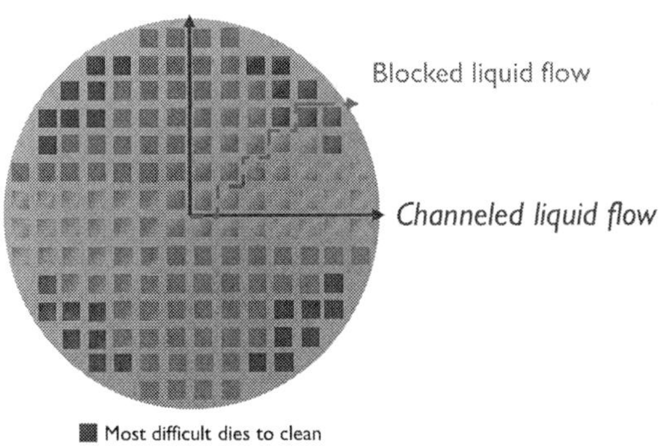


**Computed liquid film height as a function of wafer rotation speed (Q = 1l/min)**

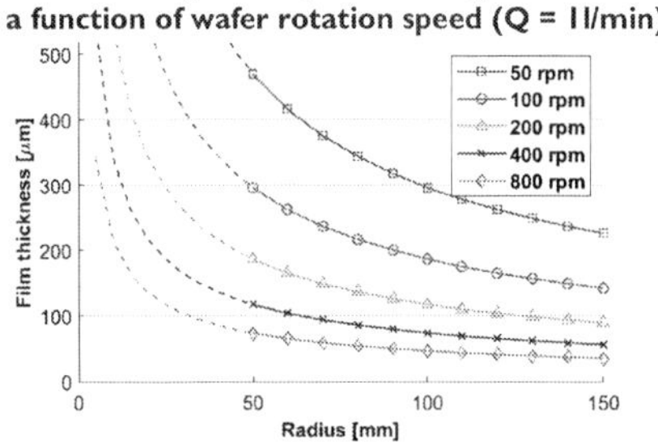


Computed based on:
*Prieling D., Computational investigation of liquid films on rotating disks, phD thesis, 2013*

- Outside the dispense zone, the expected liquid film height on a **flat** surface is < 300 μm
- Thin dies (50 um) are wetted but thick dies (775 um) oppose a resistance to the liquid flow depending on their orientation and position

➢ Die thickness impacts process uniformity

imec 10

## CLEANING OF DIES ON A CARRIER
### LIQUID SPLASHES DURING CENTRAL CHEMICAL DISPENSE ON THICK DIES

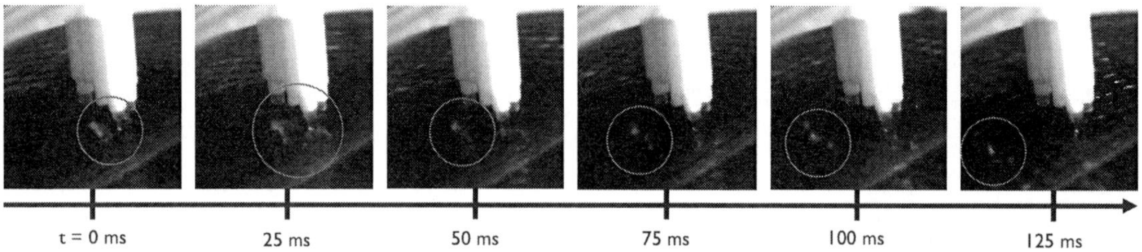


- Liquid jets and bubbles formation are observed below the dispense nozzle
  - ➢ Risk of nozzle contamination!
- Droplets formed will fall back on the wafer
- Liquid jets an bubbles are not observed when processing 50 um thick dies

imec 11

## CLEANING OF DIES ON A CARRIER
### LIQUID SPLASHES AT INCREASED ROTATION SPEED

- Splashes are visible below the nozzle
- Splashes reaches the shield
- The entire chamber is wet

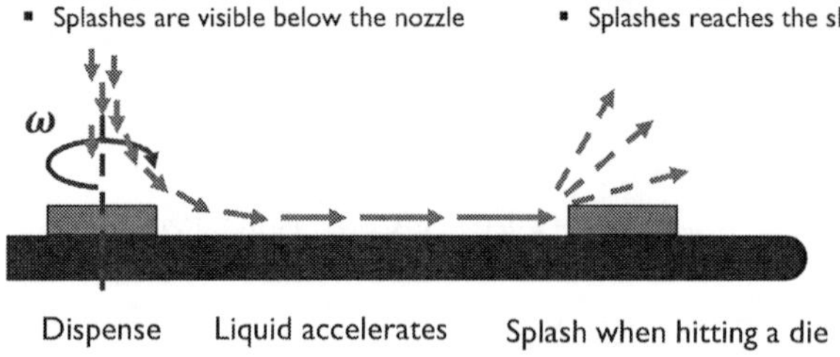


- Multiple parameters play a role in splashing (rotation speed, arm position, die population, die thickness...)
- Splashes can contaminate the tool chamber
- ➢ Rotation speed to be limited when processing thick dies to avoid splashing

imec 12

## CLEANING OF DIES ON A CARRIER
### DIE SHIFT OR RELEASE DURING PROCESS

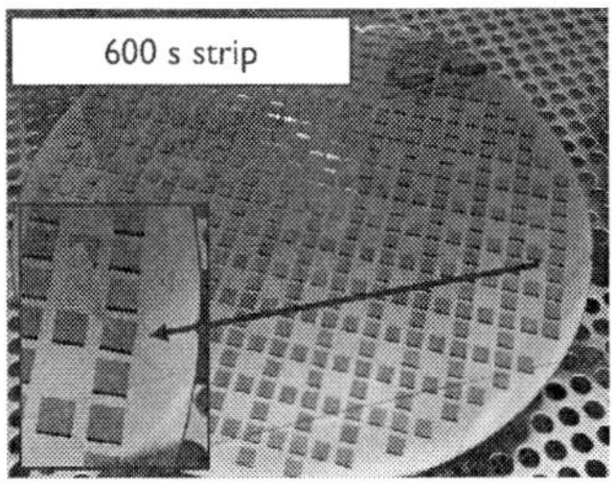


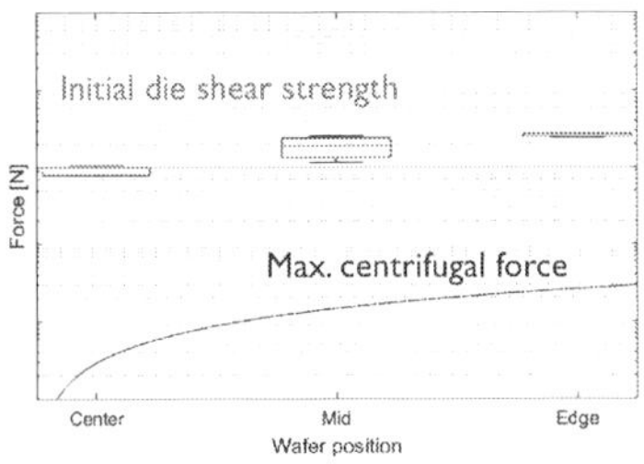


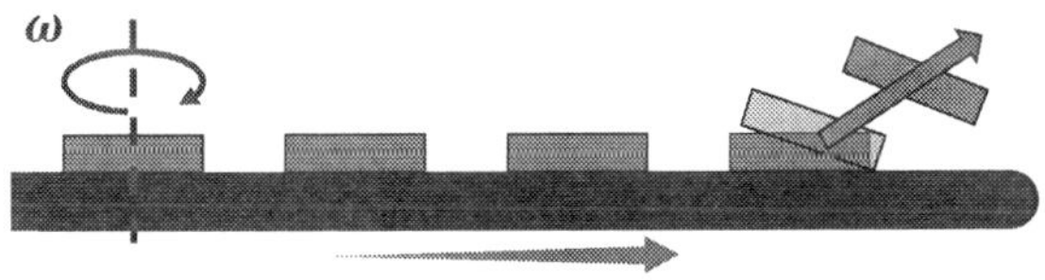


- Initial shear strength >> centrifugal force: die will not move in normal conditions
- Local defects (particles, bubbles, etc.) can lead to bond weakening
- Chemical compatibility with TBM can also affect bond strength
- Dies can shift or lift off due to centrifugal and hydrodynamic forces acting on them (drying step is the most critical, $\omega$ > 1000 rpm)
- Die shift or release will result in bonding failure and possible tool damage

imec

13

## CLEANING OF DIES ON A CARRIER
### NON-UNIFORM DRYING ON DIE SURFACE

**Standard DIW rinse**

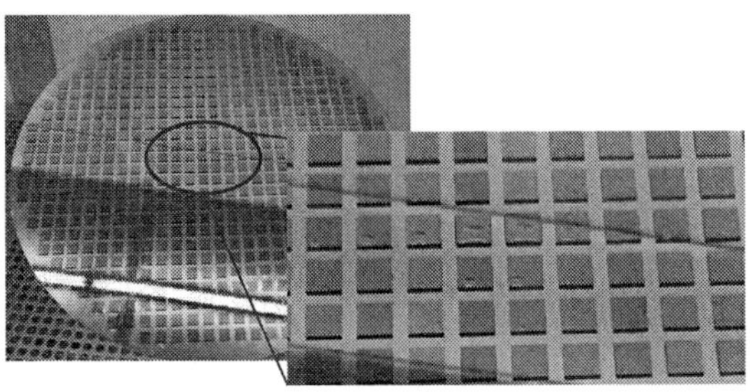

**IPA rinse + N2 blow**

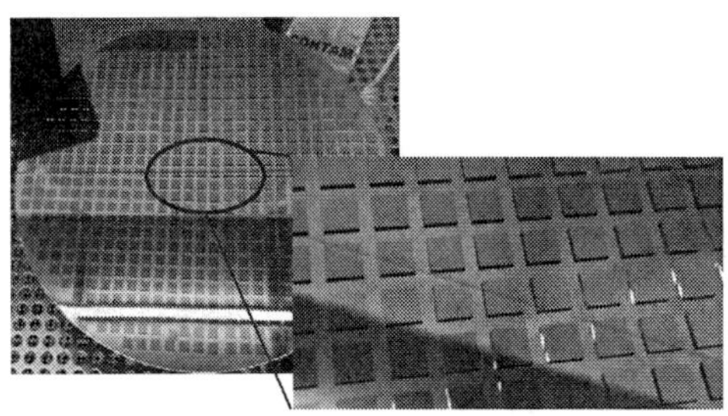

- Liquid remains trapped on top of the dies even after extended DIW and spin-drying time due to local low centrifugal forces
- Trenches between dies appears to be dry
- An IPA rinse combined with $N_2$ blow is used to dry die top surface

imec

14

## WET CLEANING CHALLENGE SUMMARY

1. Thick dies interacts strongly with the flow pattern
2. Splashes may occur when the wafer is populated with thick dies
3. Local defects may result in die shift under centrifugal and hydrodynamic forces
4. An IPA rinse/$N_2$ blow combination is required to dry the populated wafer succesfully

imec 15

## TOWARDS WET PROCESS OPTIMIZATION
### IMPROVED PROCESS PARAMETERS

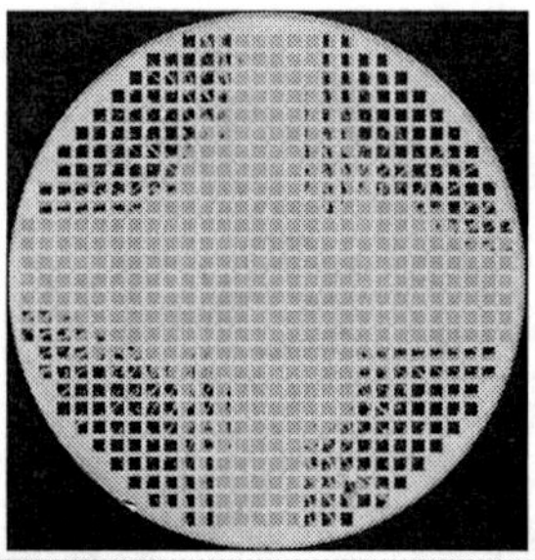
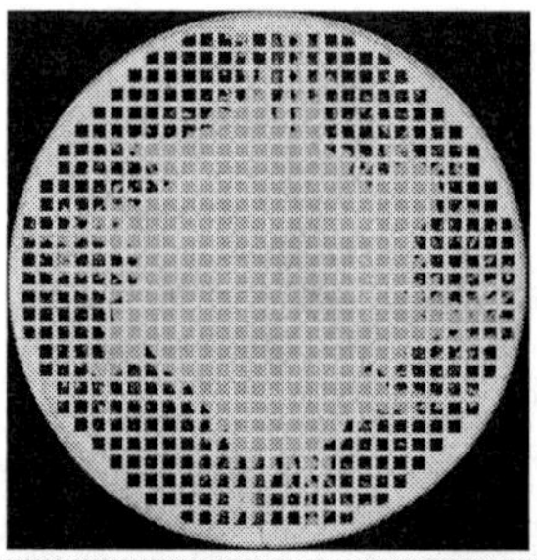
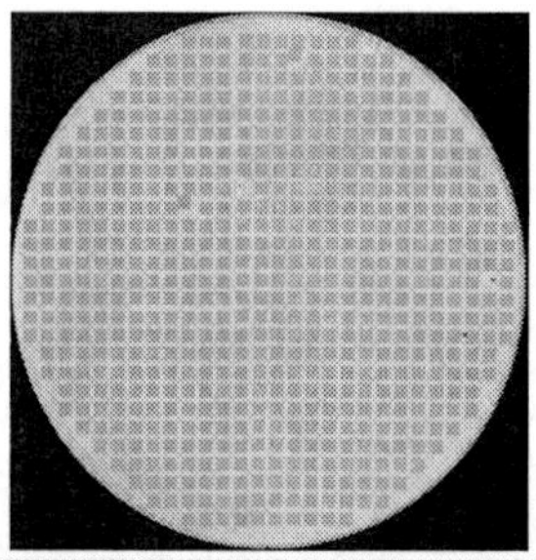

| | | | |
|---|---|---|---|
| **Thickness** | 775 um | 775 um | 775 um |
| **Population** | 100 % | 100 % | 100 % |
| **Dispense** | 3 min – central dispense | 3 min – central dispense | 3 min – tailored scan |
| **Rot. speed** | Normal | Low | Low |
| **Resist strip** | Incomplete, X-like pattern | Incomplete, radial pattern | Complete |

imec 16

## TOWARDS WET PROCESS OPTIMIZATION

### CFD SIMULATIONS TO UNDERSTAND FLUID BEHAVIOR

**Navier-Stokes Equation:**

$\nabla \cdot v = 0$ (mass conservation)

$(v \cdot \nabla)v = -\nabla p + \frac{\mu}{\rho}\nabla^2 v$ (momentum conservation)

**Volume-of-fluid (VoF) scheme:**

- 2 distinct phases (air and liquid) such that $\phi_{air} + \phi_{liq} = 1$
- Interface tracking scheme
- Identical velocity

(V) Complex 3D geometries
Detailed solution

(X) Computationally demanding

**Thin film approximation:**

- Reduced dimensionality: Film thickness $t$ is much smaller than other dimensions
- Vertical velocity is neglected
- Pressure is mainly hydrostatic

(V) Fast computation

(X) Strong hypothesis
Only valid for simple geometries

- Expected liquid film thickness?
- Impact of dies on the flow pattern?
- ...

**Improve the process with minimized risk!**

P. Vita et al., THIN FILM FLOWSIMULATION ON A ROTATING DISC, ECCOMAS, 2012

imec 17

## SUMMARY

1. Die singulation and transfer with a low defectivity level is required to enable successful die-to-wafer bonding
2. Protective layer in combination with a dedicated clean is investigated as a route to yield clean die surface
3. Thick dies reveal multiple challenges linked to liquid splashes, non-uniform process and die drying
4. The interaction between liquid flow and dies is complex and requires tailored process parameters to enable full protective layer strip
5. CFD numerical simulations offer the possibility to understand and further optimize the process

imec 18

imec

embracing a better life

imec

PUBLIC

THE SURFACE PREPARATION AND CLEANING CONFERENCE (SPCC)

SPCC 2019 | Portland, Oregon, USA | April 2-3, 2019

# Effect of Bubble on Particle Agglomeration in Tungsten (W) CMP Slurry

April 2, 2019

H.T. Kim[1], Y.A. Jeong[1], H.Y. Ryu[1], M.K. Poddar[1], N.P. Yerriboina[1], J.H. Kim[3], J.D. Park[3], M.G. Lee[3], C.Y. Park[3], S.J. Han[3], M.J. Kim[3], and J.G. Park[1, 2†]

[1] Department of Bio-Nano Technology and
[2] Materials Science and Chemical Engineering, Hanyang University, Ansan, 15588, Korea
[3] Dongjin Semichem Co. Ltd., Hwaseong, 18645, Korea

*jgpark@hanyang.ac.kr
+820314004080
Hanyang Univ., Ansan, 15588, KOREA

NEMPL Hanyang University



## Outline

**Introduction**

- W CMP Process
- Defects in CMP Process

**Motivation**

- $O_2$ Gas Bubble Generation during CMP Process
- Mechanism of Particle Agglomeration by Gas Bubble

**Experimental Materials and Procedure**

**Experimental Results and Discussion**

- Effect of $Fe(NO_3)_3$ on $H_2O_2$ Decomposition
- Effect of $Fe(NO_3)_3$ on Particle Agglomeration

**Summary**

NEMPL Hanyang University

SPCC 2019 | Portland, Oregon, USA | April 2-3, 2019

# Introduction

NEMPL
Hanyang University



## Chemical Mechanical Polishing (CMP) Process

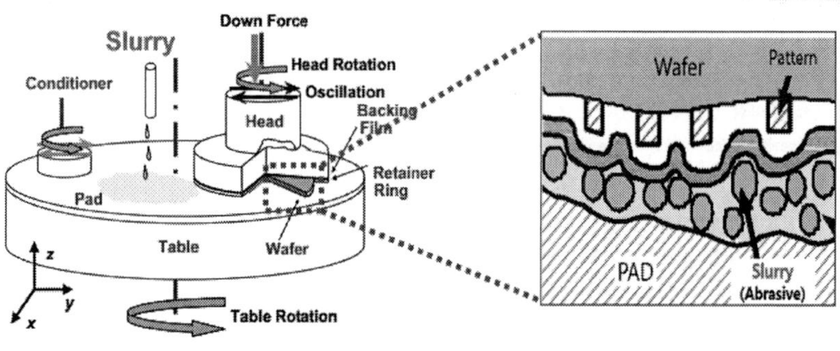


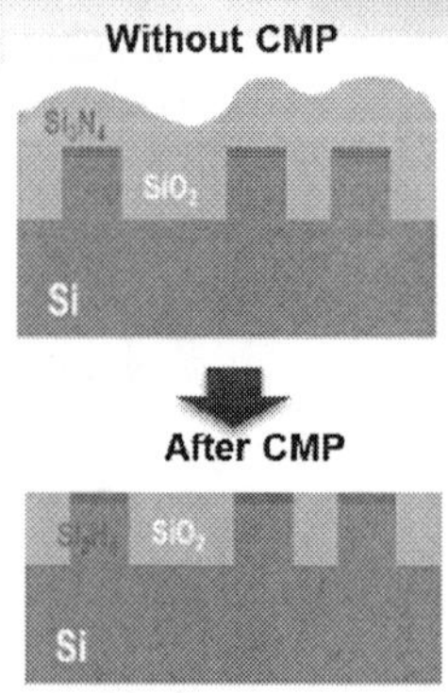


- With advancement in technology, CMP is used in STI CMP for isolation, W, Cu CMP for metallization
- CMP is critical step to planrize various surface during semiconductor process
- Slurry is the key component in CMP process

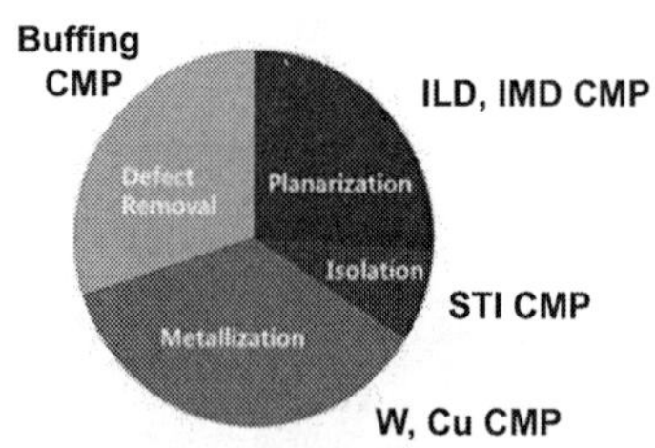

## Tungsten (W) CMP Process

### ❖ RMG Process Flow

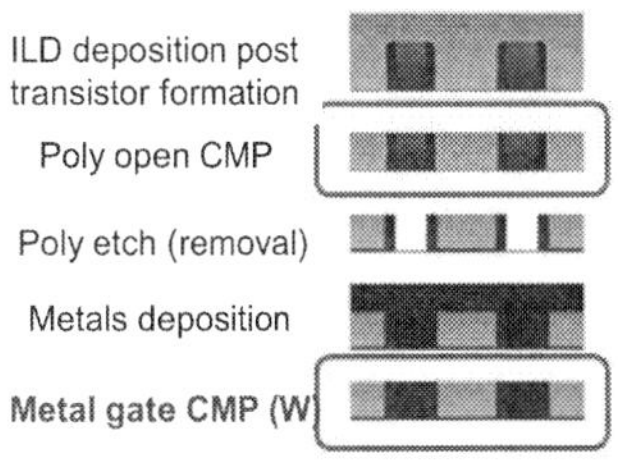


W Metal Gate

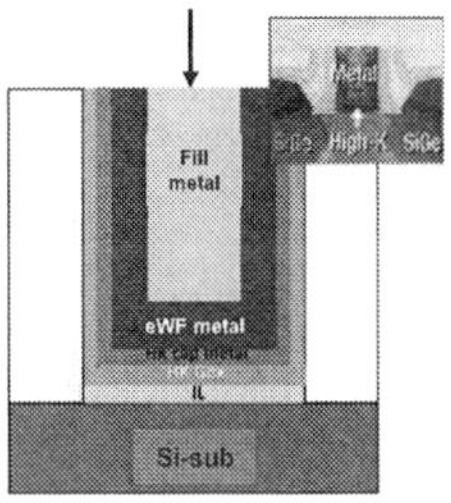


RMG : Replace metal gate
ILD : Inter-layer Dielectric
eWF : effective Work Function
HKMG : High-K Metal Gate
S/D : Source / Drain

**Advantages of HKMG:**
1) good thermal budget
2) higher strain from embedded SiGe S/D

### ❖ MOL Process Flow

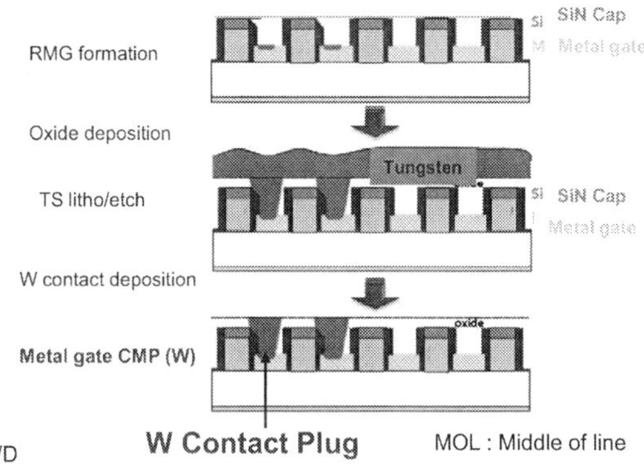


MOL : Middle of line

- The application of W CMP in the RMG(Replace Metal Gate) and MOL(Middle of Line) processes enabled semiconductor devices below 14 nm node
- CMP plays an important role during planarization of W material for the both processes

Ref: H. J. Kim et al., NCCAVS CMPUG Spring Meeting 2015



## Technology Roadmap of Metal Materials

### ❖ Technology Roadmap of Metal Materials

| | | Device | 2017 | 2018 | 2019 | 2020 |
|---|---|---|---|---|---|---|
| Production Node | | Logic | 7nm | | 5nm | |
| | | DRAM | <18nm | <16nm | | <14nm |
| | | 3D NAND | 64L / 96L | | 96L / 128L | |
| Logic | BEOL | M1 Pitch | 24nm | | 16nm | |
| | | V0 / M1 | Cu, Co | Co | Co | Co |
| | | Liner / Barrier | Co Main, Ru | | Co or Ru / Mn | |
| | MOL | M0 / Contact Plug Metal | CVD W | CVD W | Co | Co |
| | | Liner / Barrier | Ti / TiN | Ti / **FF W** | Co / TiN | Co / TiN |
| | FEOL | Gate Length | 8nm | | | |
| | | Gate Metal | W | W | W | W |
| DRAM | | Buried Wordline Contact | W | *FF W | FF W | FF W |
| 3D NAND | | Wordline Contact | W | FF W | FF W | FF W |

***FF W : Fluorine-Free W**

- W has been widely used material as a replace metal gate (RMG) and middle of line (MOL) in various semiconductor devices
- W is a promising material in the roadmap of NAND and DRAM devices.


H. J. Kim et al., *CMP challenges in sub-14nm FinFET and RMT technologies*, NCCAVS CMPUG Spring Meeting, 2016
M. Krishnan et al., *Chemical Mechanical Planarization: Slurry Chemistry, Materials, and Mechanisms*, Chem. Rev., 2010, 110(1), 178-204

# Defects in CMP Process

❖ **Scratch defect from large particle in CMP process**

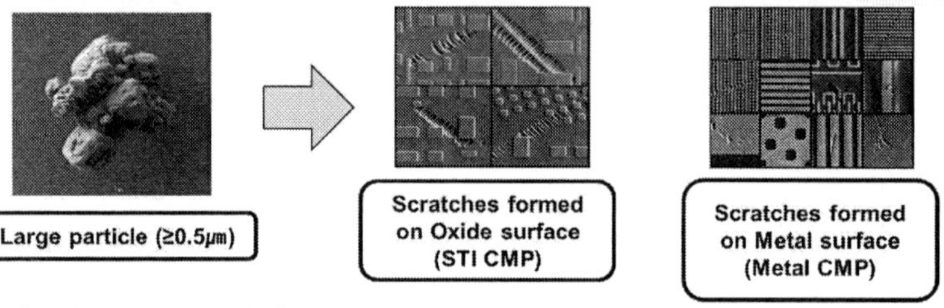


- During CMP process, defect such as scratch can deteriorate yield and reliability
- Scratch can be generated by agglomeration of particles

❖ **Particle contamination on the surface in CMP process**

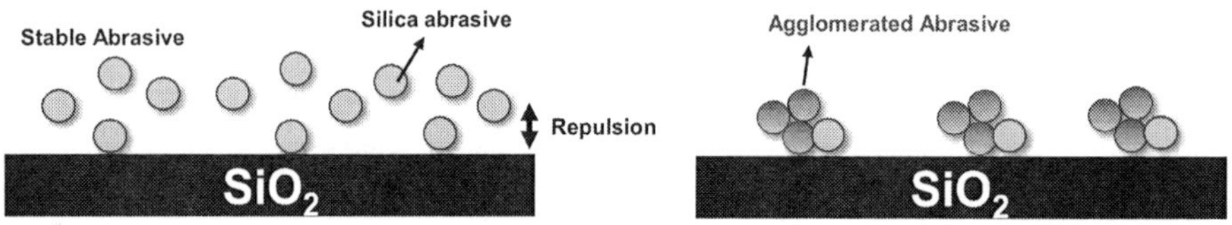


**Lower Contamination** **Higher Contamination**

- **Agglomerated large particle is one of the key factors for scratches during CMP**
- **Dispersion ability** can affect **abrasive particle defect** on surface



# Motivation

NEMPL Hanyang University

# Catalyst Reaction with $H_2O_2$

❖ **W Slurry key component : Combination of $H_2O_2$ and Metal catalyst**

❖ Fe(III)/$H_2O_2$ system, $H_2O_2$ is catalytically decomposed by Fe(III) at acidic pH

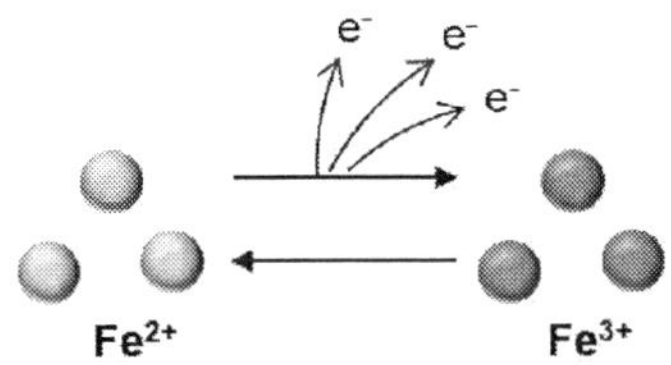


**Reaction of Fe ion catalyst**

$Fe^{2+} + H_2O_2 \rightarrow Fe^{3+} + OH^- + OH\cdot$

$H_2O_2 + OH\cdot \rightarrow H_2O + HO_2\cdot$

$Fe^{3+} + HO_2\cdot \rightarrow Fe^{2+} + H^+ + O_2$

- $H_2O_2$ Decomposition Reaction

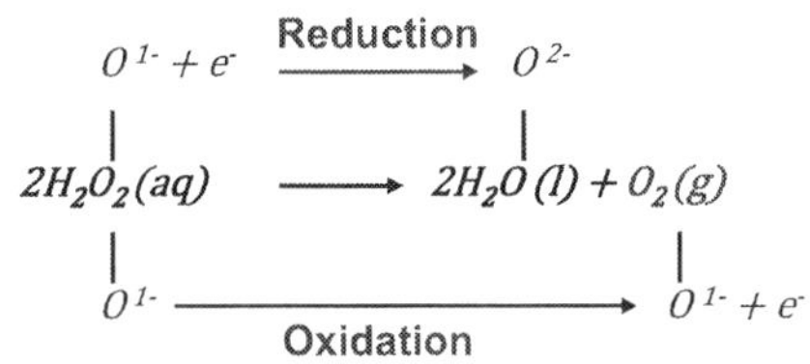


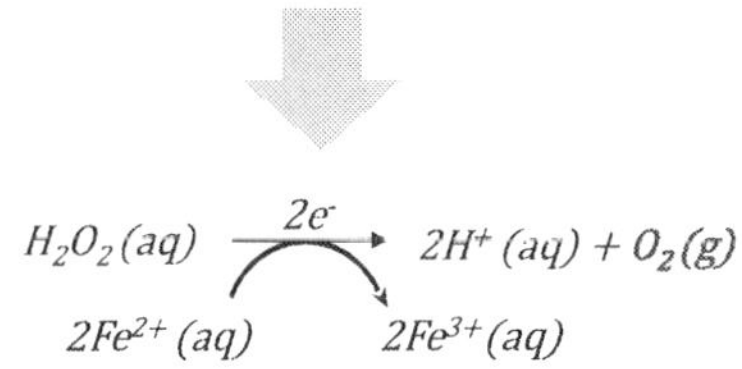


- For W CMP, usually slurry contains Fe based catalyst to promote the hydroxyl radical with high oxidative power by the decomposition of $H_2O_2$
- Effect of $O_2$ gas on particle stability was investigated

M. Pourbaix et al., *Atlas of Electrochemical Equilibria in Aqueous Solutions*, National Association of Corrosion Engineers, 1974
G. Lim et al., *Effect of oxidants on the removal of tungsten in CMP process*, Wear, 257(9-10), 2004, 863-868
C. J. Marzzacco., *The enthalpy of decomposition of hydrogen peroxide: a general chemistry calorimetry experiment*, J. Chemi. Educ., 76(11), 1999, 1517



# Mechanism of Particle Agglomeration by Gas Bubble

❖ **Two mechanisms of particle agglomeration by gas bubble**

1) Particle adhering to a static gas bubble
2) Bubble generation is giving more chances to collision, attachment between particles

**Static Bubble Condition**

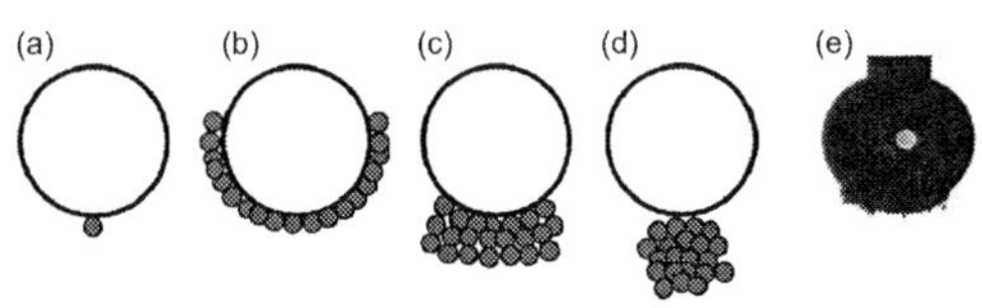


* Schematic view of particles adhering to a gas bubble: (a) single particle, (b) monolayer, (c) multilayer, (d) cluster of particles suspended through a single particle. (e) Image of an air bubble partially covered with silica particles

**Dynamic Bubble Condition**

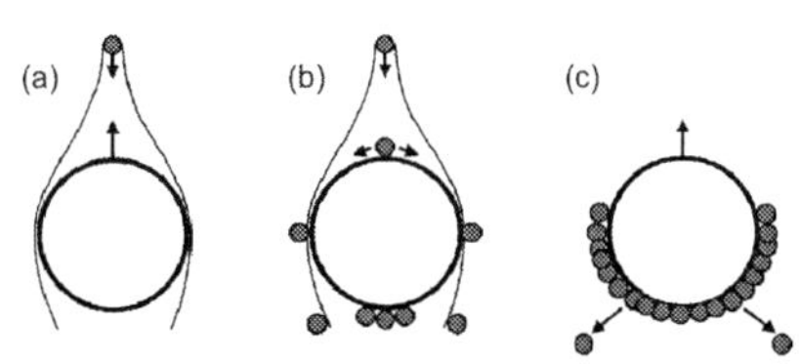


** Schematic of three successive key steps in froth flotation: (a) particle-bubble collision, (b) attachment and (c) detachment

- Particles adhere on bubble surface and it forms cluster of particles
- The $O_2$ gas bubbles generated by $H_2O_2$ decomposition might cause the agglomeration of abrasive particles in W slurry

* F. Omota et al., "Adhesion of solid particles to gas bubbles. Part 1: Modelling", Chemical Engineering Science, 61, 2006, 823–834
** F. Maoming et al., Mining Science and Technology, 20, 2010, 317-330

# Experimental Results and Discussion

NEMPL Hanyang University

SPCC 2019 | Portland, Oregon, USA | April 2-3, 2019

## Temperature Measurement (IR Camera)

**Measured Temp. vs Process Time**

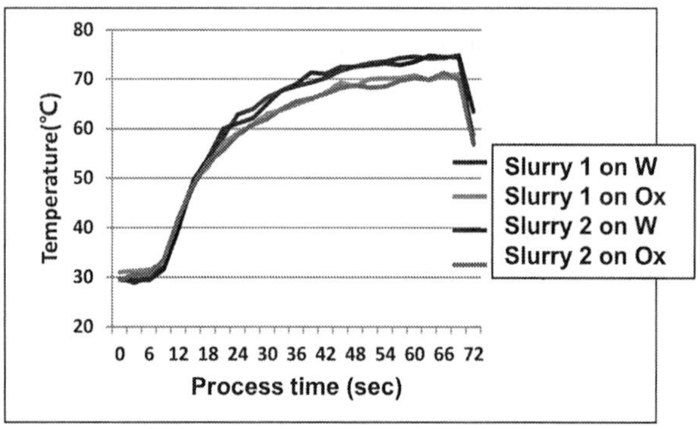


**Analyzed IR Image**

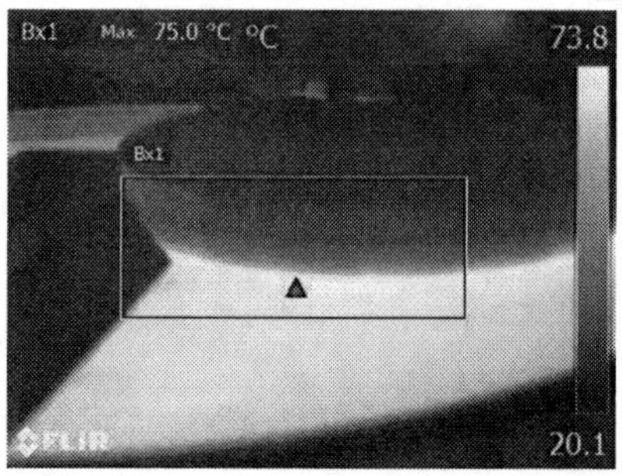


▲ : **Highest Temp. at Slurry**

**IR camera** (E85, FLIR, USA)

- The temperature of slurry during CMP process was analyzed by using IR camera
- The temperature of slurry was increased as a function of process time due to heat generated by friction between a substrate and a pad during CMP
- At high temperature, the catalyst in slurry is activated and $H_2O_2$ decomposes rapidly

SPCC 2019 | Portland, Oregon, USA | April 2-3, 2019 12/22

# Experimental Procedure

❖ **Measurement of $O_2$ gas generation by gas pressure sensor**

1) Prepare the slurry
   - 0-120 ppm $Fe(NO_3)_3$ at pH2
2) Heat the slurry to target temperature
   - 25, 60, 80 ℃
3) Add 1.5 wt% of $H_2O_2$
4) Measure the pressure change of sealed flask caused by $O_2$ gas generated from $H_2O_2$ decomposition

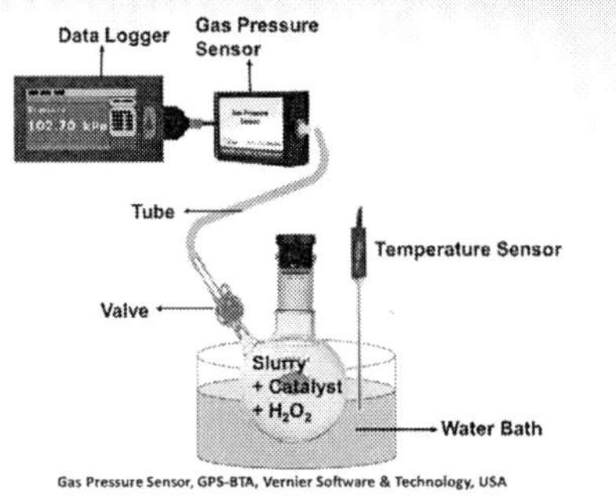


**Experimental setup to measure $O_2$ gas generation from $H_2O_2$ decomposition**

❖ **Observation of $O_2$ gas generation by high speed camera**

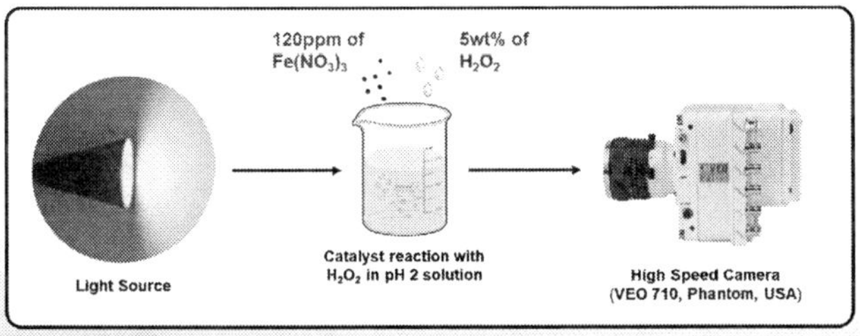


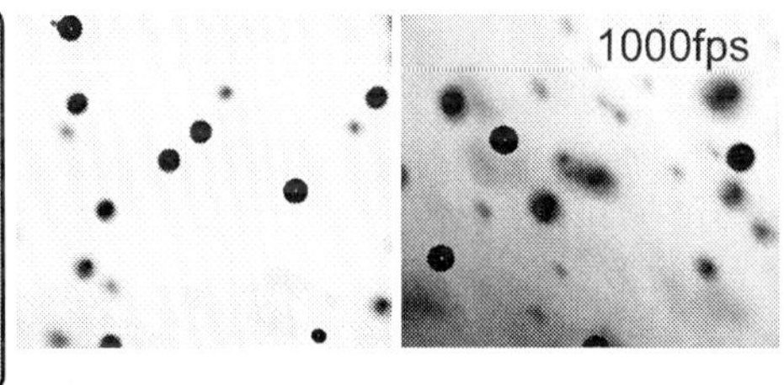




# Effect of $Fe(NO_3)_3$ on $H_2O_2$ Decomposition

❖ **Experimental conditions**

- ❑ $Fe(NO_3)_3$ : 0 and 120ppm
- ❑ $H_2O_2$: 1.5 wt%, pH: 2, Temperature: 25, 60, 80℃

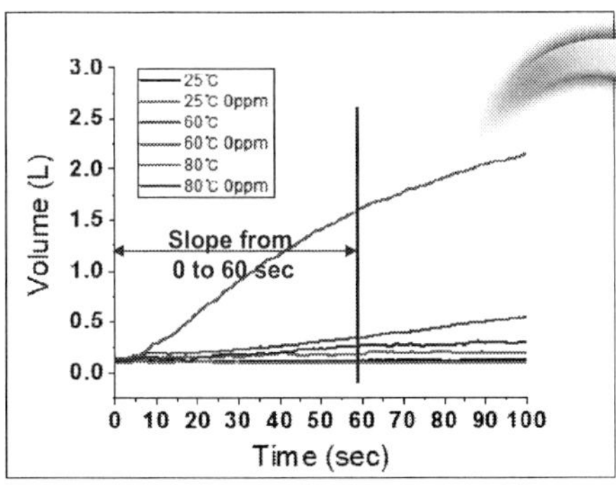


**Magnify the graph from 0 to 100 seconds**

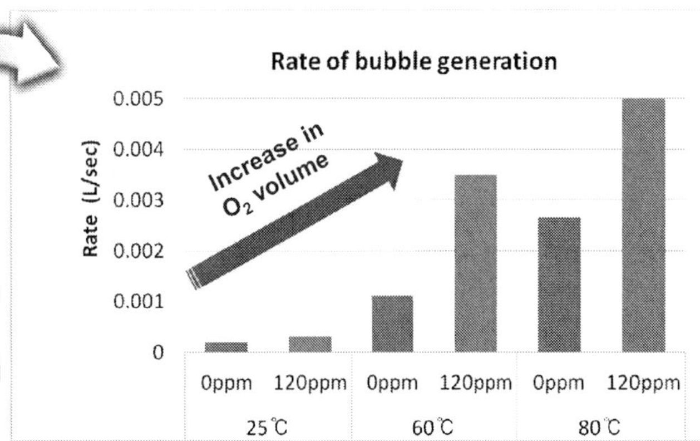


- Peroxide decomposes to produce OH radicals and oxygen gas
- With increase in temperature and Fe(III) amount, $O_2$ volume increased

## Effect of $Fe(NO_3)_3$ on $H_2O_2$ Decomposition

### ❖ Experimental conditions

- ❑ $Fe(NO_3)_3$ :0, 30, 60 and 120ppm
- ❑ $H_2O_2$: 1.5 wt%, pH: 2, Temperature: 80 ℃

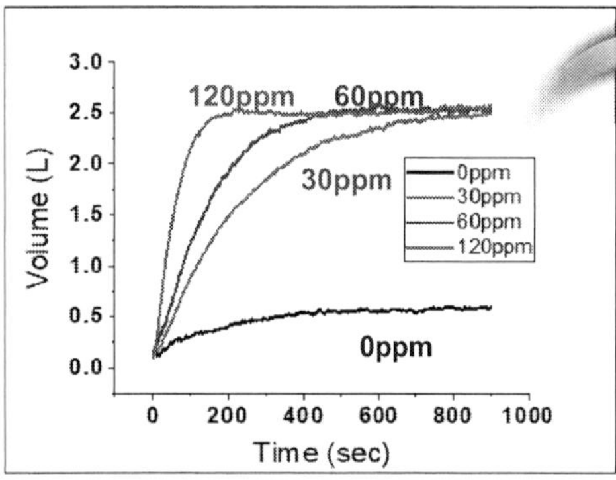


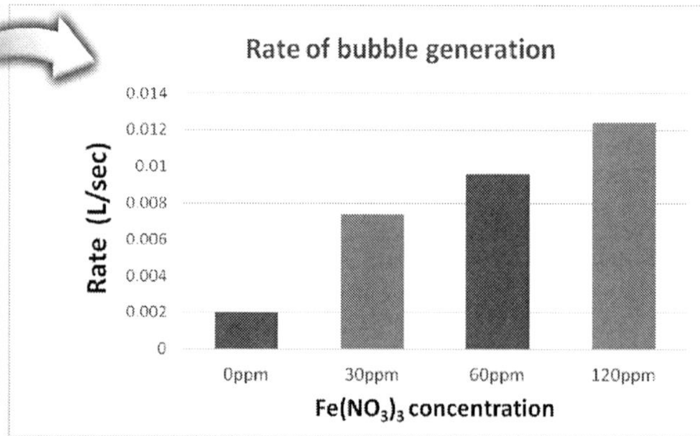


- $H_2O_2$ decomposition test at 80℃ with different $Fe(NO_3)_3$ concentration
- As Fe(III) concentration increased, the volume of $O_2$ gas also increased



## Analysis of Bubble by High Speed Camera

### ❖ Experimental conditions

- ❑ $Fe(NO_3)_3$ : 120ppm
- ❑ $H_2O_2$: 1.5 wt%, pH: 2, Temperature: 25, 60, 80℃, Filming time: 2 min

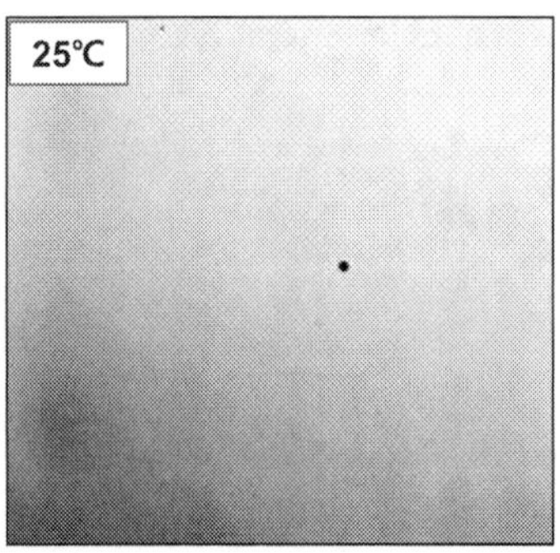


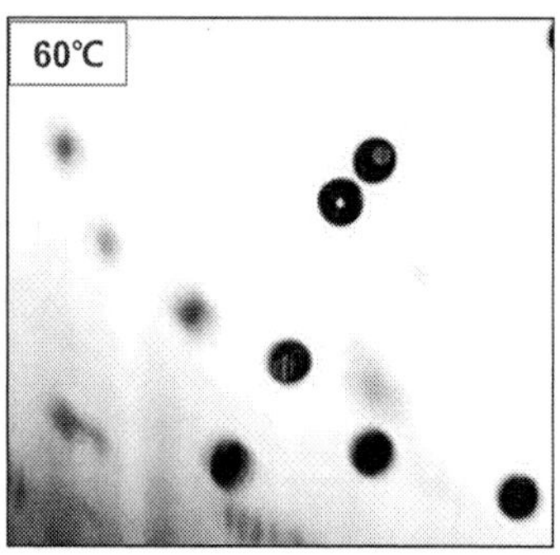


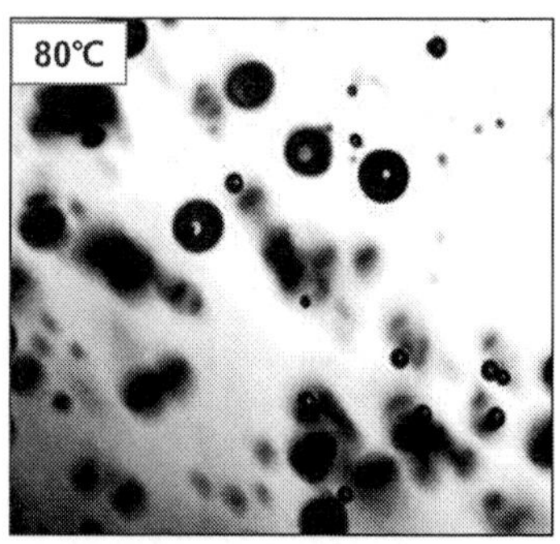


- Larger bubble size were observed with increase in temperature at high Fe(III) conc.
- This confirmed high $H_2O_2$ decomposition into $O_2$ at high temp. and high catalyst conc.

## Particle Agglomeration: Analysis with Surface Scanner

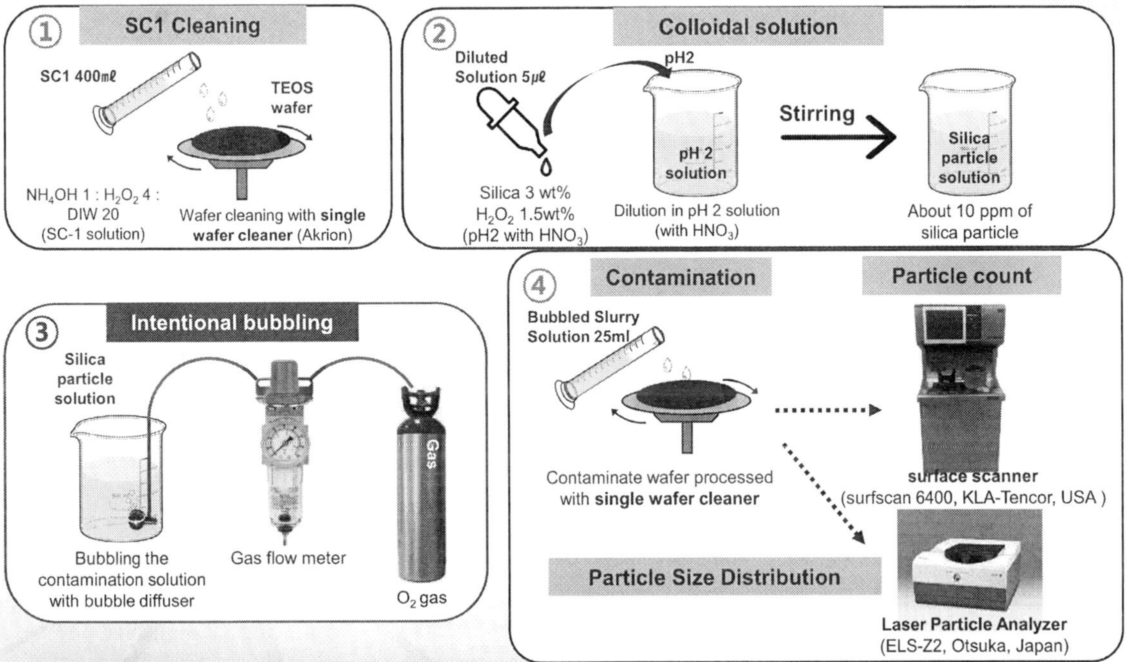




## Effect of Bubble on Large Particle Formation

### ❖ Experimental conditions

- ❑ $O_2$ gas injection with different flow rate (10min)
- ❑ $O_2$ gas injection with different flow time (100cc/min)
- ❑ Temperature: 25 ℃
- ❑ $SiO_2$: 10ppm

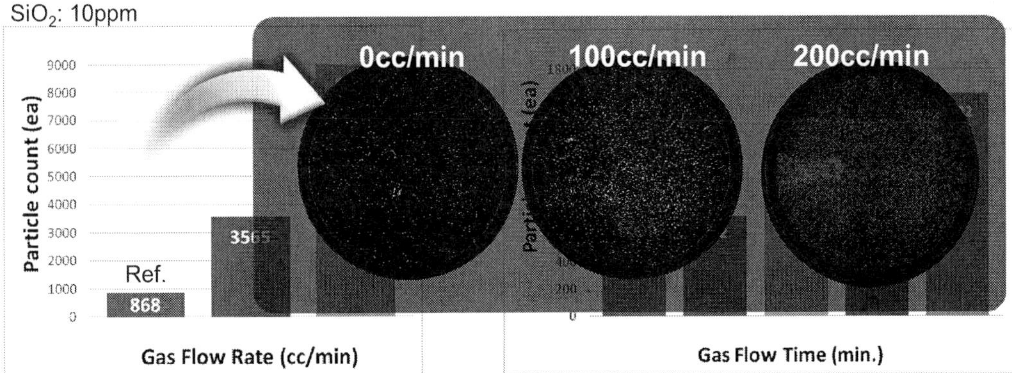


**Number of large particle (over 500nm)with gas flow rate** **Number of large particle (over 500nm) with gas flow time**

- Effect of $O_2$ gas bubble on particle agglomeration was investigated
- Particle contamination level (above 500nm) was increased with increase in gas flow rate

# Effect of Bubble on Large Particle Formation

## ❖ Experimental conditions

- ❑ Silica (3 wt%) + $Fe(NO_3)_3$ and $H_2O_2$
- ❑ $Fe(NO_3)_3$ : 120ppm
- ❑ $H_2O_2$ 5 wt%, pH2, Temperature: 25, 60, 80 ℃

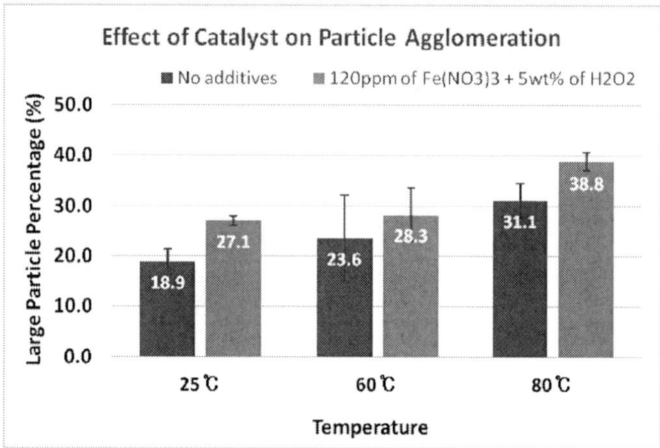


- Ratio of Large particle (> 200nm) size was analyzed
- $O_2$ bubbles at high temperature might be a prominent cause of particle agglomeration.



# Effect of Bubble on Large Particle Formation

## ❖ Experimental conditions

- ❑ $SiO_2$ (3 wt%) + w/wo $Fe(NO_3)_3$ and $H_2O_2$
- ❑ $Fe(NO_3)_3$ :120ppm
- ❑ $H_2O_2$: 1.5 wt%, pH: 2, Temperature: 25, 60, 80 ℃

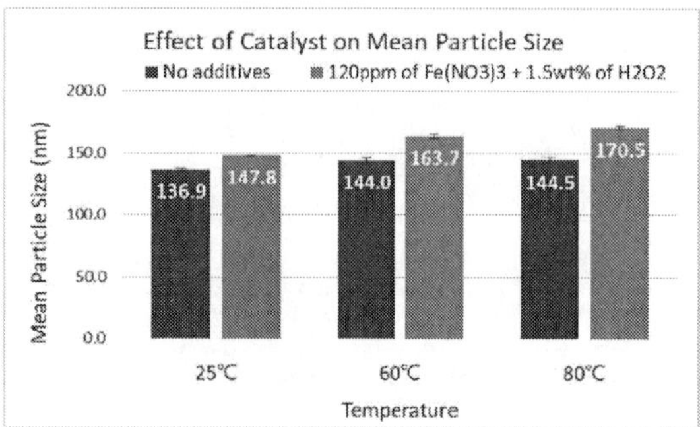


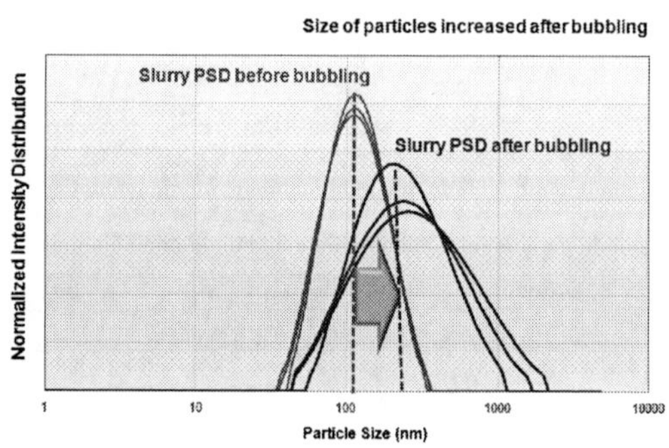


- Mean particle size increased as the temperature increases
- Increase of particle average size observed in the presence of bubbles
- This correlates similar results of $H_2O_2$ decomposition using HSC and gas pressure sensor
- $O_2$ bubbles at 80℃ could be a major cause of particle agglomeration

## Summary

- $H_2O_2$ is decomposed by $Fe(NO_3)_3$ and produce OH radical and $O_2$ gas
- $O_2$ gas bubble is generated rapidly at high $Fe(NO_3)_3$ conc. and temp.
- Bubbles may play a key role in agglomerating the particles
- Fundamental studies show that bubbling can effect defect data
- Further studies will be carried out to analyze bubble effect on defects

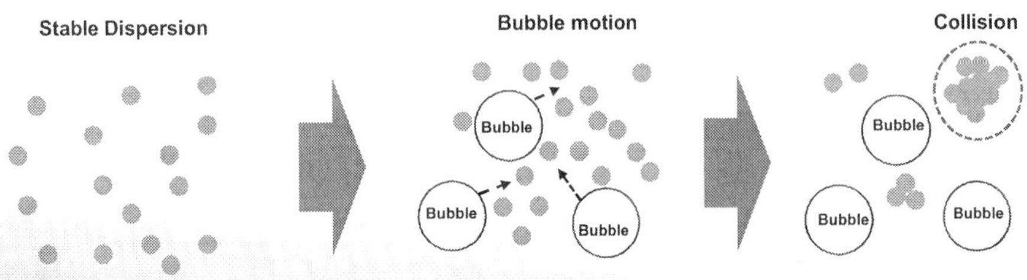

# $Fe(NO_3)_3$ Concentration Effect: Analysis with DLS

## ❖ Experimental conditions

- ❑ $Fe(NO_3)_3$ (Sigma-Aldrich): 120ppm
- ❑ $H_2O_2$: 1.5 wt%, pH: 2, Temperature: 25, 60, 80℃

**Analysis of Particle Size Distribution**

**Silica 3 wt%**
**$H_2O_2$ 1.5wt%**
**(pH2 with $HNO_3$)**

**Reacting catalyst with $H_2O_2$ for 4 mins at certain temperature in pH 2 solution with $HNO_3$**

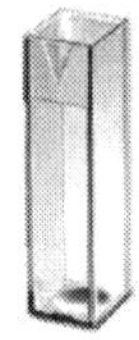

**No dilution of solution using glass cuvette at high temperatures**

**Laser Particle Analyzer**
(ELS-Z2, Otsuka, Japan)

**Heating the laser particle analyzer according to the solution temperature**

SCREEN axcelis

# Optimization of the removal of wafer backside defects arising from electrostatic clamping during ion implantation by scrubber processing with acidic chemical

SCREEN Semiconductor Solutions Co., Ltd.
○S. Ano, A. Iwasaki, J. Jeon,

Axcelis Technologies Inc.,
I. Wu, D. Kirkwood

ano@screen.co.jp, +81-80-2472-7118
480-1, Takayama-cho, Hikone, Shiga, 522-0292, Japan



SCREEN axcelis

## Agenda

- **Motivation**
  - Backside particle/defect removal requirement
  - About electrostatic clamp (ESC)
- **Experimental and Result**
  - Chemical screening for ESC defects
  - ESC defects removal by scrubber processing with acidic chemical
- **Summary**

SCREEN

## Agenda

axcelis

- **Motivation**
  - Backside particle/defect removal requirement
  - About electrostatic clamp (ESC)
- **Experimental and Result**
  - Chemical screening for ESC defects
  - ESC defects removal by scrubber processing with acidic chemical
- **Summary**



SCREEN

## Backside particle/defect removal requirement

axcelis

- **Impact of backside particles on frontside patterning**
  - Focus-spot failures in lithography become more critical in advanced logic nodes and 3D stacked memory devices.

**Wafer map of pattern defects**

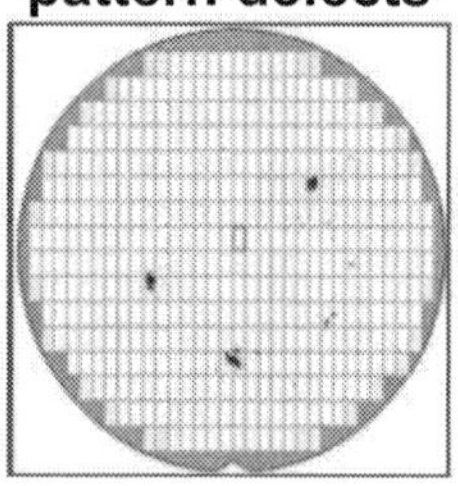

**Optical image of pattern defect**

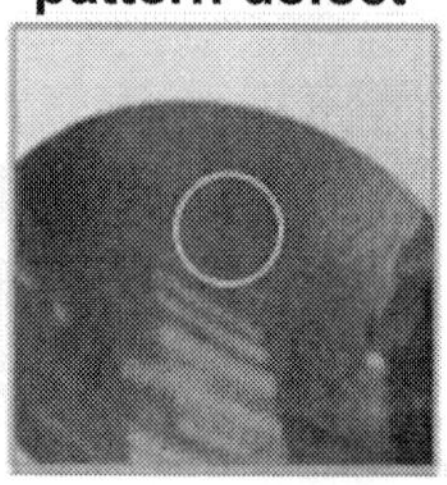

**SEM image of pattern failures**

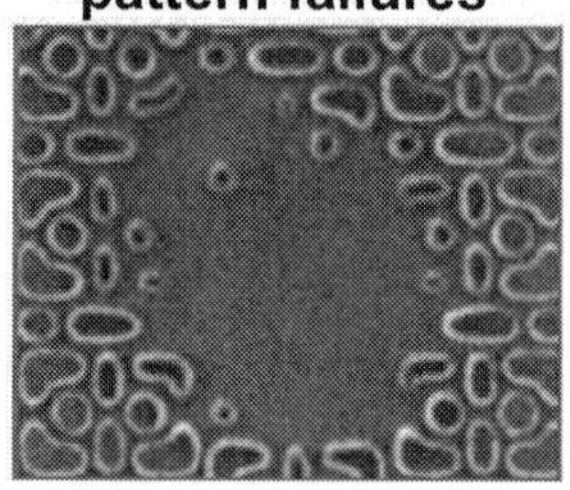

H.-S. Sohn, ECS transactions, 11 (2) 95-100 (2007)

| Year of production | 2011 | 2013 | 2015 | 2017 |
|---|---|---|---|---|
| DRAM ½ pitch (nm) | 40 | 32 | 24 | 20 |
| Critical particle size (nm) | 20 | 16 | 12 | 10 |
| Backside Defect Size (nm) | 200 | 160 | 125 | 100 |

Backside defect size requirements according to the ITRS 2011, 2013

## About electrostatic clamp (ESC)

- ESC electrode picture

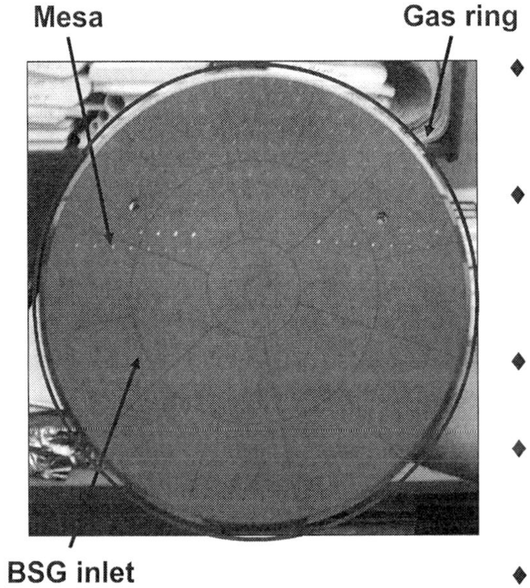


- Most ion implanter use electrostatic clamp (ESC) to hold wafer during processing
- Ion implantation requires ESC to be mechanically scanned during processing, and also to cool the wafer
- Mesa contact with wafer during processing. Material is AlN.
- Backside gas is channeled and then retained with Gas ring which is located outside of ESC electrode.
- Backside gas is purged from BSG inlet and dissipates heat from the incident ion beam.





## Typical backside defects generated by implanter

- Particle MAP after implanter

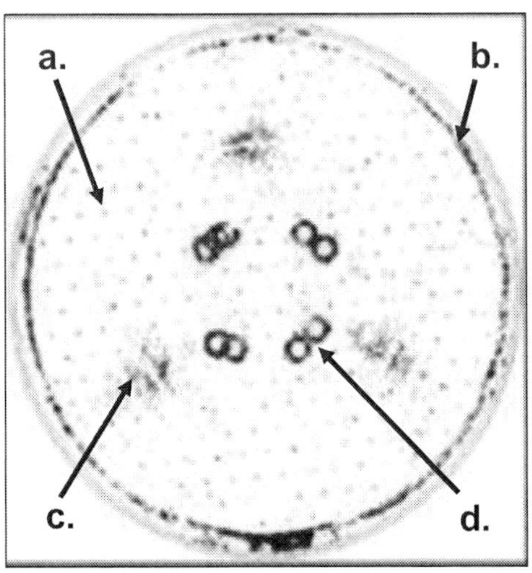


**Typical Defects**

Generate by ESC

a. Mesa
b. Gas ring
c. BSG inlet

Generate by transfer robot

d. In air end effector

SCREEN

## Identification of ESC defects

axcelis

| | Mesa | Gas ring | BSG inlet | In air end effector |
|---|---|---|---|---|
| SEM |   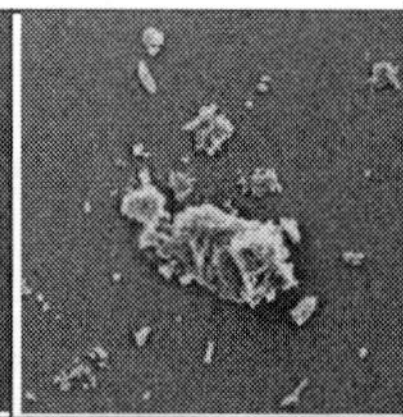 | 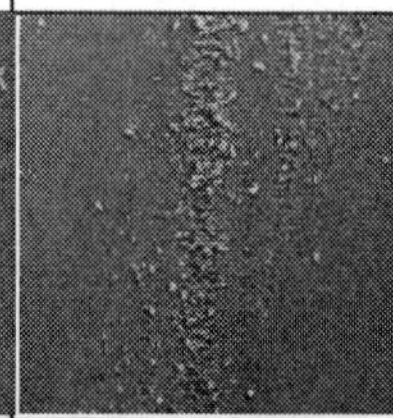 | 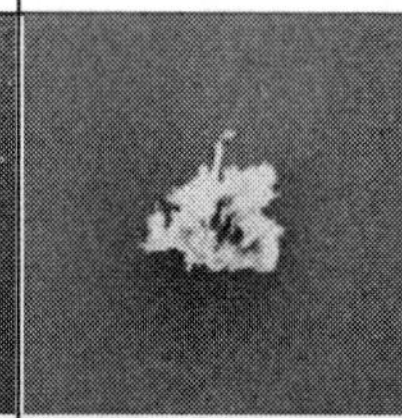 | 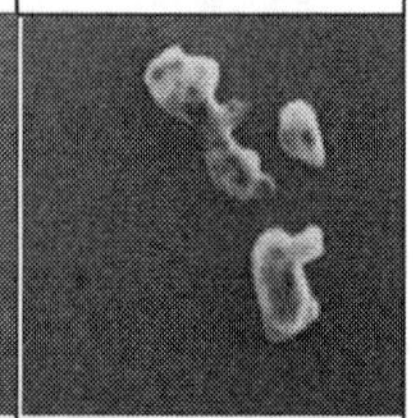 |
| EDX | 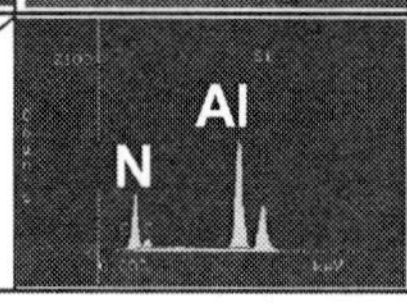  | 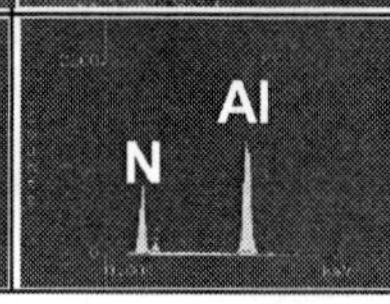  | 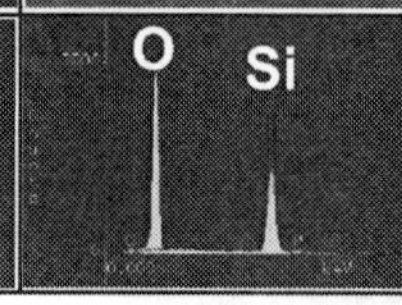  | 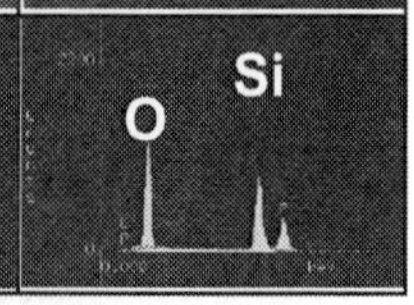  |

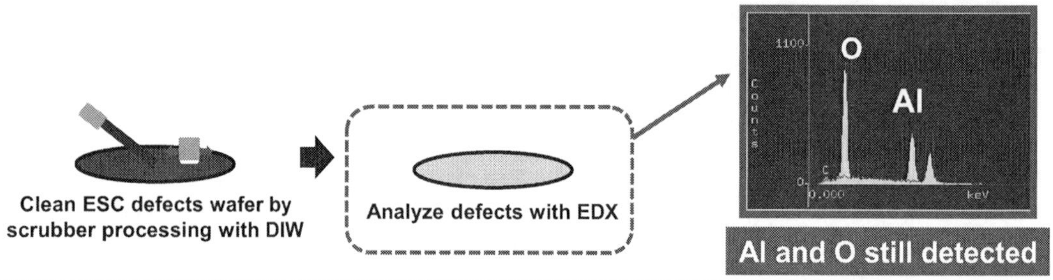




SCREEN

## Agenda

axcelis

- ♦ **Motivation**
  - Backside particle/defect removal requirement
  - About electrostatic clamp (ESC)
- ♦ **Experimental and Result**
  - Chemical screening for ESC defects
  - ESC defects removal by scrubber processing with acidic chemical
- ♦ **Summary**

SCREEN axcelis

## Chemical screening for ESC defects

- **Experimental**
  - Cleaning for SiO defects
    - Particle removal efficiency $= \frac{(pre\ particle\ count) - (post\ particle\ count)}{(pre\ \ particle\ count)} * 100\ [\%]$
      - Test chemicals; $NH_4OH$, $HCl+O_3$, $NH_4OH+O_3$ and $HF+O_3$(FOM)

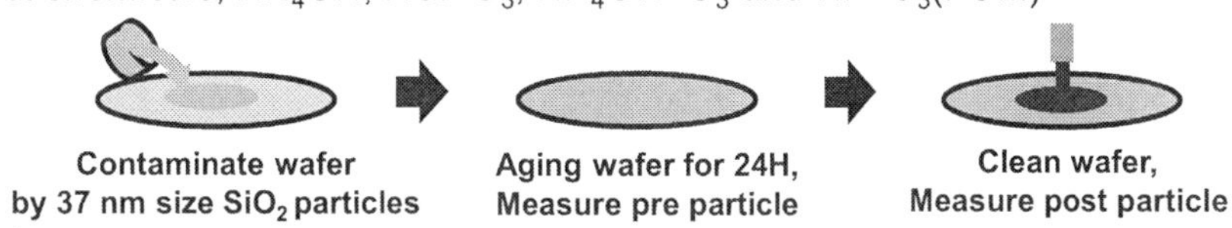


    - Si roughness

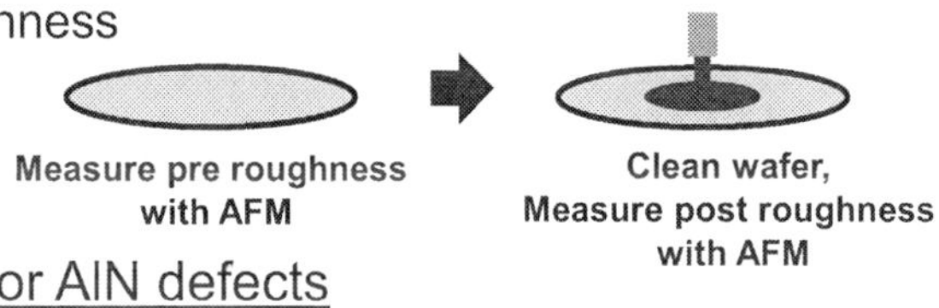


  - Cleaning for AlN defects
    - Metal removal efficiency
      - Test chemicals; SC-1, HF, SPM, SC-2 and $HF+O_3$(FOM)

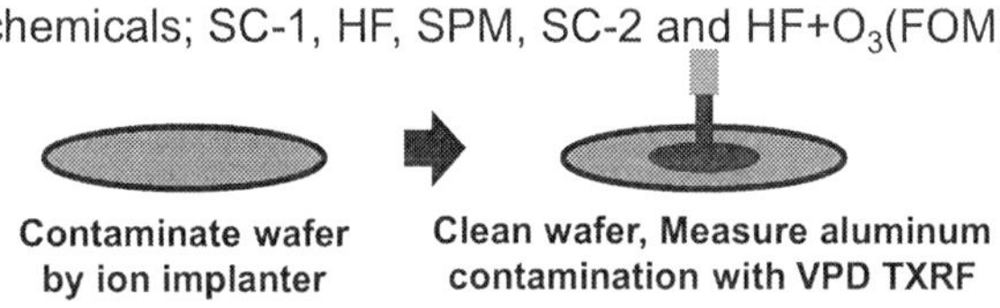




SCREEN axcelis

## Result of chemical screening for ESC defects

- **Cleaning for SiO defects**

**Particle removal efficiency**

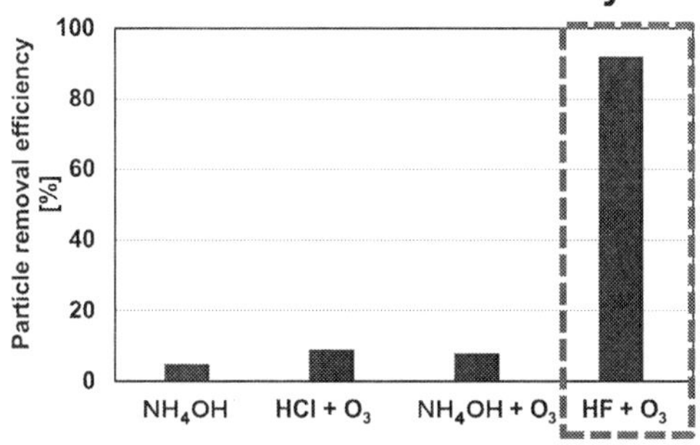


**Si roughness**

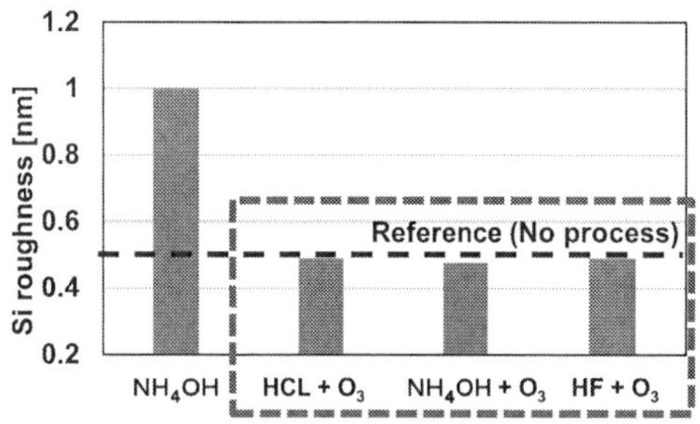


- **Cleaning for AlN defects**

**Metal removal efficiency**

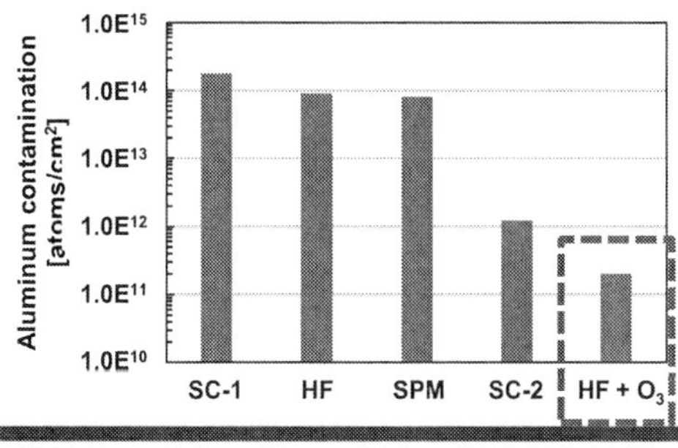


**$HF + O_3$ (FOM) is the most effective to remove SiO and AlN defects.**

SCREEN axcelis

## Agenda

- Motivation
  - Backside particle/defect requirement
  - About ESC defects
- **Experimental and Result**
  - Chemical screening for ESC defects
  - ESC defects removal by scrubber processing with acidic chemical
- Summary



SCREEN axcelis

## ESC defect removal by scrubber processing with acidic chemical

- **Experimental**
  - Implant process can generate backside particles due to electrostatic clamping. SCREEN cleans the wafers and confirms defects removal efficiency with 1.0 μm particle size, SEM inspection and EDX analysis.

**Implant wafer**

axcelis

**Clean ESC defects wafer**

SCREEN

| Test sample | | Axcelis | | | SCREEN | |
|---|---|---|---|---|---|---|
| | | Transfer | ESC clamp | Ion implantation | Scrubber processing with FOM | Scrubber processing with DIW |
| 1 | Beamless implant | V | V | | V | V |
| 2 | ESC clamp only | V | V | | V | V |
| 3 | Low Energy High Dose P+ | V | V | V | V | V |
| 4 | Mid Energy High Dose As+ | V | V | V | V | V |

SCREEN axcelis

## Result of ESC defect removal by scrubber processing with acidic chemical

### ♦ ESC defect removal efficiency

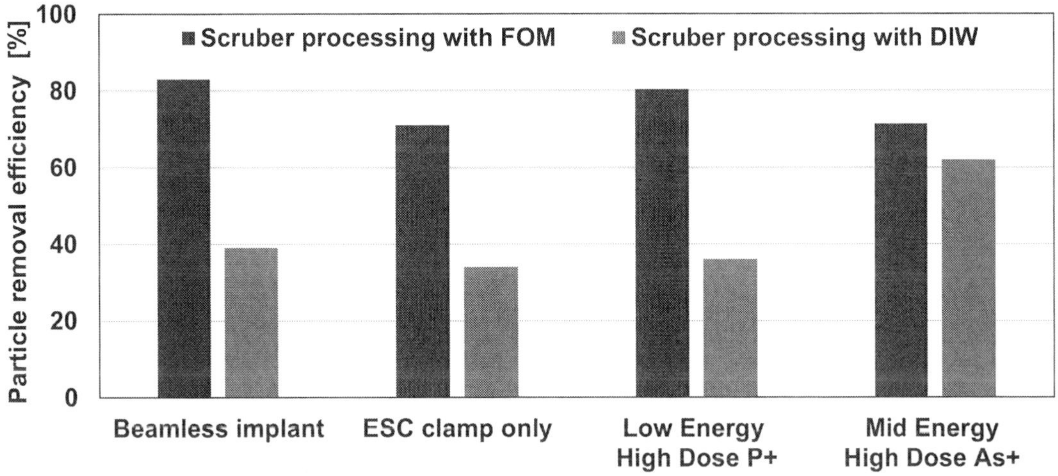


**Scrubber processing with FOM could improve 10 - 40% of ESC defect removal efficiency from scrubber processing with DIW at each test condition.**

13 SE-78-6069-L1 SCREEN Semiconductor Solutions Co., Ltd.

SCREEN axcelis

## Result of ESC defect removal by scrubber processing with acidic chemical

### ♦ SEM inspection and EDX analysis

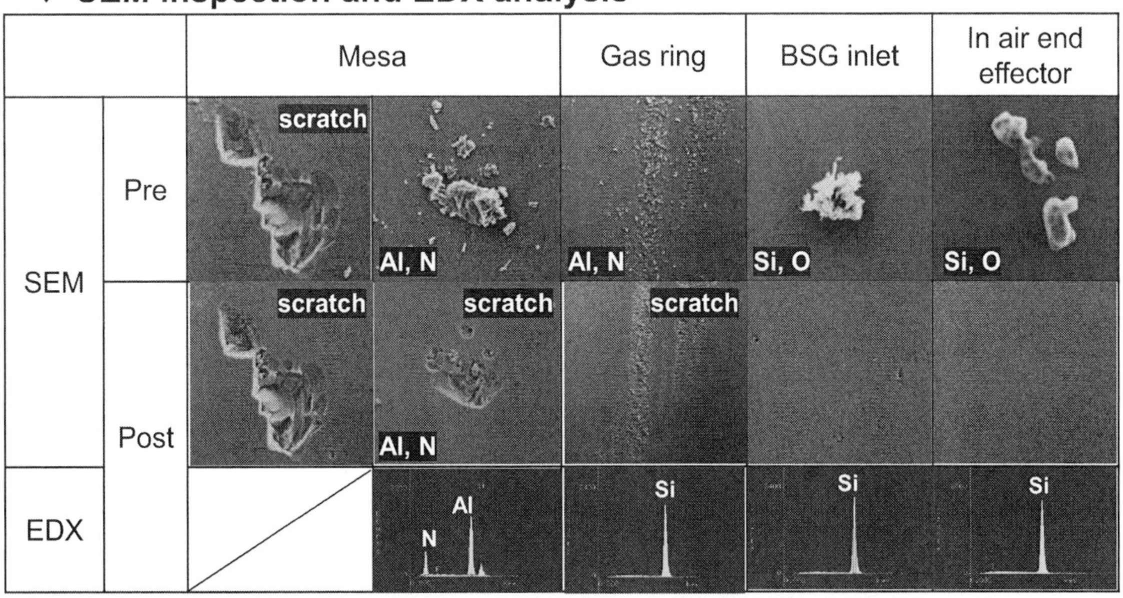


**AlN and SiO defects on wafer could be removed, but there are embedded AlN defect in scratch. Need to reduce the scratches caused by Mesa.**

14 SE-78-6069-L1 SCREEN Semiconductor Solutions Co., Ltd.

SCREEN

## Hardware approach for ESC defects and verification by scrubber processing

axcelis

- **Next experimental**
  - New type ESC to reduce scratches at mesa

| | Current type | New type |
|---|---|---|
| Mesa Material | AlN | Carbon based polymer |
| Mesa contact area | 4% (No change contact area) | |

  - Test condition
    - ESC defects removal efficiency with 1.0 μm size particle
    - SEM inspection and EDX analysis

| Test sample | | Axcelis | | | SCREEN |
|---|---|---|---|---|---|
| | | Transfer | ESC clamp | Ion implantation | Scrubber processing with FOM |
| 1 | Low Energy High Dose P+ | V | V | V | V |
| 2 | Mid Energy High Dose As+ | V | V | V | V |



SCREEN

## ESC defects comparison after implantation

axcelis

- **SEM inspection and EDX analysis (Post implant)**

| | Current type ESC | | New type ESC | |
|---|---|---|---|---|
| | SEM image | EDX | SEM image | EDX |
| Mesa | 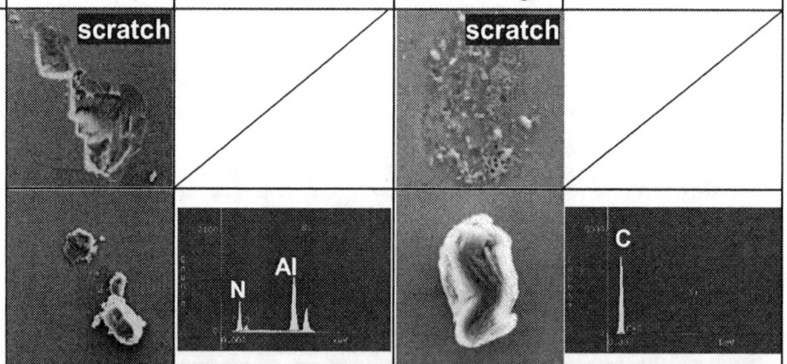 | | | |
| Gas ring | 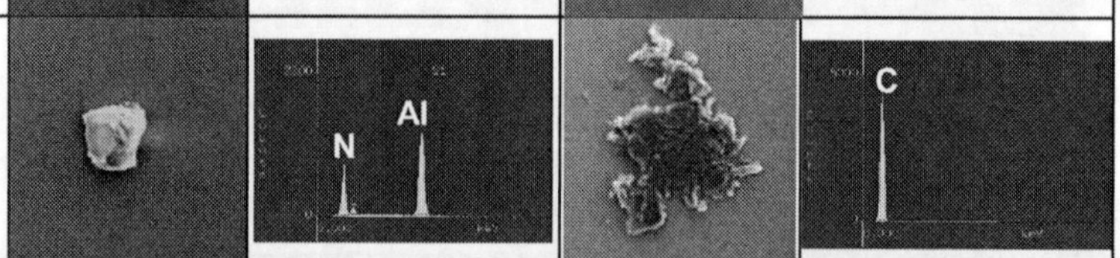 | | | |





**Scratches seem to be shallower and AlN were not detected with new type ESC, however carbon defects were observed.**

SCREEN

## Result of ESC defects removal by scrubber processing with acidic chemical

axcelis

- **SEM inspection and EDX analysis (Post backside cleaning)**

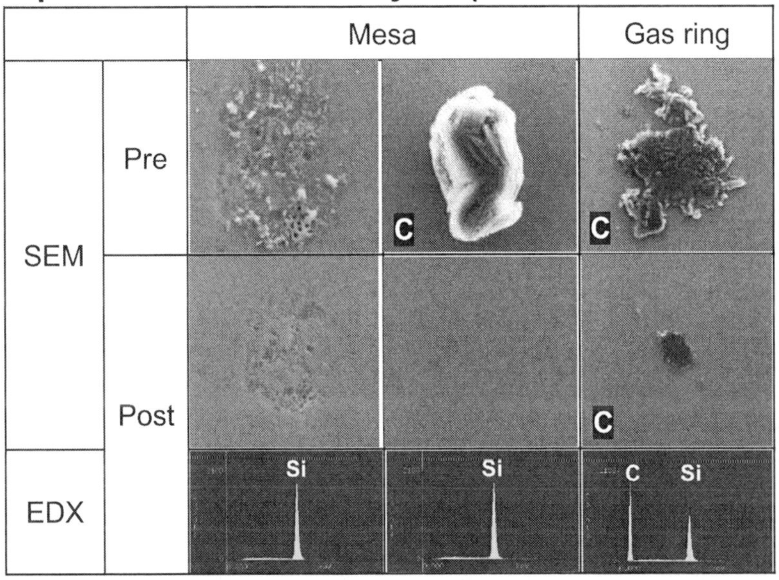


**Carbon defects could be almost removed, but there is small defects remain at Gas ring defect.**



SCREEN

## Result of ESC defects removal by scrubber processing with acidic chemical

axcelis

- **ESC defects removal efficiency**

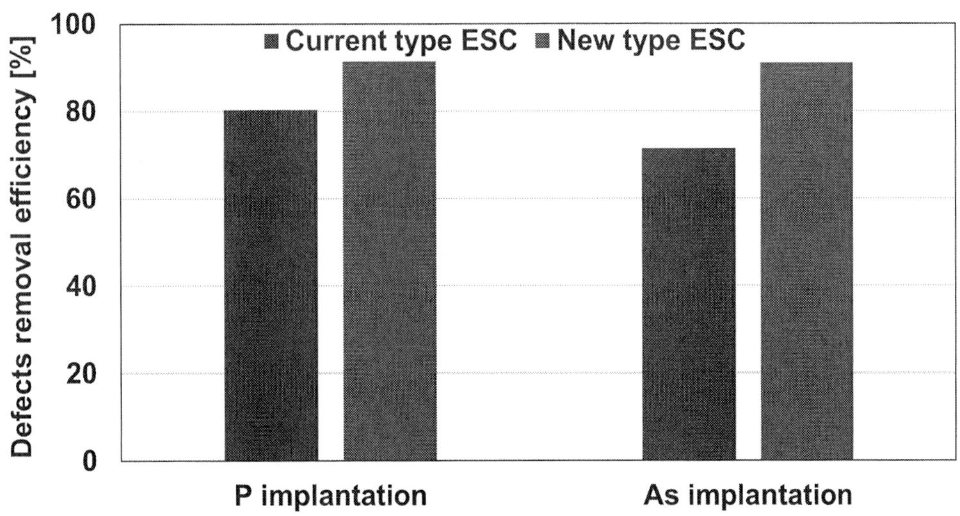


**ESC defects removal efficiency is further improved by new type ESC.**

SCREEN

## Agenda

axcelis

- Motivation
  - Backside particle/defect removal requirement
  - About electrostatic clamp (ESC)
- Experimental and Result
  - Chemical screening for ESC defects
  - ESC defects removal by scrubber processing with acidic chemical
- **Summary**



SCREEN

## Summary

axcelis

- **New scrubber processing has been developed to enable efficient removability for SiO and AlN defects, 10 - 40% improvement compared to conventional process.**
- **New type of ESC has demonstrated backside defect reduction and confirmed effectiveness against scratches related to Mesa contacts.**
- **Combination of new type ESC with new scrubber processing achieved 90% of ESC defects removal efficiency.**

SCREEN axcelis

Thank you for your attention

21 SE-78-6069-L1 SCREEN Semiconductor Solutions Co., Ltd.

SCREEN axcelis

Fit your needs, Fit your future

SCREEN

axcelis

SPCC 2019

# Oxide Regrowth Mechanism during Silicon Nitride Etching in Vertical 3D NAND Structure

Taehyeon Kim, Changjin Son, Taegun Park, and Sangwoo Lim*

Department of Chemical and Biomolecular Engineering
Yonsei University
Seoul, Korea

연세대학교
YONSEI UNIVERSITY

## Outline

YONSEI UNIVERSITY

- Introduction of selective nitride etching
- Overview of nitride etching process
- Oxide regrowth issue
  - ✓ What is oxide regrowth?
  - ✓ Where does it occur?
  - ✓ What does determine oxide regrowth?
  - ✓ Oxide regrowth mechanism
- Summary

S. Lim 2

## NAND Revenue and 3D NAND

YONSEI UNIVERSITY

NAND Revenue Forecast

| | | | | | | | | | | |
|---|---|---|---|---|---|---|---|---|---|---|
| Petabytes | 61,065 | 87,222 | 127,570 | 166,612 | 239,656 | 341,881 | 468,637 | 639,389 | 878,209 | CAGR 17-22 |
| Bit Growth | 40.8% | 42.8% | 46.3% | 30.6% | 43.8% | 42.7% | 37.1% | 36.4% | 37.4% | 39.4% |

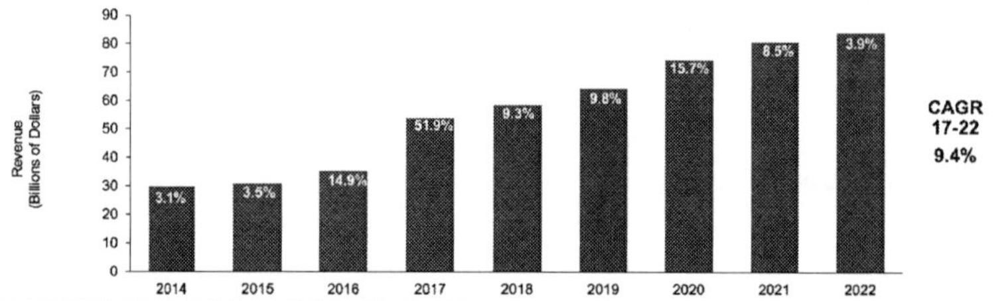


| | | | | | | | | | | |
|---|---|---|---|---|---|---|---|---|---|---|
| ASP 1GB eqv.* | 0.46 | 0.34 | 0.26 | 0.31 | 0.23 | 0.18 | 0.15 | 0.12 | 0.09 | CAGR 17-22 |
| ASP Change | -26.8% | -27.6% | -21.5% | 16.3% | -24.0% | -23.0% | -15.6% | -20.5% | -24.4% | -21.6% |

*Semiconductor Engineering*
*August 2018 - Mark LaPedus*
Source: Gartner

- 3D NAND is used for the storage applications such as solid-state drives (SSD), whose market is growing every year.
- 3D NAND provides a lower power consumption with faster performance and with longer life time.



## 3D Vertical NAND Structure

YONSEI UNIVERSITY

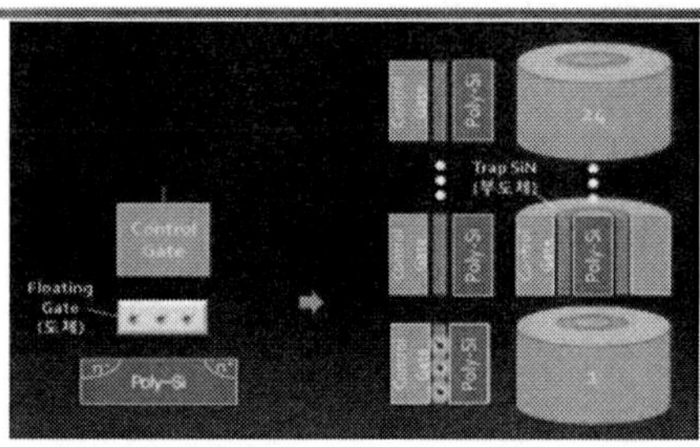


Planar NAND 3D NAND *(courtesy Samsung)*

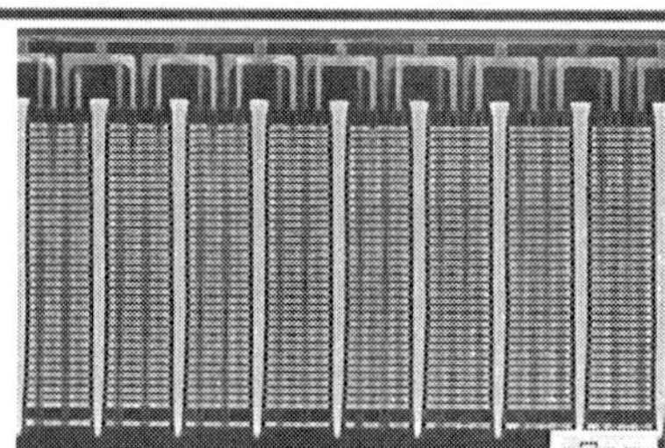


Cross section of Samsung's 86 Gbit 32-layer 2nd generation V-NAND *(courtesy Chipworks)*

- Introduction of 3D NAND increases storage capacity without shrinkage of the cells.
- As the number of stack layers increases, the memory density increases. The chip makers are racing to develop next technology generations.

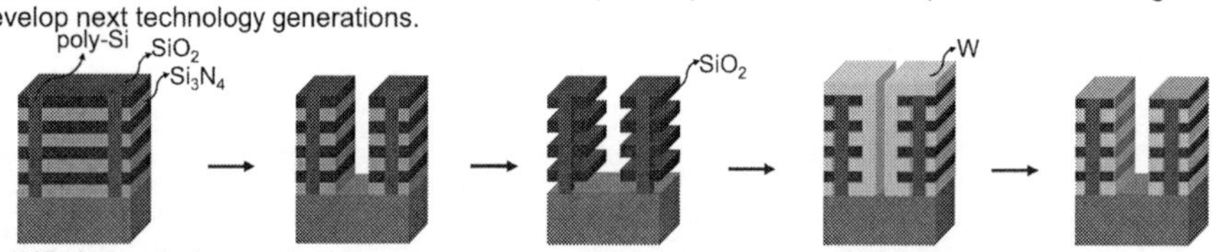


$Si_3N_4/SiO_2$ layer stack, etching and poly-Si → Dry etching of slit → Selective removal of $Si_3N_4$ → Gate dielectrics and metal deposition → Metal etching

# Etching of Nitride in $H_3PO_4$

YONSEI UNIVERSITY

ammonium phosphate · hydrated silicon dioxide

$$3Si_3N_4 + 27H_2O + 4H_3PO_4 \rightleftharpoons 4(NH_4)_3PO_4 + 9H_2OSiO_2$$

influences $SiO_2$ etching

$$\text{etching selectivity} = \frac{Si_3N_4\ \text{Etching Rate}}{SiO_2\ \text{Etching Rate}}$$

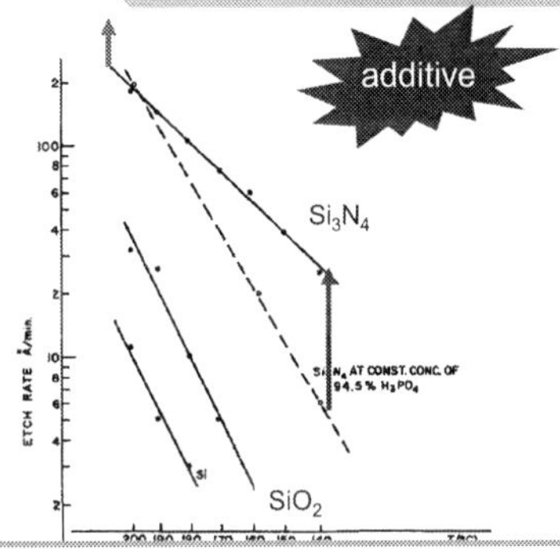


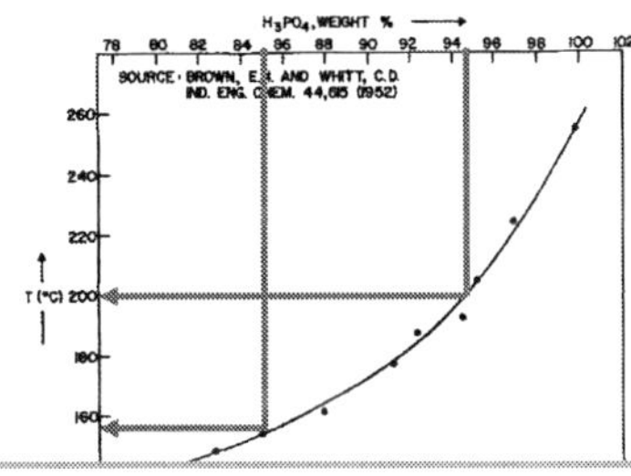


- In general, any effort to increase etching rate of $Si_3N_4$ increases etching rate of $SiO_2$ more, degrading $Si_3N_4$ to $SiO_2$ etch selectivity.
- Temp increases → $H_2O$ evaporates more → increases $H_3PO_4$ concentration → increases BP of solution → not reaching to BP → decreases etching rate
- If etching of $SiO_2$ is hindered, removal of $Si_3N_4$ is also suppressed.



# Etching Rate vs. Selectivity

YONSEI UNIVERSITY

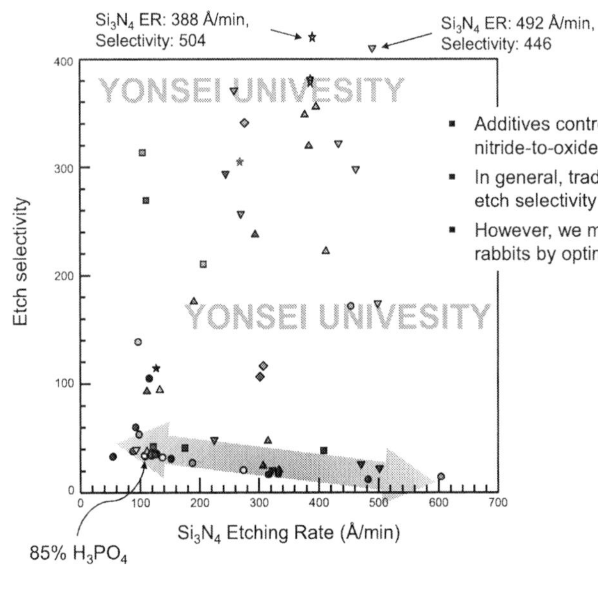


- Additives control etching rate of Si$_3$N4 and nitride-to-oxide etch selectivity.
- In general, trade-off between etching rate and etch selectivity is observed.
- However, we may be able to chase two rabbits by optimizing additives.

## Etch Behavior in 3D NAND Structure

YONSEI UNIVERSITY

with etch inhibitor addition

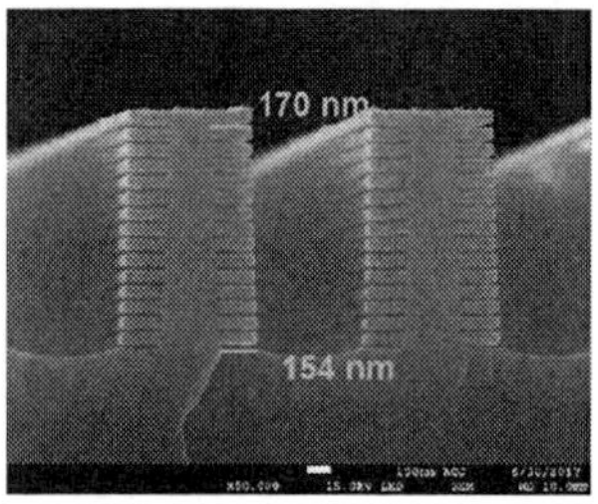


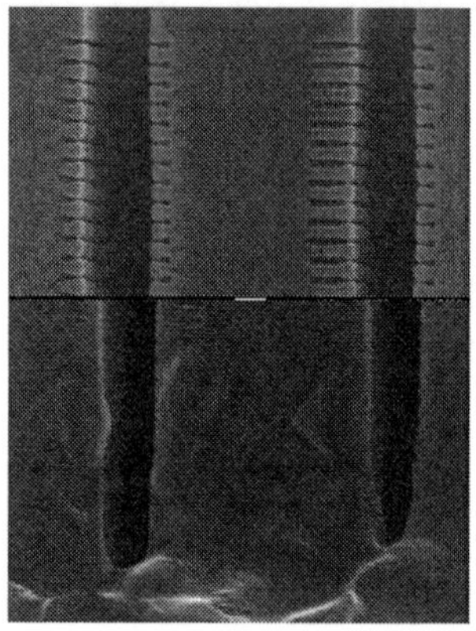


Issues are:

- Etching rates and selectivity change between blanket and 3D structure wafers.
- Etching behavior changes with the number of $Si_3N_4/SiO_2$ stack layers.
- Top to bottom etch non-uniformity becomes critical.
- Is that all?



## Nope, Oxide Regrowth

YONSEI UNIVERSITY

with etch inhibitor addition

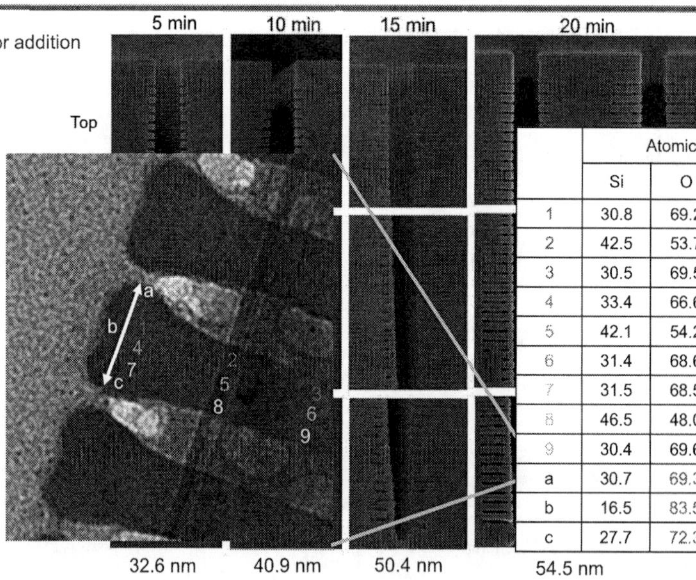


| | Atomic % | | |
|---|---|---|---|
| | Si | O | N |
| 1 | 30.8 | 69.2 | 0.0 |
| 2 | 42.5 | 53.7 | 3.8 |
| 3 | 30.5 | 69.5 | 0.0 |
| 4 | 33.4 | 66.6 | 0.0 |
| 5 | 42.1 | 54.2 | 3.7 |
| 6 | 31.4 | 68.6 | 0.0 |
| 7 | 31.5 | 68.5 | 0.0 |
| 8 | 46.5 | 48.0 | 5.5 |
| 9 | 30.4 | 69.6 | 0.0 |
| a | 30.7 | 69.3 | 0.0 |
| b | 16.5 | 83.5 | 0.0 |
| c | 27.7 | 72.3 | 0.0 |

# Why at the Bottom?

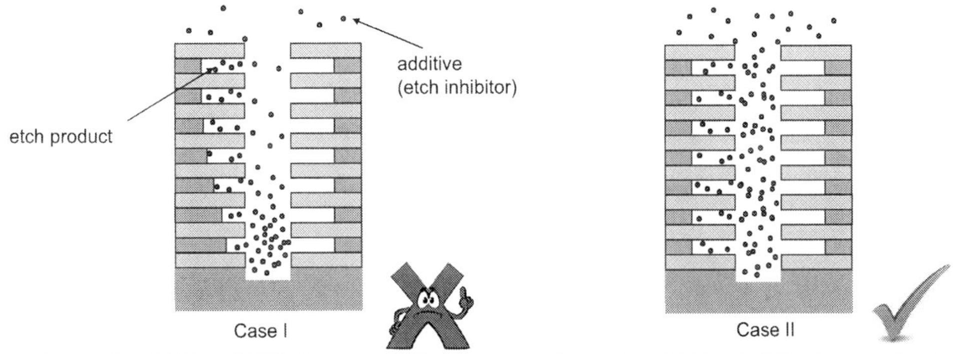


- Assumption: Additives (inhibitor) were not uniformly dissolved in the etchant, and the concentration of etching inhibitor is higher at the bottom. Etching rate is higher at the top.
- Under this condition, oxide regrowth may result mainly from the **additives distribution** at the bottom.
- However, generally etch rate was uniform at the top and the bottom.

- Assumption: Additives (inhibitor) were uniformly and completely dissolved in the etchant. Etching rate is uniform at the top and the bottom.
- Heavier oxide regrowth at the bottom is due to poor **mass transfer** of etch product from the multi-layer stack to the bulk solution.



# Why at the Corner?

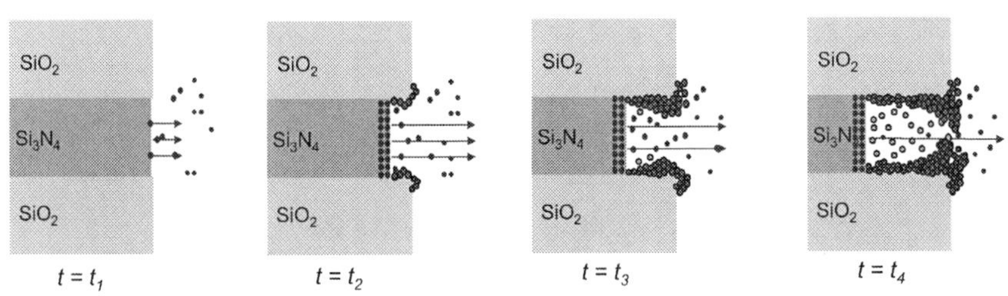


- In the beginning of etching process, diffusion of etch product to bulk solution (●) easily occurs. Only a limited portion of etch product chemisorbs on the $SiO_2$ layer (●) surface.
- As the etching proceeds, amount of nitride etch products that will chemisorb on $SiO_2$ (◎) increases. On the contrary, only limited nitride etch products will diffuse to the bulk solution.
- Oxide regrowth may be due to hydrogen bonds and Van der Waals force.

## Why at the Corner?

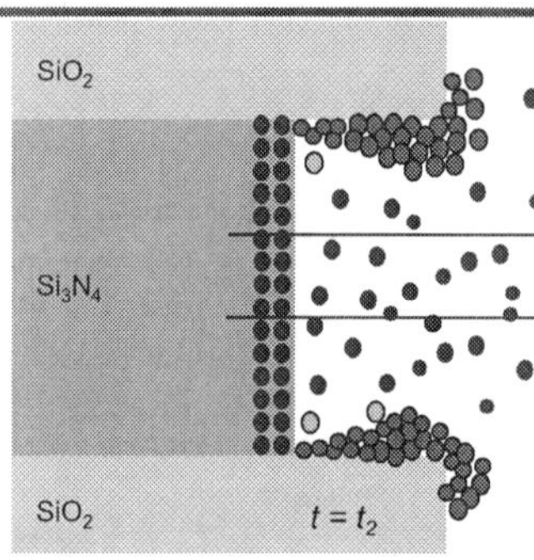


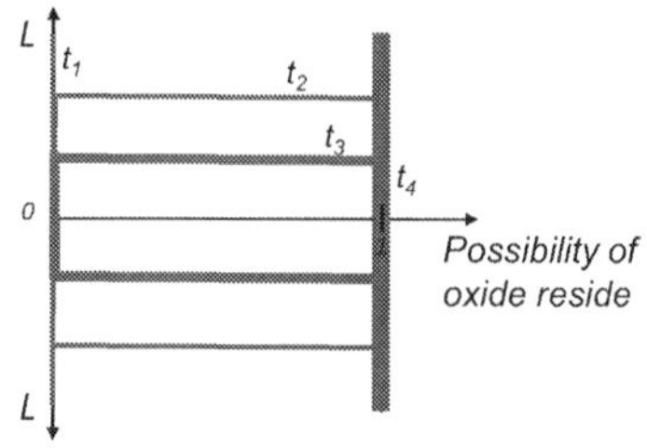


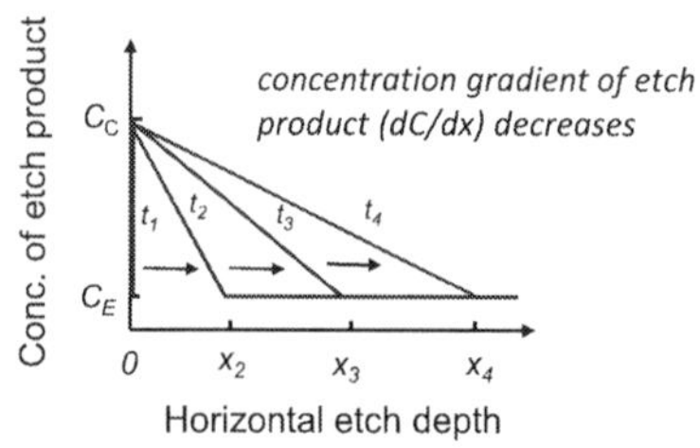


- Oxide-regrown area extends as etching proceeds.
- Entrance is exposed to the etch products for the entire period of process. Therefore, oxide regrowth is heavier at the entrance.
- As etched depth increases, concentration gradient of etch product decreases. Thus, diffusion of etch product to bulk solution is hindered and oxide regrowth easily occurs.

S. Lim

11

## Effect of Additives

$H_3PO_4$ with strong inhibitor addition @ 160 °C

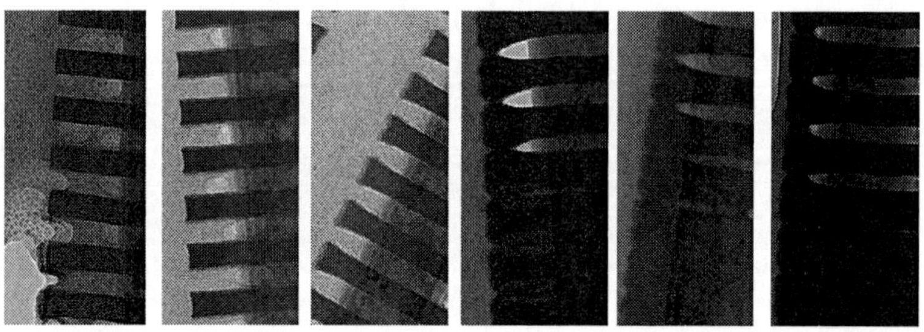

1x → 5x

- The additives concentration determines degree of oxide regrowth.

S. Lim

12

## Effect of Etch Product

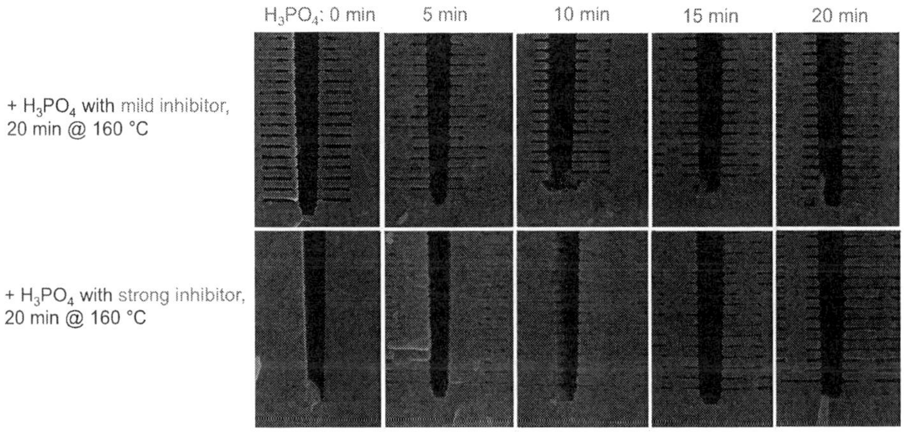


- With the increase in the first etching period in $H_3PO_4$, oxide regrowth was suppressed.
- The amount of etch product produced near the trench (i.e. etching rate) is another key factor that decides behavior of oxide regrowth.

S. Lim 13

## Additives or Etch Products?

**In general, oxide regrowth is heavier at the bottom.**

only by additives?

- Additives are completely dissolved in etchant. Therefore, there shouldn't be any concentration difference in the etchant.
- If additives were not completely dissolved, its concentration must be higher at the top of the 3D NAND structure than at the bottom and **oxide regrowth should be heavier at the top.**

only by etch products?

- There is no chance that etching rate, which is generation rate of etch product, is faster at the bottom of the 3D NAND structure.
- Therefore, if oxide is regrown only by etch products, **oxide regrowth should be heavier at the top.**

**Oxide regrowth might be brought by both additives and etch products.**

S. Lim 14

# Degree of Oxide Regrowth

YONSEI UNIVERSITY

> Degree of oxide regrowth = $k$ x inhibitor conc. x etch product generation rate (≈ ER) at a given structure

$k$ : determined by surface wetting property, reactivity, adsorption & condensation rate, and so on

- $H_3PO_4$-only: just thinning of $SiO_2$ with **no oxide regrowth because of zero inhibitor concentration**
- Certain amount of inhibitor is required to get a higher etch selectivity, and a higher etching rate is definitely needed for manufacturing.
- ***k*** **is the only factor we can control over the margins. Optimization of additives will be able to reduce it.**

> Degree of oxide regrowth = $k'$ x (1 / diffusion rate of etch product) in a given solution

$k' = k$ x etch product generation rate (≈ ER)

- The location of oxide regrowth (bottom > top, entrance > cul-de-sac) will be determined by how fast the etch products diffuse to the bulk solution.



# Additives Optimization

YONSEI UNIVERSITY

TEM images of $Si_3N_4/SiO_2$ stack layers after selective wet etching

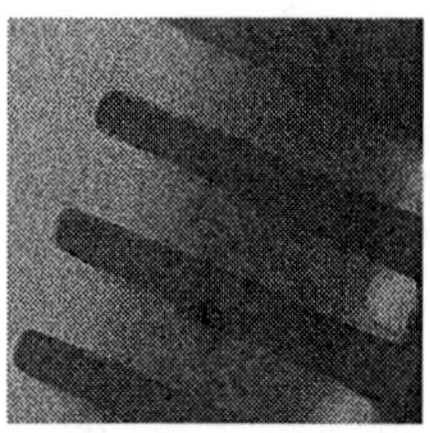

conventional $H_3PO_4$ process

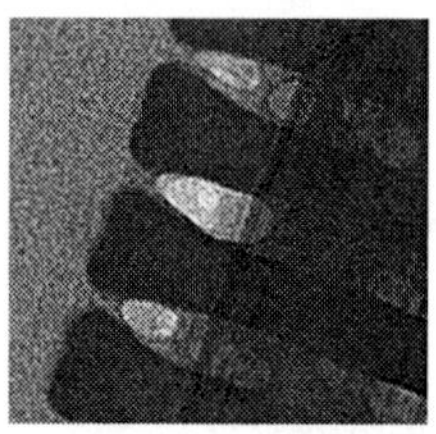

Oxide regrowth with additives

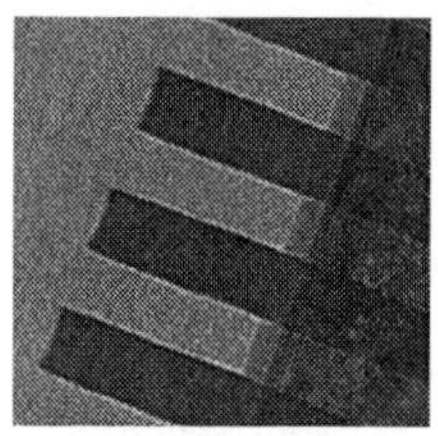

optimization of additives

## Summary

YONSEI UNIVERSITY

- In the selective nitride etching process for the manufacturing of 3D VNAND structure, not only etching rate of nitride and etch selectivity, but also oxide regrowth is a critical issue that we need to consider.
- Oxide regrowth occurs more at the bottom and the entrance of $Si_3N_4/SiO_2$ stack entrance because of the mass transfer issue.
- Additives introduced to control etching rate and etch selectivity are definitely the source of oxide regrowth. However, degree of oxide regrowth is determined by the amount of etch product produced near the trench as well as the type and concentration of etching inhibitor.
- Surface wetting property, reactivity, and adsorption and condensation rate may be controlled by optimization of additives, which will suppress oxide regrowth without loss of etching rate and selectivity.

S. Lim 17

# Phosphoric acid rinse impact on CMOS devices

L.Caron, P.Garnier

Philippe-e.garnier@st.com

STMicroelectronics - France

(+33) (0) 4 38 92 34 70

life.augmented

## Outline

2

- Defects from $H_3PO_4$
- Case study: STI / DTI SiN pullback
- Defects characterization
- Solutions

life.augmented

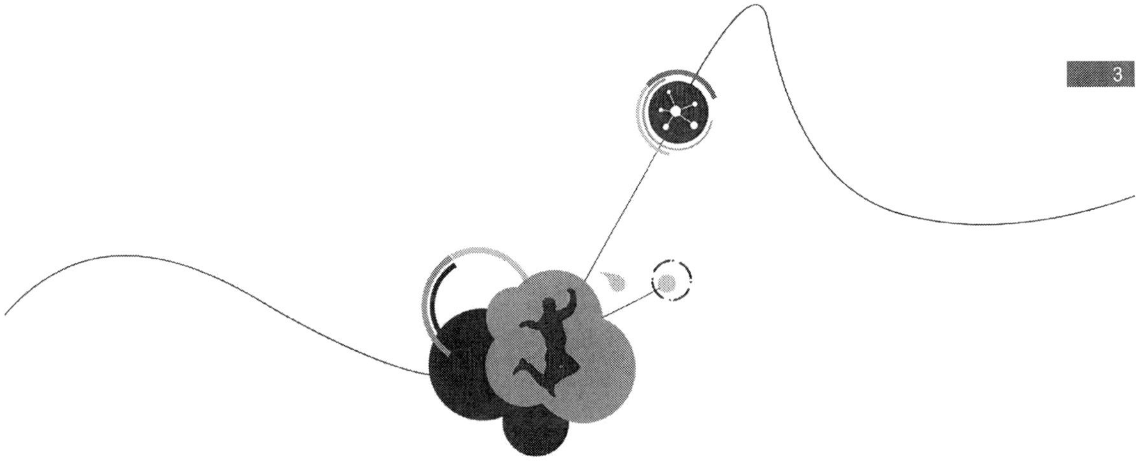

# Defects from $H_3PO_4$ bath

life.augmented

## $H_3PO_4$ wet bench: defects zoology



$H_3PO_4$ main FEOL use = SiN partial or complete removal

Usual particles

Teflon (Hardware)

Silicates (SiN etch products deposition)

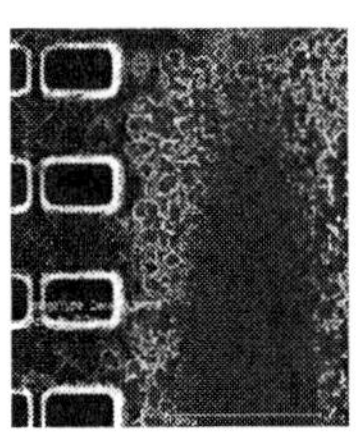

Slurry

Silica

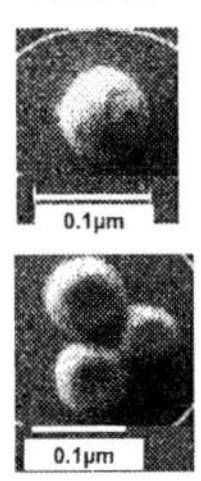


Ceria

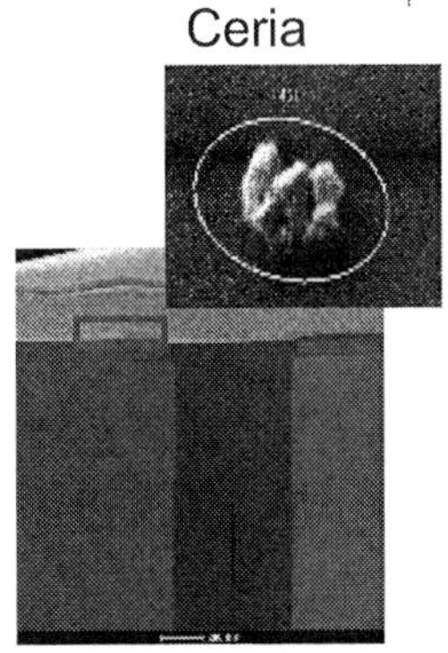

C stains

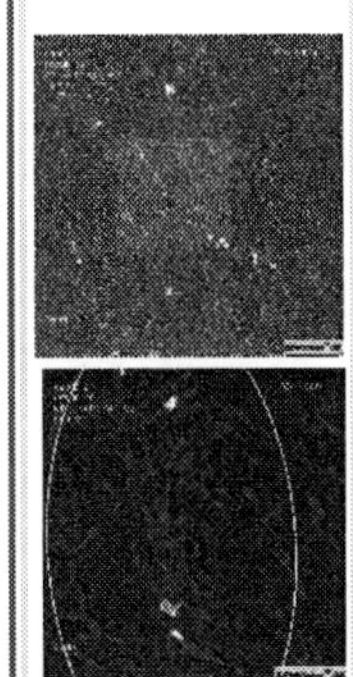

Other defects

Usual Yield killers

Size: few µm wide
Occurrence: only observed if H3PO4 used
IPA quality checked
Root cause: IPA condensation
Impact on die: unknown

life.augmented

# Defects of interest: occurences

5

POR (Process Of Record) Sequence:
HF – $H_3PO_4$ – SC1 – IPA dryer

variables:

- POR
  (observed after 5 days)

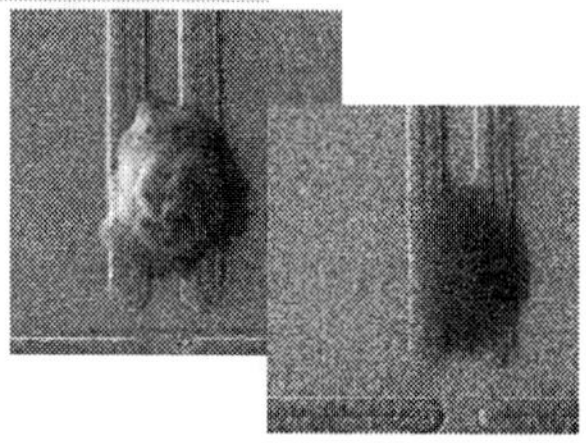

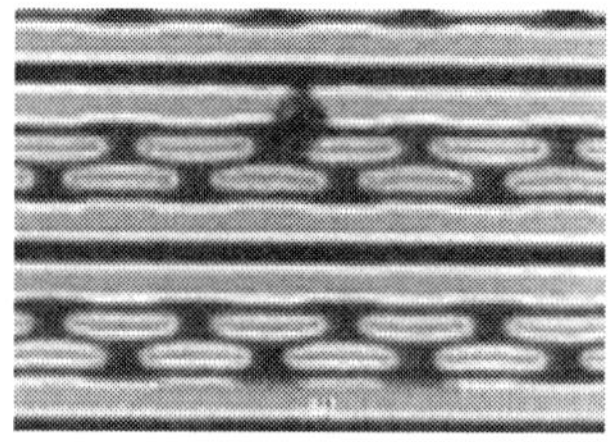

- decoupled POR sequence:
  « HF-$H_3PO_4$ - IPA dryer » then « SC1 – IPA dryer »

- POR with Shortened SC1

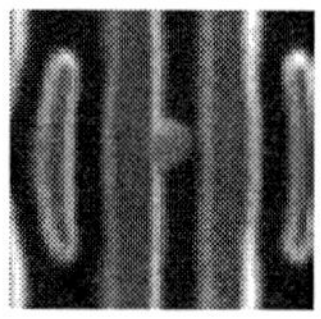

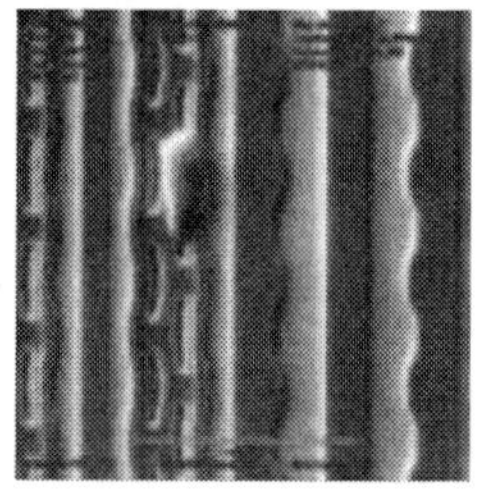

- POR with SC1 room temp i.o. 50°C

life.augmented

---

6

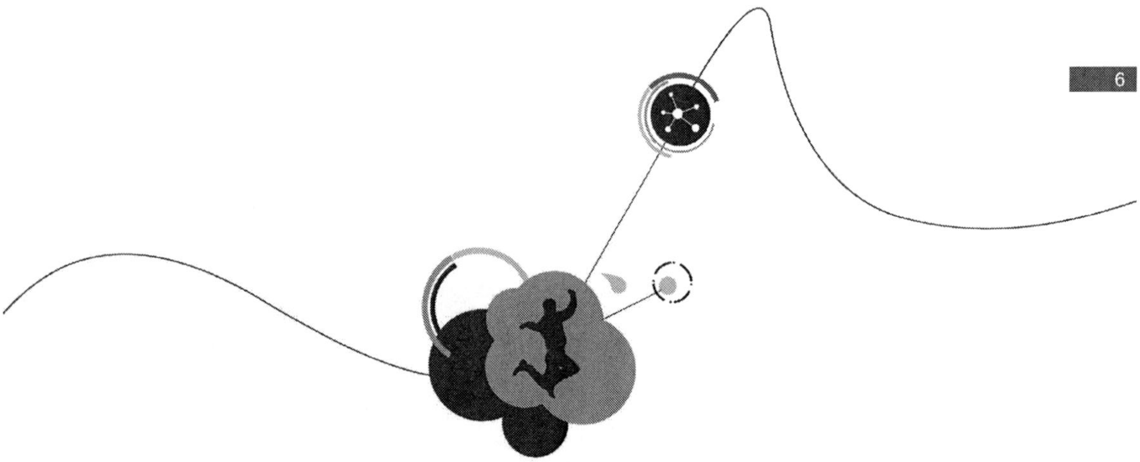

# Case study: STI / DTI SiN pullback

life.augmented

# STI / DTI integration

7

SiN pullback goal: better trench top profile to ease gapfilling w/o voids

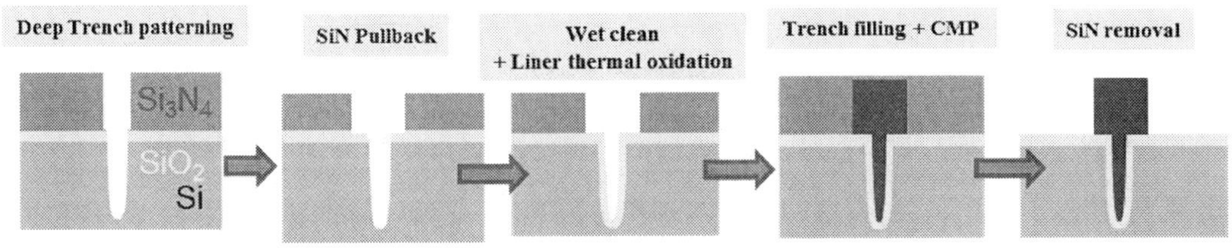


life.augmented

# STI / DTI integration

8

SiN pullback goal: better trench top profile to ease gapfilling w/o voids

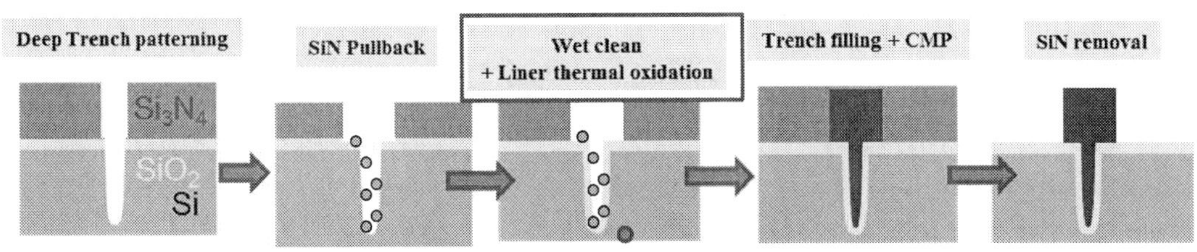

**Neither RCA nor oxidation remove defects**

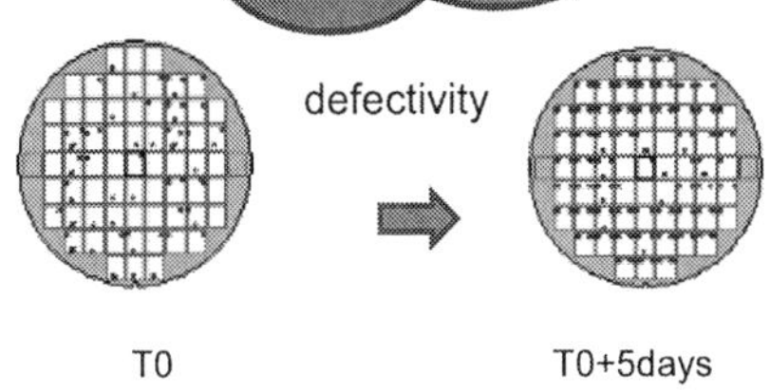


life.augmented

# STI SiN pullback: impact on device



STI $SiO_2$ Liner oxidation: RTO **temperature ramp up** conditions

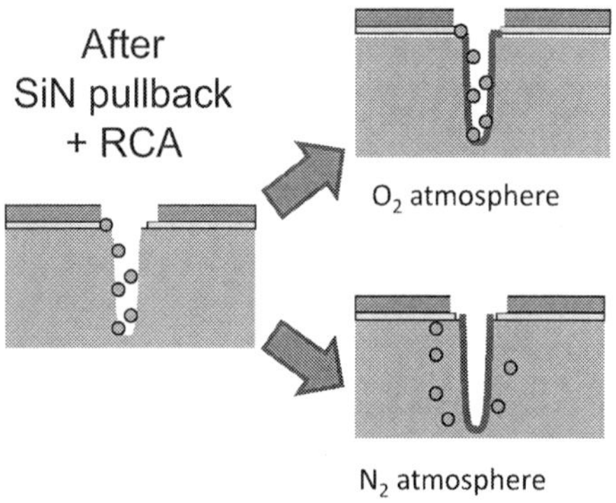


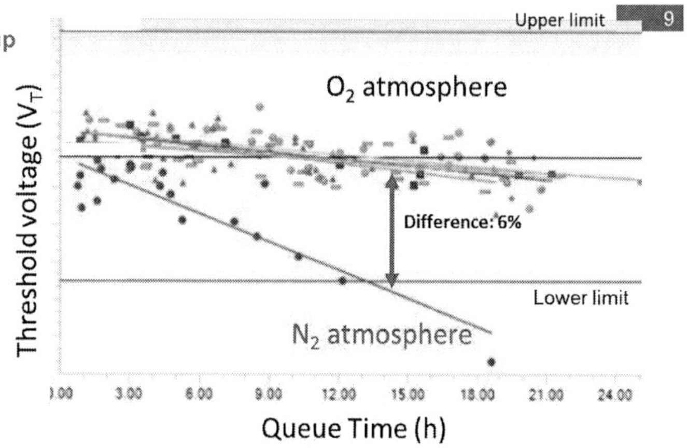


Transistor Vth shift:

- Phosphates ➔ Phosphorous counter-doping of active areas
- liner oxidation : temperature ramp up
  - N2 ramp up ➔ diffusion through Si
  - O2 ramp up ➔ P trapped by oxidation on STI edges
- Queue time influence between P contamination & liner oxidation

ST life.augmented



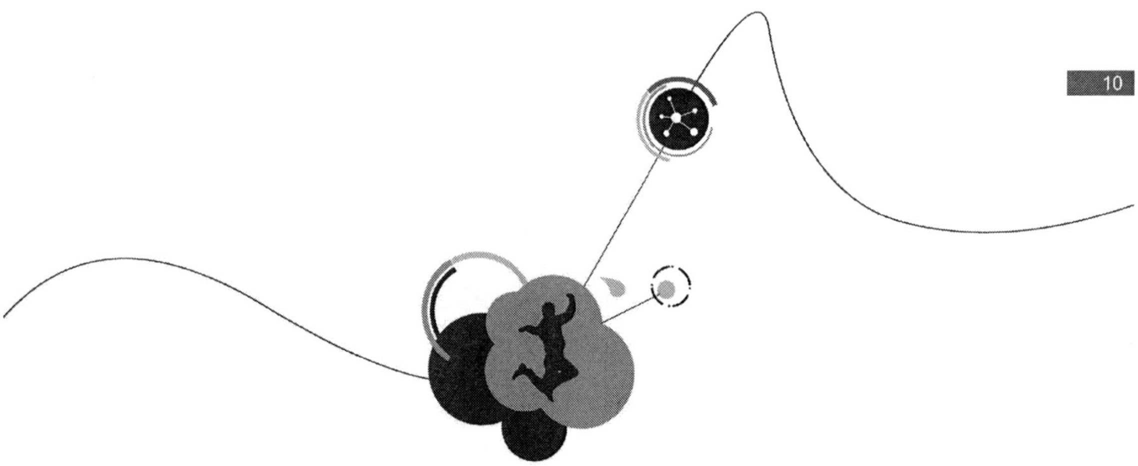

# Defects characterization:

- Intra wafer mapping
- Growth kinetics
- Influence of surface

ST life.augmented

# Defects: intra wafer mapping

11

Blanket Si wafer / HF-$H_3PO_4$ - no SC1 – IPA
SC1 = removed in many experiments to ease detection / charac of defects of interest

~ 50 000 defects > 30nm

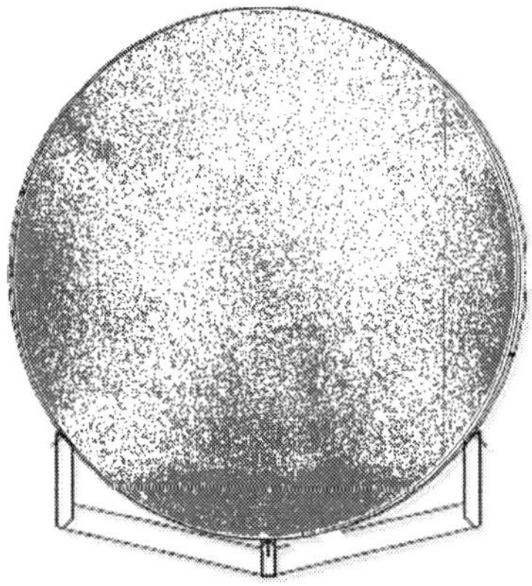

Wet bench wafer lifter

Defects:
- Localization: in worst cleaning area in bath, around, wafer lifter contact positions
- Growth: +33% amount, 5 days after contamination
- No impact on thermal oxide grown thickness

# Defects characterization on blanket wafer

3 different surfaces dipped into aged $H_3PO_4$ bath (w/o SC1)

12

**$SiO_2$**

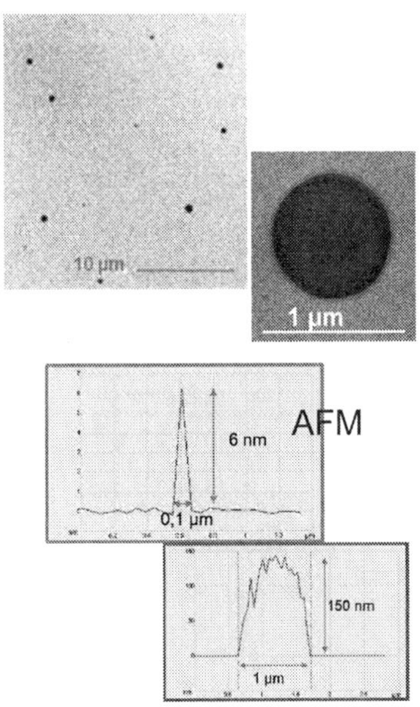


**Si surface (SiH or SiOH)**

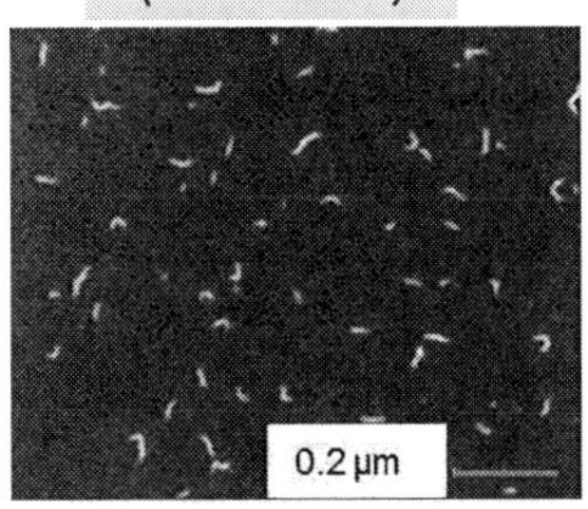


**SiN surface**

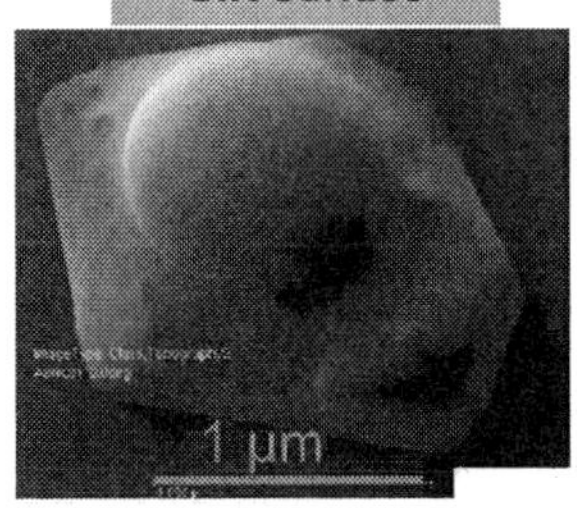


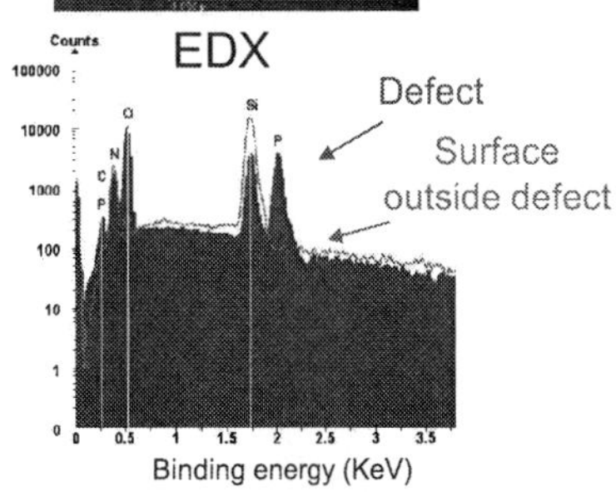


High [P]
Partially volatile
& hygroscopic

**Defects shapes & size depending on surfaces nature**

# Defects characterization on blanket wafer

3 different surfaces dipped into aged $H_3PO_4$ bath (w/o SC1)

13

Thermal SiO2 | Si surface (SiH or SiOH) | SiN surface

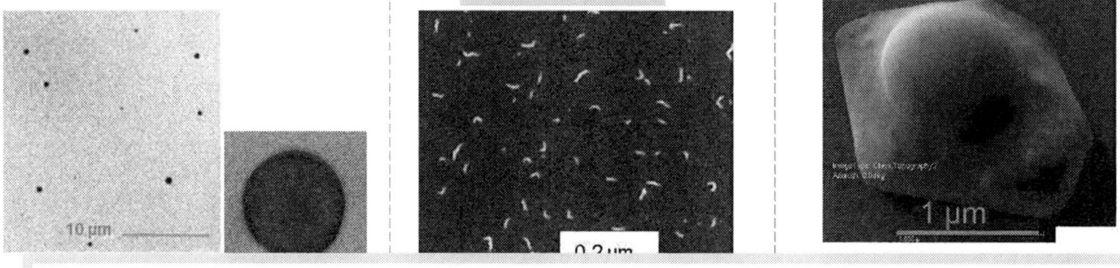


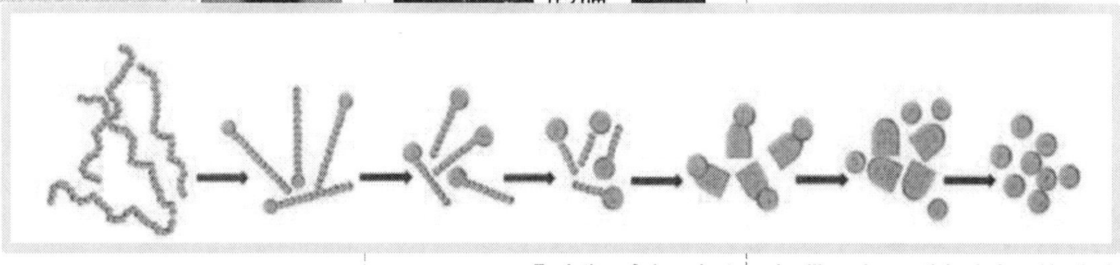

*Evolution of shape isotropy in silica microparticles induced by the base, J. Brijitta, Colloid and Polymer Science, 2017*

**Morphology could probably be explained by silica sol-gel reactions & deeper characterizations**

ST life.augmented

# Defects charac. on STI pattertned wafer

3 different surfaces dipped into aged $H_3PO_4$ bath (w/o SC1)

14

**Hydrophilic** Si surface during $H_3PO_4$ | **Hydrophobic Si** surface during $H_3PO_4$ | **Hydrophobic SiN** surface during $H_3PO_4$

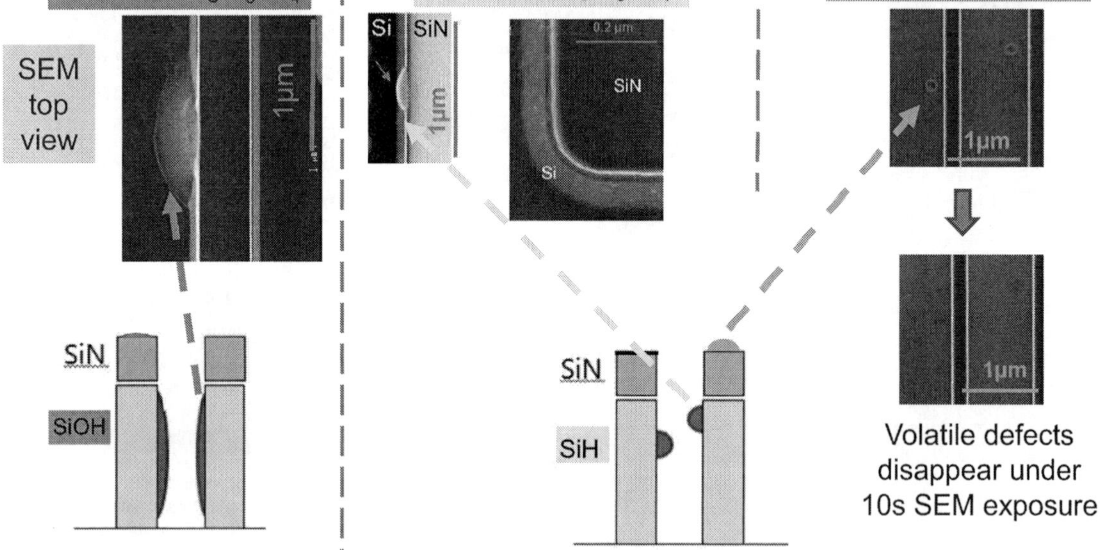


Volatile defects disappear under 10s SEM exposure

ST life.augmented

# Defect composition

15

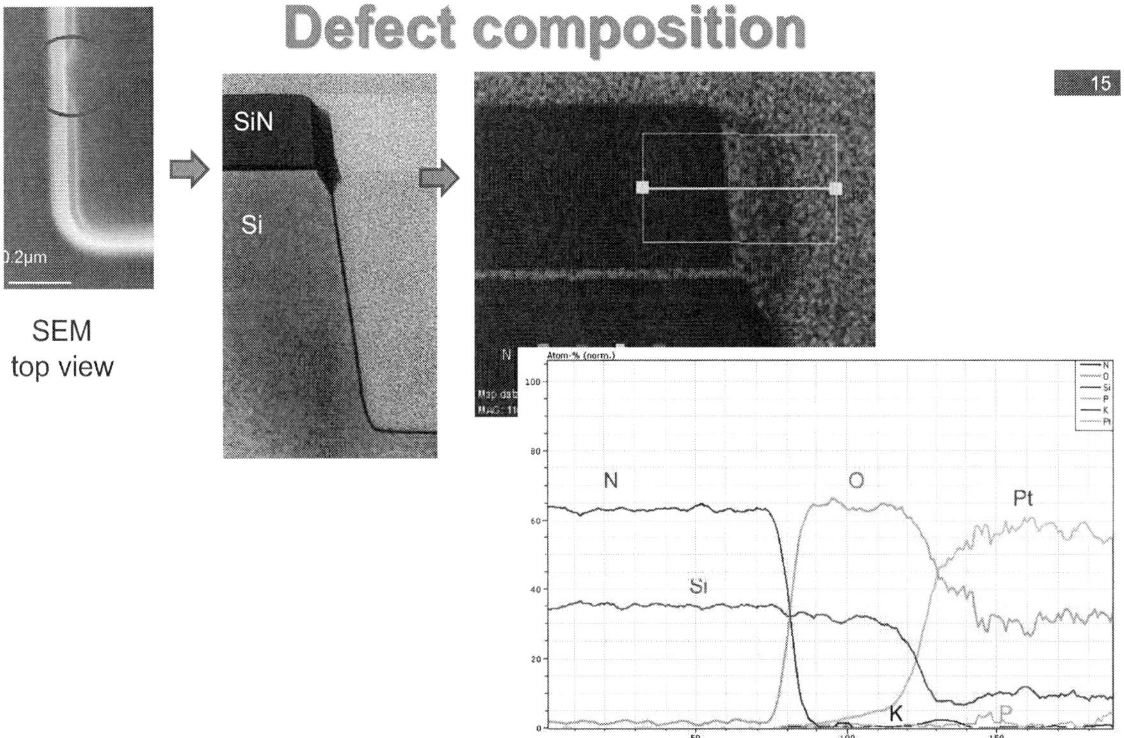


SEM
top view

**Defect = silicate with little Phosphorous**
**[P] ~ 1E+14at/cm²**

life.augmented

# Defect mechanism

16

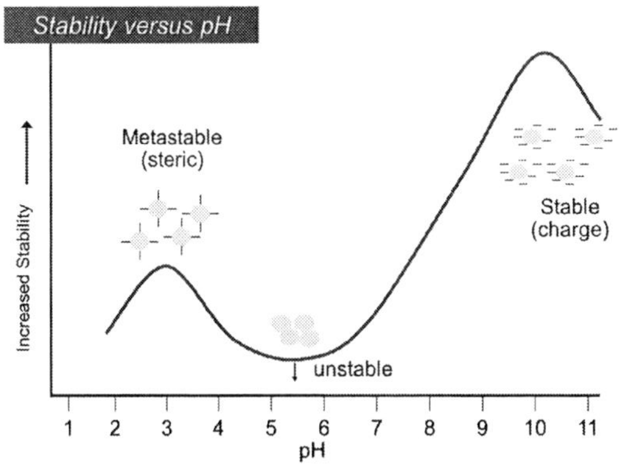


R. Iler, The Chemistry of Silica, 1979, p.367

**Hot reclaimed $H_3PO_4$ bath contains 20 – 100 ppm dissolved silicates (SiN etch by products)**

**Wafer rinse post $H_3PO_4$:**
**2 detrimental parameters lead to some phospho silicates deposition on Si**
- **Temperature decrease**
- **Quick pH change**

life.augmented

## Insufficient $H_3PO_4$ rinse = degraded device

17

White pixels amount vs $H_3PO_4$ equipment
(red > green > blue)
Sequence: HF - $H_3PO_4$ - SC1 in DTI

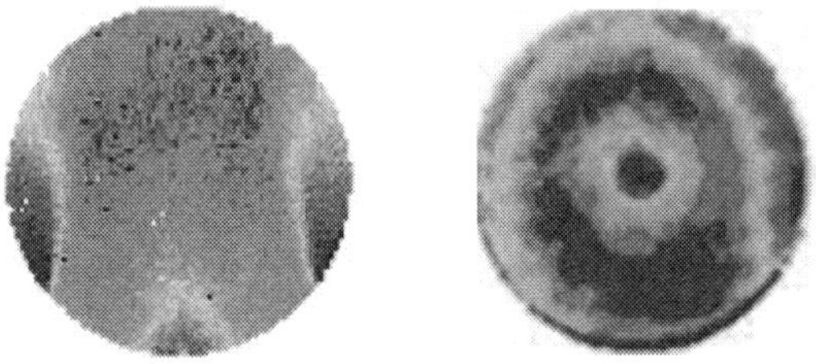

Wet bench    Single wafer

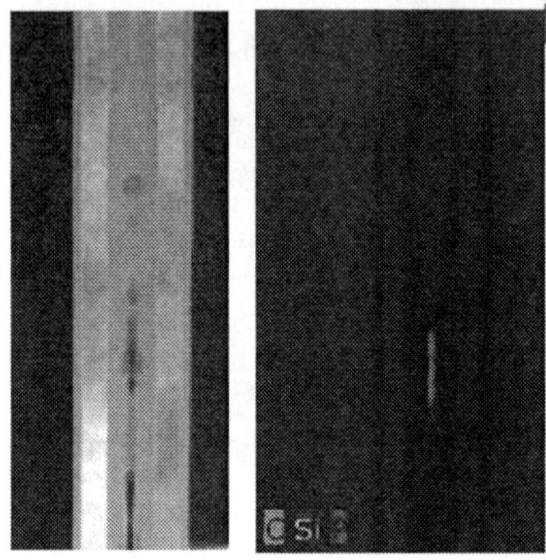


P doped silicates can infiltrate voids in Deep Trench Isolation during $H_3PO_4$ sequence
➔ severe degradation of Image sensor

**w/o adequate $H_3PO_4$ rinse optimization, phospho silicates deposit without being detected inline, leading to counter doping & device degradation**

## Solutions

18

- Alternative chemicals to $H_3PO_4$:
  Available for low SiN etch applications:
  - hot diluted HF
  - HF/EG mix (ESH concern)
  - Other formulations
- Apply HF post $H_3PO_4$ sequence, to remove silicates: very efficient but:
  - Some device integrations requires no HF to avoid excessive SiO2 consumption
  - Doubts on good defects removal in DTI contaminated voids
- $H_3PO_4$ Rinse optimizations

# Conclusions

19

- $H_3PO4$ :
  - Well known chemical
  - But can deposit phospho silicates, **whose detection & quantification isn't easy** ( [P] < LLD of all analytical tools)
  - 2 kind of defects evidenced: volatile on SiN & phospho silicates on Si / SiO2
- These silicates can counter dope silicon:
  - Threshold voltage shifts
  - Device malfunctions
- Solutions:
  - Use HF, post $H_3PO_4$ sequence where possible
  - Otherwise, improve $H_3PO_4$ rinse

life.augmented

20

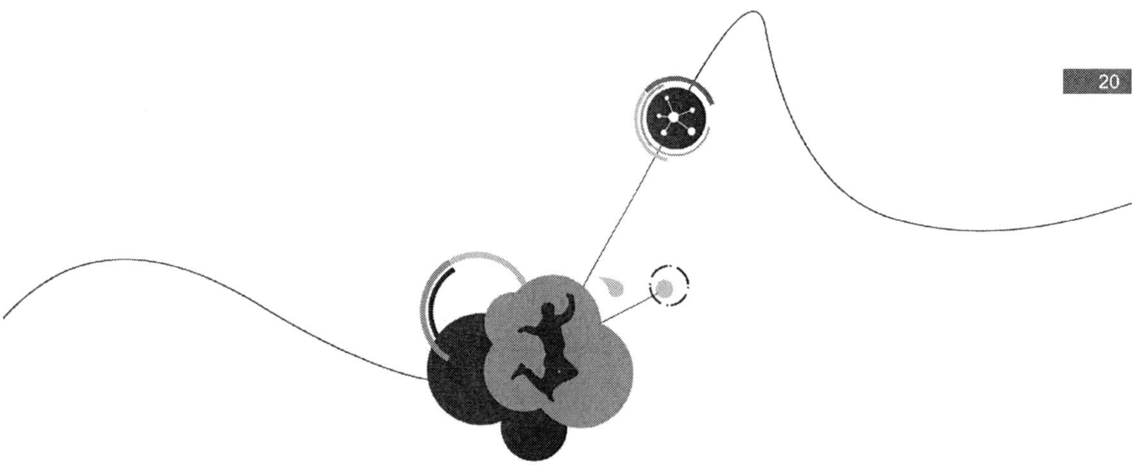

# Backup slide

life.augmented

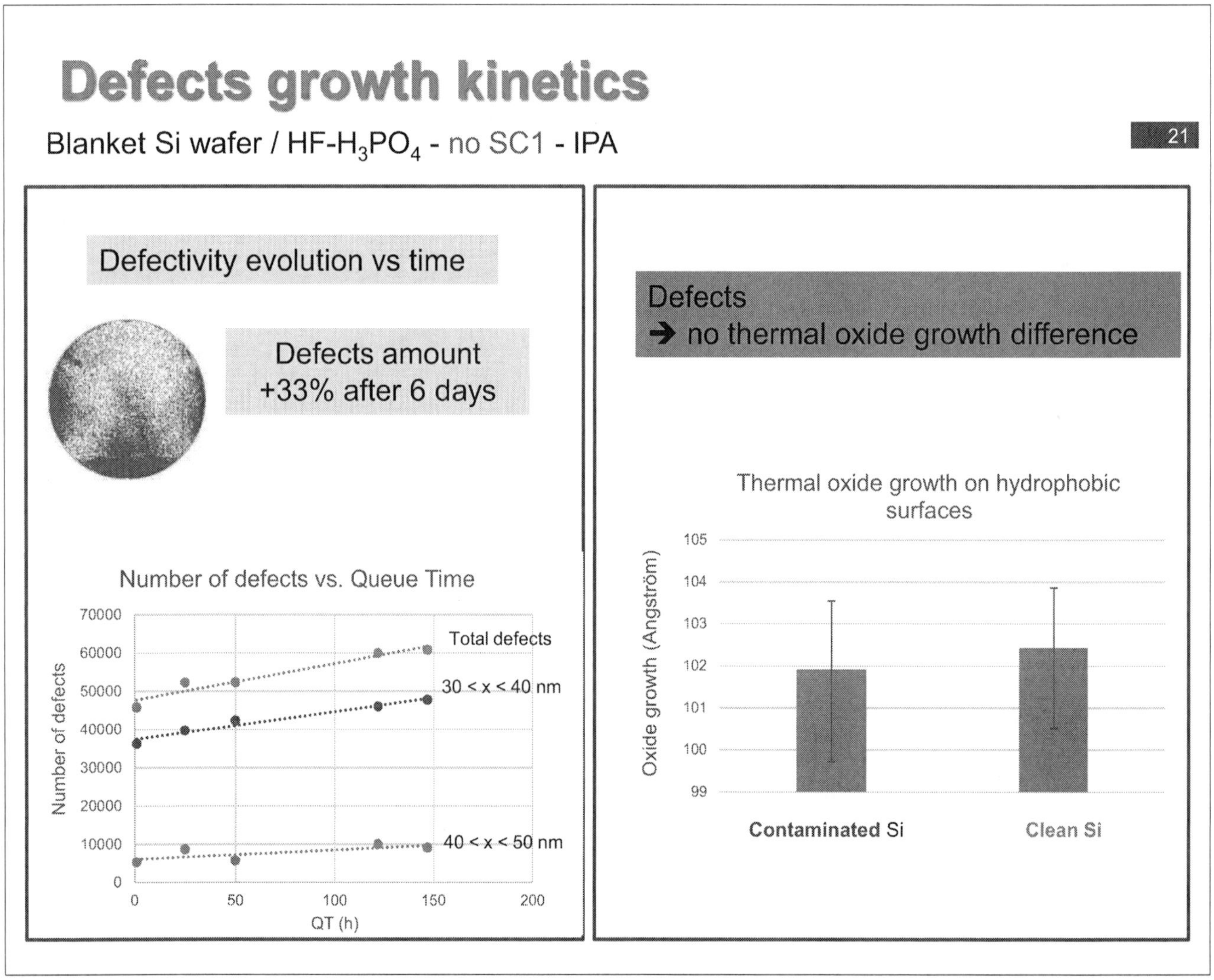
Defects growth kinetics
Blanket Si wafer / HF-H3PO4 - no SC1 - IPA
21
Defectivity evolution vs time
Defects amount
+33% after 6 days
Number of defects vs. Queue Time
Number of defects
QT (h)
Total defects
30 < x < 40 nm
40 < x < 50 nm
0
10000
20000
30000
40000
50000
60000
70000
0
50
100
150
200
Defects
➔ no thermal oxide growth difference
Thermal oxide growth on hydrophobic surfaces
Oxide growth (Angström)
99
100
101
102
103
104
105
Contaminated Si
Clean Si

# Hybrid DHF and $N_2$ Jet Spray Cleaning for Silicon Nitride and Metal Layer

Kook-Hyun An[a], Dong-Kyu Lee[c], Hun-Hee Lee[c]
Woo-Sik Choi[c], Tae-Hoon Jung[c] and Jin-Goo Park[a, b, †]

[A]Department of Bio-Nano Technology and
[b]Materials Science and Chemical Engineering, Hanyang University, Korea
[C]Samsung Electronics, Hwaseong, Korea

[†]jgpark@hanyang.ac.kr
+82031 4004080
Hanyang Univ., Ansan, 15588, KOREA





## OUTLINE

1. Research Background
   1) Research motivation : limitation of the current cleaning methods
   2) Research objectives : diluted HF jet hybrid cleaning
2. Experimental Materials/Procedure
3. Experiment Results and Discussion
   1) Particle removal efficiency of HF spray process
   2) Correlation of "Etch rate" and "Particle removal efficiency"
   3) Surface roughness of HF spray process
   4) Evaluation of patterned wafer particle removal
4. Summary

2019 Surface Preparation and Cleaning Conference
April 2-3, Portland, Oregon, USA

# Research Background

NEMPL Hanyang University



## Clean processing method conversion

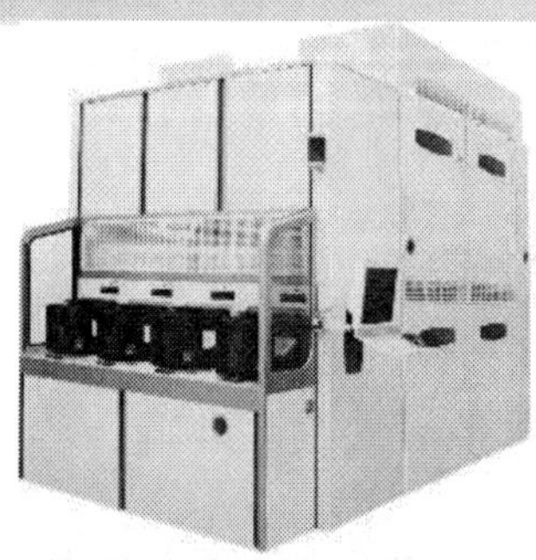

**Switch from batch to single equipment**

- Wafer size increased from 200 mm to 300 mm
- IC circuit line width is reducing : micron(μm) → nanometer(nm)

  → **Need to control small contaminants for yield improvement**

# Importance of fine particle control

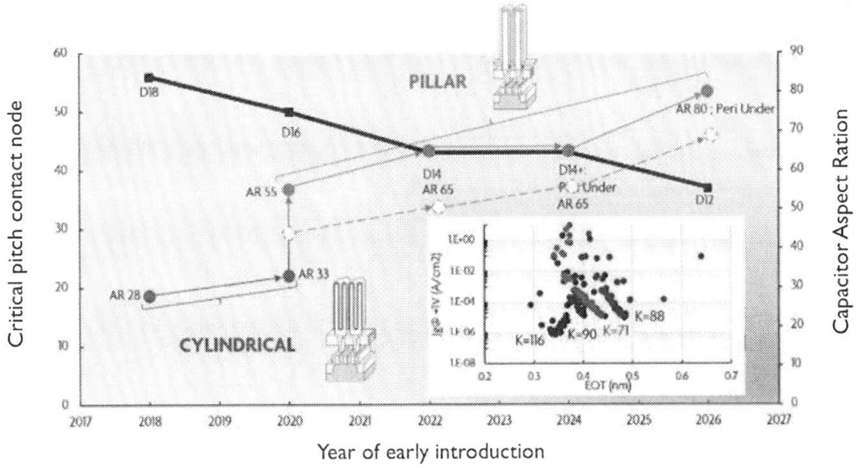


## DRAM device roadmap

- Scaling and structures are more complex
- Capacitor aspect ratio is increase

→ **Fine particle removal much more important**

Imec Magazine September 2018 Emerging memories for the zettabyte era



# Current jet spray issues

## Particle remove cleaning technology

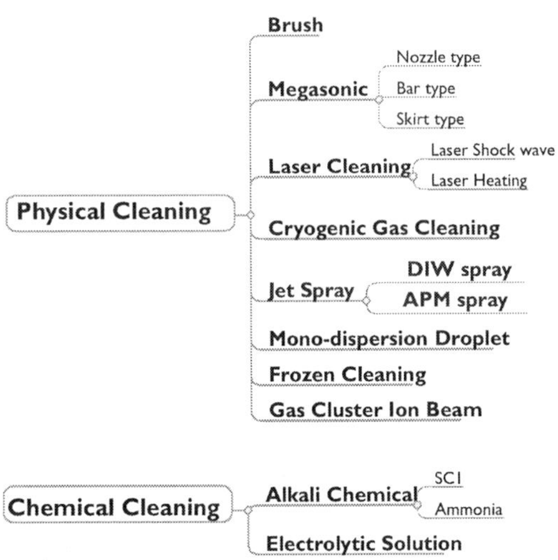


## Current issues on jet spray

- **DIW spray**

  Pattern damage occurs when physical force is increased

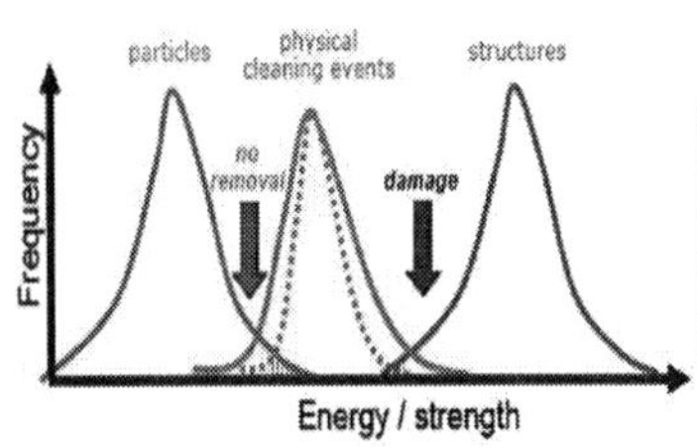


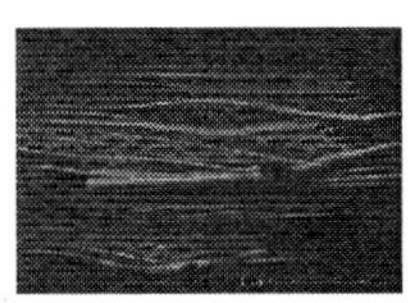

(Pattern damage)

- **APM spray**

  Corrosion defect occurs when used in metal layer

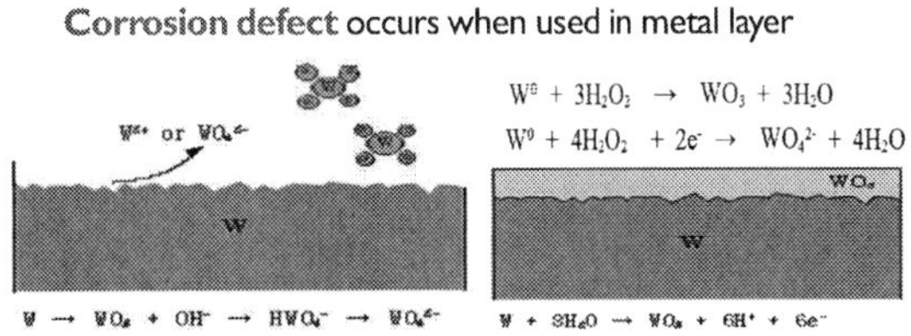


- **DIW spray and APM spray**

  Low PRE on silicon nitride film

P.W. Mertens st al, VLSI Taiwan 2006

# Research objective

## Efficiency of metal removal

- DHF : Metal contaminants removal no metal attack

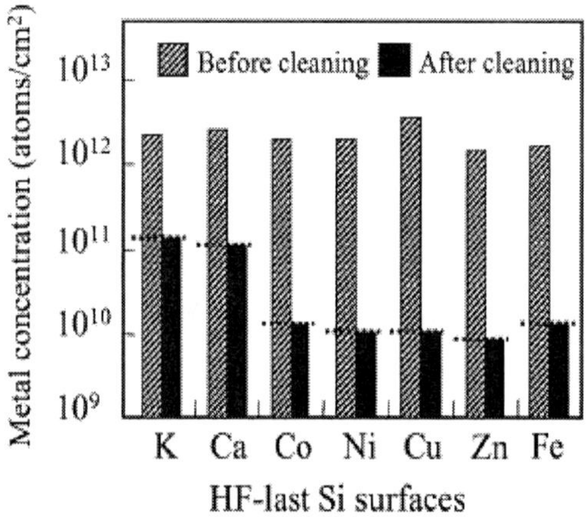


## Physical force vs. Particle removal efficiency

- HF spray : High PRE at silicon substrate

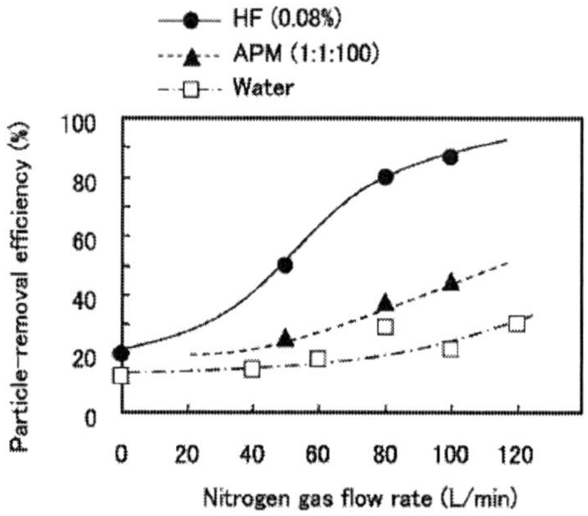


**Figure 5.** Efficiency of nitride-particle (>50 nm) removal on HF-last silicon surfaces using a mixed-liquid/nitrogen jet spray for 20 s using three different cleaning liquids as a function of the nitrogen flow rate.

**Evaluation and application of DHF Spray cleaning for SiN and metal wafers**

Hideki Hirano et al., " Damage-free ultra diluted HF/nitrogen jet spray cleaning for particle removal with minimal silicon and oxide loss."(2006).



2019 Surface Preparation and Cleaning Conference
April 2-3, Portland, Oregon, USA

# Experiment Materials And Procedure

NEMPL Hanyang University

# Experimental materials

## Contamination

- Contamination solution: silicon powder mixed with DIW
- Wafer contamination: poured a contaminant on wafer surfaces (2 min/wafer)
- Adsorption of contaminants : store in clean room 24 hours



# Experimental materials

## Materials

❖ **PRE evaluation**

- Si wafer (p-type (100))
- 90 nm SiN on Si

❖ **Etch amount**

- 90 nm thermal Oxide ($SiO_2$)
- 90 nm poly-Si
- 60 nm TiN
- 40 nm W

❖ **Field evaluation (Patterned)**

- 40 nm-scale DRAM (Poly, SiN, TiN and W)

## Analysis equipment

**Unpatterned wafer defect inspection**
(SP-5, KLA-Tencor)

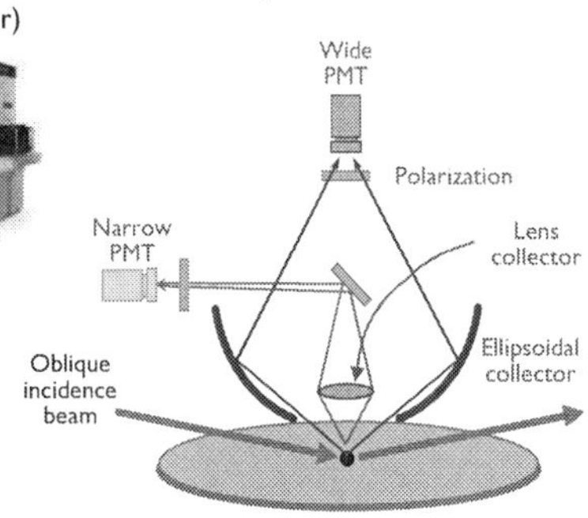


**Spectroscopy ellipsometer**
(Aleris, KLA-Tencor)

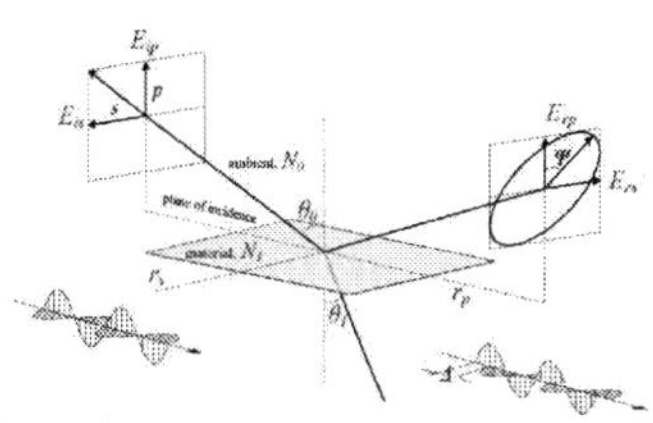

# Experimental setup

## Experimental parameters

| DIW | Diluted APM | Diluted HF |
|---|---|---|
| 1. **Temperature :** 25°C | 1. **Temperature :** 65°C<br>2. **Concentration :**<br>$NH_4OH$ : 0.3 ~ 1.7 wt%<br>$H_2O_2$ : 0.3 ~ 6.5 wt% | 1. **Temperature :** 25°C<br>2. **Concentration :**<br>DHF : 0.08 ~ 1.6 wt% |

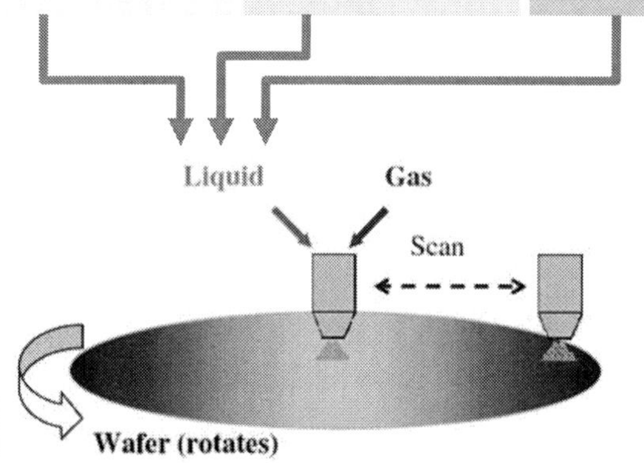


300mm singe wet cleaning equipment

## Controlled parameters

- **Nozzle**
  - Design: In-nozzle mixing
  - Wafer ~ nozzle distance: 30~50 mm
- **Gas**
  - Flow rate: 40~55 l/min
- **Scan**
  - Speed: 40~100 mm/sec
  - Frequency: 2~4 scans
  - Scan profile: 170 mm 0 mm
- **Wafer**
  - Rotation speed: 0~2000 RPM
- **Liquid**
  - Flow rate: 30~60 cc/min





# Experiment Results and Discussion

NEMPL Hanyang University

# Limit of physical cleaning (DIW-$N_2$ spray Only)

## Particle removal vs. pattern damage

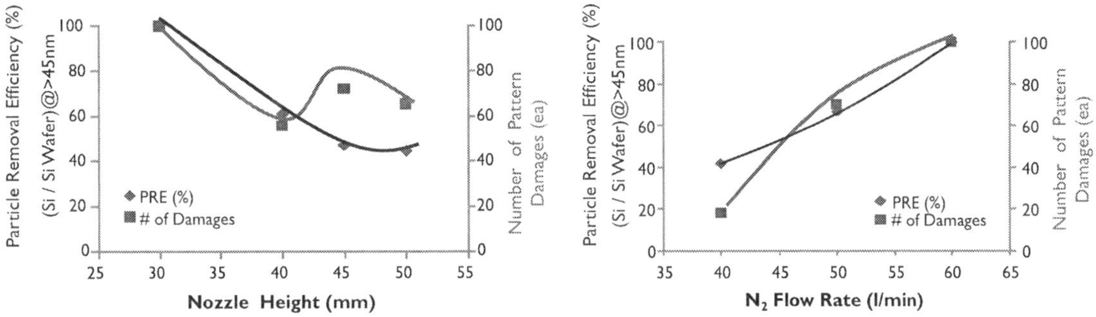


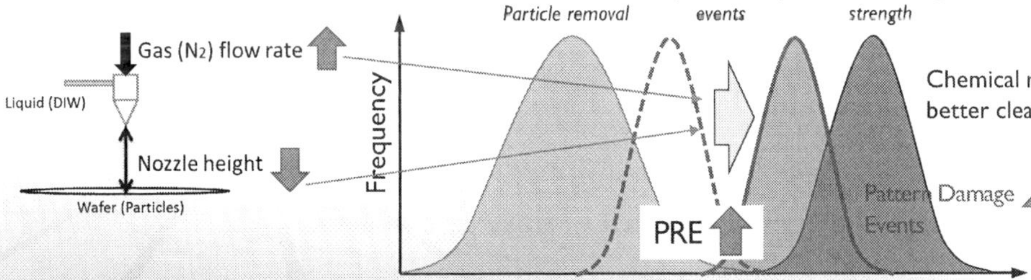


Chemical reaction needed to better cleaning performance



# PRE on SiN film in HF spray

## PRE on SiN compared to other cleaning conditions

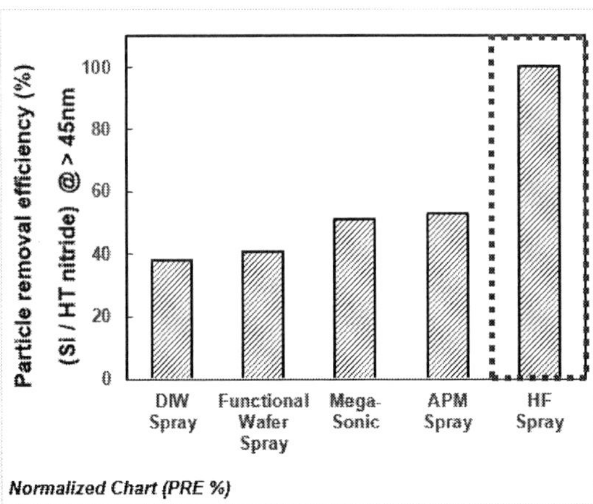


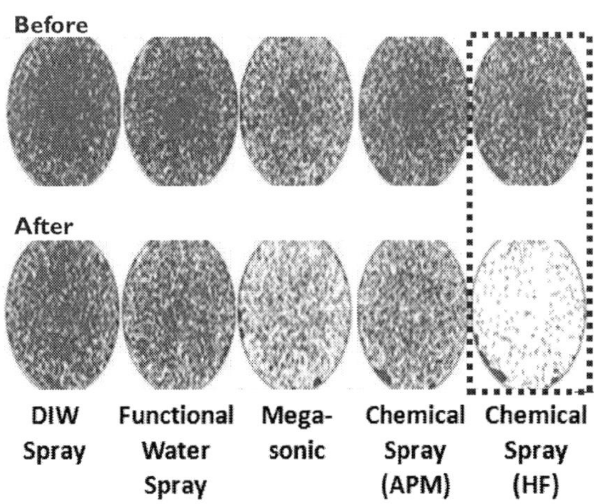


- DHF combined with $N_2$ spray showed the best PRE
- Chemical etching is required on SiN

HT Nitride 900Å Layer
Particle Source – SI Powder +DIW diluted Solution

# Dominant factor for PRE on SiN layer

**Comparison of PRE according to etch amount on SiN and oxide films**

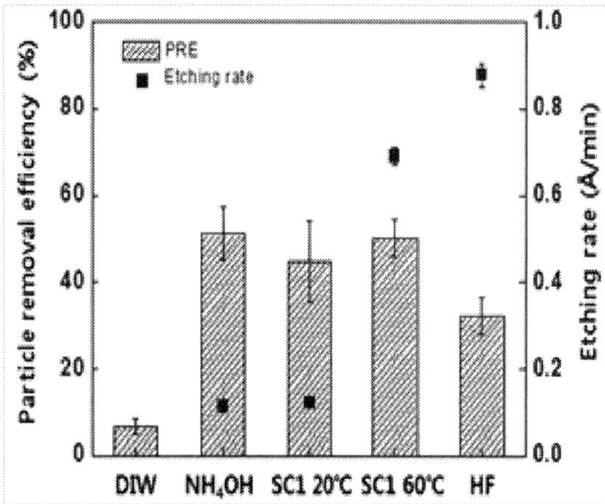


< Oxide layer >

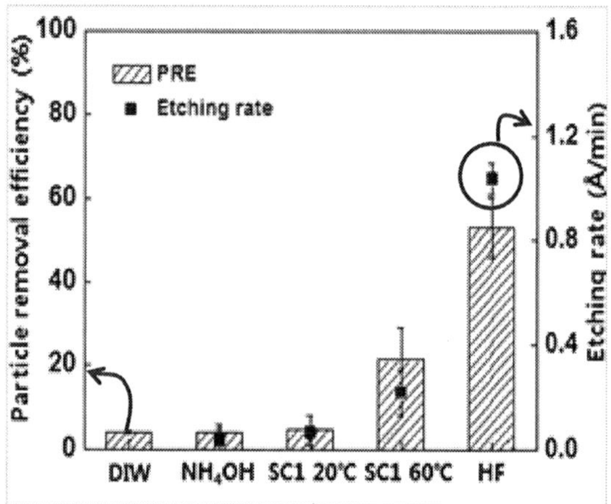


< SiN layer >

- Oxide film : No correlation between surface etch amount and PRE
- Silicon nitride film : PRE has very strong correlation with etch amount

  ➔ Support the estimated mechanism



# HF etch amount on metal layers

**Etch amount according to HF process time**

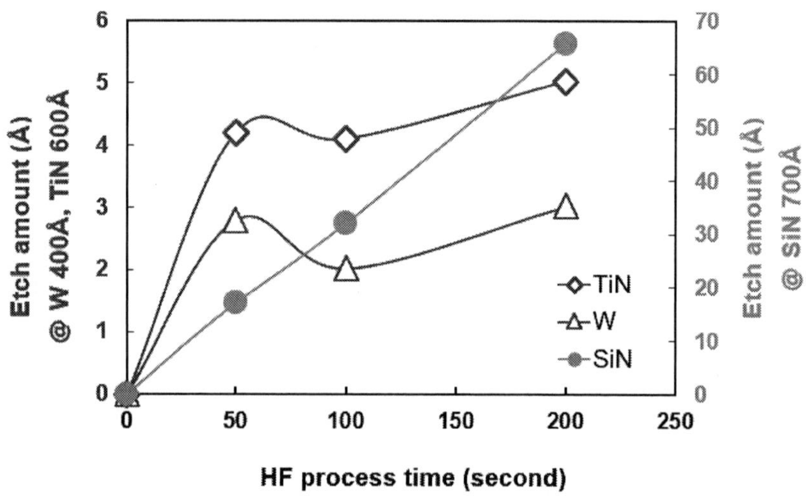


- TiN,W surface showed slight etch amount in HF

  ➔ HF spray was expected to have high PRE on TiN,W films

# PRE on W patterned wafer in HF spray

**Comparison of PRE with DIW spray & HF spray**

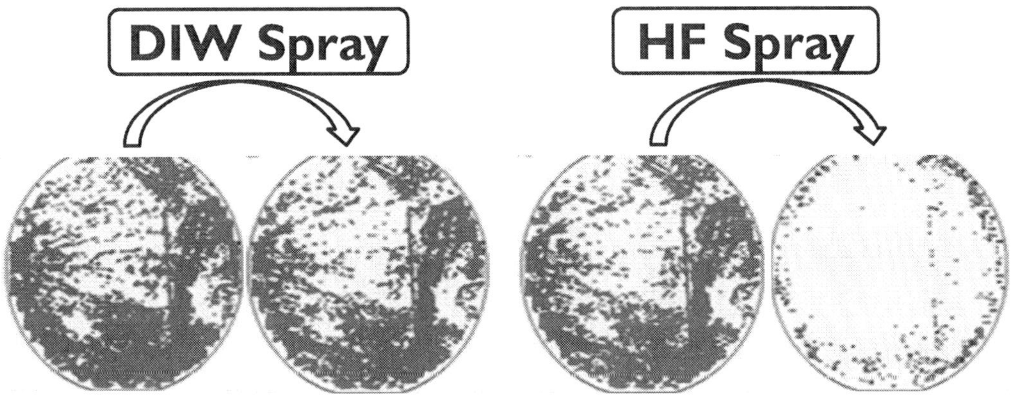


**PRE - 34.28%** **PRE - 90.23%**

DRAM 4x – W Layer / Particle : Si Powder

- HF spray showed far superior PRE on W patterned wafers
- Higher PRE was achieved by slight surface etching of the films



# Surface roughness changes at different chemical

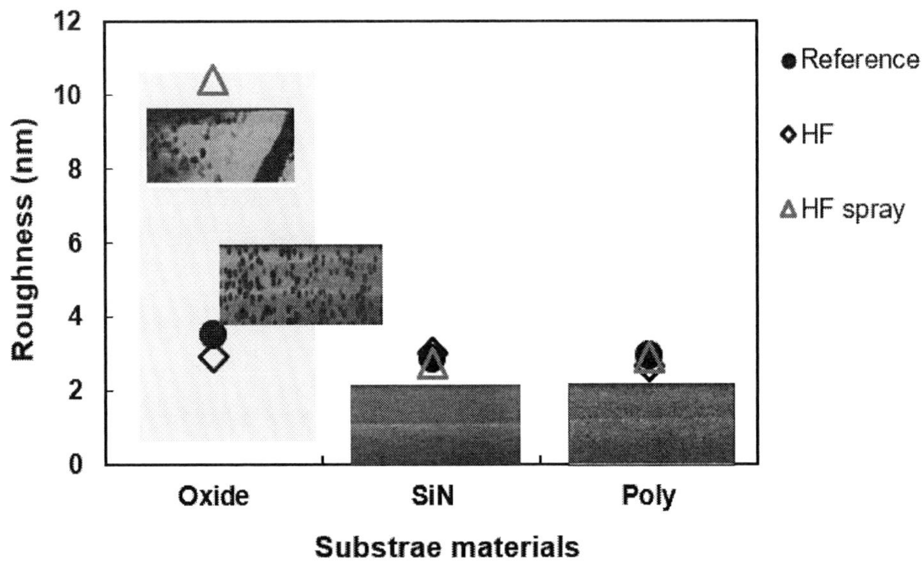


- SiN, Poly-Si layer : No difference between HF spray and reference
- Oxide layer : high roughness of 10.4nm after treated in HF spray
  → Need caution for roughness change when applying on oxide films

## Process application on patterned wafers

**PRE results (HF spray vs. conventional spray)**

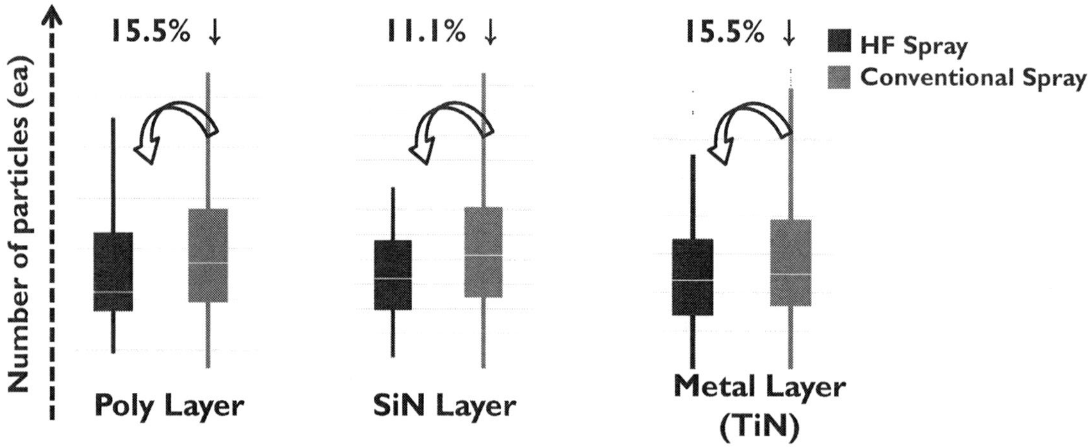


- DRAM 4x – Poly-Si, SiN, TiN layer particle trend (1month, over 3000 wafers)
- SiN layer : 11.1% improvement with HF spray
- Poly-Si, TiN layer : 15.5% improvement with HF spray



## Summary

**Need new type of cleaning method**

- Increased PRE on SiN/Metal layers
- No pattern damage on reduced design rule
- No attack on metallic films

**High particle removal performance requires**

- Chemical cleaning cooperated with physical force
- Slight etching of the surface on SiN/Metal films

**New hybrid cleaning of " Diluted HF/N2 jet spray" can be answer**

- High PRE on SiN and Metal layer without pattern attack

HANYANG UNIVERSITY
한양
1939
THANK YOU
NEMPL
NEMPL
Hanyang University
2019 SPCC, April 2-3, Oregon, USA

# Python based data analysis of ToF-SIMS

Sean Stuart
Sean.c.stuart@aero.org

April 3rd, 2019



## Introduction

- **Introduction**
- Getting the data
- Visualizing the Data
- Principle component analysis & Multivariate Curve Resolution
- Binning data for PCA + MCR
- Multi-Sample comparison
- Conclusion

2

## My background

*How and why I started using Python*

- I started using Python to make professional looking graphs
  - *To fit with publication standards (font size, figure dpi… etc.)*
  - *Excel wasn't cutting it, and other options were expensive….*

- After getting a feel for python, it was easy to expand from making graphs
  - *Modeling/simulation was relatively straightforward, just import a few packages.*

- I started working at The Aerospace Corporation in 2015, focusing on ToF-SIMS
  - *Working with **John Chaney** and **Robert Moision**, who had developed lots of python based techniques for ToF-SIMS data analysis.*
  - *Using their work I was able to expand on existing techniques, increasing the speed + resolution of the statistical analysis we performed.*

- I will show code snippets to demonstrate how easy and straightforward Python can be

3

## Toolbox for Python data analysis

*Why Python?*

- NumPy
  - *N-dimensional arrays*
  - *Linear algebra and I/O tools*
- Pandas
  - *"database" with tools for visualizing, searching, sorting data*
- HDF
  - *Hierarchical data format, developed for synchrotron data provides high performance searching /binning tools*
- Scikit Learn
  - *Machine learning tools (and "classic" statistical analysis methods)*
- Many more
  - *MCR, output to PowerPoint, read your email, make a user interface….*

***All this, and more for free!***

4

## *ToF-SIMS*

*Time of Flight Secondary Ion Mass Spectrometry*

- ToF-SIMS is an imaging mass spectrometry technique, with a long history in the semiconductor industry.
- Pros:
  - *Extreme surface sensitivity, sampling the top atomic / molecular layer*
  - *Generally non destructive*
  - *High sensitivity, generally ppm sometimes ppb*
  - *Capable of detecting and identifying elements and chemicals from hydrogen to large polymers simultaneously.*
- Cons
  - *Generates large complex datasets*
  - *UHV technique, generally expensive*

***Large complex datasets created in ToF-SIMS require statistical tools, and lots of visualization methods to understand and communicate the data.***

5

## ToF-SIMS
## Time-of-flight secondary ion mass spectrometry

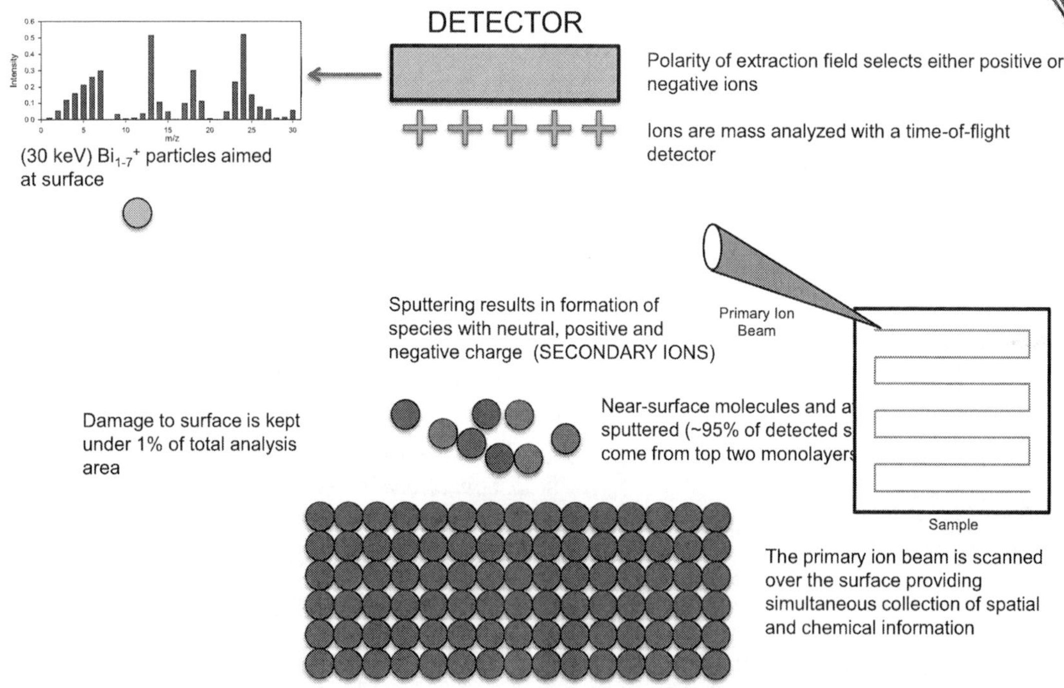


6

# Time-of-flight (ToF) Mass Spectrometry

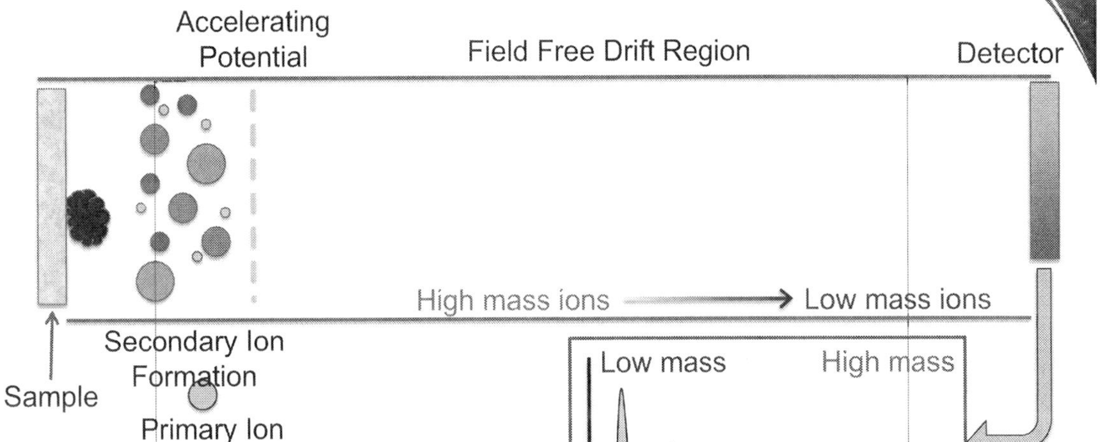


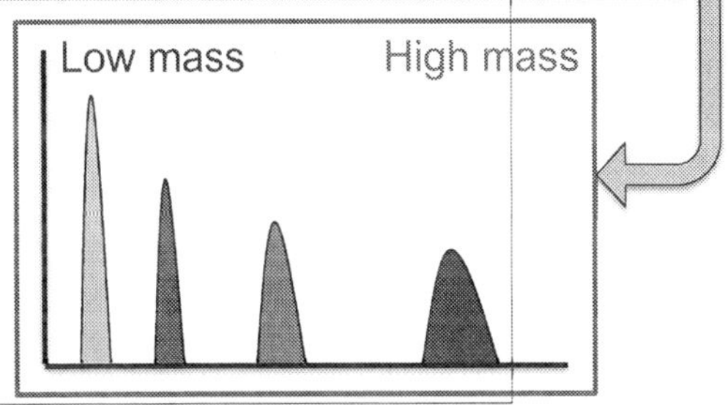


- Requires a pulsed source
- High mass resolution
- All secondary ions are detected

$$time \propto \sqrt{\frac{mass}{charge}}$$

7

## *Surface Sensitivity*

*Thin layers of contamination will dominate the signal*

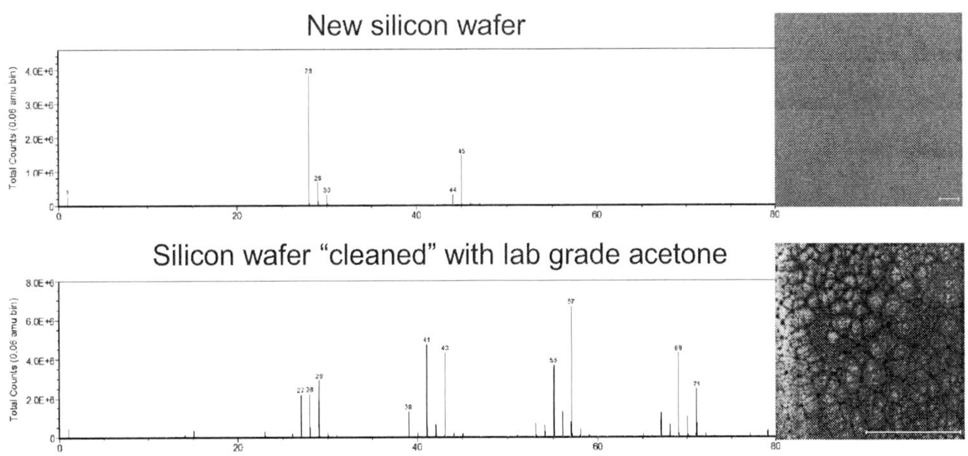


***Extreme surface sensitivity makes detecting monolayer contaminants easy. (in this case acetone residue is likely much more than a monolayer).***

8

# ToF-SIMS Depth Profiling

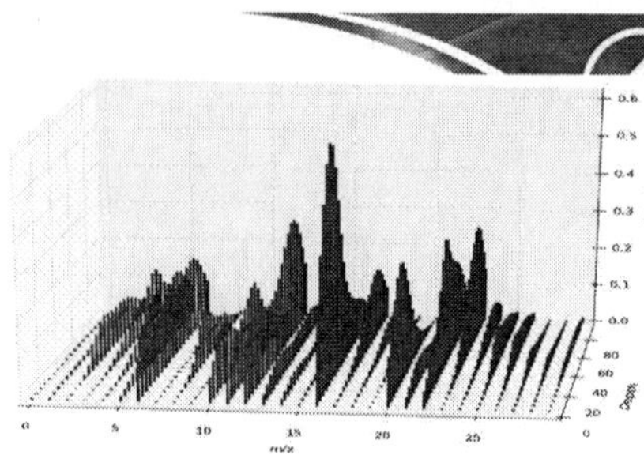

- During depth profiling, analysis ($Bi^+$) and sputtering ($Cs^+$ ,Ar, O) beams are alternated
- This provides chemical information as a function of depth
- The resulting data looks like a series of slices, which each slice containing a spectral signature of that layer

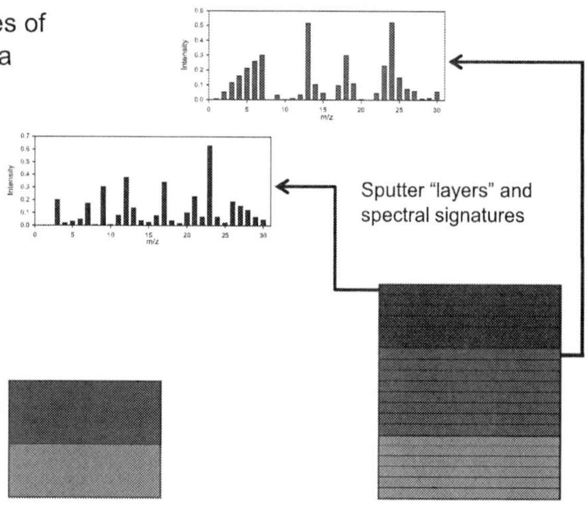


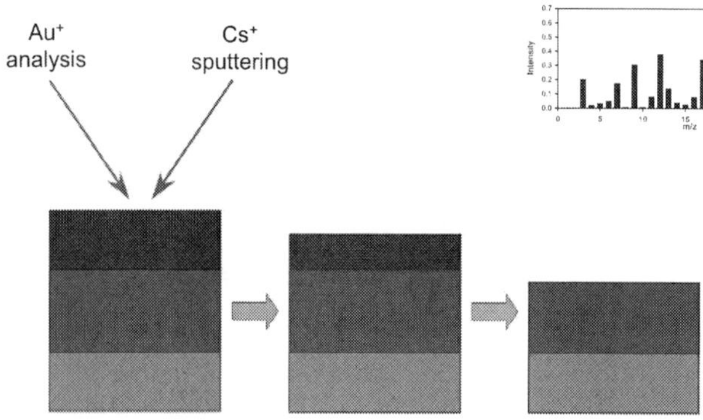


9

# *Getting the Data*

- Introduction
- **Getting the data**
- Visualizing the Data
- Principle component analysis & Multivariate Curve Resolution
- Binning data for PCA + MCR
- Multi-Sample comparison
- Conclusion

10

## Getting the data

*Generally the hardest part of data analysis..*

- Easy:
  - *Text files, CSV*
- Moderate:
  - *XML, HDF, non proprietary data formats*
- Difficult:
  - *Proprietary binary files*

```
directory = r"C:\Data\folder"
os.chdir(directory)

filename = r'E0g 003 4 comp pos.csv'
print(filename)

datafile=open(filename,'r')
numpyimport = numpy.genfromtxt(datafile,skip_header=9,delimiter = ',')
```

```
datalist = []
for x in range(21):
    measurment = All_data[1+x*202:202*(x+1)]
    print(measurment[0])
    datalist.append(np.genfromtxt(measurment))
```

```
n_pts = 0
#Now start the normal python routine
phased_profile = False
phased_sputter_time = 0; phase_num = 0
with open( in_name, "rb" ) as f:
    header_length = int(header.get("HeaderSize"))
    f.seek( header_length, 0 )
    if(header.get("SputterInterval", False)!= False):
        phased_profile = True
        phased_sputter_time = 0
    else:
        phased_profile = False
    token = 1; pixel = 0; channel = 0; n_pts = 0
    while token != 0:
        prev_token = token
        buffer = f.read(2)
        buf = buffer
        s_ptr = <unsigned short*>&buf[0]
        token = deref(s_ptr)
        if( token == 0 ):
            break;
        #For mosaic data. Include extra points for stuff
        if( token == 6 ):
            for i in range( 1, 8 ):
                f.read(2)
        if( token == 8 ):
            f.read(2)
        #Read 4 bytes for the number of data bytes
        f.read(4)
        buffer = f.read(4); buf = buffer
        l_ptr = <unsigned long*>&buf[0]
        num_data_bytes = deref(l_ptr)
        #timer_ticks, num_data_bytes = struct.unpack( "=II", f.read(8))
        if( token == 4 and prev_token == 3 and num_data_bytes%6 == 0):
            phase_num += 1
            f.read(num_data_bytes)
        elif( token == 4 and (num_data_bytes%6)!=0):
            cur_loc = f.tell()
            char = ''; num=''; str=''; len=0; read_num=False
            while char != '\x01':
                cur_loc = f.tell()
                char = struct.unpack( "1c", f.read(1))[0]
                str = str + char
                if( read_num == True ): num = num + char
                if( char == ':' ): read_num = True
            time = float( num[:-1])
            phased_sputter_time += time
            f.seek( cur_loc )
        else:
            buffer = f.read(num_data_bytes)
            buf = buffer
            #n_bytes = int(num_data_bytes/6)
```

***Translating data from one format to another can be challenging... but usually worth it***

11

# Visualizing the Data

- Introduction
- Getting the data
- **Visualizing the Data**
- Principle component analysis & Multivariate Curve Resolution
- Binning data for PCA + MCR
- Multi-Sample comparison
- Conclusion

12

## *Visualizing data*

*Mass, location, depth*

- ToF Sims records a mass spectra at every pixel in the data.
- We can create an Image of each peak in mass spectra
  - *Including as a function of depth*

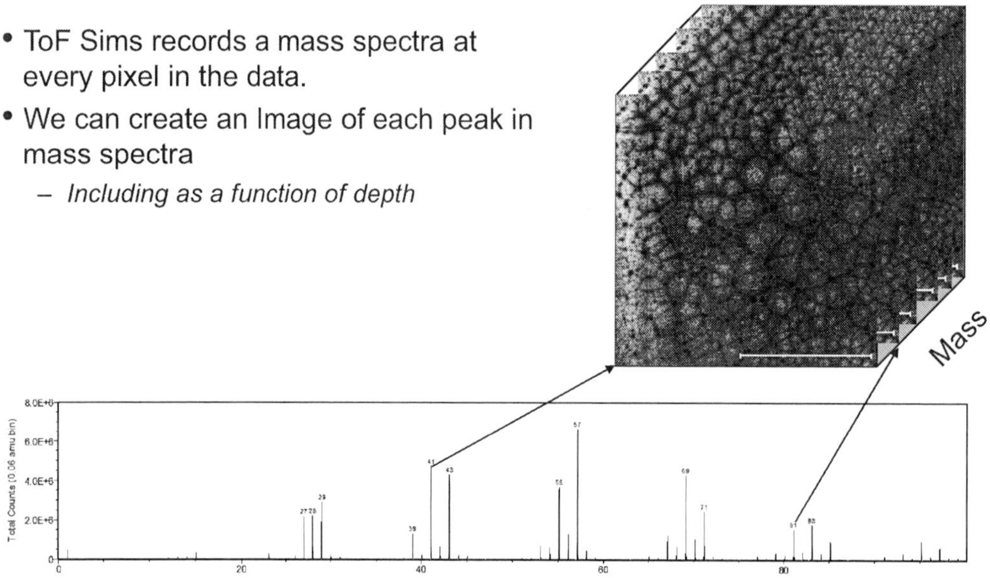


13

## *Data selection with HDF5*

*Opened using PyTables*

In the HDF data file, we can store information about each ion detected in the ToF-SIMS measurement:

```
class data_dt(tb.IsDescription):
    px = tb.UInt32Col()
    ch = tb.UInt32Col()
    time = tb.UInt32Col()
    px_x = tb.UInt16Col()
    px_y = tb.UInt16Col()
    m_z = tb.Float32Col()

dt = np.dtype([('px', np.uint32), ('ch', np.uint32), ('time', np.uint32),
              ('px_x', np.uint16), ('px_y', np.uint16), ('m_z', np.float32) ])
```

Making an image of a specific mass involves a search, basically asking: Give me all of the ions with mass between <lowmass> and <highmass>. Then count up how many are in each pixel, oh and arrange the pixels into an image:

```
def makeMapofMassRange(raw_dat,lowmass,highmass,image_size = (256,256),scale = 1):
    search_str = '((m_z>={})&(m_z<={}))'.format(lowmass,highmass)
    px_map = [x['px'] for x in raw_dat.where(search_str)]
    px_map = np.bincount(px_map,minlength=image_size[0]*image_size[1])

    if scale != 1:
        image = np.reshape(px_map,(image_size[0],image_size[1]),order='F')
        image = resizebox(image,scale)
        px_map = np.reshape(image,-1)
    return px_map
```

***HDF provides efficient, parallel data searches.***

14

## Depth Profile Video

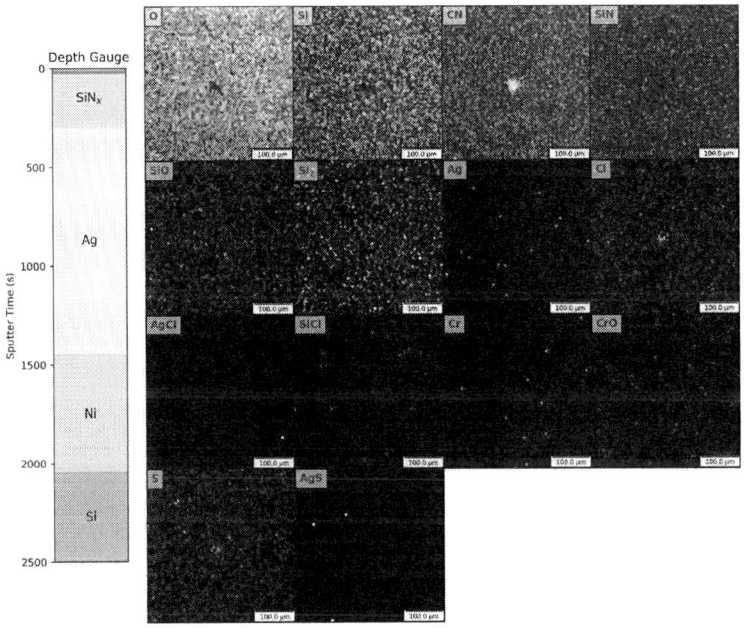


***By automating data visualization (with Python) you can generate enough images for a video***

15

## Why use Python for data visualization?

Control over every detail.

- The software provided with the instrument might have most of what you need but…
  - *Making changes to fonts, aspect ratios, graph styles is difficult or impossible (sometimes publications have strict requirements on figures)*
  - *Making lots of identical graphs can be tedious.*
  - *Extracting 100's of images (for a video) is not worth the time.*

- Once you have access to the data in Python, making a new visualization is about as time consuming as figuring out how to use a new feature of a program… and you are not limited by what is already implemented.
- If you have the code to make a graph, I find it much faster to use python.

16

# PCA and MCR

- Introduction
- Getting the data
- Visualizing the Data
- **Principle component analysis & Multivariate Curve Resolution**
- Binning data for PCA + MCR
- Multi-Sample comparison
- Conclusion

17

## Deconvolution of two similar signatures using PCA

*Using positive and negative correlations to separate components*

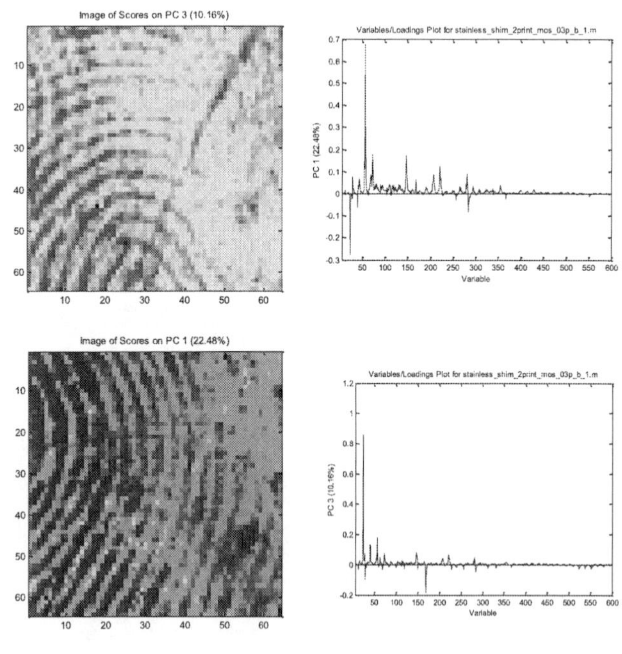


- Two overlapping thumb prints (from two people) are distinguished chemically.
- Positive and negative correlations can be used to distinguish regions
- Both prints overlap but are clearly resolvable with different spectral signatures
- Without PCA analysis these chemical differences would be extremely difficult to detect.

18

# ToF SIMS analysis background

*Data processing with multivariate analysis (MVA)*

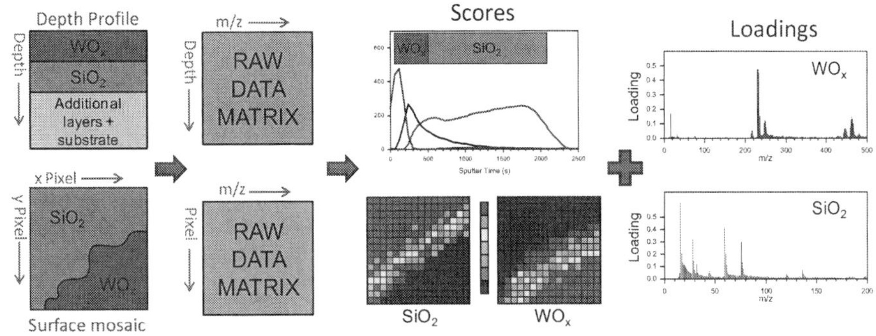


- A given data set contains mass, pixel position and time
  - *Easily exceeding 9,000,000 data points*
  - *Single files may be 4 GB*
- Multivariate curve resolution (MCR) and principle component analysis (PCA) are well suited to ToF-SIMS.
  - *Methods describe data using scores and loadings*
  - *Scores resemble ion images or depth profiles*
  - *Loadings resemble mass spectra*

19

## ***Principle Component Analysis***

*Dimensionality reduction*

- PCA is as an orthogonal linear transformation, which transforms data into vectors that model the variance in the data.

- The PCA components look like mass spectra but with negative values, the negative values are peaks anticorrelated with the component.

- The scores of each component represent the spatial correlation of each component, it is an image of where a particular correlation spectra is located on the sample.

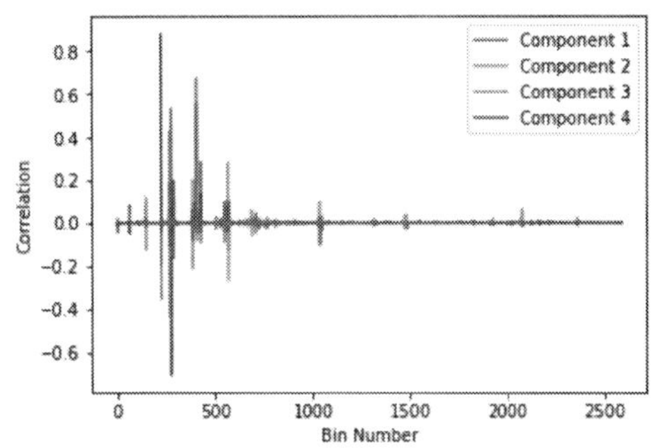


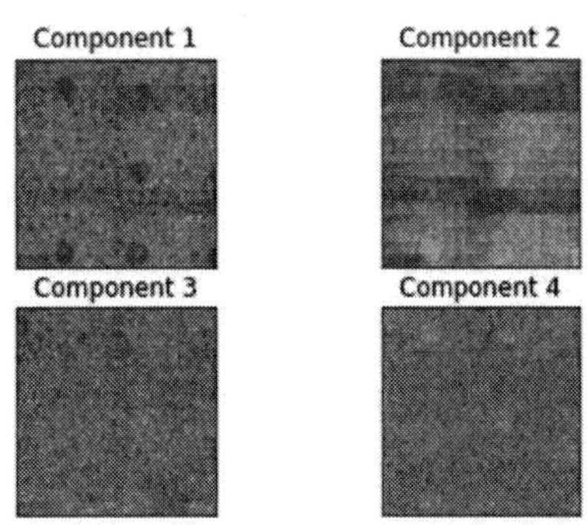


***PCA is a useful statistical technique for ToF-SIMS analysis, but produces somewhat non-intuitive and non physical results.***

20

## *Principle Component Analysis with Python*

*Dimensionality reduction*

- Data = principle components X score
- The data is provided in a matrix with each row containing a mass spectrum of a particular pixel.
- When the results are visualized, the pixels are just "wrapped" back up into a 2-D array

```
In [3]:  pca = PCA(n_components=13)

In [4]:  fit = pca.fit(data)
         pca_score = pca.explained_variance_ratio_
         print('data fitted')

         data fitted
```

21

## *PCA*

*Dimensionality reduction*

- We also get an explained variance ratio, the amount of data contained in each principle component
- The blue line plots this value as a function of principle component number (orange line is the derivative)
- We can see that there are diminishing returns past 8 components
- Sometimes a interesting chemical feature is found which is not a statistically significant part of the data but shows a new contamination source. (ie a speckle of sodium/potassium from breathing on a sample)

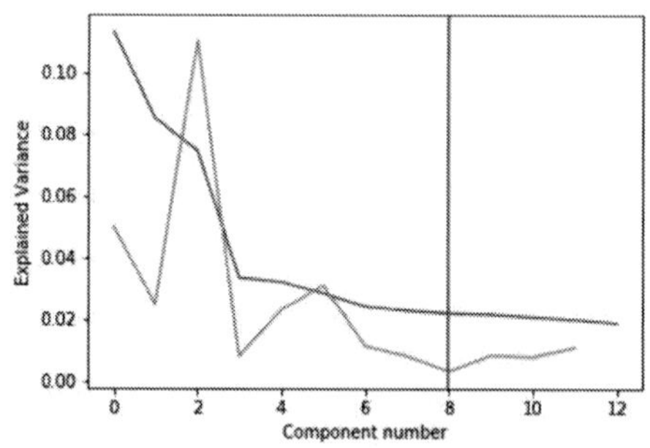


***Explained Variance ratio decreases with each component, this can suggest how many components are necessary to model the dataset accurately.***

22

## *Multivariate Curve Resolution with pyMCR*

*PCA with constraints*

- Using MCR is not significantly more difficult to implement than PCA but it often gives better results.
- First it does not produce correlations and anticorrelations, just a spectra of individual components
- It has been implemented for Python: https://pypi.org/project/pyMCR/ with useful constraints (normalization, non negativity ect)
- Solves:
  - *Data = Concentration * Spectra$^T$*
- The method requires an "initial guess" of the spectra or Concentration. This can be random, or the results of another algorithm (ie PCA)

```
mcrals = McrAls(max_iter=5, tol_err_change = 5e-4 ,st_regr='NNLS',
                c_regr=NNLS(), c_constraints=[ConstraintNonneg(),
                                              ConstraintNorm()])
```

```
insp = pca.components_[:components]
insp[insp < 0] = 0
insp[insp >1] = 1
```

```
mcrals.fit(data, ST= insp, verbose = True)
```

```
Iter: 1 (C)     mse: 1.4795e-01
Iter: 1 (ST)    mse: 3.4401e-02
Iter: 2 (C)     mse: 3.2413e-02
Iter: 2 (ST)    mse: 3.1585e-02
Iter: 3 (C)     mse: 3.1207e-02
Iter: 3 (ST)    mse: 3.0981e-02
Iter: 4 (C)     mse: 3.0850e-02
Iter: 4 (ST)    mse: 3.0758e-02
Change in err below tol_err_change (2.2305e-04). Exiting.
```

***MCR is a good choice for ToF-SIMS data.***

23

## *MCR results*

*8 component fit*

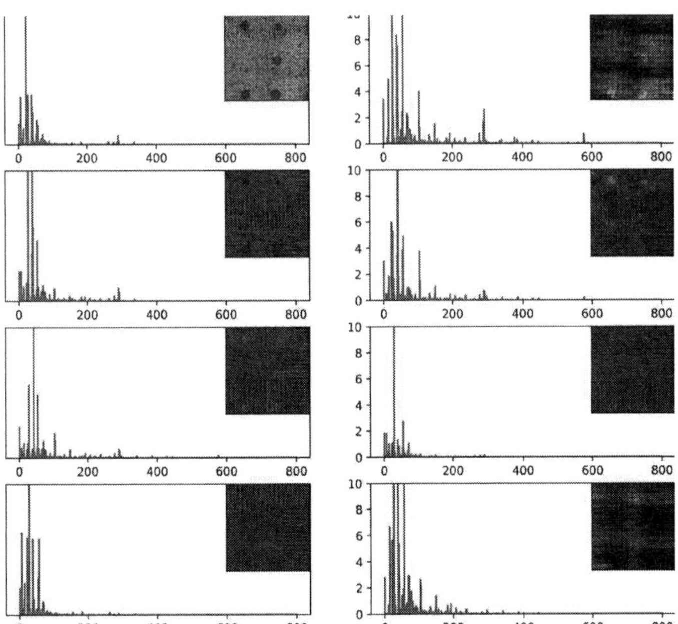


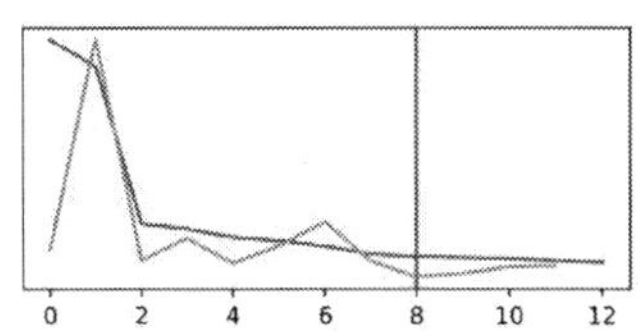


***Overfitting can make multiple components without significant (or meaningful) differences...***

24

## MCR results

6 component fit

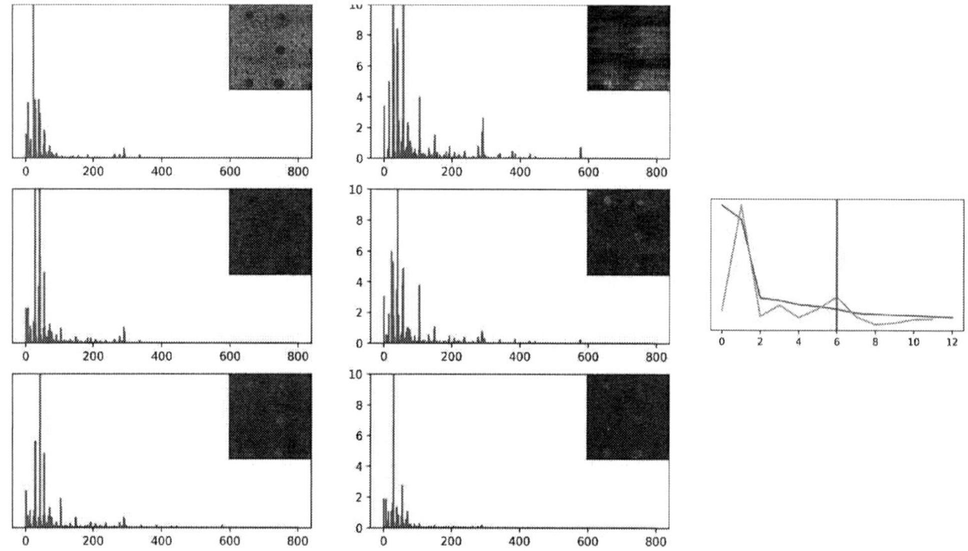


***6 components show most of the features of interest, perhaps this is enough?***

25

## MCR results

4 component fit

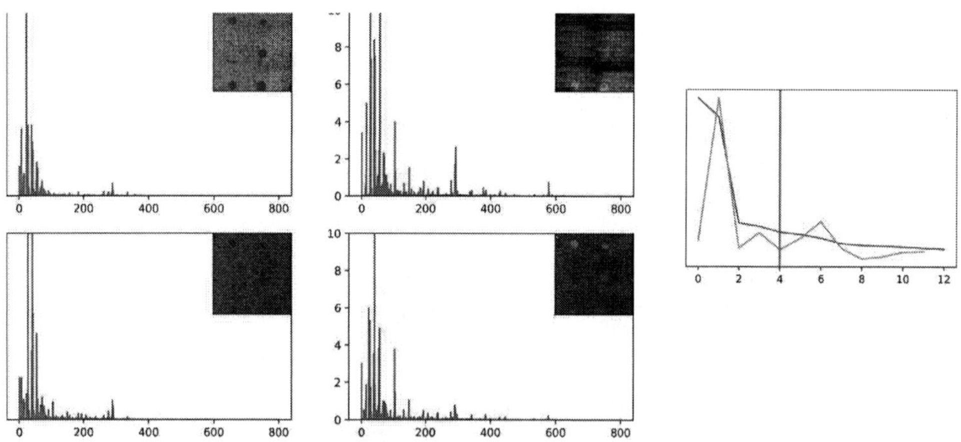


***Underfitting may miss details in the data that are important (ie small features buried in the dataset with unique chemistry)***

26

## *MCR results*

*11 component fit*

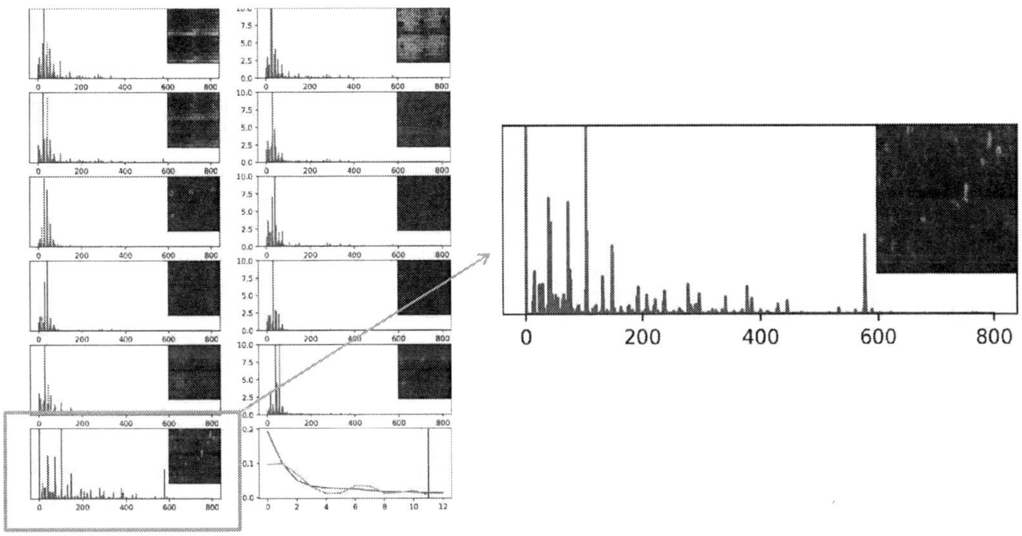


***Underfitting may miss details in the data that are important (ie small features buried in the dataset with unique chemistry)***

27

# *Binning data*

- Introduction
- Getting the data
- Visualizing the Data
- Principle component analysis & Multivariate Curve Resolution
- **Binning data for PCA + MCR**
- Multi-Sample comparison
- Conclusion

28

## *Computational constraints*

- The size of the data matrix increases the computational time dramatically.
- Since the size mass x pixels, we can use less pixels or less masses
- Then it is a classic database search problem, find all particles from location (x,y) with mass within the desired range

```
def makeMapofMassRange(raw_dat,lowmass,highmass,image_size = (256,256),scale = 1):
    search_str = '((m_z>={})&(m_z<={}))'.format(lowmass,highmass)
    px_map = [x['px'] for x in raw_dat.where(search_str)]
    px_map = np.bincount(px_map,minlength=image_size[0]*image_size[1])

    if scale != 1:
        image = np.reshape(px_map,(image_size[0],image_size[1]),order='F')
        image = resizebox(image,scale)
        px_map = np.reshape(image,-1)
    return px_map
```

***Function to get a map of a specific range***

29

## *Binning mass spectra*

*How many bins are enough?*

- Binning ToF-SIMS data is necessary for PCA and MCR analysis.
- The original data has a resolution of approximately 1/1000th of an atomic mass unit (amu, about the mass of a proton). Keeping that resolution would require 2 million bins.
- By binning the mass spectra to 1/10 amu, most peaks can be segregated (ie Mg and $C_2$)
- Binning further speeds up the analysis, but will combine different peaks into one value.
- We can also manually define peaks, but this is time consuming and may neglect peaks that might be hiding in the noise.

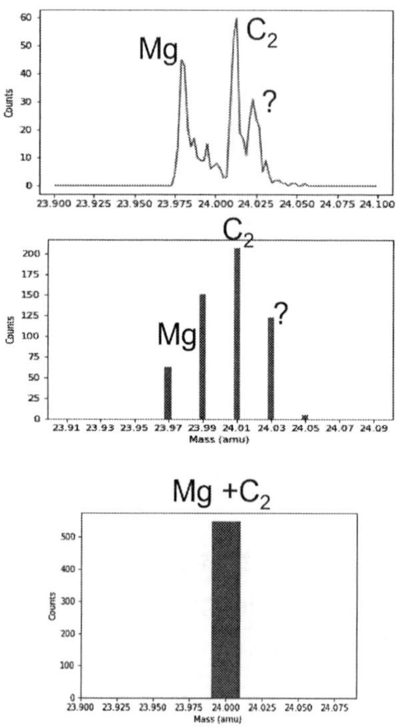


***Classic problem of processing time vs resolution.***

30

## *Reduction spatial resolution*

- For statistical analysis, the data is a matrix of size # mass spectra bins x # of pixels.
- Normally we collect 256x256 pixel images.... But for larger areas we can easily collect 2560x2560 pixel images by stitching multiple images together into a mosaic.
- The size of the data matrix can greatly increase the processing time for MCR, for example a 1000x1000 matrix takes <10 minutes
- Decreasing resolution is relatively simple, just add the mass spectra of neighboring pixels to define a new, combined pixel.

32x32 pixels

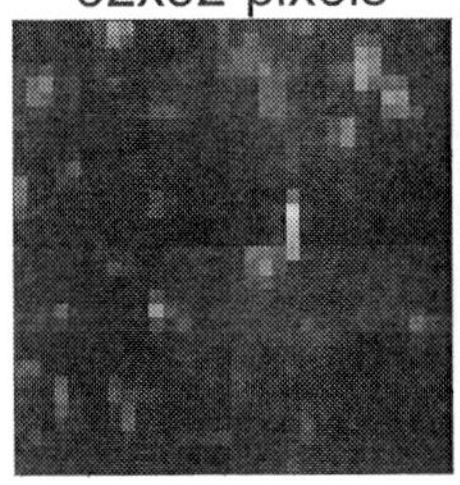

512x512 pixels

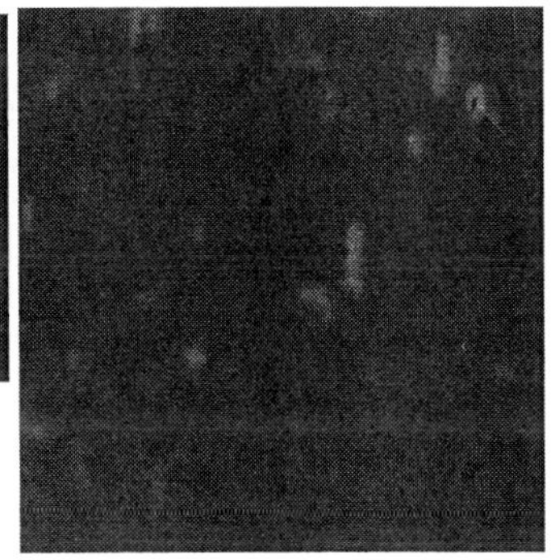

```
def resizebox(array,scale):
    scale = int(scale)
    shape = array.shape
    x_pixels = int(shape[0]/scale)
    y_pixels = int(shape[1]/scale)
    reducedimage = np.zeros((x_pixels, y_pixels))
    for newpx_x, x in enumerate(range(0,int(shape[0]),scale)):
        for newpx_y, y in enumerate(range(0,int(shape[1]),scale)):
            reducedimage[newpx_x,newpx_y]=np.sum(array[x:(x+scale),y:(y+scale)])
    return reducedimage
```

31

# *Multi-Sample Comparison*

- Introduction
- Getting the data
- Visualizing the Data
- Principle component analysis & Multivariate Curve Resolution
- Binning data for PCA + MCR
- **Multi-Sample comparison**
- Conclusion

32

## *Multi-sample comparison*

- Comparing multiple samples with MCR (or PCA) can identify variations in a process
- Comparable spectra from multiple samples can be selected (such as at an interface as shown)

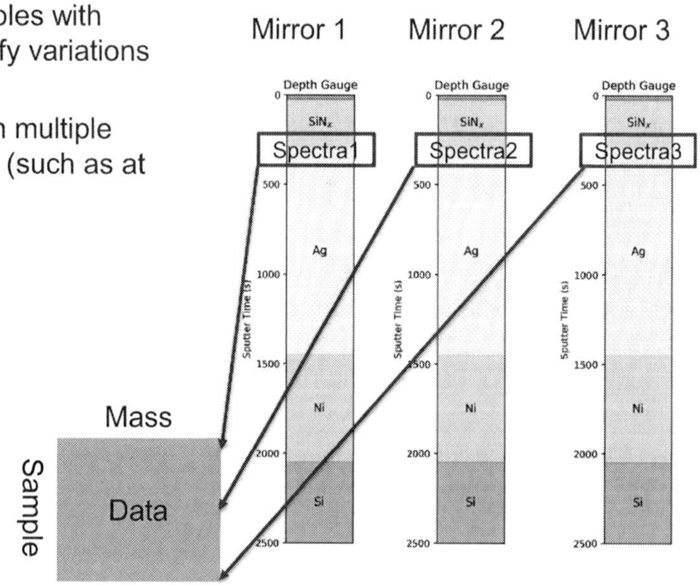


***You can also include data from other complementary measurements to correlate chemical data to electrical, optical… etc..***

33

# *Conclusion*

- Introduction
- Getting the data
- Visualizing the Data
- Principle component analysis & Multivariate Curve Resolution
- Binning data for PCA + MCR
- Multi-Sample comparison
- **Conclusion**

34

## *Python for ToF-SIMS data analysis*

*Access to state of the art methods*

- Cons of using Python for data analysis
  - *Learning curve can be significant*
  - *No graphical user interface (unless you make one yourself)*
  - *Proprietary data formats can be challenging to import*
- Pros:
  - *State of the art tools and methods are free, with huge community improving and expanding what's available.*
  - *You can automate your workflow, eliminating tedious tasks.*
  - *Can run on your laptop, server, supercomputer*
  - *Your options for data analysis is limited only by time and imagination.*



35

BALAZS™ nanoanalysis

# Optimization of Single Nanoparticle ICP-MS Analysis for Controlling Nanoparticles in Process Chemicals and UPW used in Wet Surface Cleaning

**Lisa M. Mey-Ami***, Jinjin Wang, and Fuhe Li

Air Liquide – Balazs NanoAnalysis
46409 Landing parkway, Fremont, CA 94538, USA

*Lisa.Mey-Ami@airliquide.com
(510) 624-4019

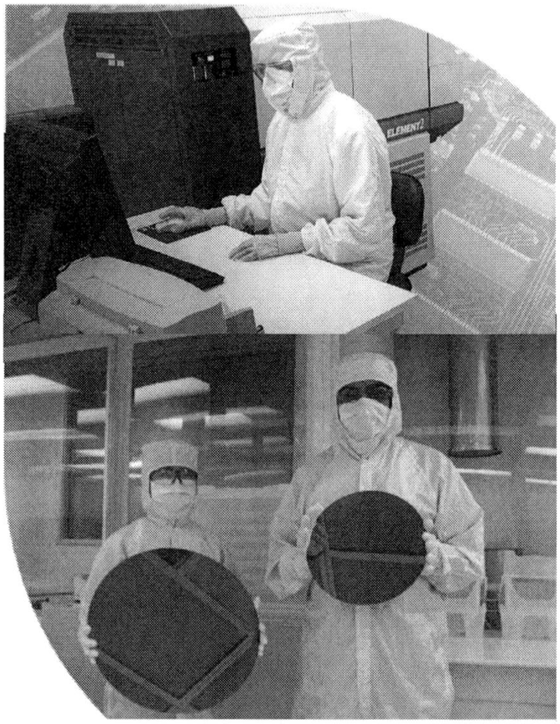



## Introduction

- Nanoparticles are defined as particles with at least one dimension in the nano-scale (1-100 nm) range and they are either natural occurring or engineered.
- Natural occurring particles are often present as contaminants in all aspects of the chip fabrication process, including etching, lithography, chemical mechanical planarization (CMP), deposition, and implantation.
- Nanoparticle contaminants will have adverse impact on <10 nm device fabrication process and detrimental effect on production yield.
- Improvements in our SP-ICPMS methods since last year will be shown for UPW and process chemicals used in wet surface cleaning (e.g. sulfuric acid).
- Particle counts for aqueous surface extracts of various OEM parts (e.g. quartz ring, ceramic ring and Al coupon) will be presented

# Why Nanoparticle Analysis is needed

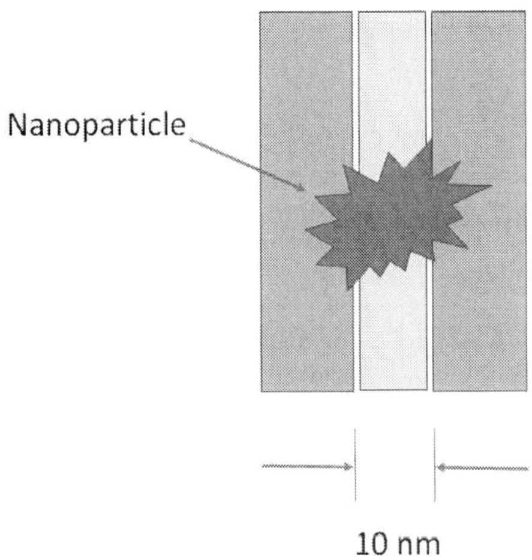


10-100 nm particles are now critical. Monitoring & controlling single nm and near 1 nm particles will soon be required.

**Semiconductor manufacturing processes**

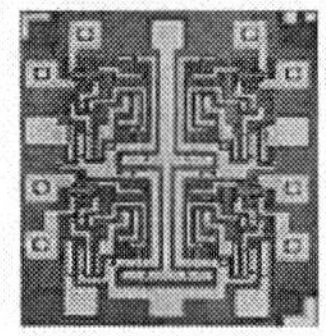

10 µm – 1971
6 µm – 1974
3 µm – 1977
1.5 µm – 1982
1 µm – 1985
800 nm – 1989
600 nm – 1994
350 nm – 1995
250 nm – 1997
180 nm – 1999
130 nm – 2001
90 nm – 2004
65 nm – 2006
45 nm – 2008
32 nm – 2010
22 nm – 2012
14 nm – 2014
10 nm – 2017
**7 nm** – *~2019*
5 nm – *~2021*



# Nanoparticle Analysis Fundamentals

ICP-MS has been used regularly for the determination of trace metals in UPW, process chemicals and extraction solutions from OEM tool parts. The detection limits are often as low as sub parts per trillion (ppt). Since trace metals are completely dissolved and aspirated into the ICP plasma continuously, producing a constant signal intensity, only total elemental concentration is determined.

Continuous signal from dissolved ionic analyte

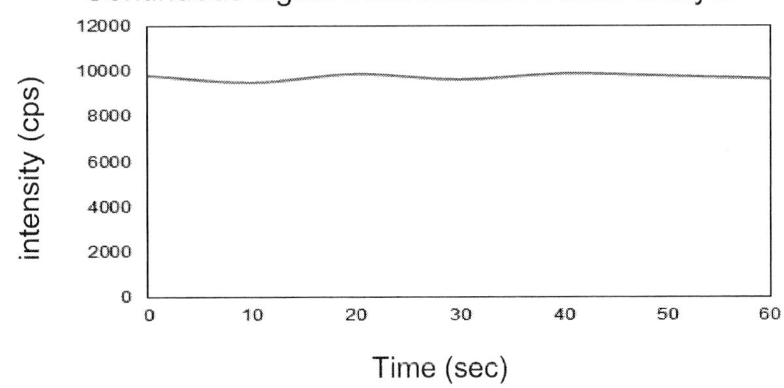

# Nanoparticle Analysis Fundamentals

Using time resolved data acquisitions with short dwell times, the transient signal of each individual nanoparticle can be differentiated from the steady-state ICP-MS signal generated by dissolved ionic species. This unique feature coupled with fast electronics enables ICP-MS to characterize nanoparticles down to single nm sizes, producing results in the form of particle concentration, particle median size, and size distribution information.

Each nanoparticle gives a transient signal, collected at point/0.1 ms

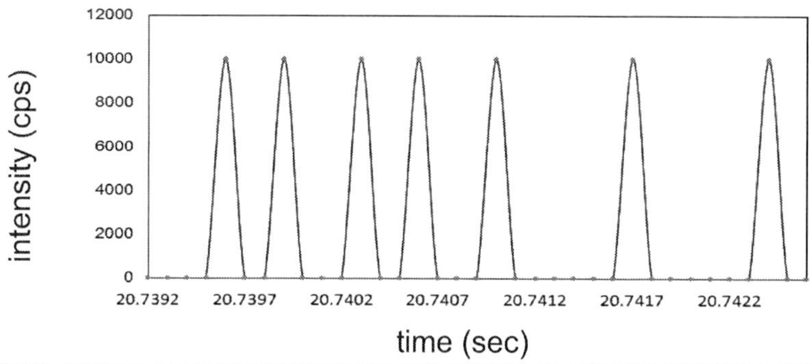




# Particle Size Distribution: Fe vs. Si

(similar concentration of particles but different median particle size)

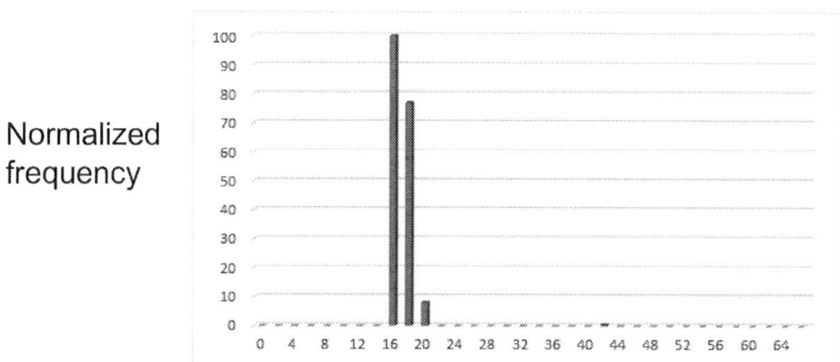


**Fe**

(75953 particles/mL)

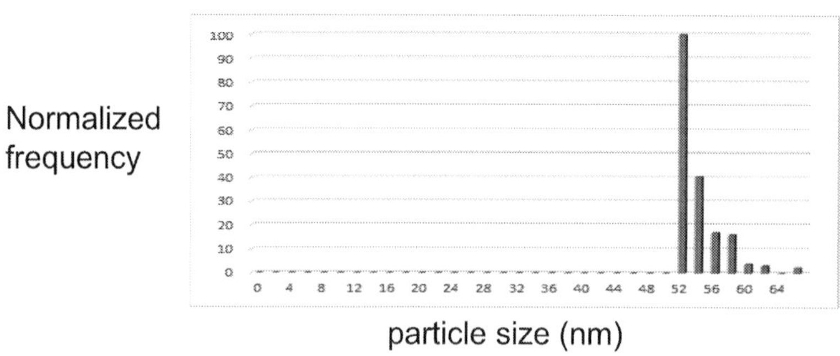


**Si**

(62420 particles/mL)

## Typical Nanoparticle Analysis Figures of Merit

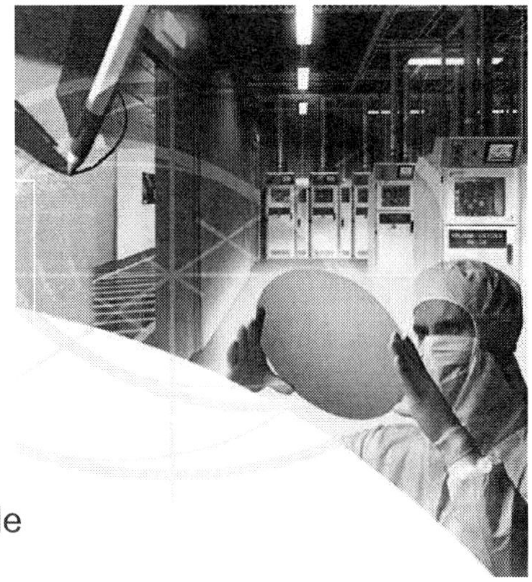

- Particle Concentration (number of particles/mL)
- Particle Size Distribution
- Median Particle Size; nm
- Elemental Composition of Particles
- BED (background equivalent particle diameter; nm)



## Analysis of a UPW Sample Collected from a 10 nm Inline Filter (2018 vs 2019)

| Element | Particle Concentration (particles/mL) 2018 | Particle Concentration (particles/mL) 2019 |
|---|---|---|
| Mg | 4143 | ND |
| Al | 567,613 | 200 |
| B | 8286 | 200 |
| Fe | 75,953 | ND |
| Si | 62,420 | 5000 |
| Zn | 6629 | ND |
| Cr | 99,053 | ND |
| V | 268 | ND |
| Ag | 34,000 | ND |

## Parameters contributing to the lowered particle concentrations

Some elements difficult to assess owing to background contribution and mass interferences...Consequently..

- ICP-MS conditions modified and optimized (e.g. reaction gas and different isotope selected)
  - Molecular and isobaric interferences avoided (e.g. $^{48}Ti$ vs $^{32}S^{16}O$ in sulfuric acid)
  - Signal/ background noise improved resulting in low BED (some single digit)
- Background contribution lowered (e.g. glassware modification)
- Cleaning/leaching methods of glassware improved



## Analysis of a 96% Sulfuric Acid sample Using Modified SP ICP-MS method

| Element | Particle Conc (particles/ mL) | BED (nm) | Median Size (nm) |
|---|---|---|---|
| Mg | 390,000 | 12 | 22 |
| Al | 10,000,000 | 9 | 15 |
| B | ND | 19 | —— |
| Fe | 640,000 | 6 | 13 |
| Si | 860,000 | 19 | 50 |
| Ti | ND | 12 | —— |
| Zn | 22,000 | 11 | 31 |
| Cr | 21,000 | 5 | 15 |
| V | ND | 7 | 18 |
| Ag | ND | 4 | 12 |

## Reproducibility of the Analysis of a 96% Sulfuric Acid Sample

| Replicate | Aluminum (particles/mL) | Iron (particles/mL) | Magnesium (particles/mL) |
|---|---|---|---|
| 1 | 1205612 | 2258615 | 504652 |
| 2 | 1508783 | 2558635 | 448637 |
| 3 | 1051614 | 2485782 | 541224 |
| 4 | 1287939 | 2347514 | 435879 |
| 5 | 995802 | 2159785 | 537894 |
| 6 | 1132575 | 2685763 | 495112 |
| Average | 1197054 | 2416016 | 493900 |
| S.D. | 185105 | 196516 | 44062 |
| % RSD | 15 | 8 | 9 |



## Laser Particle Counter (LPC) Results From Surface of OEM Parts: Only Total Particles Reported

| Particles/cm² | >0.1 –1 µM | >1 - 2 µM | >2 - 5 µM |
|---|---|---|---|
| Quartz Ring | 100.9 | 159.8 | 50.2 |
| Aluminum Coupon | 70.1 | 11.5 | 1.4 |
| Ceramic Ring | 867.1 | 178.2 | 2.6 |

## SP-ICPMS Results From the Surface of OEM Parts: Concentrations (particles/cm$^2$)

| Element Tested | Al | Mg | Mo | Ti | Zn | Ni |
|---|---|---|---|---|---|---|
| Quartz ring | 347,512 | 48,709 | ND | 296 | 3063 | 1778 |
| Al coupon | 111,867,151 | 79,921 | ND | ND | 57,699 | 2,465,754 |
| Ceramic Ring | 62,599,369 | 378,326 | ND | 89 | 876,918 | 2,619,337 |



## Aluminum (Al) Containing Nanoparticles Found From the Surface of Various OEM Parts

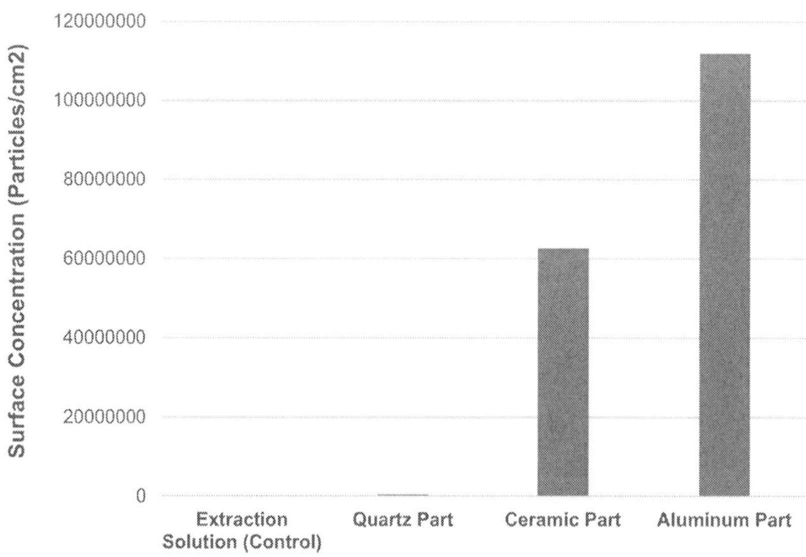

## SP-ICPMS Results From the Surface of OEM Parts: Concentrations (particles/$cm^2$)

| Element Tested | Cu | Cr | Fe | Co | Si |
|---|---|---|---|---|---|
| Quartz ring | 2272 | ND | 45,547 | 1087 | ND |
| Al coupon | 154,283,928 | 198 | 143,458 | ND | 11,376 |
| Ceramic Ring | 5,550,154 | 1159 | 1,108,231 | 179 | 137,281 |



## Iron (Fe) Containing Nanoparticles Found From the Surface of Various OEM Parts

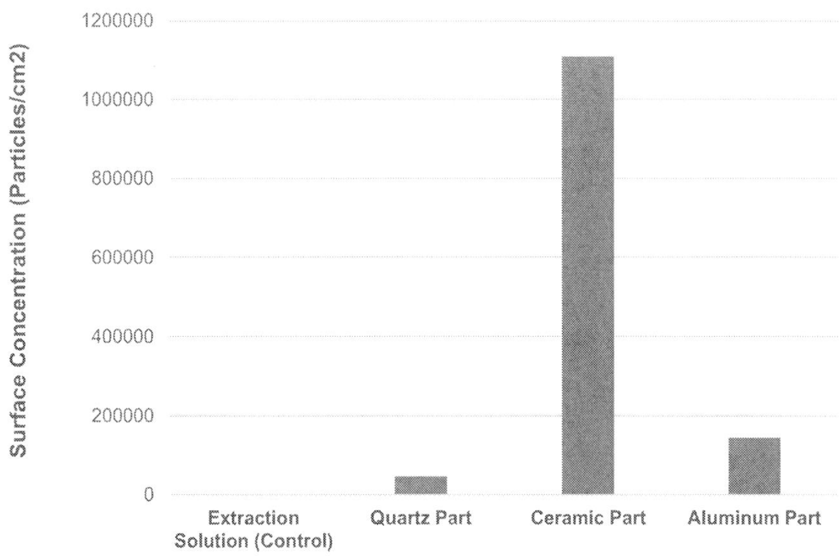

# Conclusion

- SP ICP-MS methods for determining nanoparticles in UPW and process chemicals have been improved
- SP ICP-MS background is low and consistent
- SP ICP-MS results are reproducible
- SP ICP-MS has successfully been used to measure nanoparticles extracted from surface of OEM parts used in etching and deposition chambers
- SP ICP-MS complements laser particle counter (LPC) technique by providing concentration and size distribution information of nanoparticles ranging from single nm to 100 nm.

UNISERS
UNIVERSAL LIQUID ANALYZER

# Universal Surface Enhanced Raman Spectroscopy for the metrology of High-Purity liquids


Ali Ozhan Altun, Ph.D. ETHZ
Co-founder & CEO of UNISERS AG
altun@unisers.ch +41 78 933 92 60
www.unisers.ch


SPCC-2019
Portland, OR
03.04.2019

## Outline

- **Introduction: Raman → SERS → UNISERS**
- **Application UNISERS to high-purity liquids**
- **Conclusions**

www.unisers.ch

2

# Raman Spectroscopy

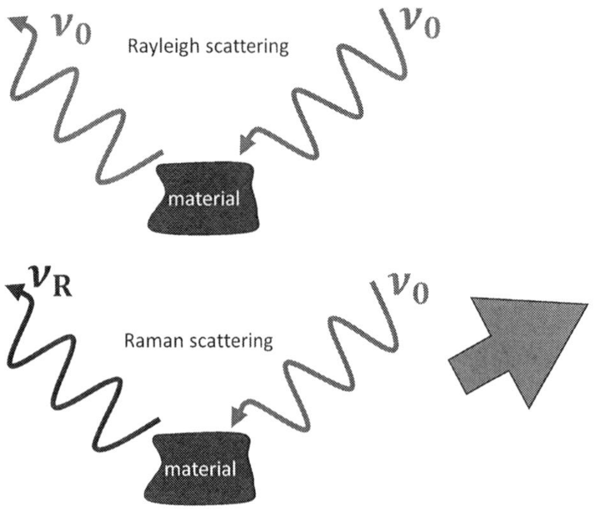


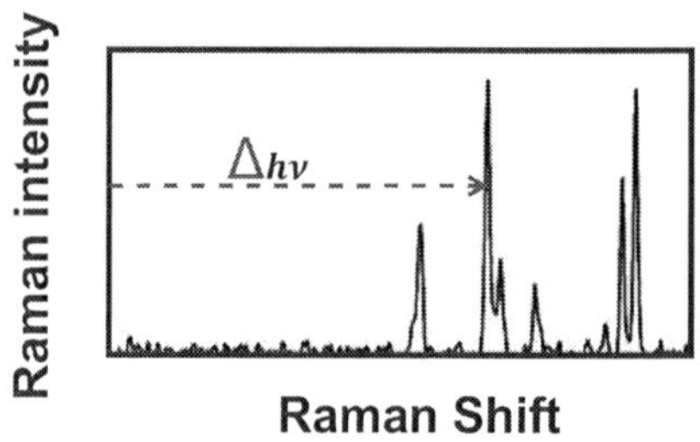


**"Molecular fingerprint spectra"**

***Used for material identification***

3

# Surface Enhanced Raman Spectroscopy (SERS)

**Raman scattering is a million times weaker than Rayleigh scattering**

$$Raman\ signal \propto (Near\ field\ enhancement\ of\ the\ substrate)^4$$

**Example SERS substrate:**
**Bow-tie structures**

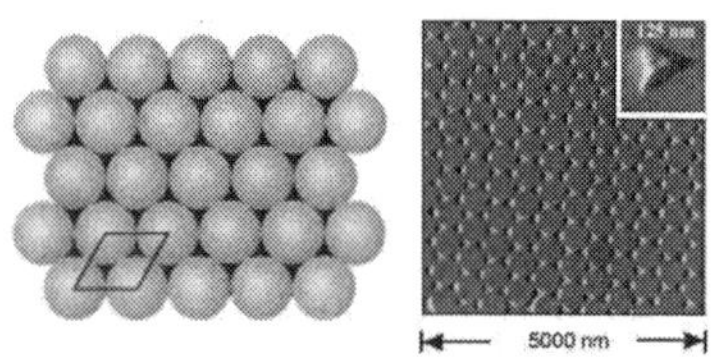


Haynes *et al.*, J. Phys. Chem. B (2001) **105**, 5599.

~$10^8$ to $10^{10}$ Raman enhancement

More than 12,000 patents with key word "Surface enhanced Raman"

But no extensive industrialization due to lack of **reliability** and **universality**

4

# SERS reliability and universality problem

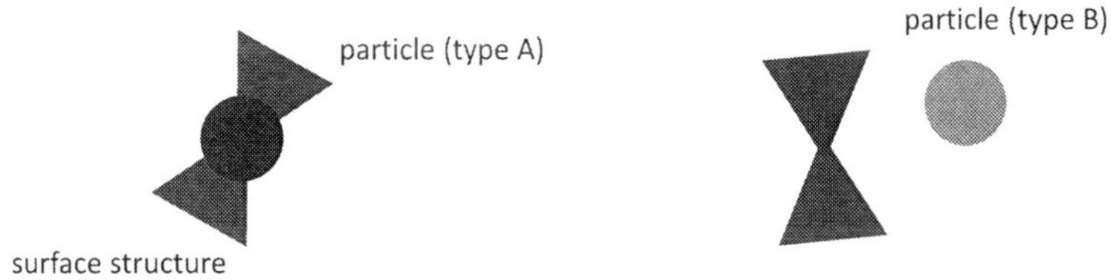


SERS requires a surface coupling of the particle to enhance Raman

Using a **functionalization layer** for reliable surface coupling works only for some particles

Only suitable to quantify some impurities

www.unisers.ch

5

# **Universal** Surface Enhanced Raman Spectroscopy (UNISERS) has a consistent surface coupling

| Parameter | Raman | SERS* | UNISERS** |
|---|---|---|---|
| **Detection of trace-impurity** | ✗ | ✓ | ✓ |
| **Universality** | ✓ | ✗ | ✓ |
| **Multiple trace-impurity analysis** | ✗ | ✗ | ✓ |

*SERS: Surface enhanced Raman spectroscopy
**UNISERS: Universal surface-enhanced Raman spectroscopy

**Quantifying trace impurities becomes feasible with a reliable surface coupling**

www.unisers.ch

6

# Quantification and identification of non-volatile impurities

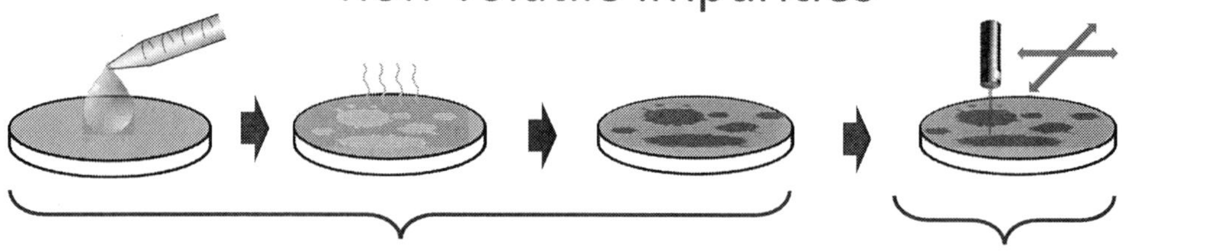

**Impurity-immobilization on UNISERS substrate**

- Amplification of Raman signal
- Control of cross-contamination

**Raman-scanning**

- Identification and quantification of multiple impurities pixel by pixel

**Universal methodology for different impurities and different liquids**

www.unisers.ch

7

# Analysis of impurities with Raman bands

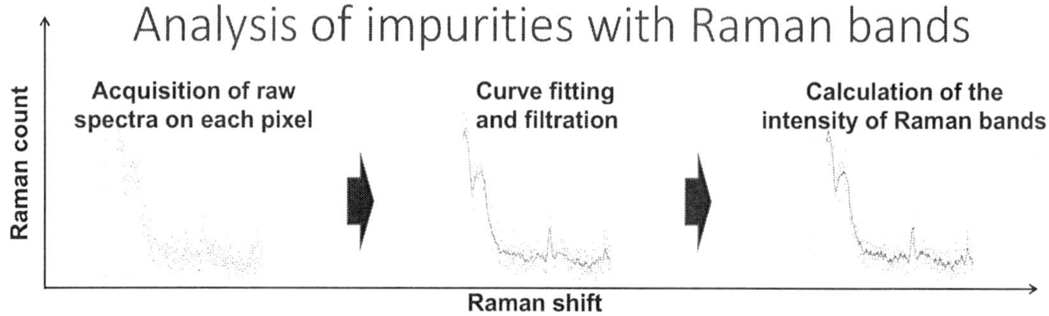


Intensity plot of detected Raman bands of overall scan

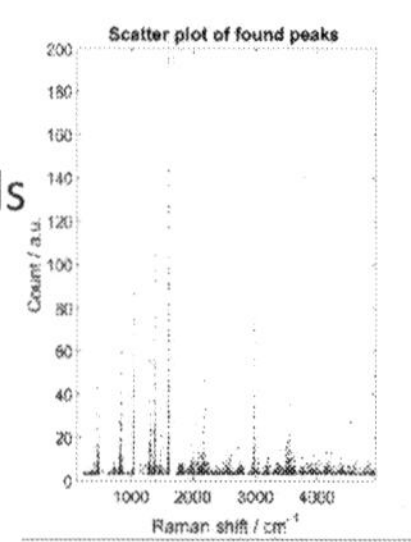


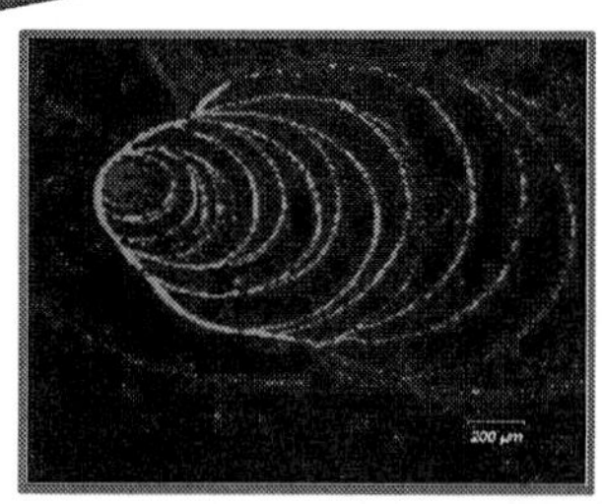


Mapping of Raman bands

www.unisers.ch

8

# Monitoring and analysis

**Total areal coverage** of Non-Volatile Impurities

**Analysis** of specific Non-Volatile Impurities

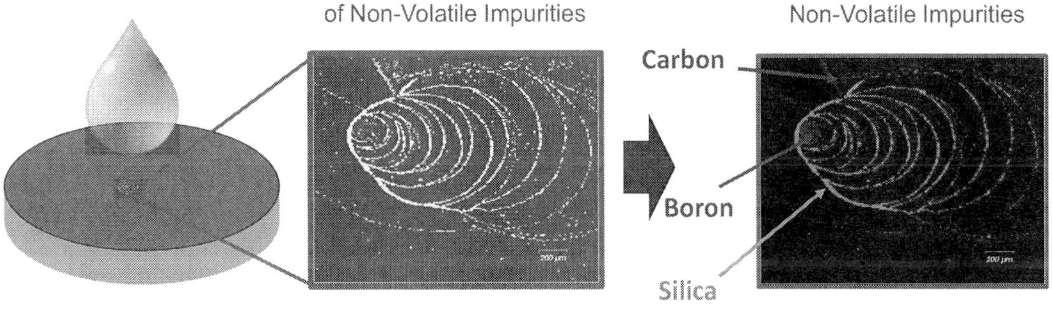


fast Raman scan

slow Raman scan

Total analysis time: 5-10 mins 30-60 mins

www.unisers.ch

9

# Quick monitoring mode

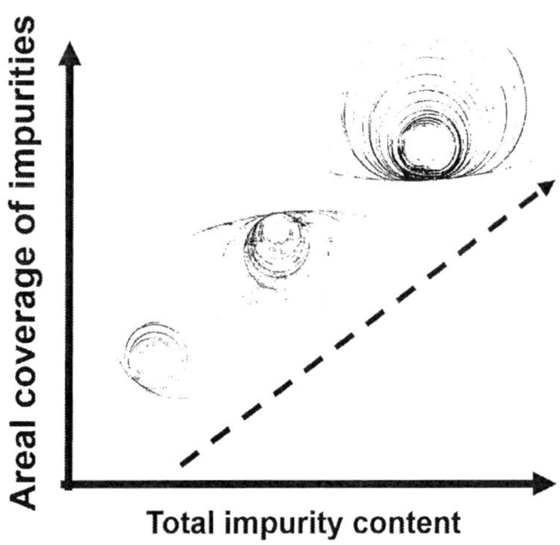


www.unisers.ch

10

# Application to High-Purity liquids

Detect the areal coverage of:

- Total non-volatile impurities
- Total silica
- Total non-volatile carbon

**Multi-parameter monitoring with a single test**

in multiple liquids:

- UPW
- Acids like. $H_2SO_4$ and HCl
- All solvents (such as IPA)

**Multi-liquid monitoring with same methodology**

www.unisers.ch

11

# Detection of total areal coverage of non-volatile impurities for acids

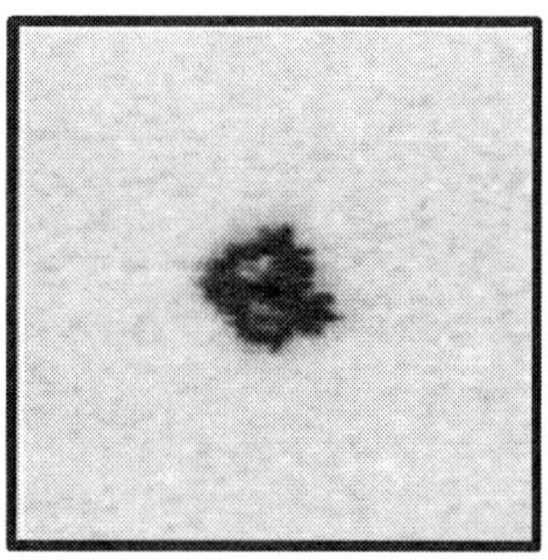

**$H_2SO_4$**

**0.5 mm$^2$ total non-volatile impurity**

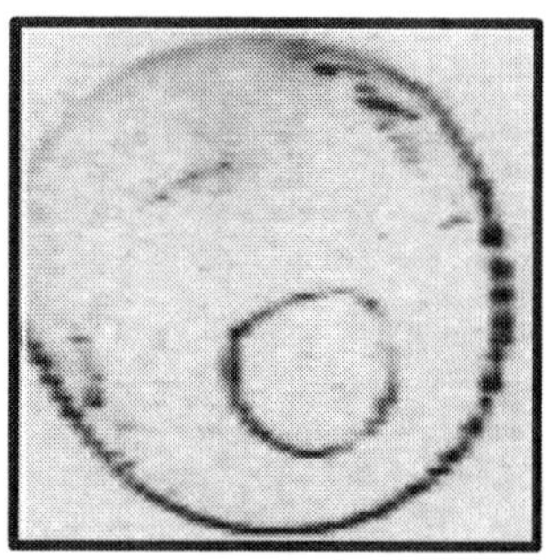

**HCl**

**0.82 mm$^2$ total non-volatile impurity**

www.unisers.ch

12

# Detection of the areal coverage of silica and carbon in $H_2SO_4$

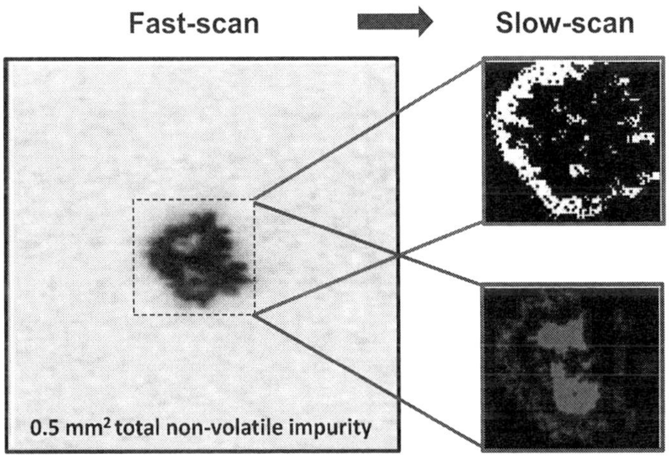


*(Raman bands are used only for material identification. Areal coverage data is used for quantification.)*

**0.11 mm²**
**Silica areal coverage**

**0.46 mm²**
**Carbon areal coverage**

www.unisers.ch

13

# Detection of the areal coverage of silica and carbon in HCl

**Slow-scan only a portion of area**

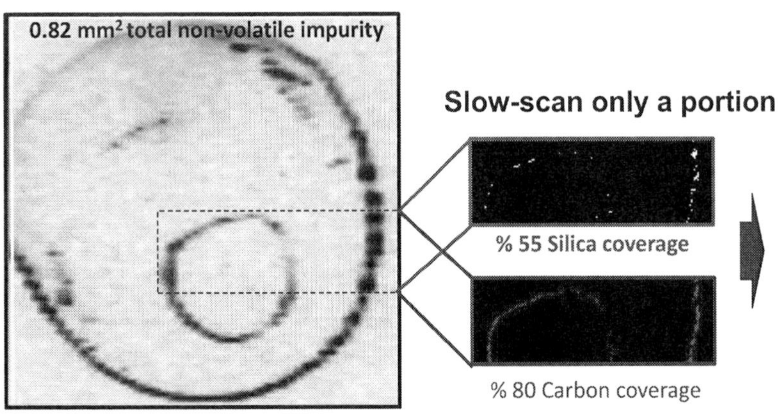


**0.45 mm²**
**Silica areal coverage**

**0.66 mm²**
**Carbon areal coverage**

www.unisers.ch

14

# Detection of the areal coverage for solvents

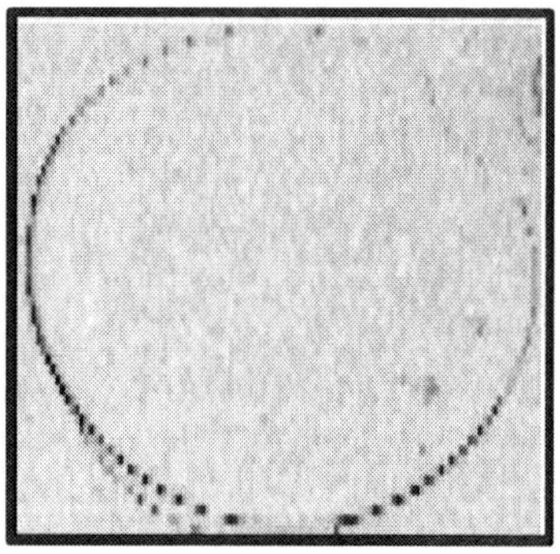

IPA

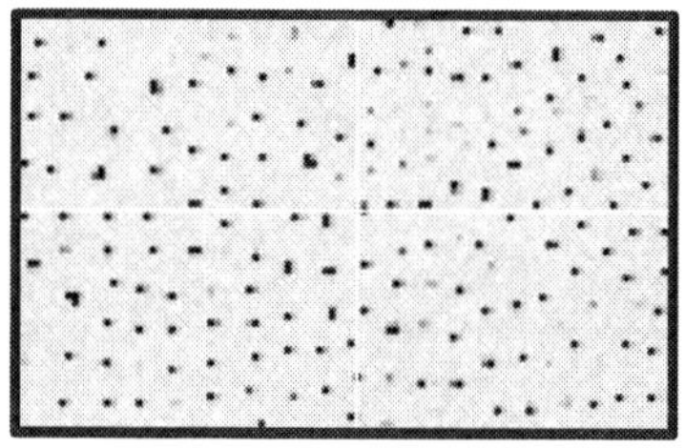

Acetone

www.unisers.ch

15

# Estimation of silica concentration in UPW

*(Both Raman band intensity & areal coverage is used for quantification)*

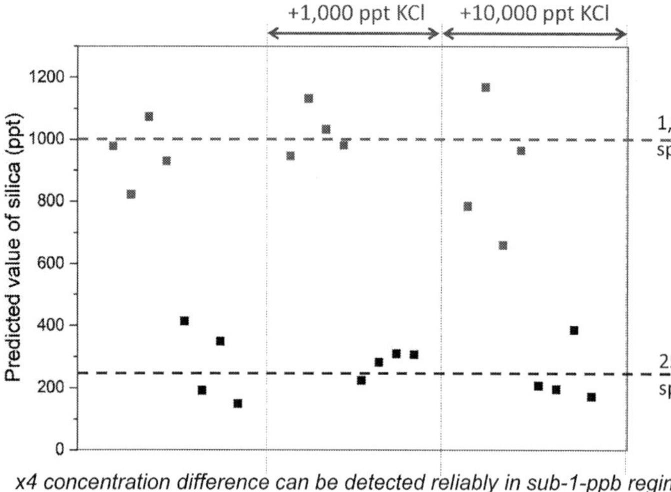


*Sodium silicate is used in spiking.

*x4 concentration difference can be detected reliably in sub-1-ppb regime*

www.unisers.ch

16

# Detection limit vs analysis time

**Detection of sub-10-picogram impurity deposit on the surface is possible**

Detection limit can be improved by increasing the sample volume.

**100 µl sample** **1400 µl sample**

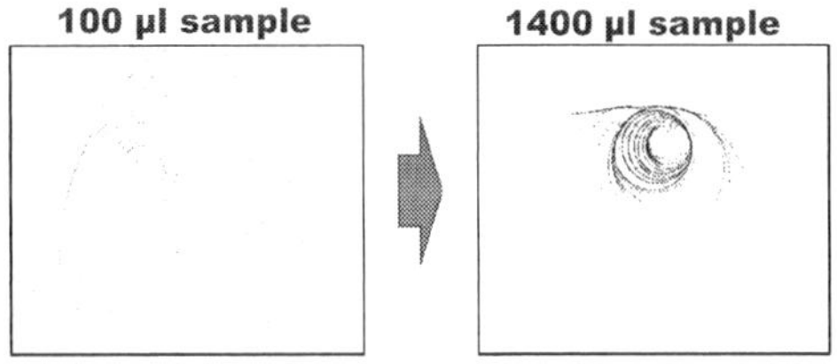

*Larger sample volume*
*+) More detailed analysis*
*+) Improved detection limit*
*-) Longer analysis time*

www.unisers.ch

17

# Monitoring with relative values

***Example case:***

**Time 1** **Time 2**

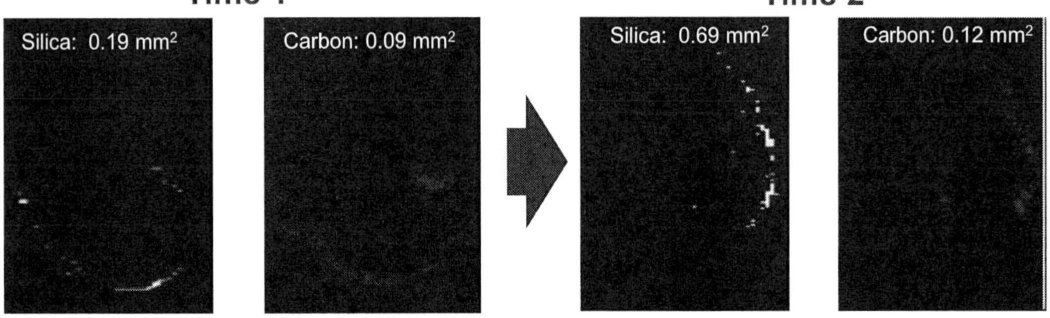


**Silica/carbon=2.1** **Silica/carbon=5.8**

**INCREASE IN SILICA CONCENTRATION**

www.unisers.ch

18

## Monitoring with relative parameters allows **cross-referencing**

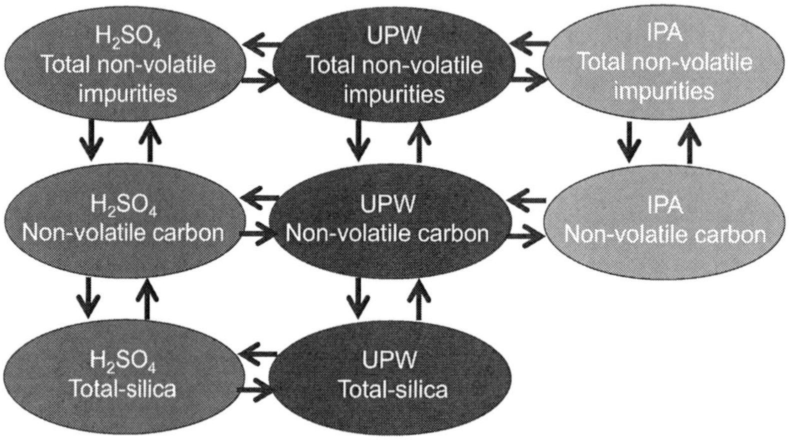




## Direct correlation with process DEFECTIVITY allows to **quickly detect the cause of yield loss**

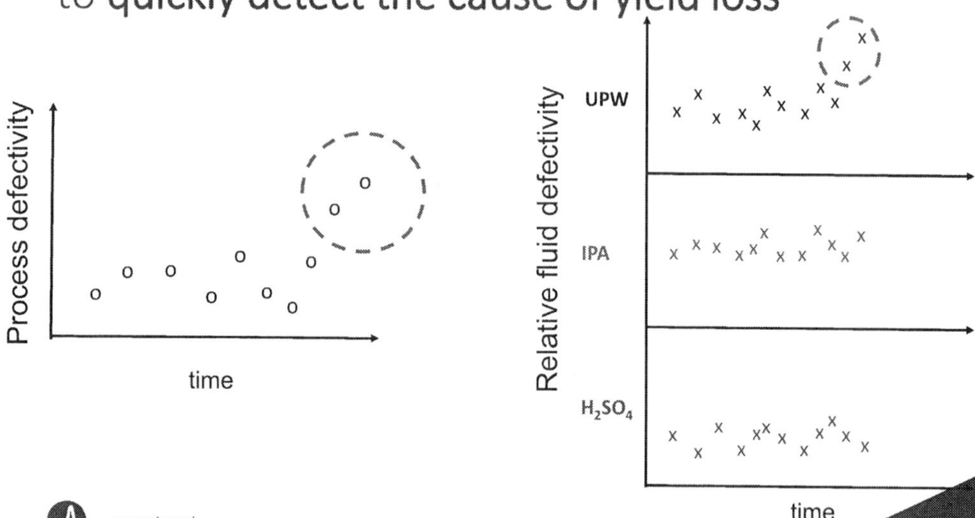

# Conclusion: Root-cause analysis of defectivity

Process defectivity — What is the defectivity of single-wafer-cleaning station?

↓

Fluid defectivity — Which liquid is the main source of defects?

↓

Component defectivity — What is contaminating UPW?

www.unisers.ch

21

# APPENDICES

www.unisers.ch

22

# UNISERS pipeline

### Impurities

- **Reactive silica vs colloidal silica**
- Polar vs nonpolar carbon
- Organic vs inorganic carbon (graphitic carbon)
- Molecular / colloidal carbon
- Total chloride salts
- Boric acid
- Total conductive impurities
- …

### Sample fluids

- Other solvents
- $H_2SO_4$+$H_2O_2$
- TMAH
- …

www.unisers.ch

23

# Big-Data: Improved root-cause analysis

**UNISERS usage as liquid monitoring**

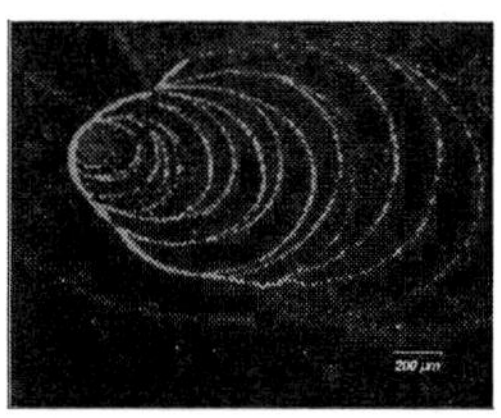

**Leverage existing fab information like wafer inspection results**

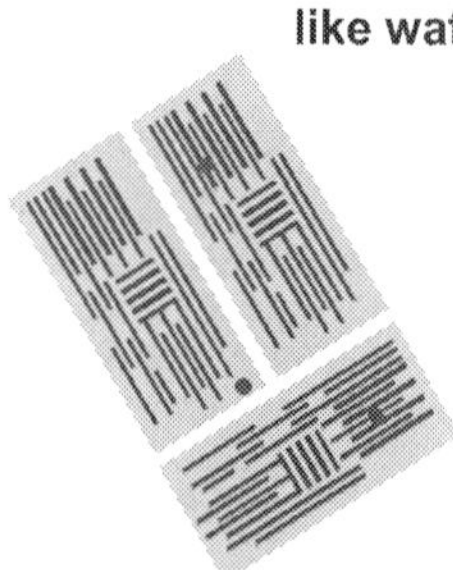

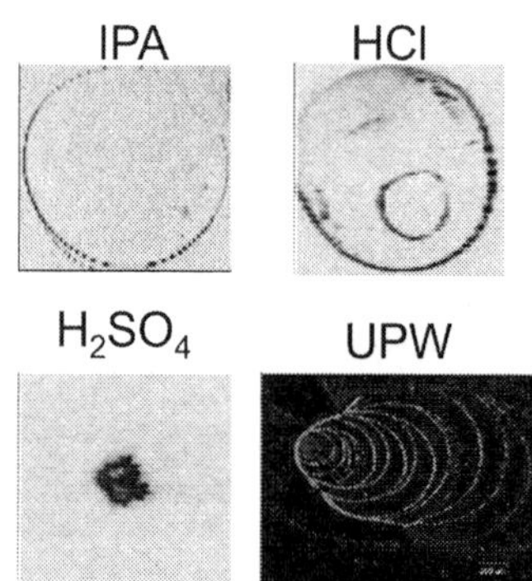


- Optical metrology
- SEM
- others

www.unisers.ch

24

# Modular analysis setup allows flexible integration

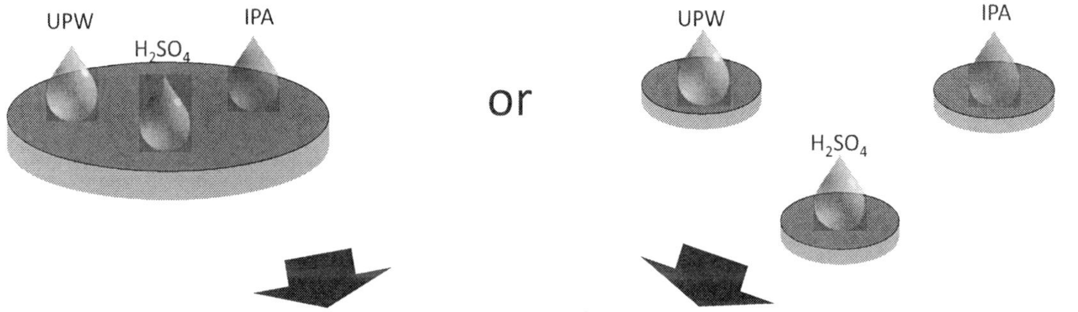


**IMPURITIES ARE ANALYZED WITH THE SAME MACHINE**

www.unisers.ch

25

**As universal liquid metrology UNISERS can facilitate an overall process improvement by identifying the bottlenecks in the process liquids**

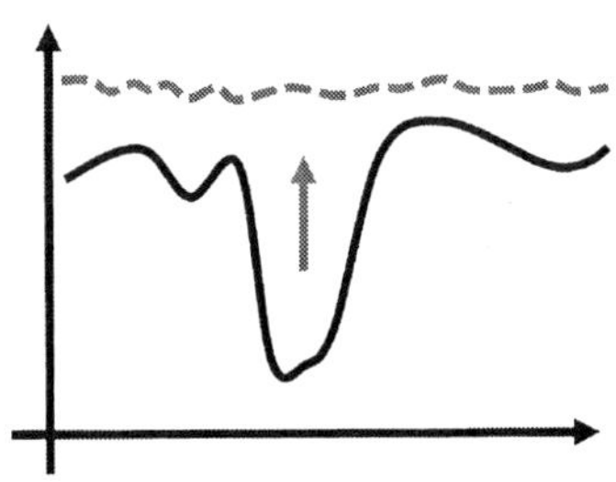

**Improved cumulative yield**

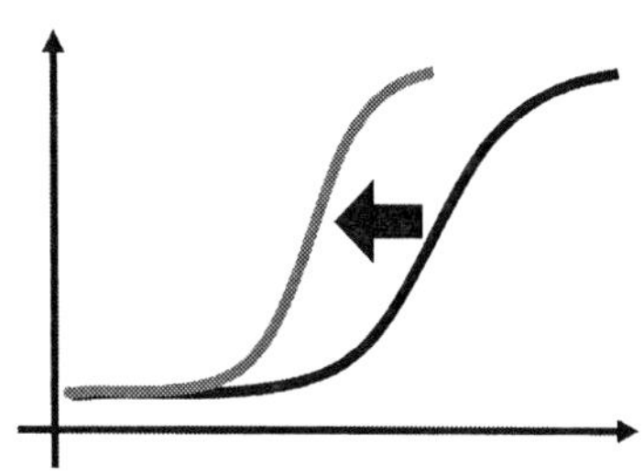

**Faster ramping**

www.unisers.ch

26

Hilltop Economics

# A Slower Global Economy with a Recession Watch in 2019

## Implications for the Semiconductor Outlook

Surface Preparation and Cleaning Conference (SPCC)
April 3, 2019

Duncan Meldrum
Chief Economist, Hilltop Economics, LLC

duncan.meldrum@hilltopeconomics.com
610-709-7951

**A global outlook based on Consensus Economics, Inc. CONSENSUS FORECASTS®*



Hilltop Economics

# Outline

- **Forecast with key themes:**
  - **Slower growth forecast from the Consensus**
  - **Higher uncertainty**
  - **Rising risk of recession**
- **Recession Risk Dashboard**
- **Semiconductors:**
  - **Near term risk**
  - **Long term strength**

Hilltop Economics

SPCC 2019 Forecast

# Summary Overview: 2019-2020

- Now in the *mature phase of the business cycle with decelerating growth and a rising risk of recession*
  - Healthy but more cautious consumer
  - Investment moderating despite need
  - Low inflation, diverging interest rates, currency unsettled
  - Higher economic policy uncertainty:
    - Domestic fiscal and monetary policy (China, US, EU, Japan)
    - Trade
    - Geopolitics
- Less-healthy backdrop for semiconductor industry

*Duncan Meldrum*
*SPCC April 3, 2019*

3

Hilltop Economics

MARCH 2019

# Consensus: Growth Weakens to Trend in 2019-20

**World* Real GDP Growth**

**2018: 3.0% 2019: 2.7% 2020: 2.7%**

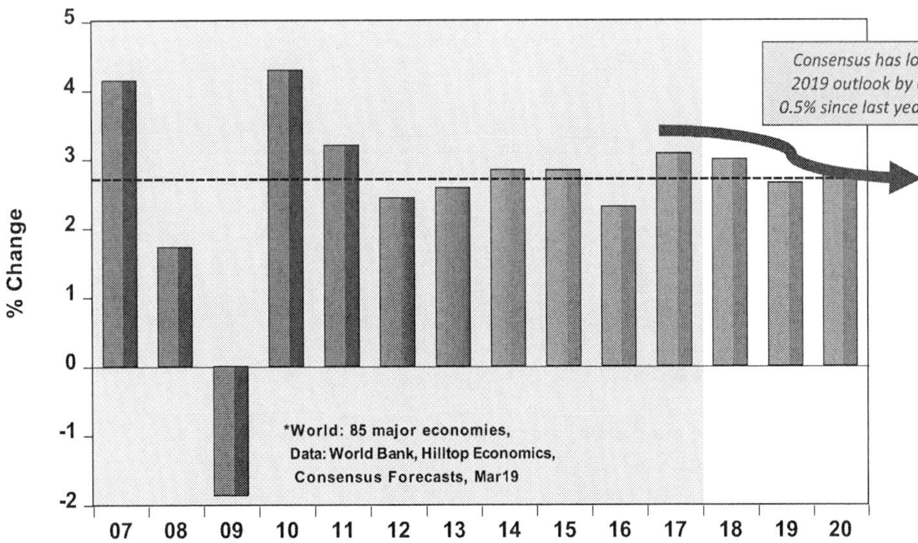


*Duncan Meldrum*
*SPCC April 3, 2019*

4

Hilltop Economics

~60% of GDP

# Global Consumer Healthy

- Unemployment rates are low – below equilibrium
- Wages are stable to rising
- Inflation is low – at or below policy maker targets
- Oil prices are not a big problem
- Consumer debt is under control
- Confidence has slipped off the peaks of 2018

*Duncan Meldrum*
*SPCC April 3, 2019*

5

Hilltop Economics

~60% of GDP

## World* Real Consumption

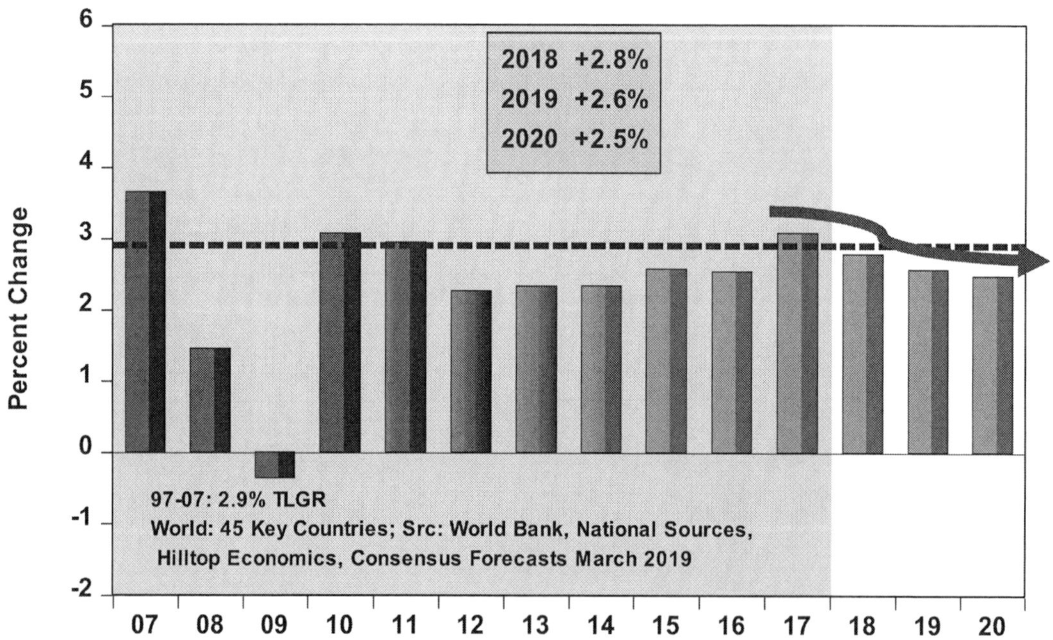


*Duncan Meldrum*
*SPCC April 3, 2019*

6

Hilltop Economics

~25% of GDP

## Investment Should Be Growing at this Point: Needed to Drive Cycle Higher

- **The largest developed economies are growing ABOVE their potential rate of growth (i.e. above capacity)**
  - **Investment typically grows more rapidly when potential is reached**
- **Revenue growth OK**
  - Real GDP growth (volume) now at 2.7%
  - Some inflation (price) on average from ~1% - 2%
  - "Revenue" now ~4 ½ %
- **Interest rates generally remain low**
- **Modest cost pressures in most regions**
- **Profitability still positive**

*Duncan Meldrum*
*SPCC April 3, 2019*

7

Hilltop Economics

~25% of GDP

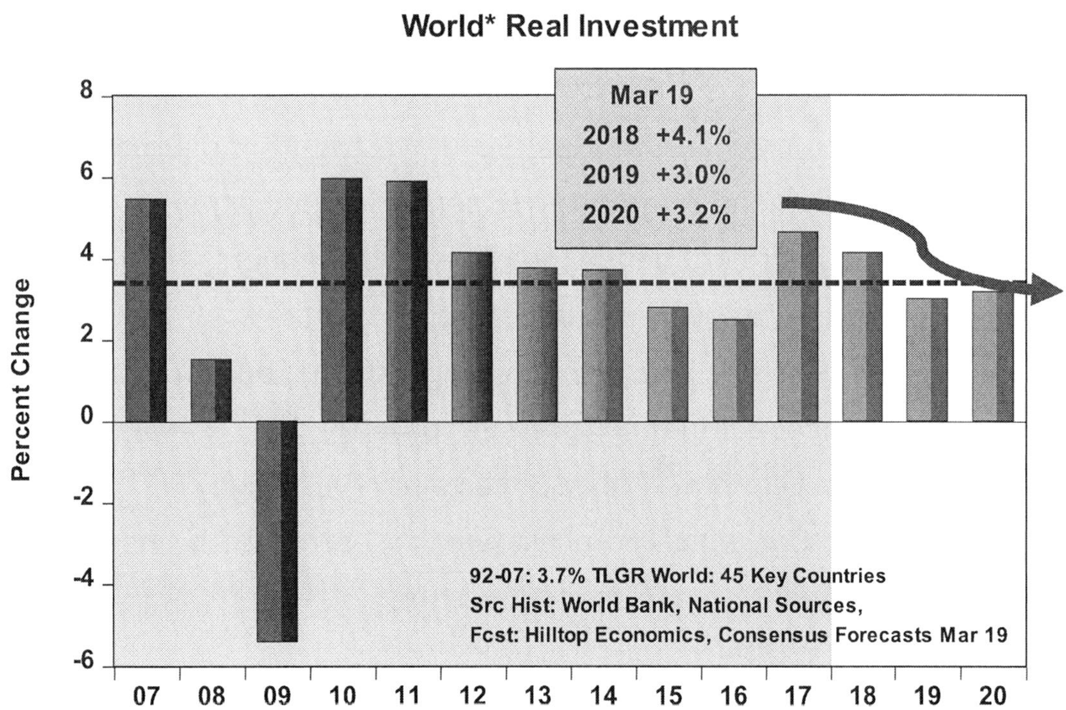


*Duncan Meldrum*
*SPCC April 3, 2019*

8

Hilltop Economics

## Investment Should Be Growing at this Point: But Forecast to SLOW – What's the Problem?

- ***Risk aversion** is the problem again*
- ***Economic policy uncertainty** volatile and high on a global level*
    - Trade wars (shifting international institutional arrangement)
    - US (post-stimulus fiscal policy, post-govt shutdown, monetary policy, immigration)
    - Europe (Brexit, Italy, France, EU-wide e.g. immigration)
    - China stimulus impact, international situations
    - Japan tax increases
    - ***Geopolitical concerns***
- ***Forecasts on a downward trajectory***
- ***Interest rates reflecting the uncertainty***

*Duncan Meldrum*
*SPCC April 3, 2019*

9

Hilltop Economics

## Recession Watch Dashboard: Cautious (Yellow)

- Output momentum (Real GDP, Investment, Ind. Production)
- Employment momentum (Unemployment rates)
- Consumer confidence measures (US & majors)
- Equities (recovered & but still a bit volatile)
- Profits & Small Business Earnings (strong, but off peak)
- Oil Prices & Inflation (pressure eased)
- Global Economic Policy Uncertainty (Very high)
- World Consensus Forecast (weaker each month since April)
- U.S. Yield Curve (close to inverting)
- Interest Differentials (US vs EU rising)

*Recession watch: leading indicators that may help identify turning points that precede a downturn in the next 18 to 24 months*

*Duncan Meldrum*
*Business of Cleans April 1, 2019*

10

*Hilltop Economics*

# Uncertainty Peaked at Crisis Levels in January 2019

## Global Economic Policy Uncertainty

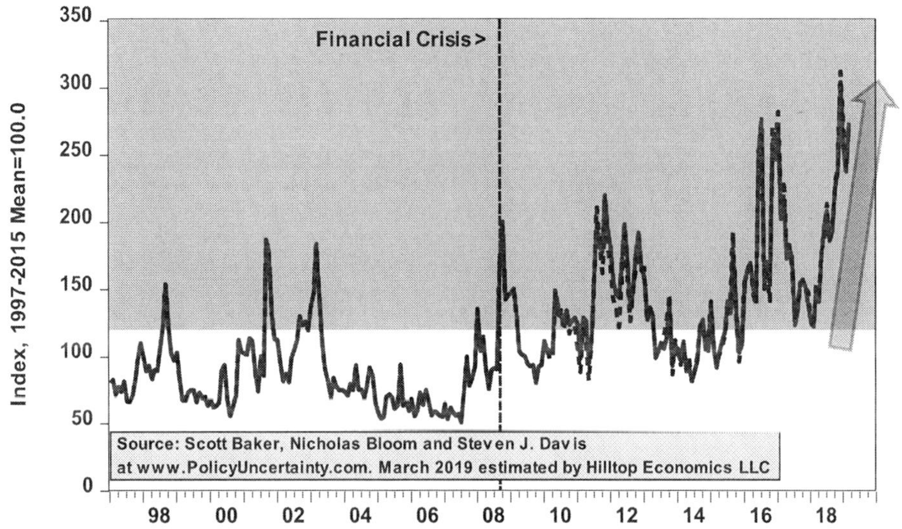


*Uncertainty over future economic policy skyrocketed back to levels high enough- in the grey zone on the graph- to dampen investment by the middle of 2018.*

*Duncan Meldrum*
*SPCC April 3, 2019*



*Hilltop Economics*

# Uncertainty UP Everywhere (Even U.S.)

Economic Policy Uncertainty: U.S. & Europe

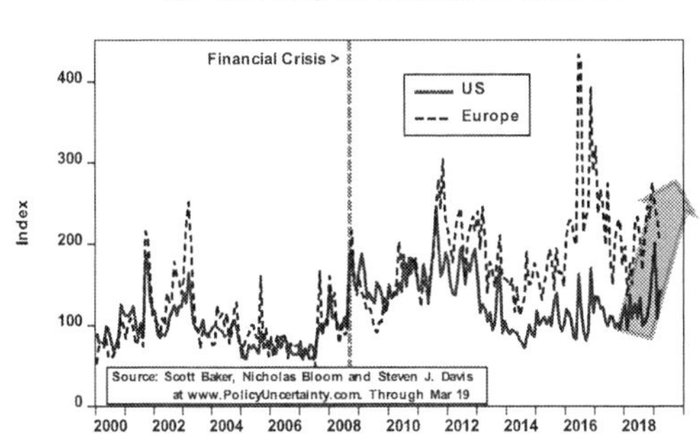


Economic Policy Uncertainty: Japan

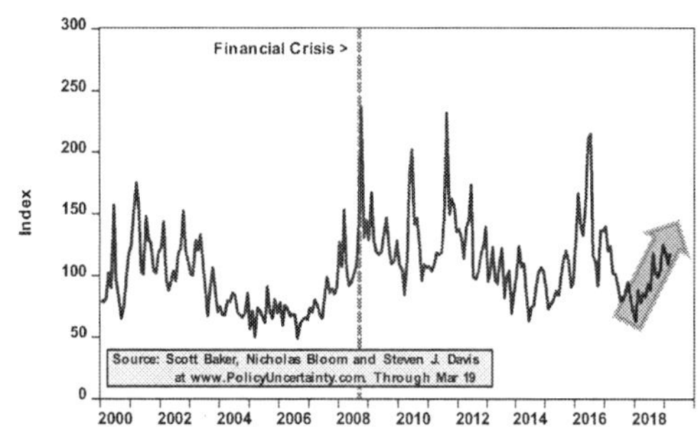


UK Economic Policy Uncertainty

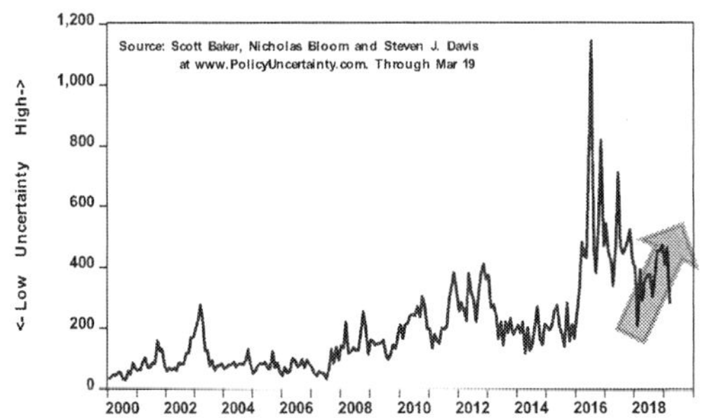


Economic Policy Uncertainty: China

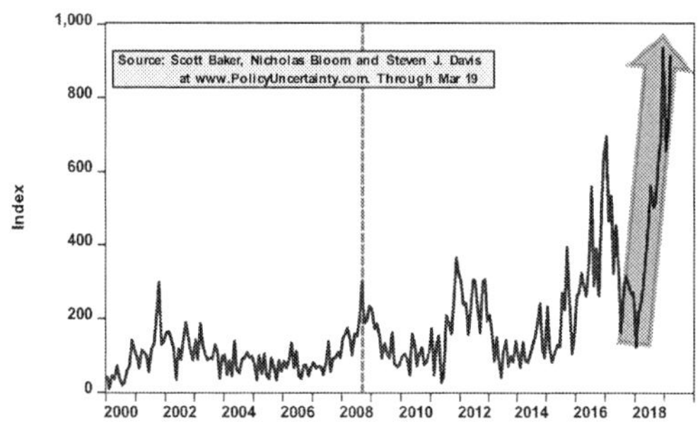


*Duncan Meldrum*
*SPCC April 3, 2019*

Hilltop Economics

## Global Real GDP Forecasts for 2019 Lowered Eleven Months in a Row

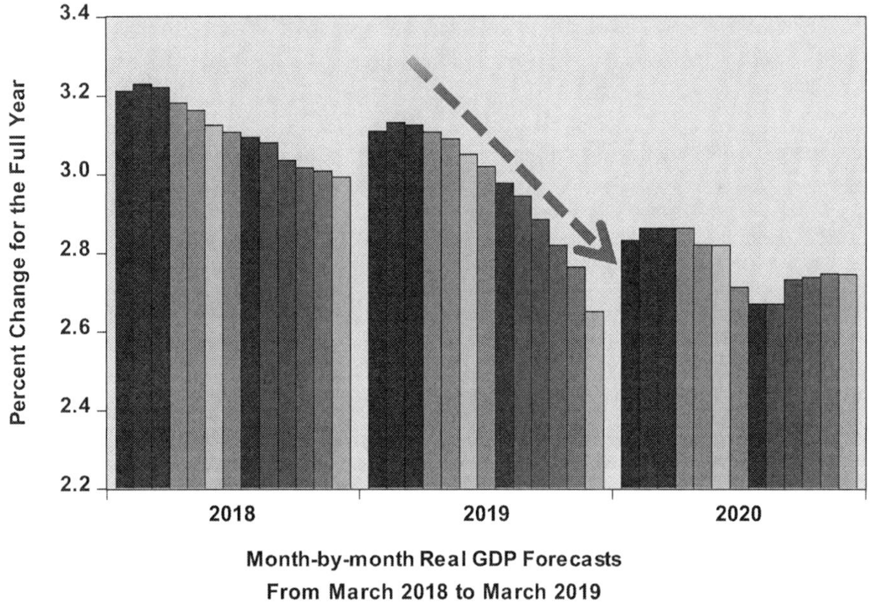


Duncan Meldrum
SPCC April 3, 2019

13

Hilltop Economics

## Yield Curve on Verge of Signaling Recession

***Uncertainty & fear of recession driving U.S. longer-term rates lower; Consensus (the arrows) too optimistic as of the end of March***

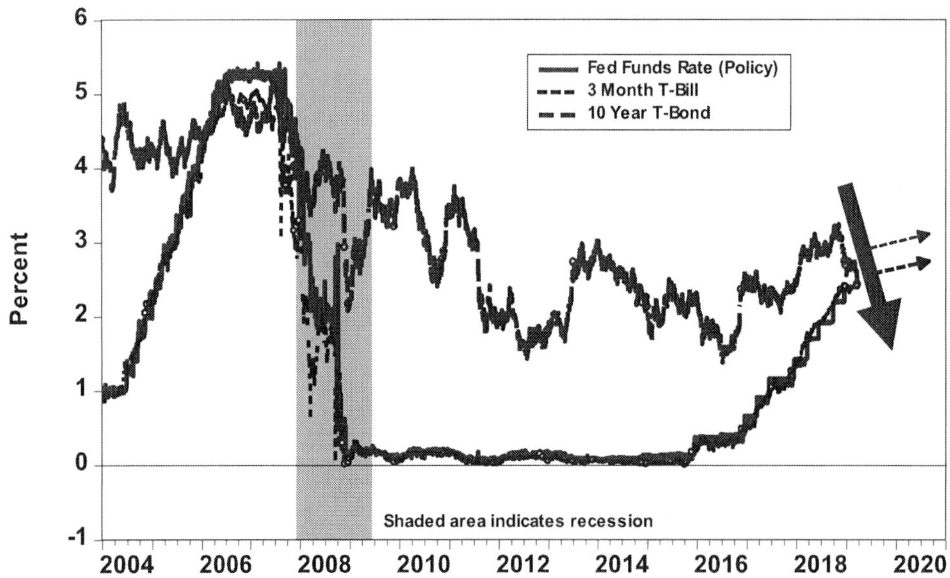


Duncan Meldrum
SPCC April 3, 2019

14

*Hilltop Economics*

# U.S. Yield Curve Foreshadows Recession

***Seven (eight?) inversions, seven recessions in the past 50 years***

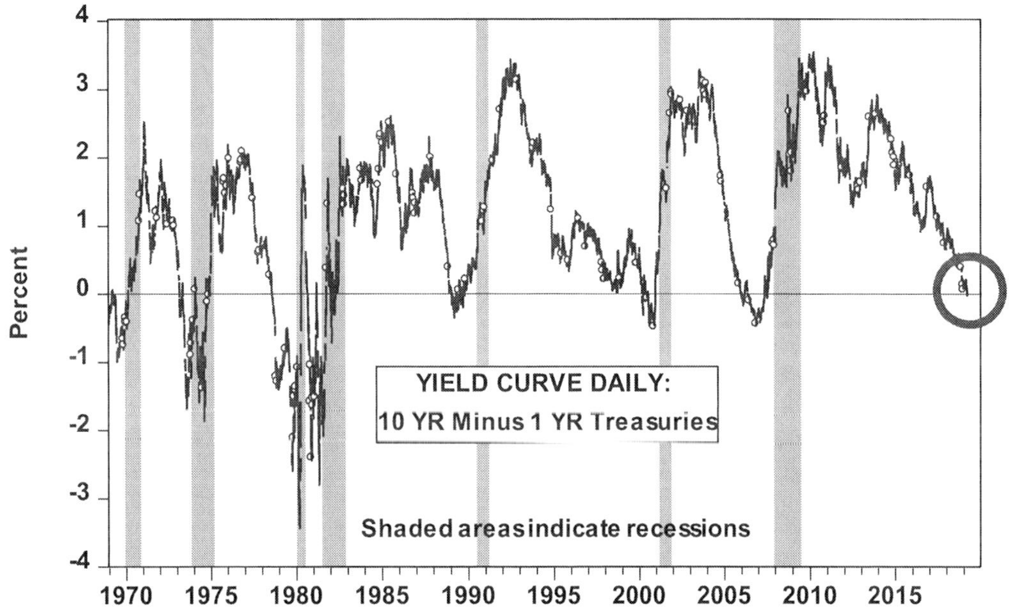


*Duncan Meldrum*
*SPCC April 3, 2019*

15

*Hilltop Economics*

# Different International Interest Rate Trends

***A risk to financial flows, currencies, and a contributor to uncertainty***

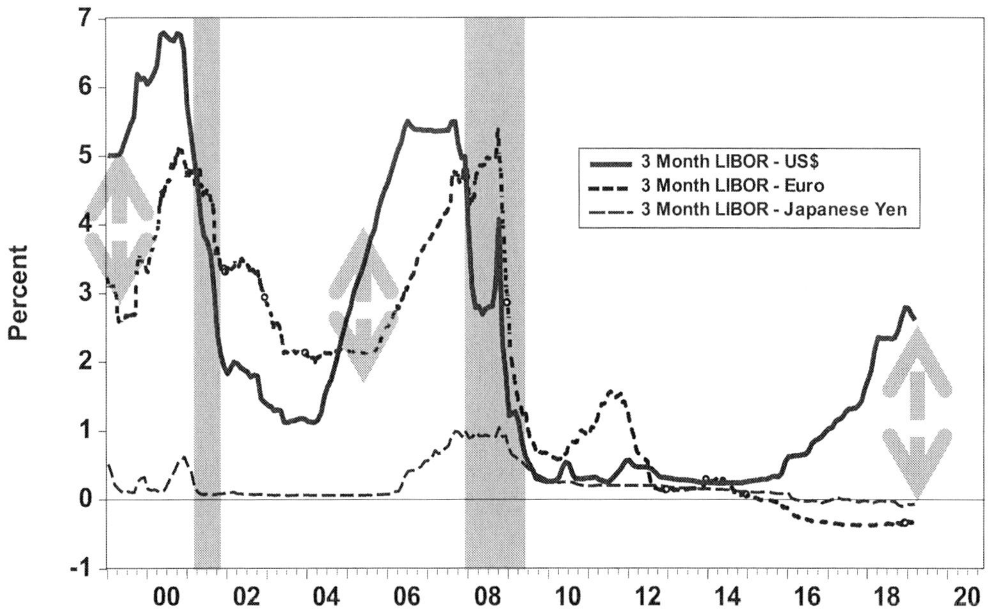


*Duncan Meldrum*
*SPCC April 3, 2019*

16

*Hilltop Economics*

# Summary of Semiconductor Driving Forces in 2019

- Demand drivers from the economic outlook weakening
  - Slower investment growth weakens growth in:
    - Servers
    - PCs
    - Industrial applications
  - Moderating consumer growth dampens growth in:
    - Mobile phones
    - PCs
    - Autos
  - Evidence of inventory builds
  - Falling prices for key products & recently expanded capacity
  - Evidence of slowing trade, declining revenue
- RESULT: Moderate decline in MSI in 2019

- LONG-RUN STAYS VERY HEALTHY: AI/5G/ML and continued expansion of the digitalization of the economy

*Duncan Meldrum*
*SPCC April 3, 2019*

17

*Hilltop Economics*

2019 MSI Demand Outlook

# U.S. Consumer Spending on Tech Goods Weakening

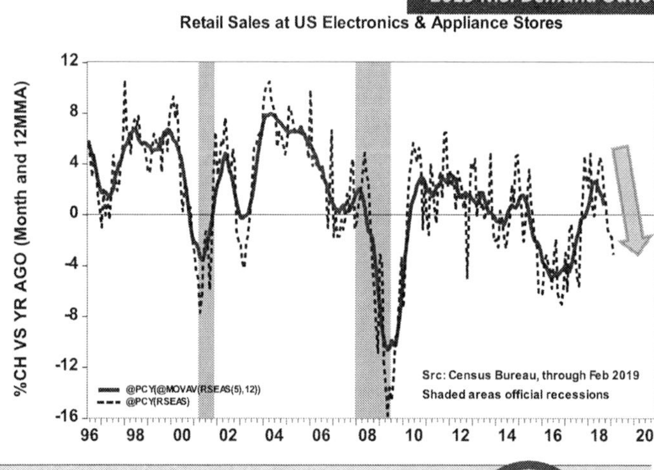


| U.S. Consumer Spending on Technology | Jun 18 | Jul 18 | Aug 18 | Sep 18 | Oct 18 | Nov 18 | Dec 18 | Jan 19 | Feb 19 |
|---|---|---|---|---|---|---|---|---|---|
| PCE (year % ch.) | 5.1 | 5.4 | 5.4 | 4.7 | 4.9 | 4.7 | 3.8 | 3.7 | -- |
| PCE Technology Goods (year % ch.) | 5.8 | 6.6 | 6.6 | 6.6 | 5.6 | 4.2 | 1.0 | 2.6 | -- |
| | | | | | | | | | |
| Information Processing Equip (year % ch.) | 6.5 | 7.3 | 7.5 | 7.6 | 6.6 | 5.3 | 1.4 | 3.6 | -- |
| _PCs and Peripherals (year % ch.) | 6.9 | 7.6 | 7.9 | 8.1 | 7.2 | 6.0 | 1.8 | 4.3 | -- |
| *_Software and Accessories (year % ch.)* | *6.3* | *7.0* | *7.1* | *7.2* | *6.2* | *4.7* | *1.2* | *3.0* | -- |
| _Calculators, etc. (year % ch.) | 5.2 | 6.6 | 7.3 | 6.5 | 5.5 | 5.8 | 2.0 | 4.8 | -- |
| _Telephone and Fax Equip (year % ch.) | 3.6 | 4.7 | 4.0 | 3.5 | 2.4 | 0.6 | (0.6) | (1.0) | -- |
| | | | | | | | | | |
| Retail Sales: Electronics & Appliance Stores (year % ch.) | 3.3 | 4.5 | 3.9 | 1.9 | 1.0 | (1.2) | (1.6) | (1.7) | (3.1) |

*Duncan Meldrum*
*SPCC April 3, 2019*

18

# Stable PC Outlook, Minimal Growth

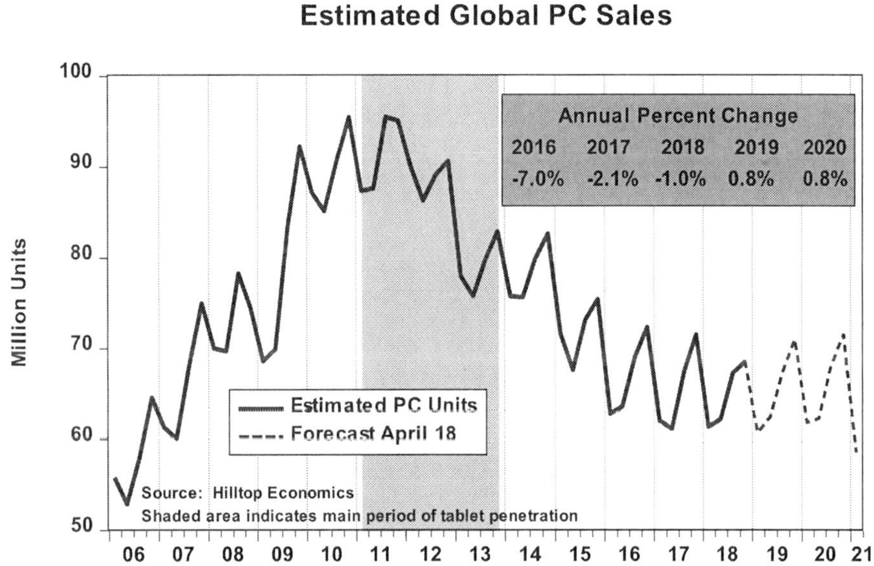


Duncan Meldrum
SPCC April 3, 2019





# Global Smartphones Grow Slowly

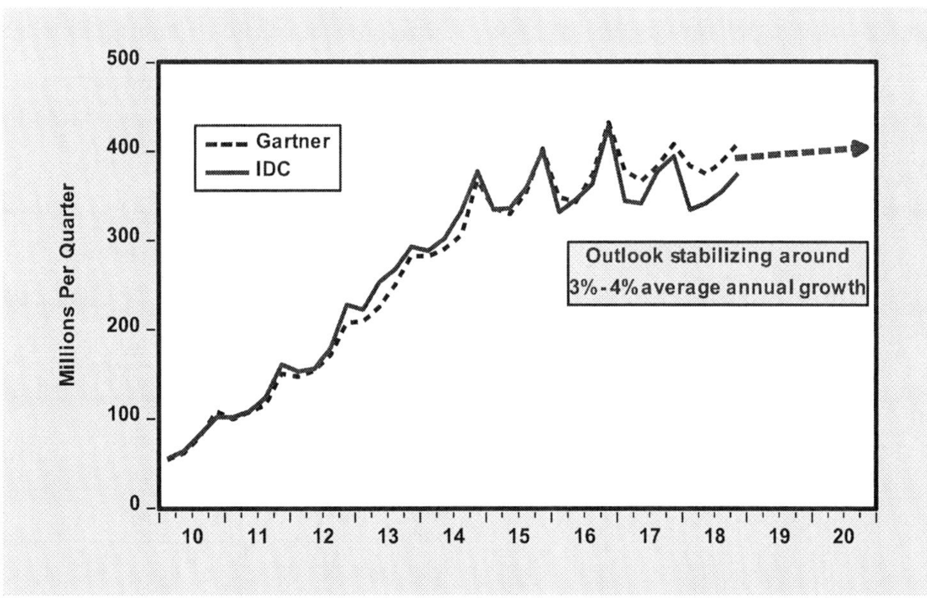


Duncan Meldrum
SPCC April 3, 2019

Hilltop Economics

MSI Demand Outlook

# Servers

***Likely heading to low single-digit unit growth temporarily with the slipping in investment growth – 5G, AI, ML will sustain growth after 2020***

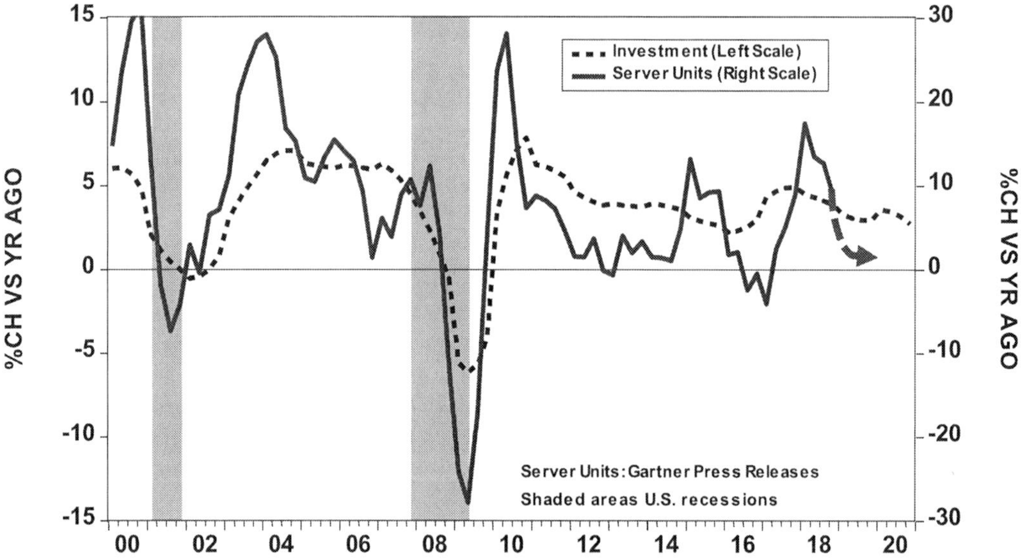


*Duncan Meldrum*
*SPCC April 3, 2019*

21

Hilltop Economics

2019 MSI Demand Outlook

# Hints of Inventory Building (Upstream and Downstream)

US Computers & Peripheral Devices

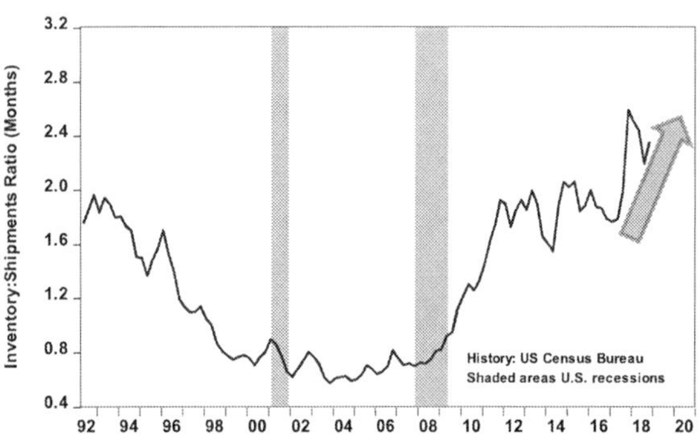


Information and Communications Electronics Equipment
Inventory-Shipments Ratio (Indexed)

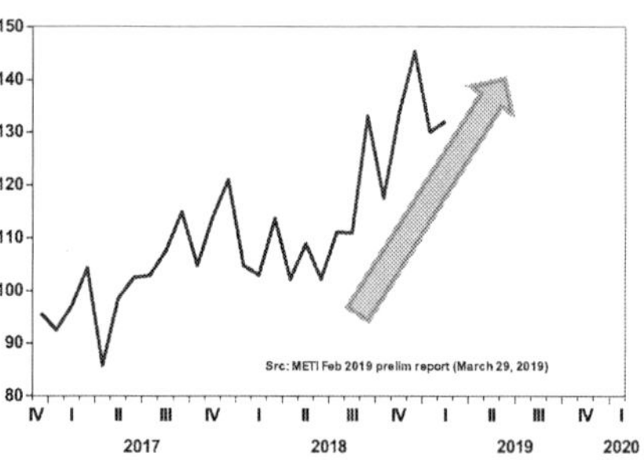


Japan (METI) Wafer Data
Inventories Divided by Shipments

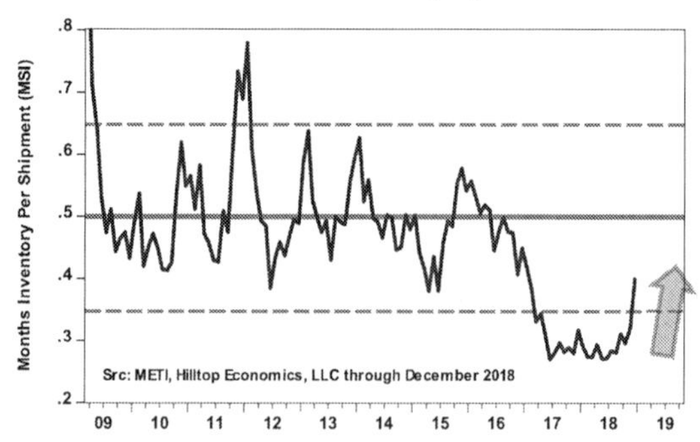


*Duncan Meldrum*
*SPCC April 3, 2019*

22

# Lots of Semiconductor Measures Turning Down

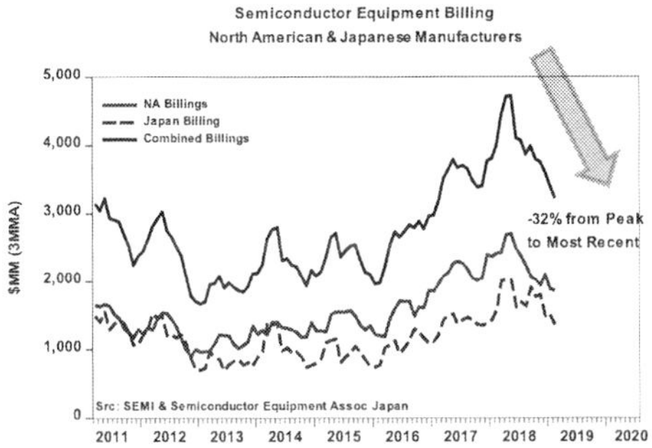


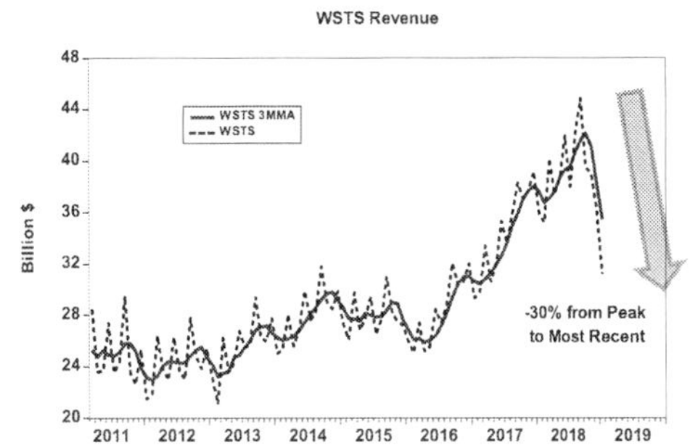


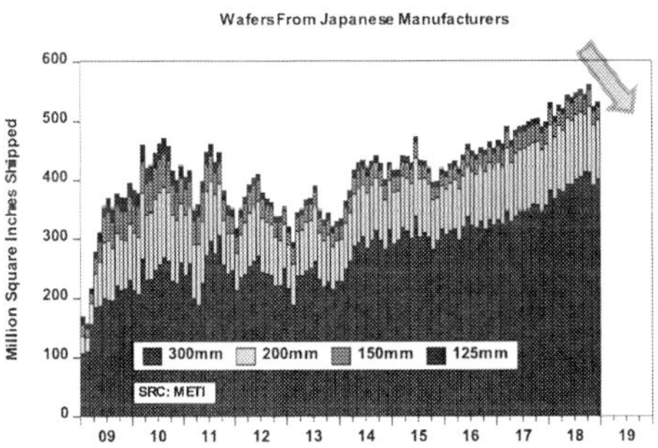


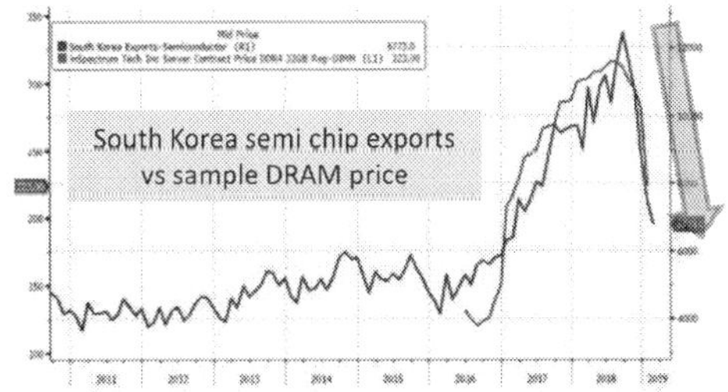


As a reminder, South Korea releases its trade data earlier than most other major economies and is a key link in the global supply chain, making it a bellwether for trade.





# Weaker Final Demand Pulls MSI Down in 2019

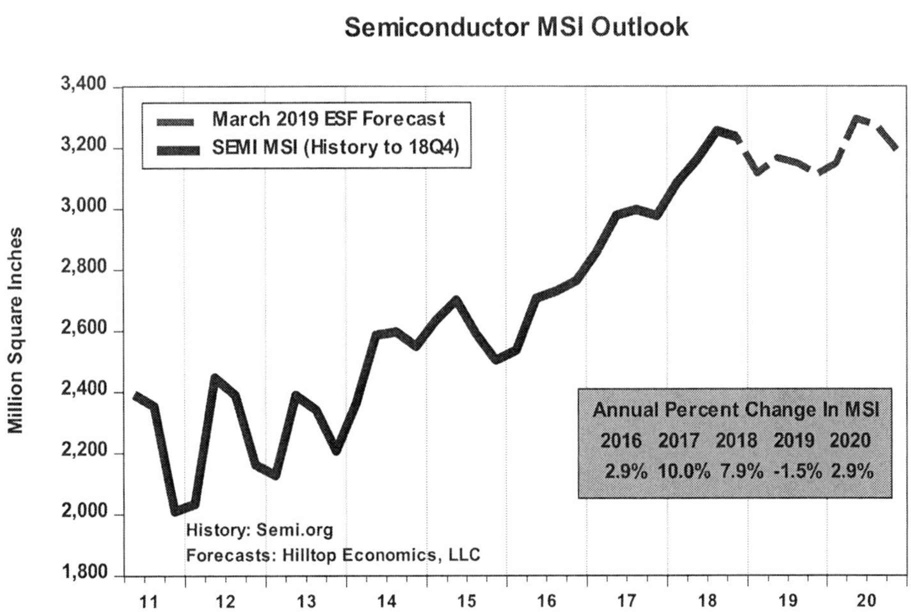


| Annual Percent Change In MSI | | | | |
|---|---|---|---|---|
| 2016 | 2017 | 2018 | 2019 | 2020 |
| 2.9% | 10.0% | 7.9% | -1.5% | 2.9% |

Hilltop Economics

Semiconductors beyond the cycle

## Semiconductors VS. World Real GDP

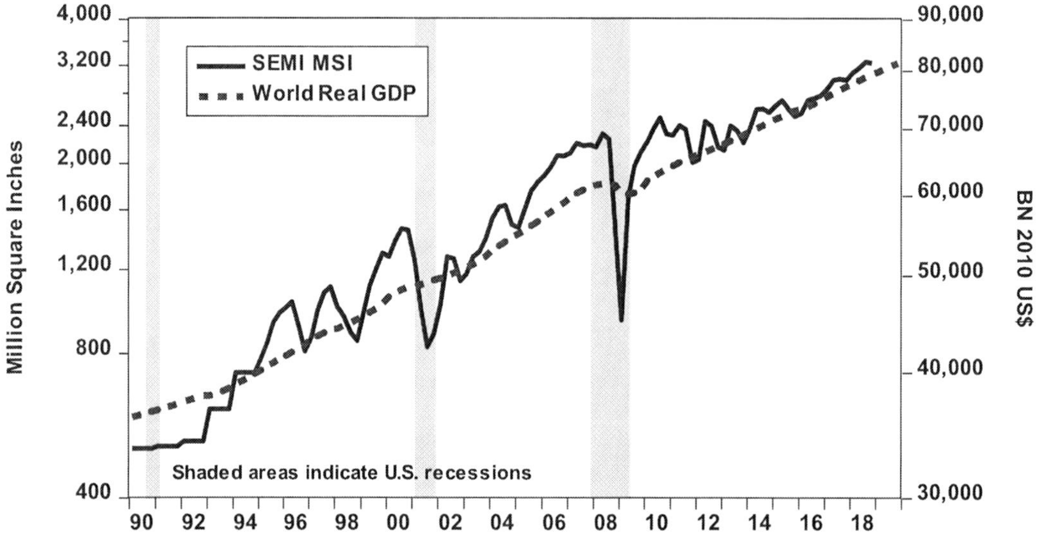


Duncan Meldrum
SPCC April 3, 2019

25

Hilltop Economics

Semiconductors beyond the cycle

## Semiconductor Penetration of the Global Economy

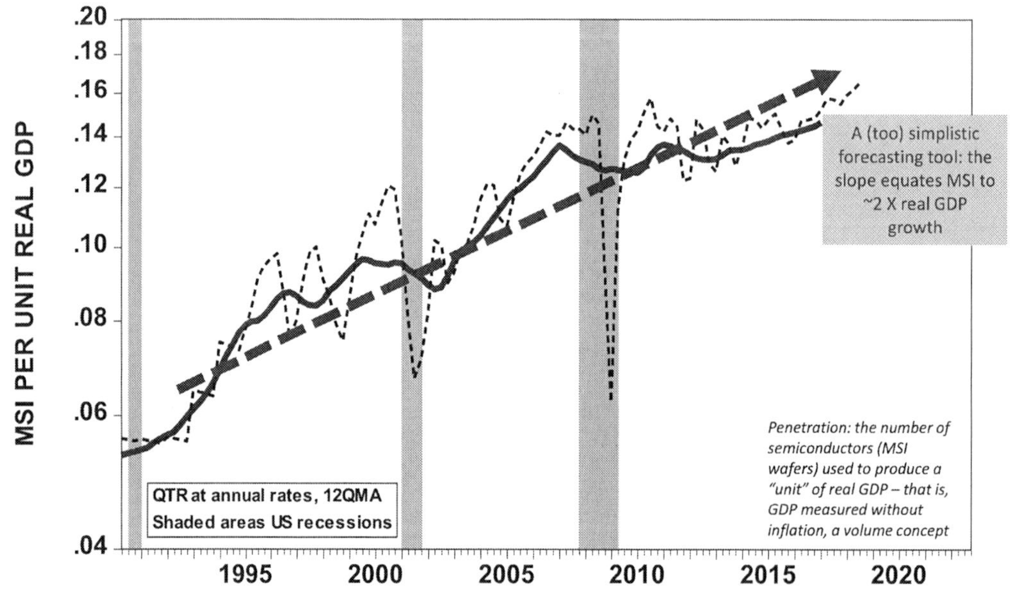


Duncan Meldrum
SPCC April 3, 2019

26

Hilltop Economics

**Semiconductors beyond the cycle**

## Semiconductor Penetration of the Global Economy

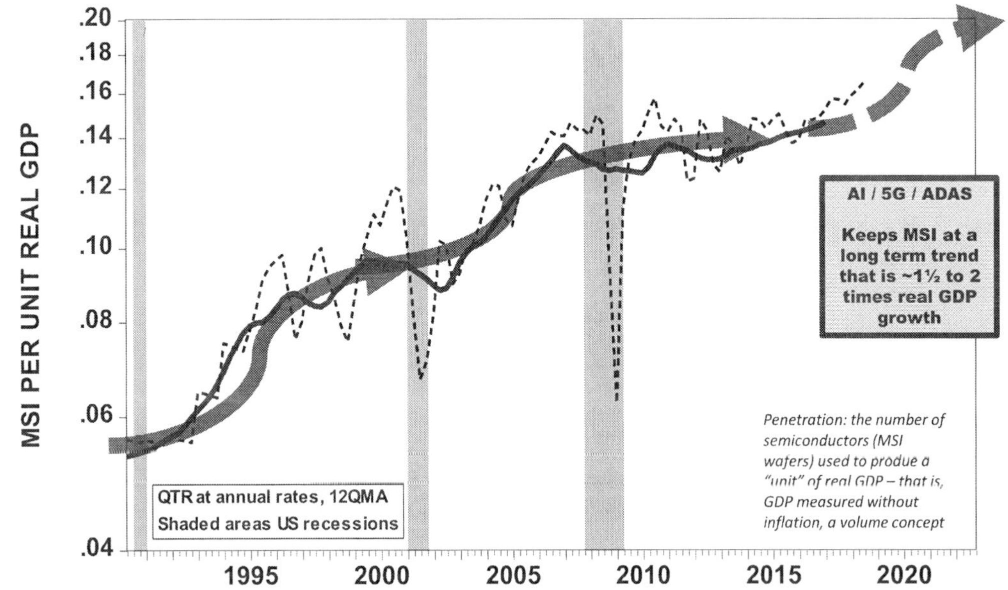


*Duncan Meldrum*
*SPCC April 3, 2019*

27

Hilltop Economics

**SPCC 2019 Forecast**

# Summary: 2019-20

- The world economy has moved into a decelerating phase of the business cycle.
- Consensus expects slightly below-trend growth in key final demand measures that drive semiconductors (investment, consumption).
- ◯ Recession risks rising (economic policy uncertainty).
- Semiconductors slide in 2019, but surge back to a healthy trend (1.5 to 2 times real GDP growth) once past the down cycle.
- Growth Rates:

| Annual % Change | 2017 | 2018 | 2019 | 2020 |
|---|---|---|---|---|
| World Real GDP | 3.1 | 3.0 | 2.7 | 2.7 |
| Investment | 4.7 | 4.1 | 3.0 | 3.2 |
| Consumption | 3.1 | 2.8 | 2.6 | 2.5 |
| *SEMI MSI* | *10.0* | *7.9* | *(1.5)* | *2.9* |

*Duncan Meldrum*
*SPCC April 3, 2019*

28

Hilltop Economics

# Duncan Meldrum
# Chief Economist
# Hilltop Economics, LLC


**duncan.meldrum@hilltopeconomics.com**

**610.709.7951**


*Duncan Meldrum*
*SPCC April 3, 2019*

29

Hilltop Economics

# Jobs Markets Support Consumption

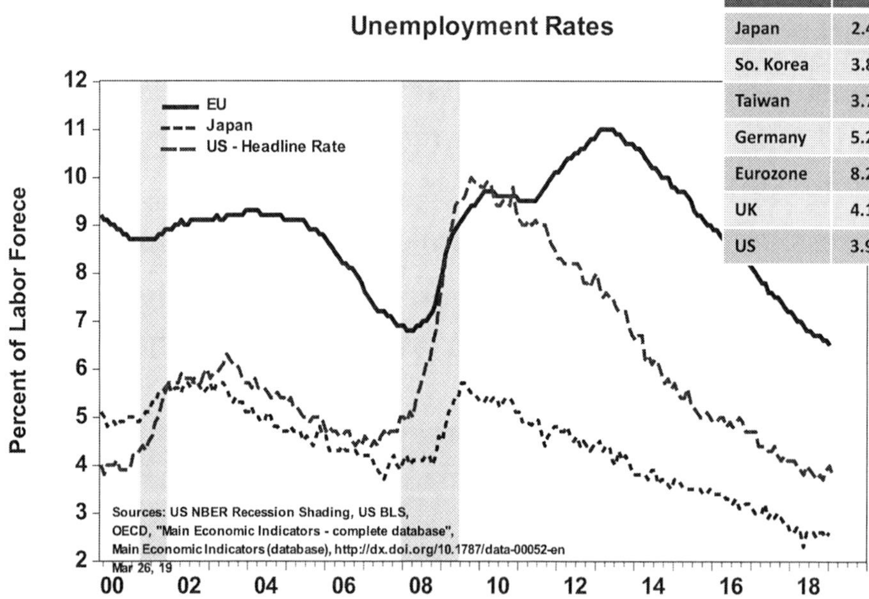


| % Unemp. | 2018 | 2019 | 2020 |
|---|---|---|---|
| Japan | 2.4 | 2.4 | 2.4 |
| So. Korea | 3.8 | 3.9 | 3.9 |
| Taiwan | 3.7 | 3.7 | 3.8 |
| Germany | 5.2 | 4.9 | 4.8 |
| Eurozone | 8.2 | 7.8 | 7.6 |
| UK | 4.1 | 4.1 | 4.2 |
| US | 3.9 | 3.7 | 3.7 |

*Duncan Meldrum*
*SPCC April 3, 2019*

30

*Hilltop Economics*

# Consumer Confidence Indexes Are High

***But off the 2018 peaks, with the exception of China***

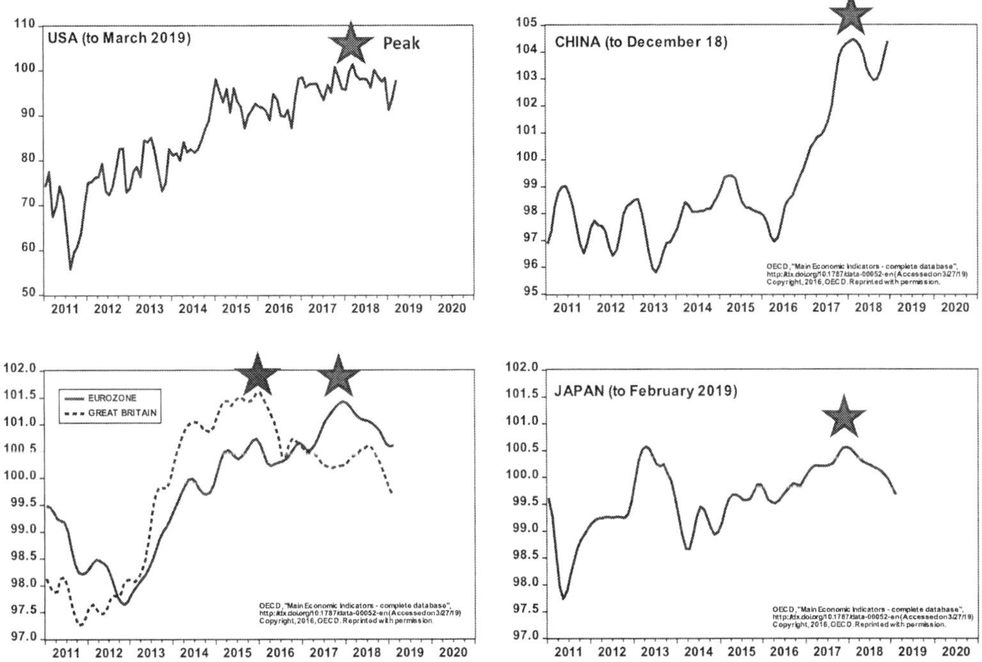


*Duncan Meldrum*
*SPCC April 3, 2019*

31

*Hilltop Economics*

# CPI Inflation Weakens in 2019

***Despite tightness in labor markets and closed output gaps***

| CPI % Change | 2016 | 2017 | 2018 | 2019 | 2020 |
|---|---|---|---|---|---|
| China | 2.0 | 1.6 | 2.1 | 2.1 | 2.1 |
| Japan | (0.1) | 0.5 | 1.0 | 0.7 | 1.0 |
| South Korea | 1.0 | 1.9 | 1.5 | 1.3 | 1.6 |
| Taiwan | 1.4 | 0.6 | 1.4 | 0.8 | 1.1 |
| Germany | 0.5 | 1.8 | 1.8 | 1.5 | 1.6 |
| France | 0.2 | 1.0 | 1.9 | 1.3 | 1.5 |
| Italy | (0.1) | 1.2 | 1.2 | 0.9 | 1.2 |
| UK | 0.7 | 2.7 | 2.6 | 2.2 | 2.2 |
| US | 1.3 | 2.1 | 2.4 | 1.8 | 2.2 |

Hilltop Economics, Based on Consensus Forecasts, March 19
2018 actual vs. forecast: China: -0.2, Germany: +0.1, US & Japan no error

*Duncan Meldrum*
*SPCC April 3, 2019*

32

*Hilltop Economics*

# Oil Prices

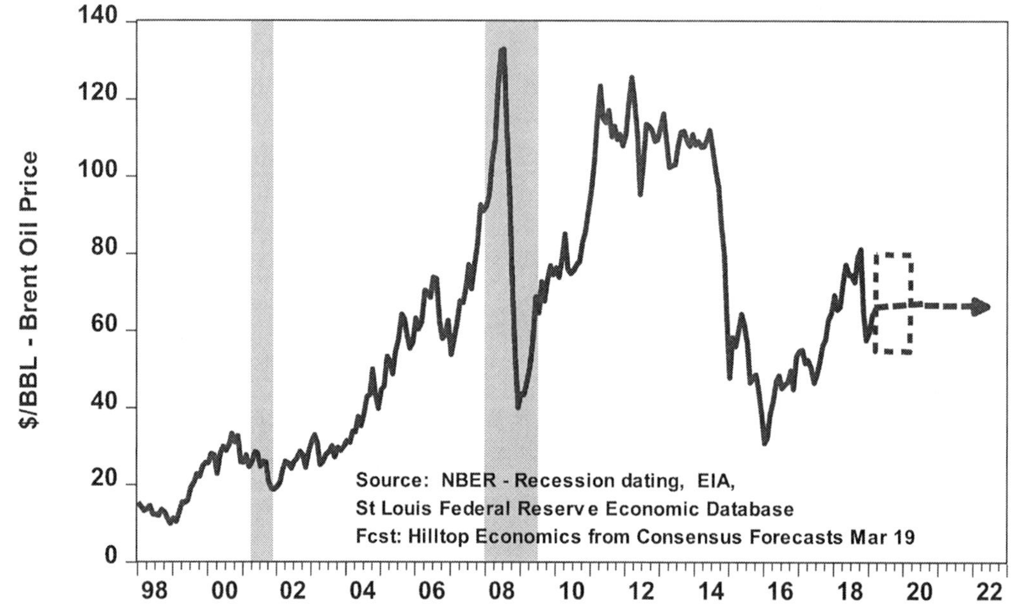


*Duncan Meldrum*
*SPCC April 3, 2019*

33

*Hilltop Economics*

# Equity Prices Well Above December Lows as of End-March, but Volatility Unsettling

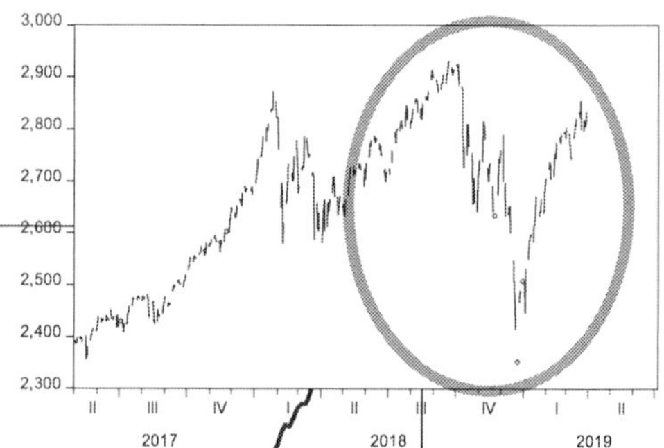


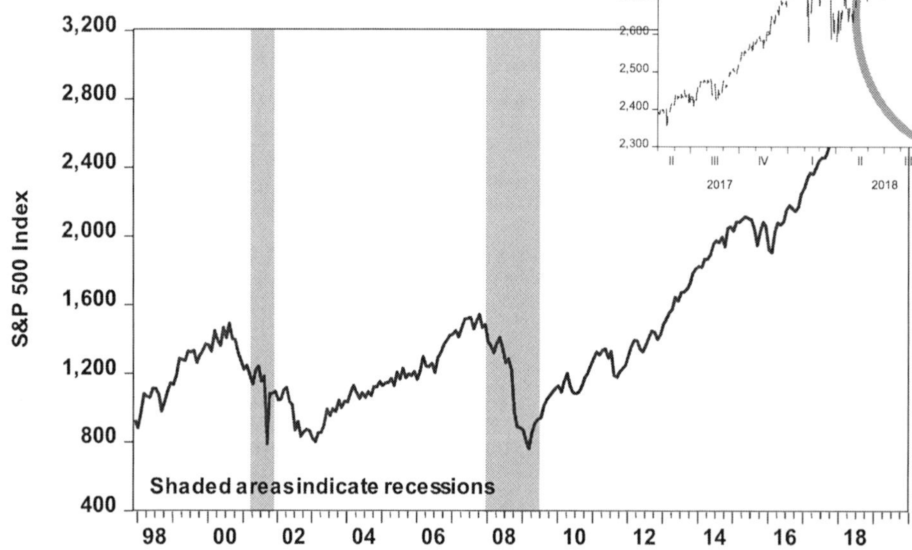


*Duncan Meldrum*
*SPCC April 3, 2019*

34

34

*Hilltop Economics*

## Consensus: Profits Grow in Line with Revenue At Least Through 2020

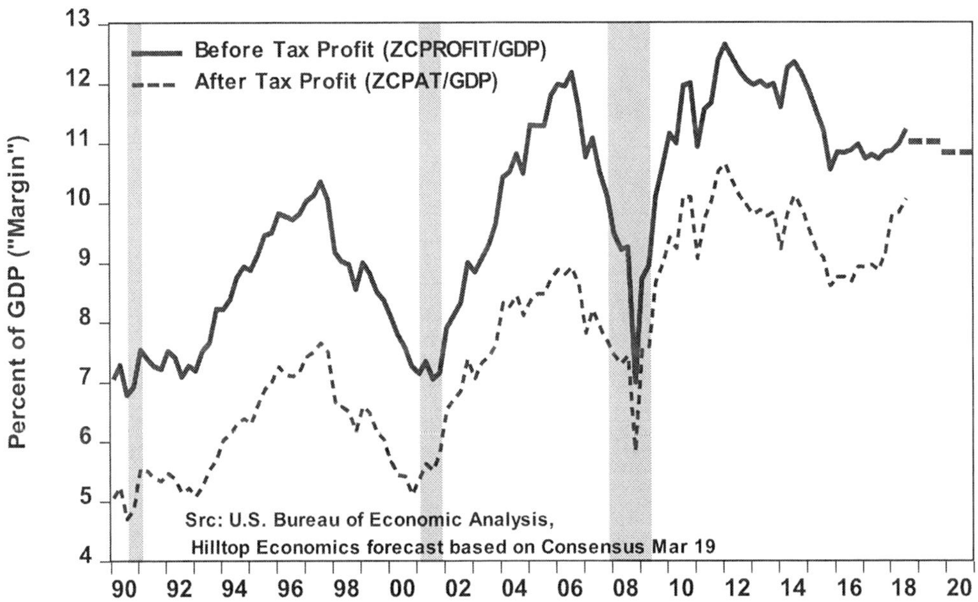


*Duncan Meldrum*
*SPCC April 3, 2019*

35

*Hilltop Economics*

## Dollar Strengthened, Trade War an Issue

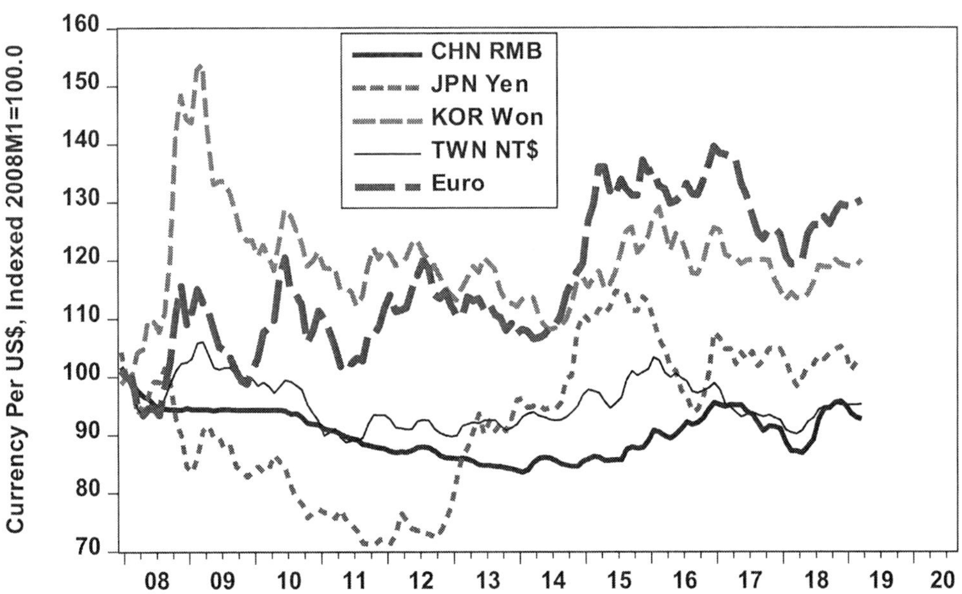


*Duncan Meldrum*
*SPCC April 3, 2019*

36

Hilltop Economics

# Metrics

- **Real GDP**: Gross Domestic Product (final "real" or "volume" demand for goods and services, in the largest 85 countries measured in Billions of 2010 US$ - that is, with inflation removed)
  - **Real consumer spending** (for key 45 countries in Consensus Economics)
  - **Real business investment** (for key 45 countries in Consensus Economics)
  - *WHY Real GDP and its final demand components? Semiconductors have become so ubiquitous, used in almost every product and in support of almost every service, that broad final demand measures best explain MSI trends.*
- **Million Square Inches (MSI)** of silicon wafers shipped to semiconductor manufacturers (Source: SEMI.ORG)
  - *WHY MSI? Best volume measure of semiconductors, to relate to the volume of demand as measured by real GDP and other units of products*
- WSTS **semiconductor revenue** (World Semiconductor Trade Statistics) - $000 sales by major region reported by WSTS.ORG
- Semiconductor **Equipment Billings** (~90% of worldwide semiconductor equipment sales)
  - Reported for North America by SEMI.ORG
  - Reported for Japan by SEAJ.ORG

*Duncan Meldrum*
*SPCC April 3, 2019*

37

Hilltop Economics

# Semiconductor Data Concerns

- Disconnect between MSI and WSTS revenue
- U.S. Government shutdown
- METI discontinuing monthly wafers (?)
- MSI quarter-to-quarter distortions due to increases in long-term contracts (?)

*Duncan Meldrum*
*SPCC April 3, 2019*

38

# Particle adsorption model on silicon wafer for clarifying required ultrapure water quality


*Kurita Water Industries Ltd.*

○ Yoichi Tanaka, Yoshiaki Ide, Hideaki Iino, Minoru Uchida




KURITA

## Introduction

KURITA

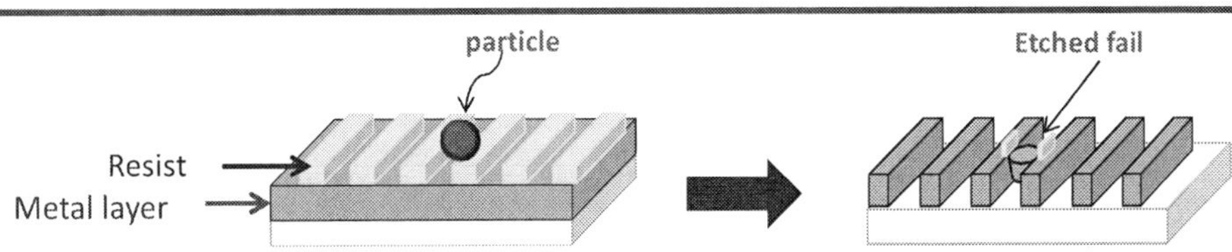


| Year of Production | 2018 | 2019 | 2020 | 2021 |
|---|---|---|---|---|
| Critical Metal [ppt] | <1.0 | <1.0 | <1.0 | <1.0 |
| Other ion [ppt] | <50 | <50 | <50 | <50 |
| Critical Particle size [nm] (Non electric active) | 18 | 14 | 14 | 12 |
| Number of Particles[pcs/L] | 10,000 | 10,000 | 10,000 | 10,000 |

(IRDS 2017)

**Proper control of the particles issue is important.**
**IRDS shows the critical particle sizes and the numbers.**

# Introduction

KURITA

## Previous study

① Particle adsorption between liquid and wafer (such as silicon, germanium and patterned wafer) has been studied.(1997～2015)

Concern

- Some studies are based on dip experiment
- Data are mostly on high concentration of particles
- No criteria for particle sizes and numbers for UPW have been shown

② Required UPW quality were suggested by relation between particles in UPW and silicon wafer. (ex. *ECS* 2015)

Concern

- Data are mostly on high concentration of particles
- No prediction procedure for small particle sizes has been proposed



3

# Introduction

KURITA

## To improve more ・・・ (Kurita idea)

Practical !

**Approach to predict adsorption of small particles at low concentration is needed.**

## In this research,

①Model equation was derived based on the relation between the number of particles in UPW and the number of particles on the silicon wafer **focused on the low particle concentration** in UPW.

②**Requirements for** water quality in terms of particles especially of **smaller sizes were predicted.**

# Introduction

KURITA

**Kurita possesses the method of detecting the number of 10 nm particles and it's composing elements in UPW.**

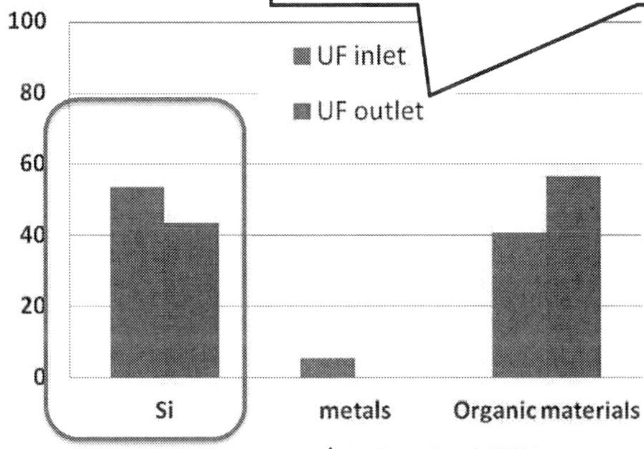


(Y. Tanaka UPW micro 2015)

Behavior of $SiO_2$ particle is selected as 1st priority to research.

**Si and organic matter are observed at both Inlet and outlet of UF.**



# Step 1: Correlation between particle in UPW and added defects

KURITA

**Method:**

6 inch (150mm) wafer
Particle; 120 nm $SiO_2$ particles, 100 pcs/mL
Pretreatment; 30ppm $O_3$ last
Procedure; With Particle 20～300 sec (1000rpm) → Dry 30sec (500rpm).

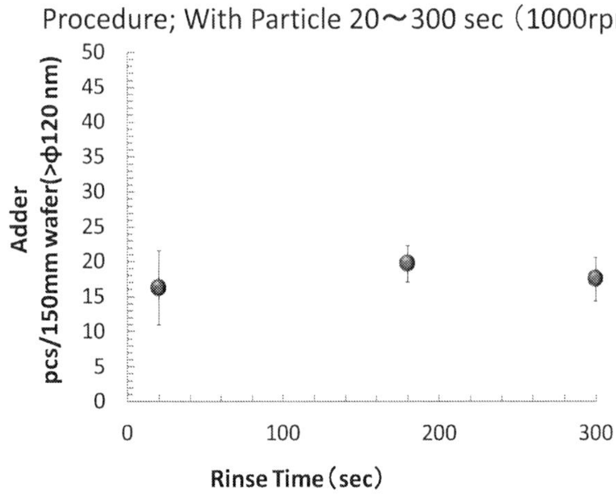


**Result:**

Adsorption on the wafer seems saturated as the rinse time gets longer.

**It seems like an equilibrium adsorption**

## Step 1: Correlation between particle in UPW and added defects

KURITA

### Method:

6 inch (150mm) wafer

Pretreatment; 30ppm $O_3$ last

Procedure; With Particle $SiO_2$ 500pcs/mL 20 sec (1000rpm) → UPW Rinse (Rinse time 0 sec, 20 sec, 60 sec, 300 sec) 1000rpm→ Dry 30sec (500rpm).

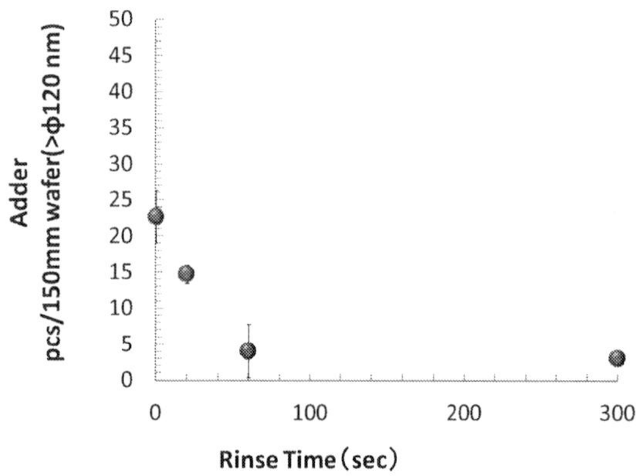


### Result:

① Adsorption on the wafer gets decreased as the rinse time of UPW gets longer.

② Finally, the adsorption level reaches as low as treated with the UPW only.

**It seems like an equilibrium adsorption**



**What type ??**

## Step 1: Correlation between particle in UPW and added defects

KURITA

### Method:

6 inch wafer

Particle; 120 nm $SiO_2$ particles, 0～500 pcs/mL

Pretreatment; 30ppm $O_3$ last

Procedure; With Particle 20 sec (1000rpm) → Dry 30sec (500rpm).

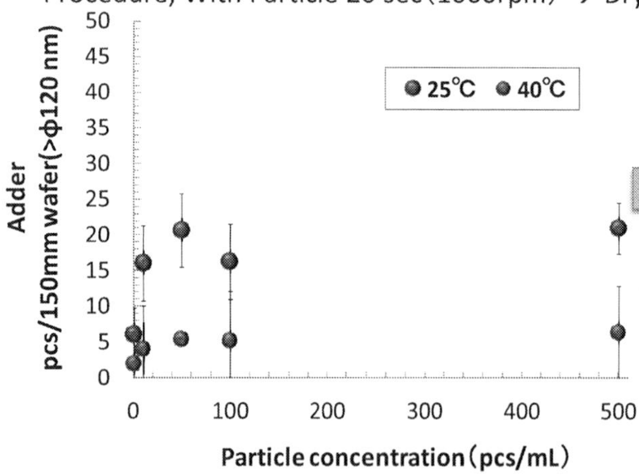


### Result:

Maximum adsorption on wafers changes as the temperature changes

**It is not a Langmuir type, and the results indicate that all adsorption sites are not homogeneous.**



8

## Step1: Model image

Image of model

⇒ Easily adsorbable sites are occupied first.

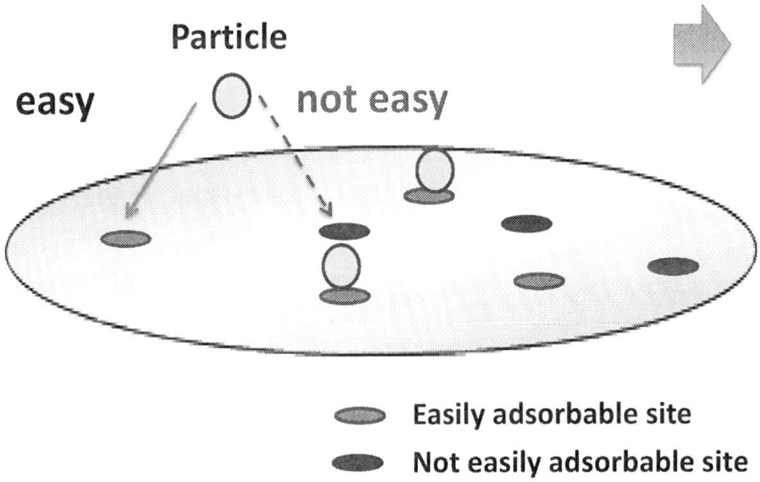


$$v = a \ln(K_A C)$$



## Step 1: Verification of the model

Verification of the model    $V = a \ln(C) + a \ln(Ka)$

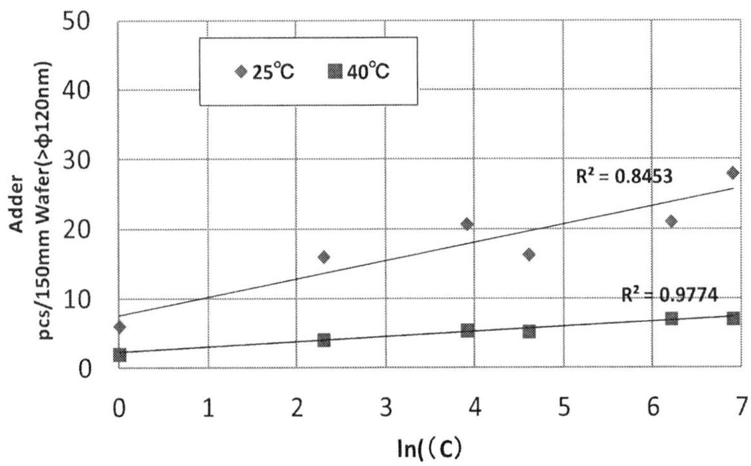


Linearity was obtained for different temperatures, which verifies the equation.

# Approach

STEP 1

**Determine the formula of particle adsorption on wafer**

$$v = a \ln(K_A C)$$

V; Required criteria of particle number on wafer
C; Required UPW quality for particle number
$K_A$; Equilibrium constant
a; Coefficient

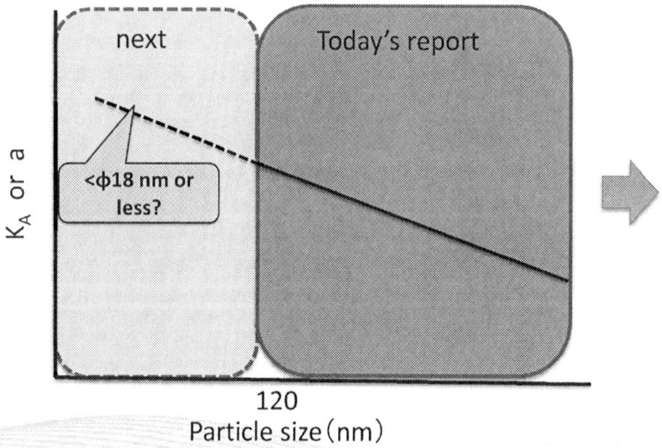


**STEP 2**
**Extrapolate from experimental data to forecast parameters for small particle sizes**

**STEP3**
**Calculate required UPW quality for particles using the formula and parameters**





# STEP 2: Experiments for prediction

Method:
6 inch wafer
Particle; $SiO_2$ particles (120 nm, 500 nm, 1000nm)
Pretreatment; 30ppm $O_3$ last
Procedure; With Particle 20 sec (1000rpm) → Dry 30sec (500rpm).

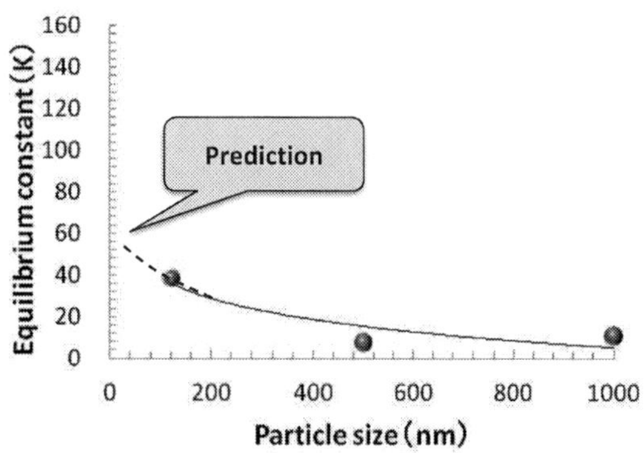


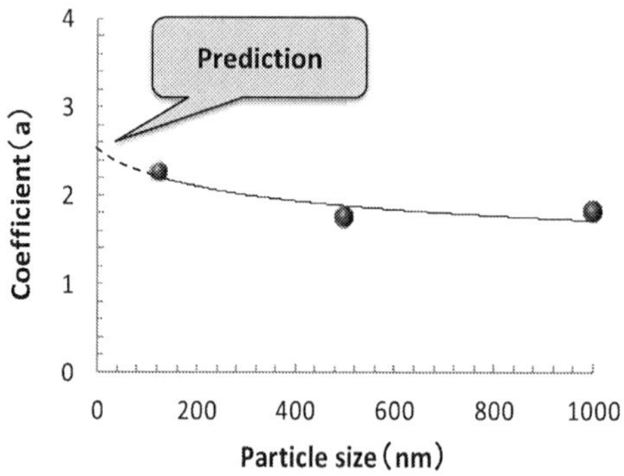


Predict the value of each factor for small particle sizes

## STEP 3: Predict the UPW quality

KURITA

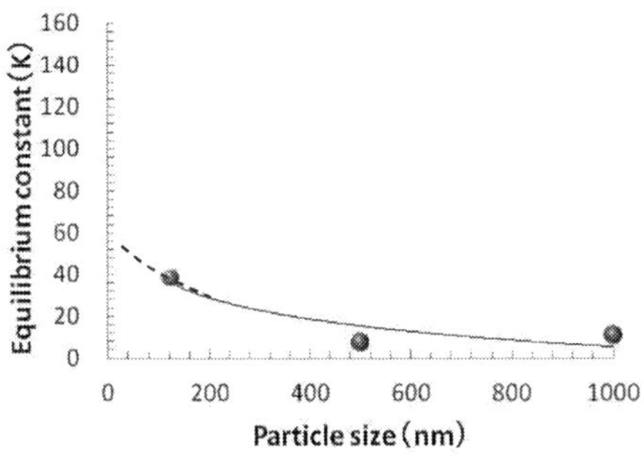


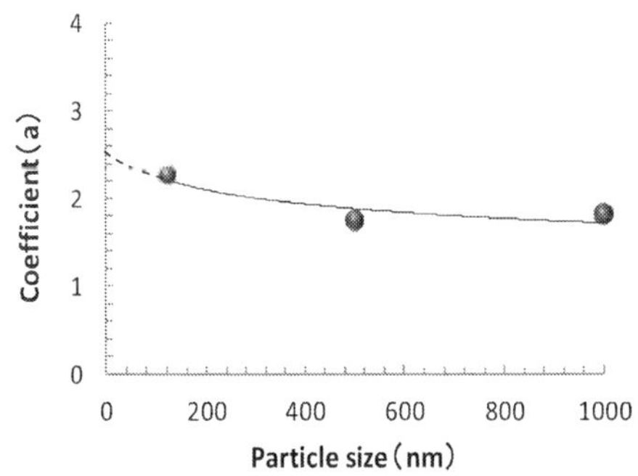


IRDS

ex) If we need to control adder 30pcs/ 300 mm wafer@18 nm up

$$v = a\ln(K_A C)$$

**<u>Number of particles of ϕ18 nm or under in the UPW should be controlled below 1,000 pcs/L</u>**

***We need to check whether the same equilibrium adsorption will be applied at small particle size.**



13

## Approach

KURITA

**STEP 1**

**Determine the formula of particle adsorption on wafer**

$$v = a\ln(K_A C)$$

V; Required criteria of particle number on wafer

C; Required UPW quality for particle number

$K_A$ ; Equilibrium constant

a; Coefficient

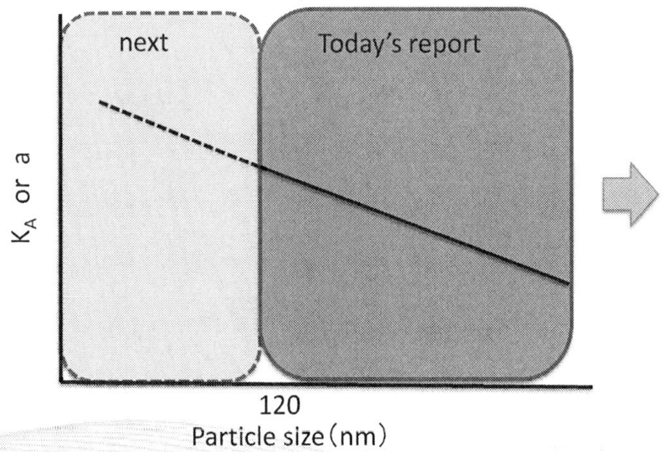


**STEP 2**
**Extrapolate from experiment to forecast parameter for small particle size**

**STEP3**
**Required UPW quality for particle number is calculated by parameter and formula**



14

## Summary

- Equilibrium model was obtained for particle adsorption
- Kurita suggests an approach to predict the required UPW quality.
- To control adder 30 pcs/300mm wafer of 18 nm or larger, particles in the UPW should be controlled less than 1,000 pcs/L at 18 nm.
- Further research is needed to improve the accuracy of prediction for small particle sizes.



15

KURITA

Thank you for your attention.

Contact:
Yoichi Tanaka (Kurita)
yoichi.tanaka@kurita.co.jp
+81-280-54-2607

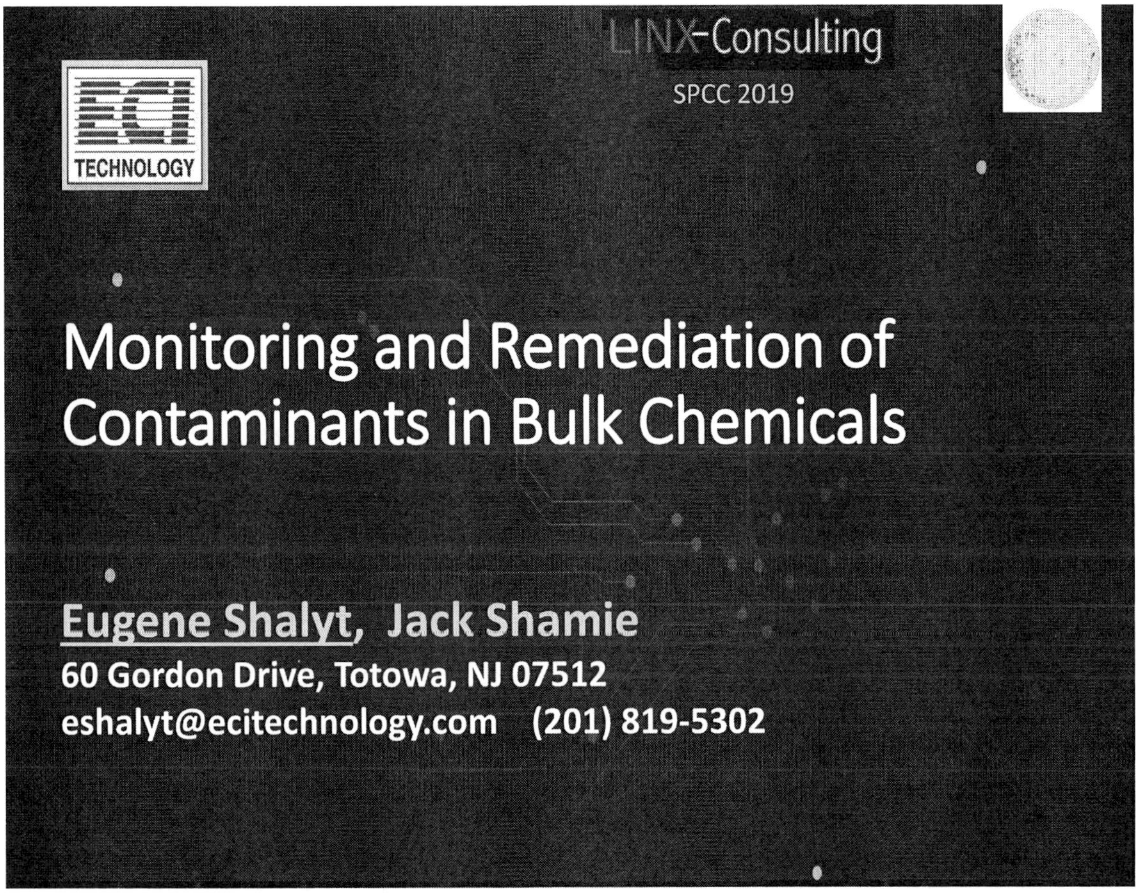
LINX-Consulting
SPCC 2019
ECI TECHNOLOGY
Monitoring and Remediation of Contaminants in Bulk Chemicals
Eugene Shalyt, Jack Shamie
60 Gordon Drive, Totowa, NJ 07512
eshalyt@ecitechnology.com (201) 819-5302

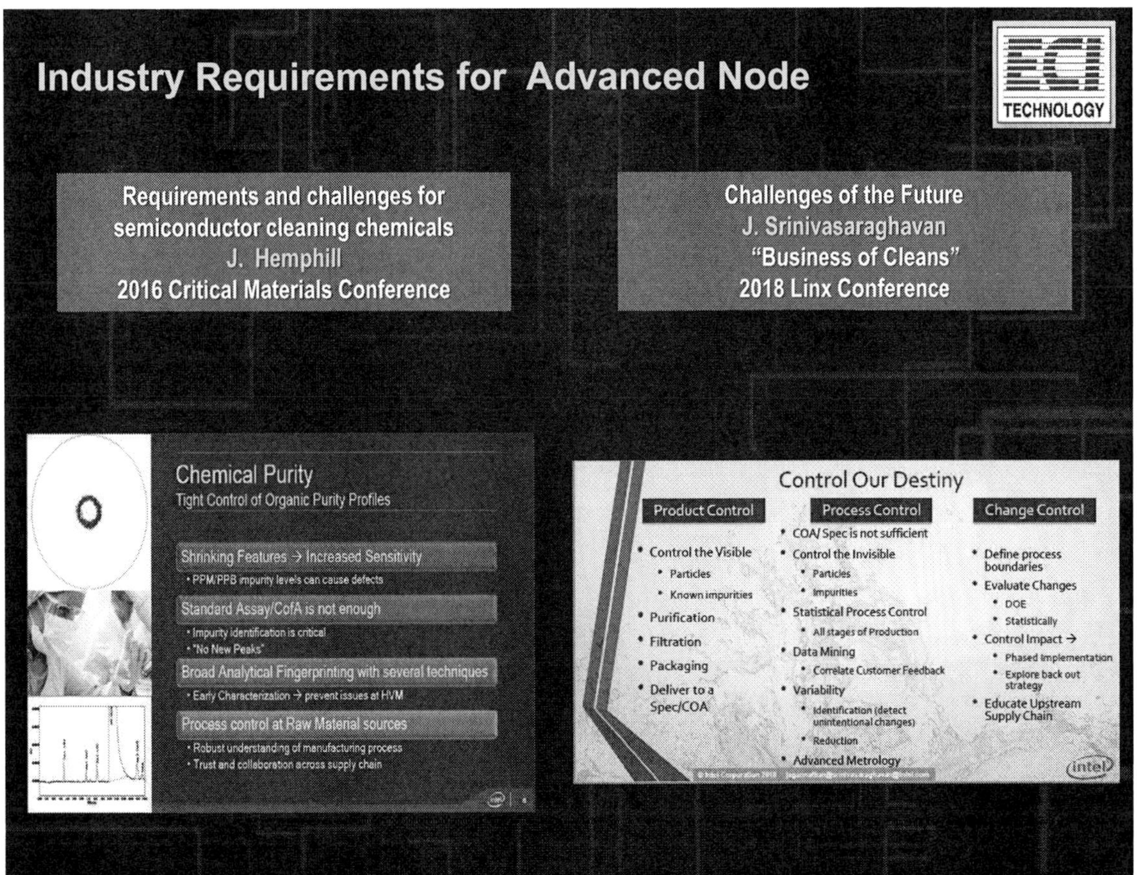
Industry Requirements for Advanced Node
ECI TECHNOLOGY
Requirements and challenges for semiconductor cleaning chemicals
J. Hemphill
2016 Critical Materials Conference
Challenges of the Future
J. Srinivasaraghavan
"Business of Cleans"
2018 Linx Conference
Chemical Purity
Tight Control of Organic Purity Profiles
Shrinking Features → Increased Sensitivity
• PPM/PPB impurity levels can cause defects
Standard Assay/CofA is not enough
• Impurity identification is critical
• "No New Peaks"
Broad Analytical Fingerprinting with several techniques
• Early Characterization → prevent issues at HVM
Process control at Raw Material sources
• Robust understanding of manufacturing process
• Trust and collaboration across supply chain
Control Our Destiny
Product Control
• Control the Visible
• Particles
• Known impurities
• Purification
• Filtration
• Packaging
• Deliver to a Spec/COA
Process Control
• COA/ Spec is not sufficient
• Control the Invisible
• Particles
• Impurities
• Statistical Process Control
• All stages of Production
• Data Mining
• Correlate Customer Feedback
• Variability
• Identification (detect unintentional changes)
• Reduction
• Advanced Metrology
Change Control
• Define process boundaries
• Evaluate Changes
• DOE
• Statistically
• Control Impact →
• Phased implementation
• Explore back out strategy
• Educate Upstream Supply Chain
intel

# Generalized Diagram of Semiconductor Processing

TECHNOLOGY

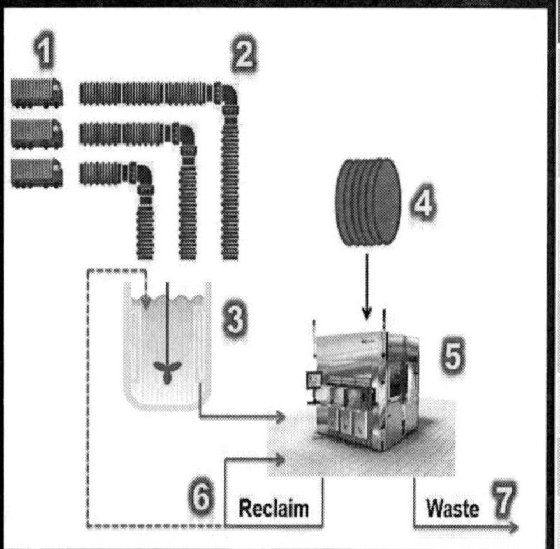


| Test point | Description | Concern |
|---|---|---|
| 1 | Raw materials as manufactured by chemical supplier | Does CoA applies to EVERY container? Degradation in transit? Does CoA cover ALL parameters? |
| 2 | Raw material as it passed through facility distribution system for bulk chemical | Degradation/Contamination of chemical in transit e.g. $O_2$, $CO_2$ exposure, particle and metal characterization |
| 3 | Process blend of individual chemicals | Was blend prepared properly? Compensation for degradation of components |
| 4 | Wafer | Outside of scope of this presentation |
| 5 | Process equipment parameters | Outside of scope of this presentation |
| 6 | Reclaim | Exact degradation of components for proper replenishment |
| 7 | Waste | Compliance |

# Case Study: Bulk Chemical X

TECHNOLOGY

## Problem Statement

- End-user reported problems loosely related to lots of raw materials

## Investigation

- End-user lab compared good and bad lots but could not find difference

# ECI In-house Tests

NIR Spectroscopy

Cyclic Voltammetric Stripping

Raman Spectroscopy

E-Chem Impedance

ICP

Conductometric Titration

Potentiometric Titration

UV-Vis Spectroscopy

HPLC

Surface Tension

Ion Chromatography

Specific Gravity

Dissolved Oxygen

OCP (Open Circuit Potentiometry)

ECI screened Good and Bad Samples by variety of analytical methods

# ECI Test Results

## Results

- Most analytical methods including ICP-MS with **ppt** sensitivity did not show difference between "Good" and "Bad"
- Surprisingly, cyclic voltammetry electrochemical method with sensitivity in **ppm** showed large difference indicating presence of Red/Ox active contaminant

## ECI Electrochemical Method

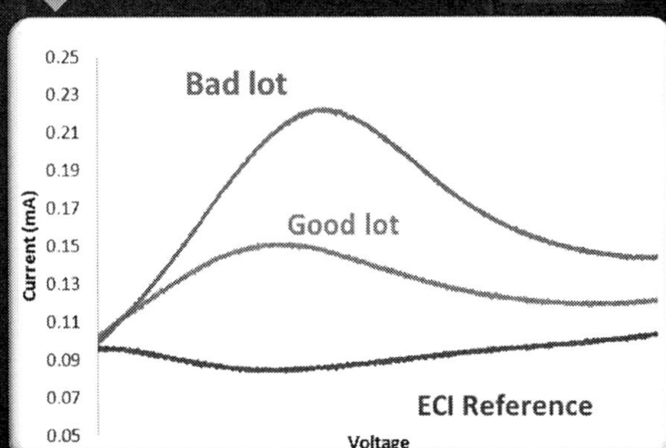

# Cyclic Voltammetry Method

ECI TECHNOLOGY

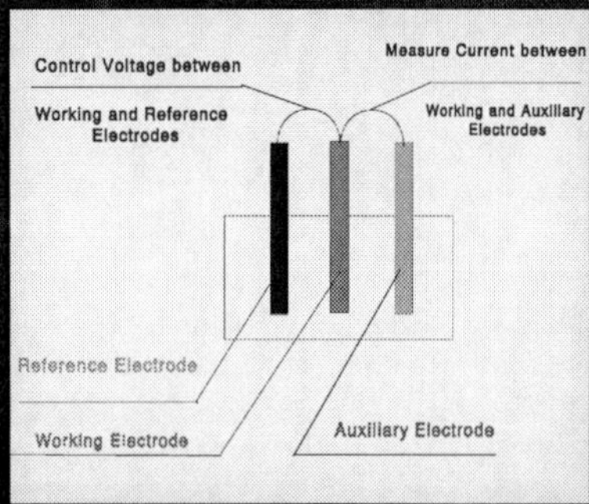


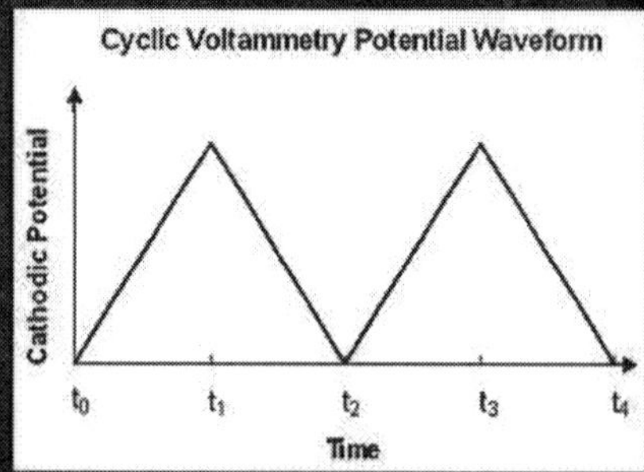


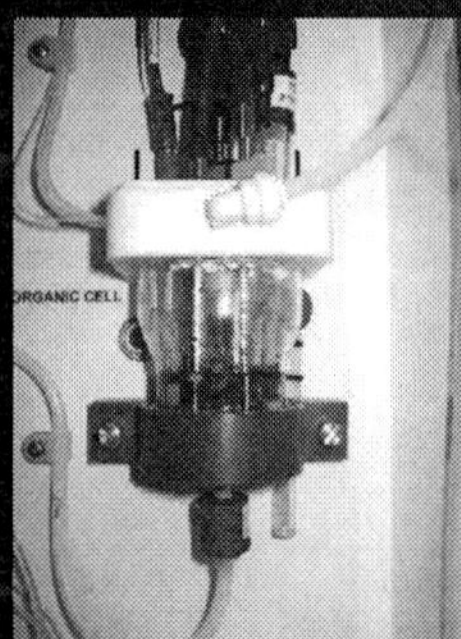

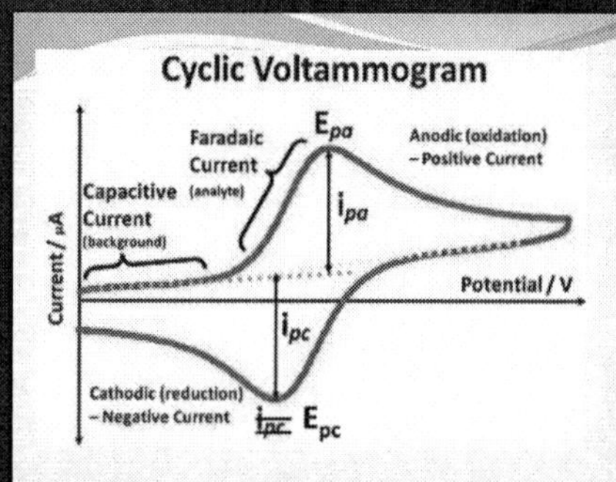


# Identifying Contaminant

ECI TECHNOLOGY

ECI Electrochemical Method

| Compound | Match | Stability in Bulk Chemical |
|---|---|---|
| $A_xB_yC_z$ | Clean bulk Chemical | Stable |
| $A_xB_yC_{z-1}$ | Contaminated bulk chemical | Stable |
| $A_xB_{y+1}C_{z-1}$ | | Unstable |
| $A_xB_{y+1}C_{z+4}$ | Clean bulk chemical | Stable |

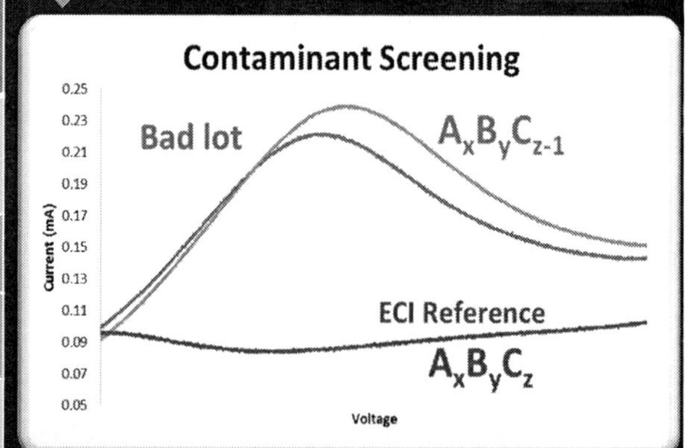

# Analytical Method

Cyclic Voltammetry analytical method was optimized to provide repeatable and accurate measurement of contaminant in ppm concentration range

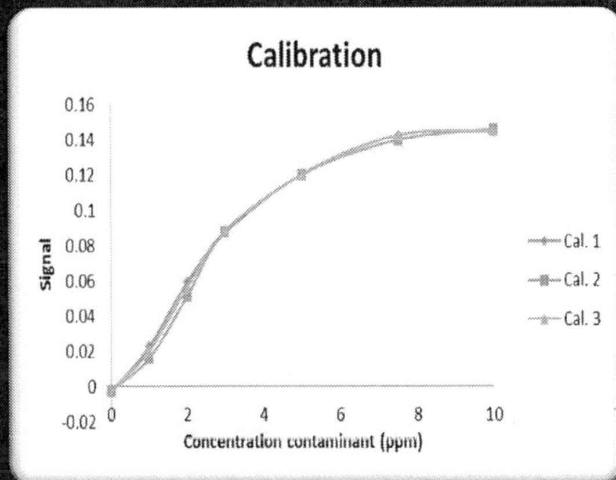


# Remediation of Contaminated Chemical

Contaminant lacks one "C" atom as compared to good product

Can contaminated chemical be remediated by treatment with C ?

Brief sparging with C gas removed contaminant signature from both "bad lot" and simulated contaminant standard

$$A_xB_yC_{z-1} + 1/n\,C_n \Rightarrow A_xB_yC_z$$

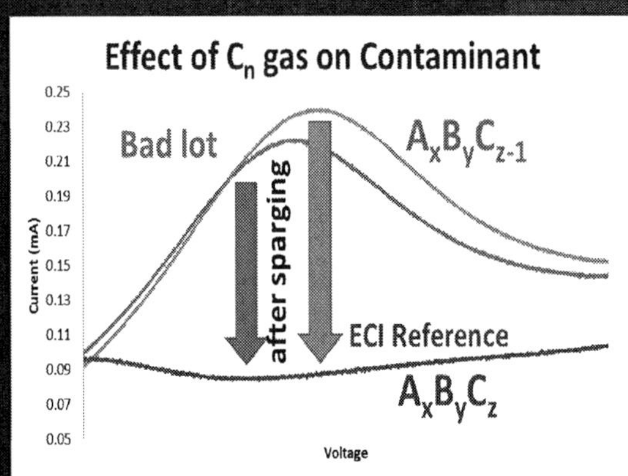

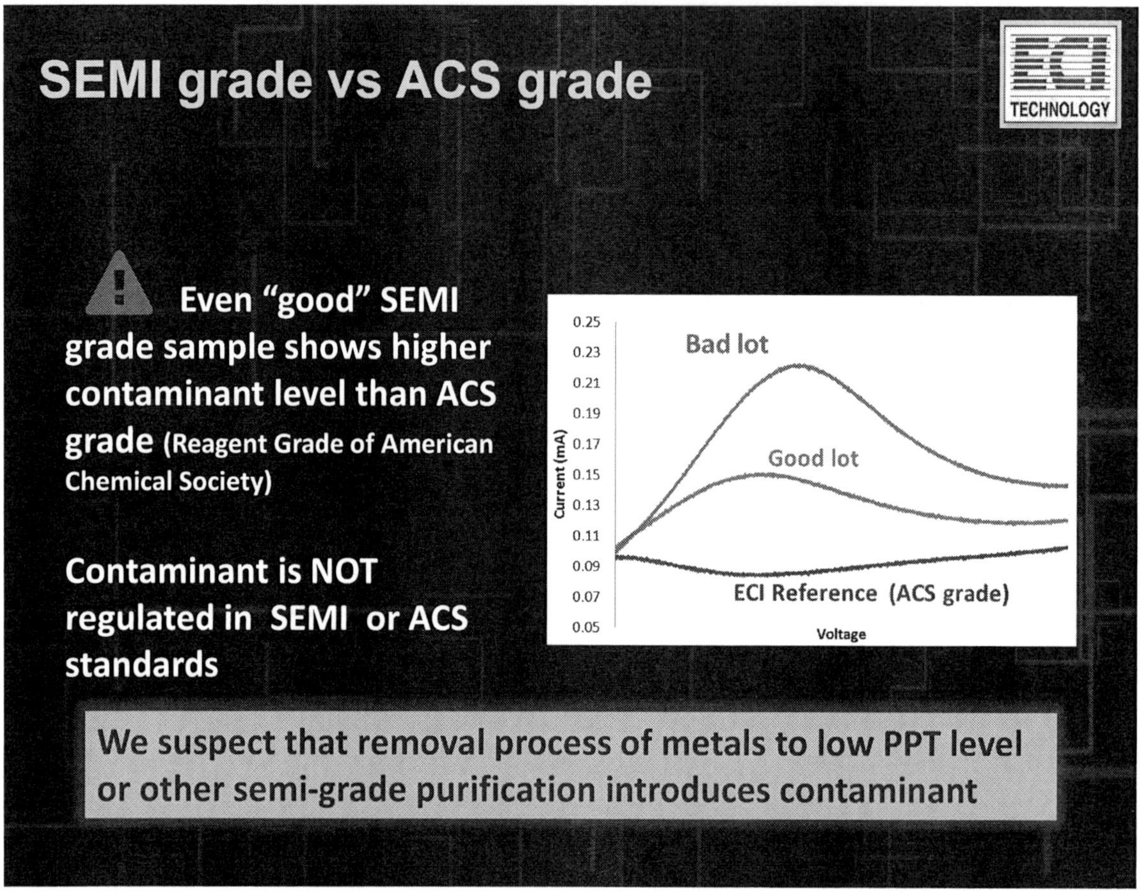
SEMI grade vs ACS grade
TECHNOLOGY
Even "good" SEMI grade sample shows higher contaminant level than ACS grade (Reagent Grade of American Chemical Society)
Contaminant is NOT regulated in SEMI or ACS standards
0.25
0.23
0.21
0.19
0.17
0.15
0.13
0.11
0.09
0.07
0.05
Current (mA)
Bad lot
Good lot
ECI Reference (ACS grade)
Voltage
We suspect that removal process of metals to low PPT level or other semi-grade purification introduces contaminant

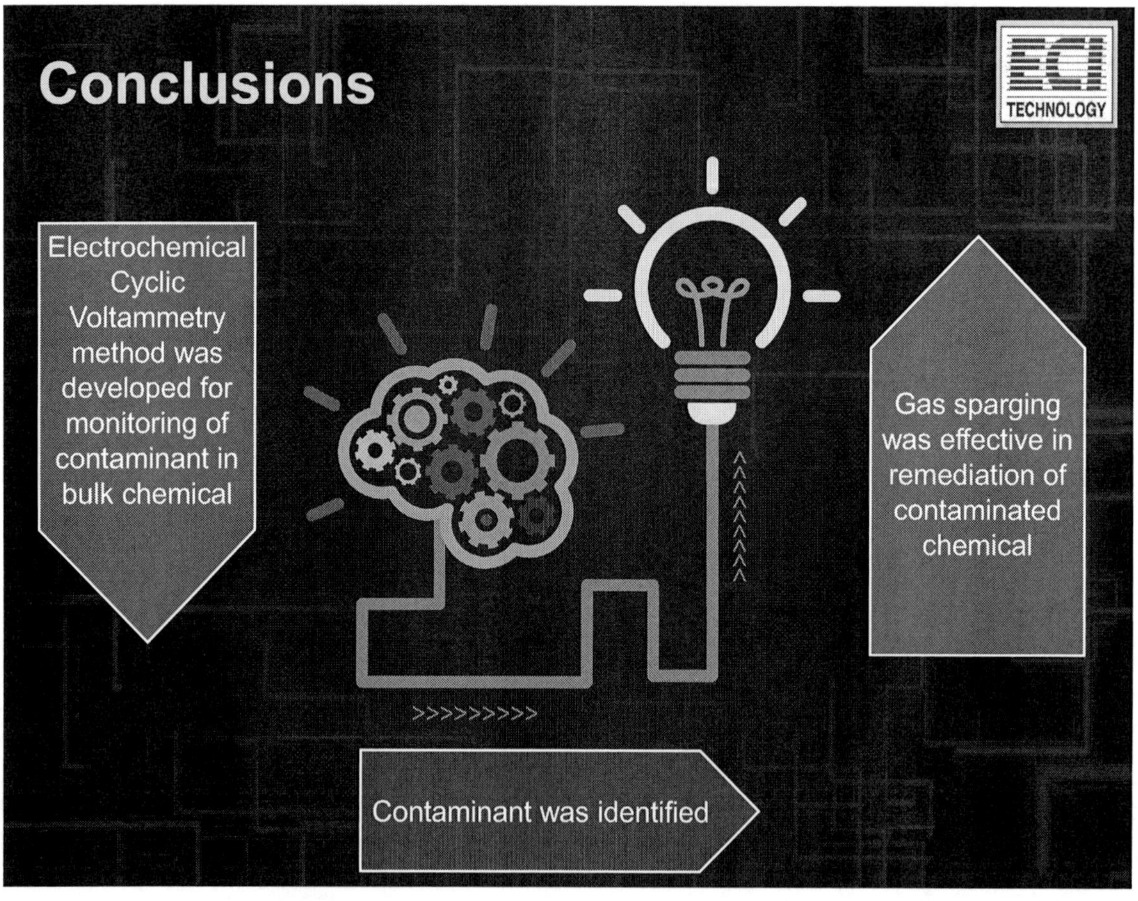
Conclusions
TECHNOLOGY
Electrochemical Cyclic Voltammetry method was developed for monitoring of contaminant in bulk chemical
Gas sparging was effective in remediation of contaminated chemical
Contaminant was identified

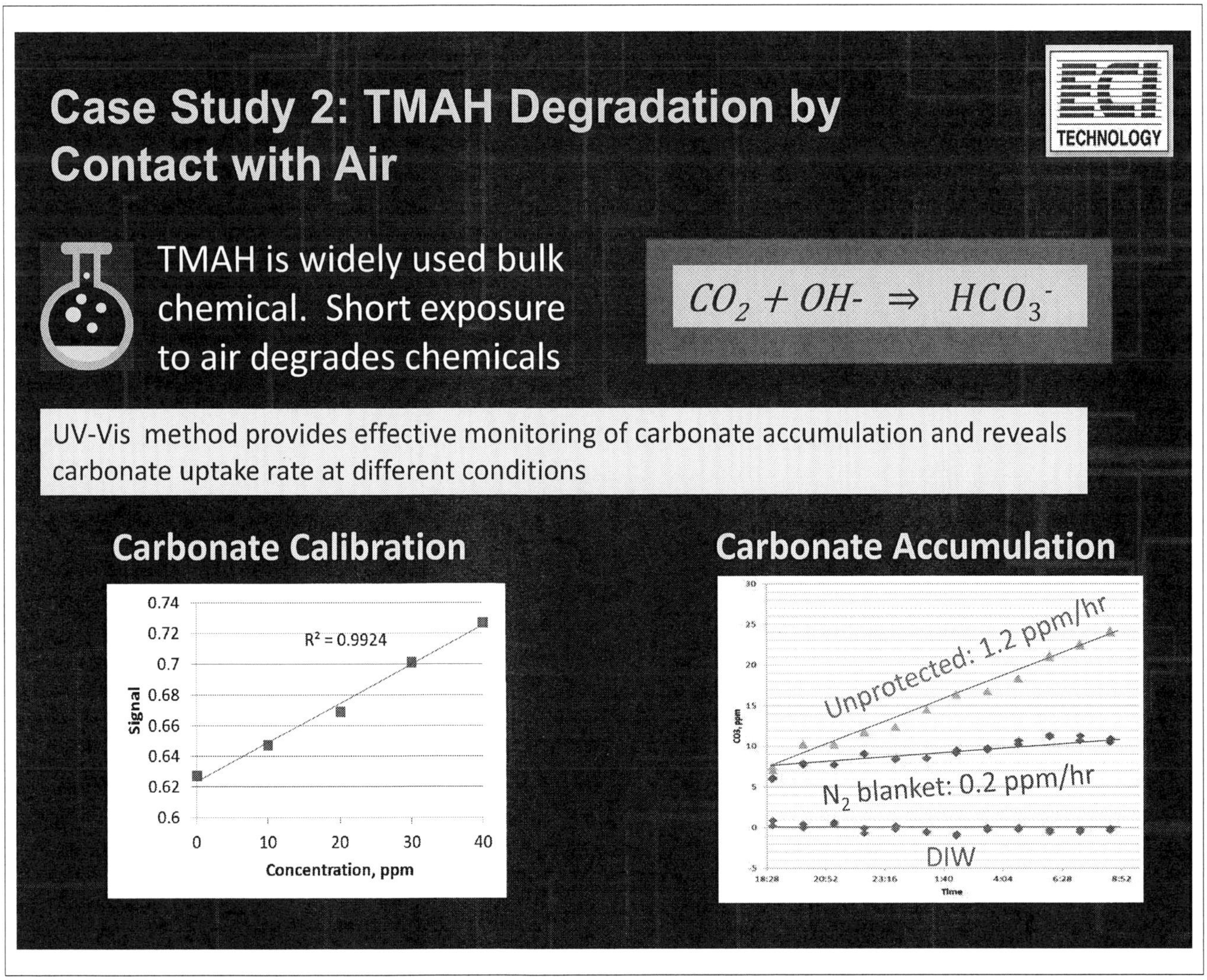
Case Study 2: TMAH Degradation by Contact with Air
TECHNOLOGY
TMAH is widely used bulk chemical. Short exposure to air degrades chemicals
$CO_2 + OH^- \Rightarrow HCO_3^-$
UV-Vis method provides effective monitoring of carbonate accumulation and reveals carbonate uptake rate at different conditions
Carbonate Calibration
$R^2 = 0.9924$
Signal
Concentration, ppm
Carbonate Accumulation
Unprotected: 1.2 ppm/hr
$N_2$ blanket: 0.2 ppm/hr
DIW
Time

Entegris

# New Methods to Reduce Variation in Bare Wafer Particle Inspection Results

Entegris: Harvey Tang, Sheryl Chou, Steven Hsiao

KLA: Vincent Wang, Oscar Lee

Harvey.Tang@entegris.com



## Agenda

Entegris

- Background
- tpM2M (Tighter Pre-scan Map-2-Map) Method
- DIOO (Dual Incidence Oblique Oblique) Method
- Applying tpM2M + DIOO to Filter Evaluations

# Background



Entegris

# Background

Entegris

As contaminant concentration reduces in advanced wafer processes, there is a higher likelihood to see significant variation in on-wafer defectivity as there are fewer defects to compare from pre-inspection to post-inspection

It is increasingly difficult to compare processes, materials or equipment when they produce low baseline particle counts

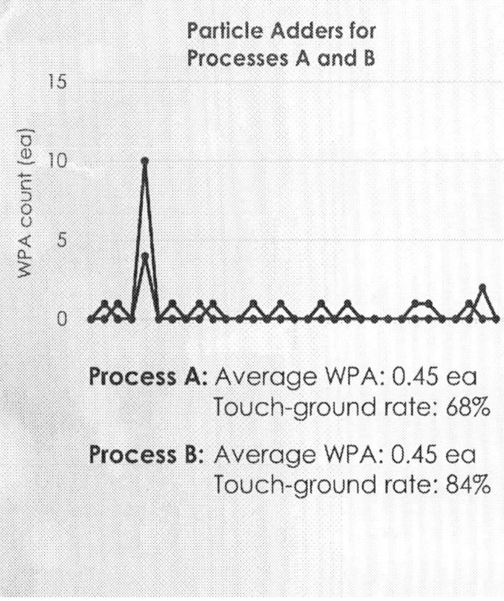


**Process A:** Average WPA: 0.45 ea
Touch-ground rate: 68%

**Process B:** Average WPA: 0.45 ea
Touch-ground rate: 84%

# Many factors influence filter choice and defectivity performance

Entegris

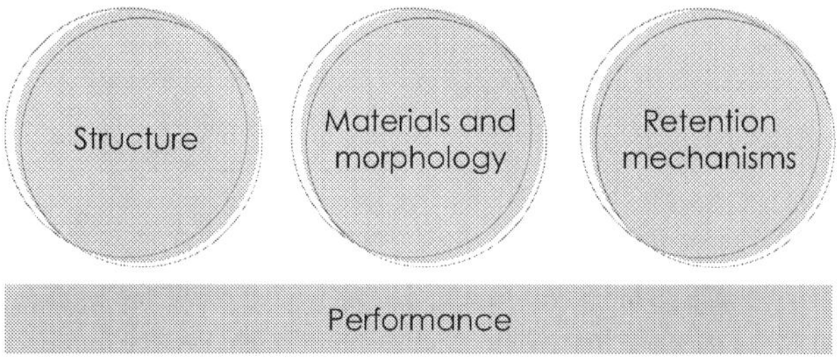


**Compatibility**
**+**
**Morphology and flow path**
**+**
**Retention mechanisms**
**=**
**Total performance**

**In order to best characterize filter performance in an application, metrology results need to provide meaningful, actionable results.**



# Fishbone diagram for process result

Entegris

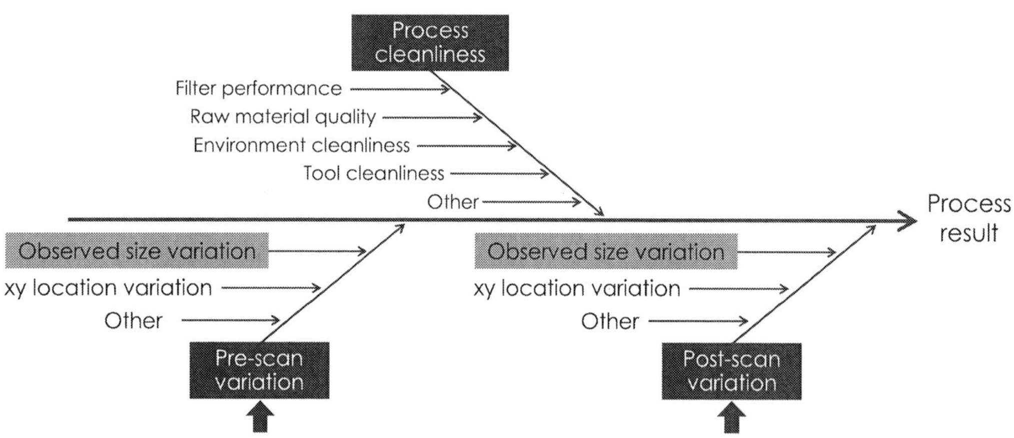


Reduced variation by "**tpM2M**"
(tighter pre-scan Map-to-Map)

Reduced variation by "**DIOO**"
(Dual Incidence Oblique Oblique)

# Metrology variation (pre-scan and post-scan) of a Surfscan SP3

Entegris

#1 Wafer
(processed DIW before)

23 nm pre-scan → No process → 23 nm post-scan

Overlap 2 maps and calculate adder

**Ideally, there should be no adder**

**Standard Map-2-Map Adder w/o wafer processing @ 23 nm (10 repeat scan on same wafer)**

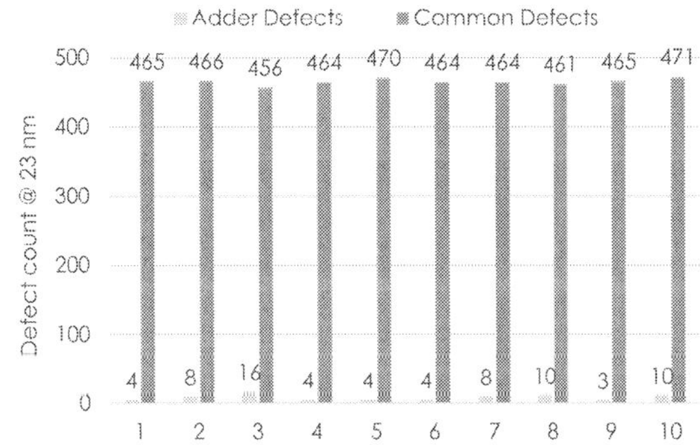


**Ideally, we would expect 0 adders in result, but some adders showed at sensitivity threshold**

7 | SPCC 2019

# Metrology variation (pre-scan and post-scan) of a Surfscan SP3

Entegris

#1 Wafer
(processed DIW before)

23 nm pre-scan → No process → 23 nm post-scan

Overlap 2 maps and calculate adder

**Ideally, there should be no adder**

**Adder Defects Size Distribution**

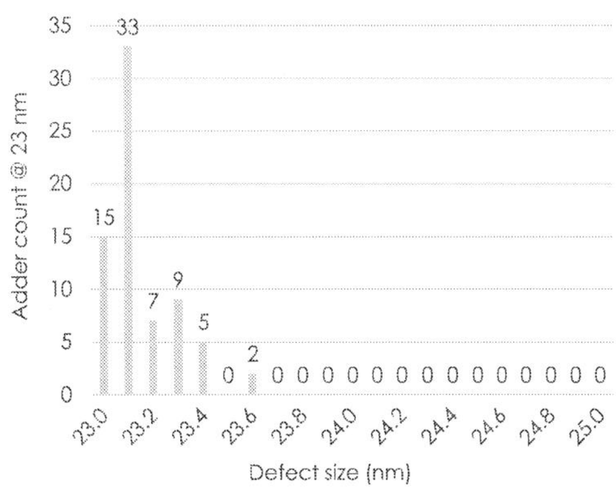


**Ideally, we would expect 0 adders in result, but some adders showed at sensitivity threshold**

8 | SPCC 2019

# Size deviation in a bare wafer inspection system

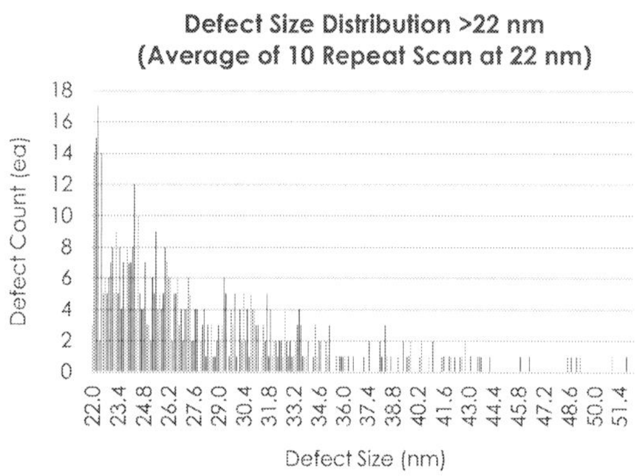


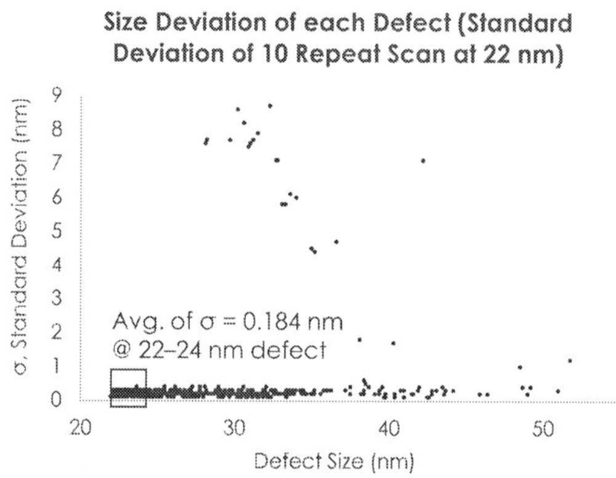


**Standard deviation of the observed size is a constant, and equals 0.184nm**

9 | SPCC 2019

# Observed size follows Gaussian distribution

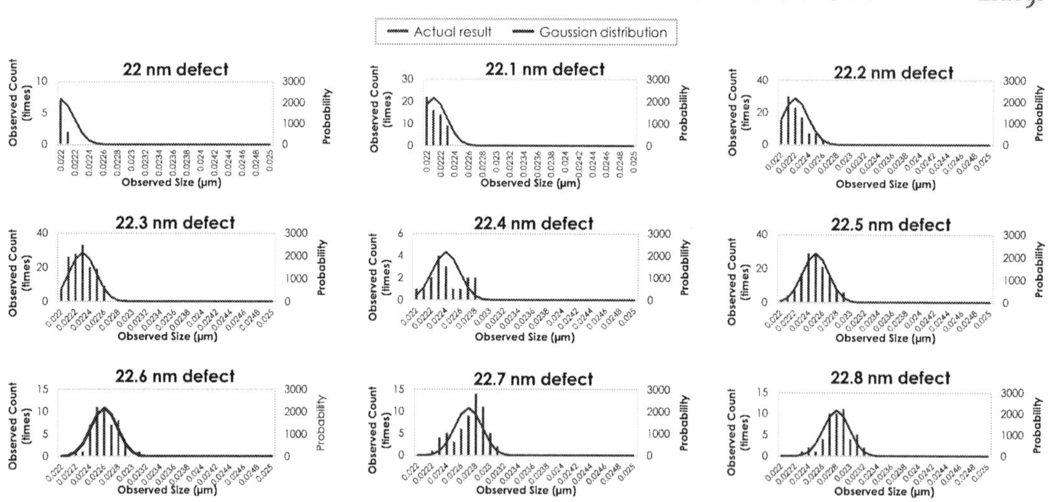


**The observed size in the 10 scans followed Gaussian distribution**

*Defects uncaptured in certain scans are not included into the plot

10 | SPCC 2019

# Probability for a defect to be captured

Entegris

**After showing the result from a bare wafer inspection system follow Gaussian distribution, the probability for a defect to be captured can be calculated.**

$$\text{Probability} = 1 - 0.5*(1 + \text{erf}[(\text{Threshold Size} - \text{Actual Size})/(\sigma *\sqrt{2})]$$

For example, a recipe threshold of 23nm is used to capture a 23nm defect.

$$\text{Probability} = 1 - 0.5*(1 + \text{erf}[(\mathbf{23} - \mathbf{23})/(\mathbf{0.184} *\sqrt{2})] = 50\%$$

Probability of seeing this defect is 50%



# tpM2M Method and Model

(Tighter Pre-scan Map-2-Map)

Entegris

# tpM2M (tighter pre-scan Map-2-Map) method and model

Entegris

## tpM2M method

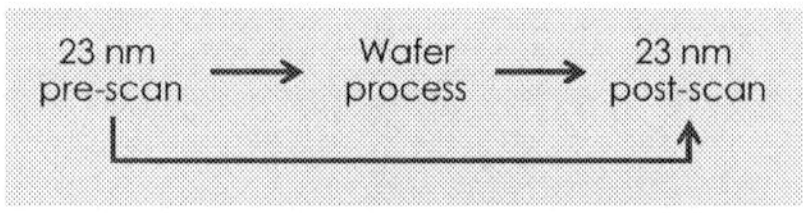


**sM2M Adder (Standard Map-2-Map)**

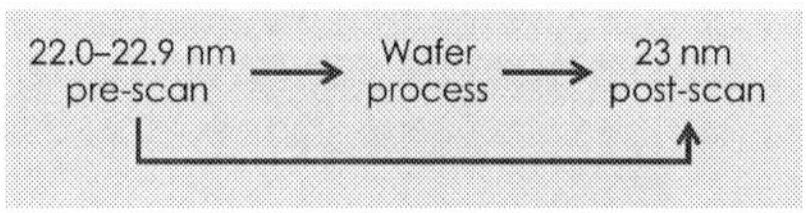


**tpM2M Adder**

**Fewer defects were missed in pre-scans by tpM2M method (fewer false adders)**

## tpM2M modeling

$$f(x) = \sum_{D=22.0}^{24.0} \left(\frac{1}{2} + \frac{1}{2}\mathrm{erf}\left(\frac{T_{pre} - D}{\sigma\sqrt{2}}\right)\right) * \left(\frac{1}{2} - \frac{1}{2}\mathrm{erf}\left(\frac{T_{post} - D}{\sigma\sqrt{2}}\right)\right) * ND$$

- First factor: Probability that a size=D defect CANNOT be seen in pre-scan
- Second factor: Probability that a size=D defect CAN be seen post-scan
- $ND$: Count of size=D defect

$T_{pre}$ = Recipe Threshold at pre-scan (22.0–23.0)
D = Diameter of the defect
σ = Standard deviation in size (Assume = 0.184 = constant)
$N_D$ = Count of size = D defect

Assumptions made:

1. All false adder defects came from observed size deviation (e.g. adder caused by x-y deviation is neglected)
2. Observed size follows Gaussian distribution
3. Standard deviation of the observed size is constant and equals 0.184 nm



# tpM2M result

Entegris

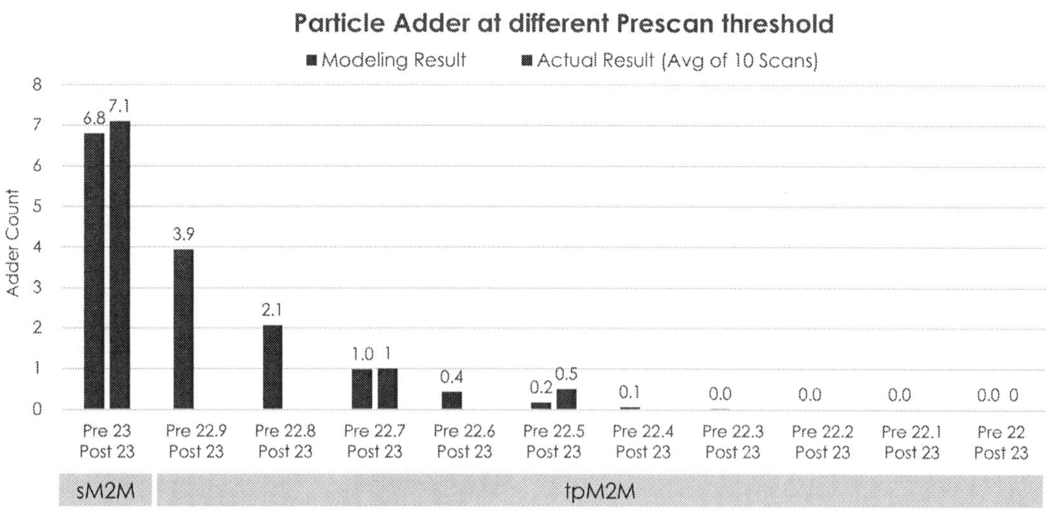

# DIOO Method and Model

(Dual Incidence Oblique Oblique)



Entegris

# DIOO method and model

Entegris

**DIOO method**

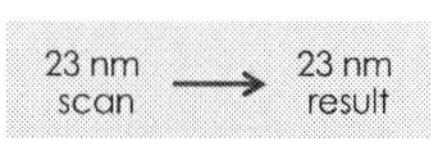


**Standard**

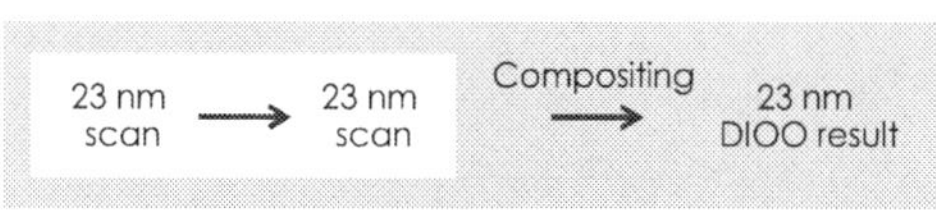


**DIOO Method**

**More true defects were captured in post-scans by DIOO method**

**Probability for a defect to be seen at different recipe threshold**

| | Recipe Threshold | |
|---|---|---|
| | 23 | |
| Actual Defect Size | Standard | DIOO |
| 22.0 | 0.0% | 0.0% |
| 22.1 | 0.0% | 0.0% |
| 22.2 | 0.0% | 0.0% |
| 22.3 | 0.0% | 0.0% |
| 22.4 | 0.1% | 0.20% |
| 22.5 | 0.3% | 0.60% |
| 22.6 | 1.5% | 2.98% |
| 22.7 | 5.2% | 10.13% |
| 22.8 | 13.9% | 25.87% |
| 22.9 | 29.3% | 50.02% |
| 23.0 | 50.0% | 75.00% |
| 23.1 | 70.7% | 91.42% |
| 23.2 | 86.1% | 98.07% |
| 23.3 | 94.8% | 99.73% |
| 23.4 | 98.5% | 99.98% |
| 23.5 | 99.7% | 100.0% |
| 23.6 | 99.9% | 100.0% |
| 23.7 | 100.0% | 100.0% |
| 23.8 | 100.0% | 100.0% |
| 23.9 | 100.0% | 100.0% |
| 24.0 | 100.0% | 100.0% |

Standard = $1 - 0.5*(1+\text{ERF.PRECISE}[(\text{Threshold Size}-\text{Actual Size})/(\sigma*\sqrt{2})]$

DIOO = $1 - \{0.5*(1+\text{ERF.PRECISE}[(\text{Threshold Size}-\text{Actual Size})/(\sigma*\sqrt{2})]\}^2$

# DIOO method and model

Entegris

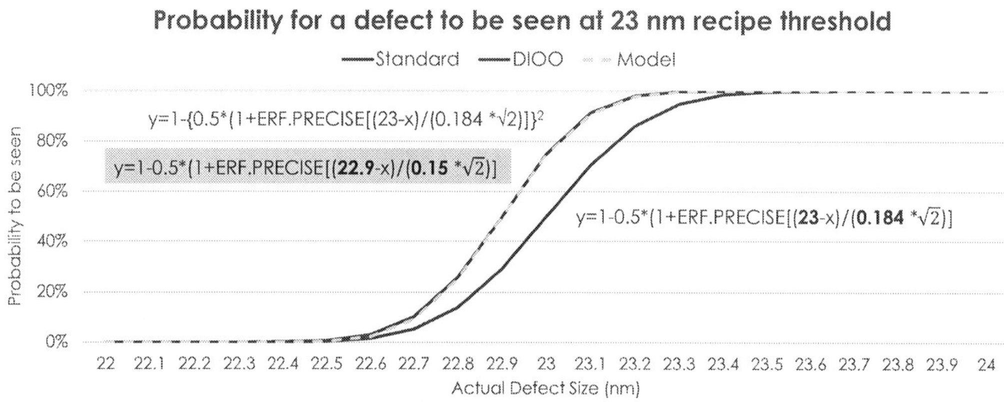


**DIOO probability curve fit the normal distribution curve with u = 22.9 and σ = 0.15**

**We can consider the standard deviation of observed size is reduced from 0.184 nm to 0.15 nm**

17 | SPCC 2019

# DIOO method and model

Entegris

Wafer #2 was scanned 10 times at 22 nm recipe threshold

The average and standard deviation in size of each defect were used to calculate the theoretical count of 100 scans (left) at 23 nm recipe threshold

The actual wafer result of 15 scans at 23 nm (right) pretty much aligned with the prediction

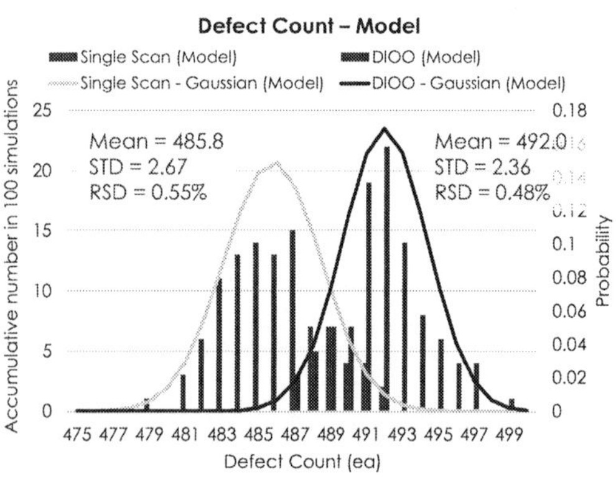


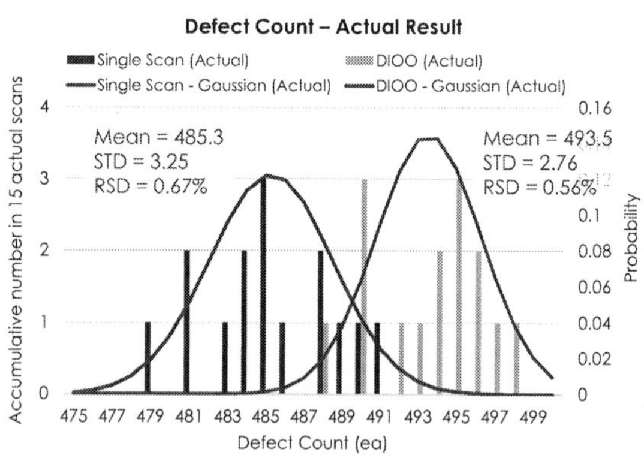


18 | SPCC 2019

# Example of applying tpM2M plus DIOO method in Wet Process Monitor



Entegris

# Applying tpM2M + DIOO on filter evaluation

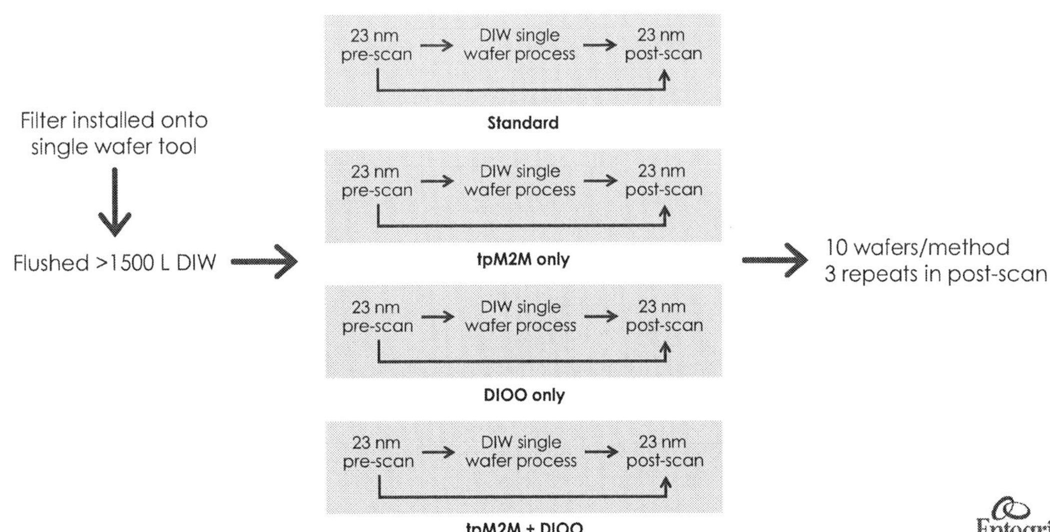




Entegris

# Results of applying tpM2M and DIOO – average

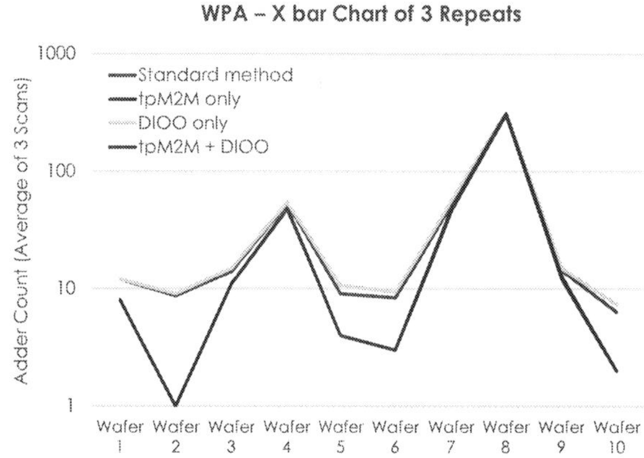


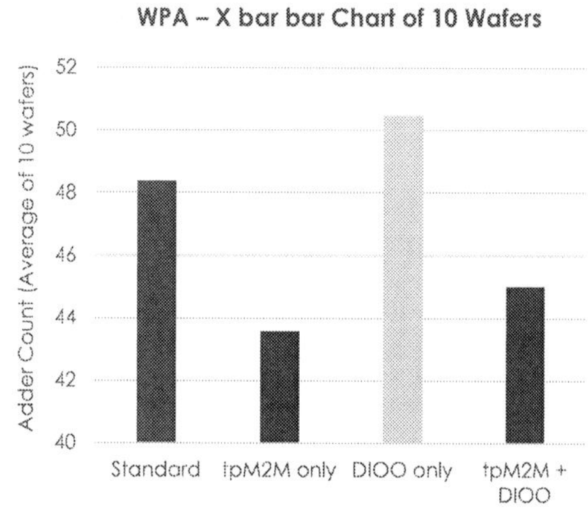


**tpM2M reduced false adder and gives a lower count**
**DIOO slightly increased sensitivity and gives a higher count**

Entegris

21 | SPCC 2019

# Results of applying tpM2M and DIOO – range

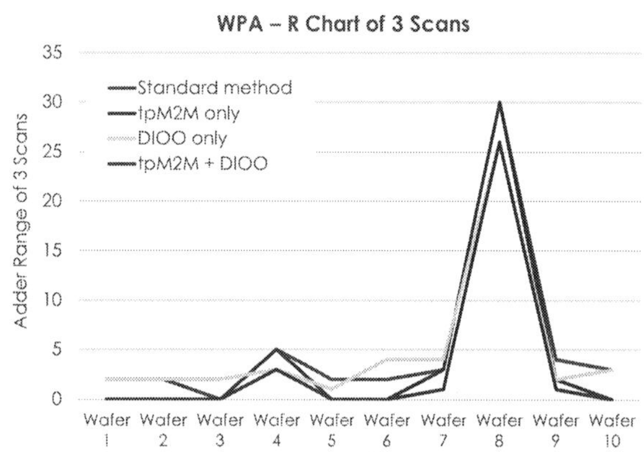


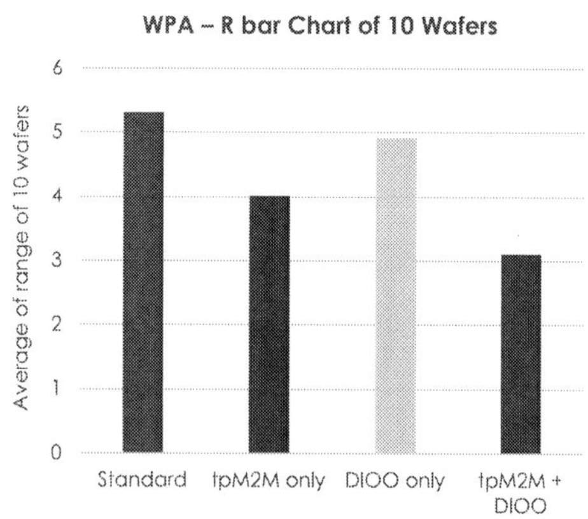


**tpM2M and DIOO method both reduced the WPA count variation of this DIW process (tpM2M more than DIOO)**
**A combination of applying both tpM2M and DIOO gives the best result**

Entegris

22 | SPCC 2019

## Summary

Entegris

- Reducing variation generated by on-wafer inspections is essential, so we can make better decisions regarding equipment/material
- Results from an unpatterned wafer defect inspection system (e.g. Surfscan) follows Gaussian distribution
- tpM2M and DIOO methods were introduced with a model to reduce the variation
- A combination of tpM2M and DIOO methods is suggested to give the best result when evaluating a process/ equipment/material

23 | SPCC 2019

## Thank you!

Entegris

24 | SPCC 2019

SPCC 2019 – Portland, USA

Minh-Phuong TRAN, CEA-Leti, *minh-phuong.tran@cea.fr*
Hervé FONTAINE, CEA-Leti, *herve.fontaine@cea.fr*
Paola GONZÁLEZ-AGUIRRE, Entegris, *paola.gonzalez@entegris.com*
Carlos BEITIA, CEA-Leti, *carlos.beitia@cea.fr*
Sung-In MOON, Entegris, *sungin.moon@entegris.com*
Jorgen LUNDGREN, Entegris, *jorgen.lundgren@entegris.com*

# Contamination kinetics of airborne ammonia on chromium-coated wafer in cleanroom conditions

2-3 April, 2019

**CEA–Leti**, University Grenoble Aples, 17 rue des Martyrs, 38054 Grenoble, France
**Entegris**, SAS, Parc Centr'Alp Ouest, 196 rue du Rocher de Lorzier, 38430 Moirans, France

## OUTLINE

1. **Introduction**
2. **Experimental setup**
3. **$NH_3$ kinetics on Cr-coated wafer**
   - **Deposition kinetics vs. [$NH_3$] and %RH**
   - **Desorption kinetics vs. %RH**
4. **Mathematical deposition model - Applications**
5. **Conclusion**

## INTRODUCTION

Entegris

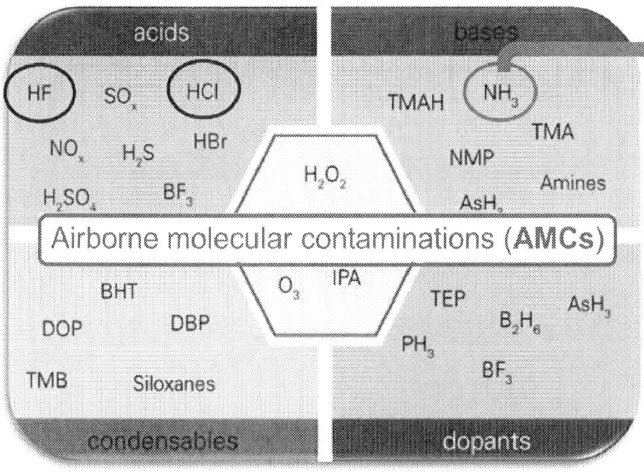


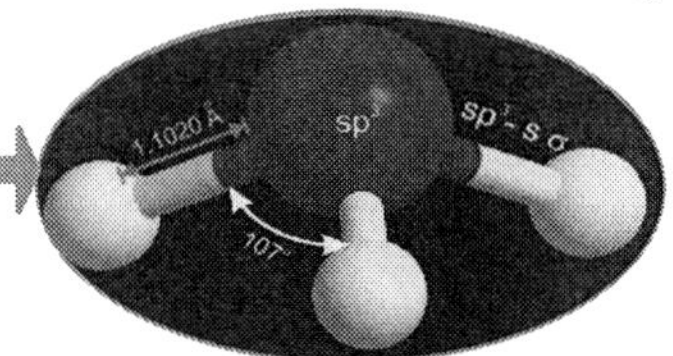


- Footing on DUV resist
- Salt $(NH_4)_2SO_4$, $NH_4Cl$ on lens masks, wafers

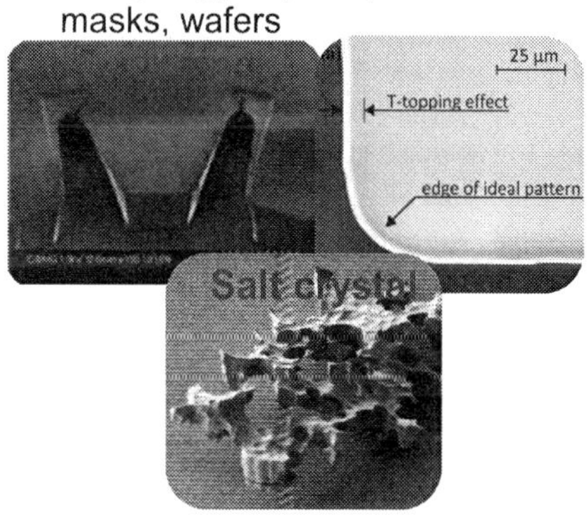


$NH_3$ sources

- Wet clean/Wet etch: Salt, $NH_4OH$, $NH_3/NH_4Cl$
- CVD deposition: TiN – Tetrakis $C_8H_{24}N_4Ti)/NH_3$
- CMP: basic slurries
- Material out-gassing, traffic & Industrial pollution

**Yield losses**

**Limit AMCs effect**

SPCC 2019, Portland USA | 3

## INTRODUCTION

Entegris

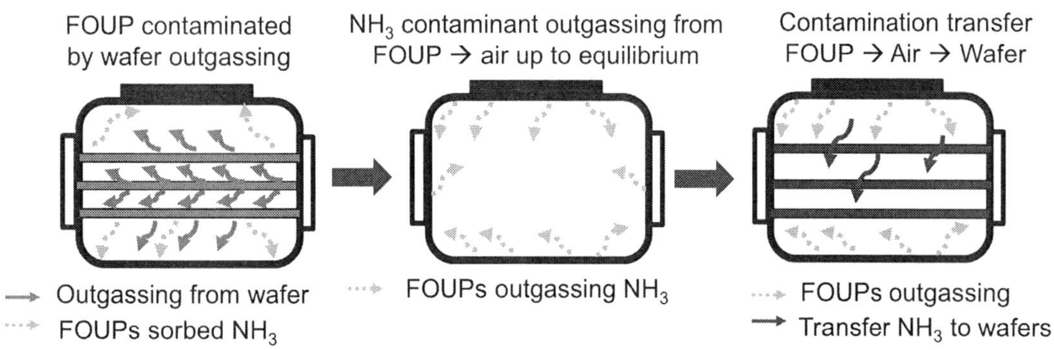


- Knowledge of $NH_3$ conta/outgassing
- Cross-contamination FOUPs to wafers ($NH_3$)
  - ✓ **Test-vehicle**
  - ✓ **Mathematical deposition model**

P. González-Aguirre *et al.*, *Microelectron. Eng.*, (2019), 205, 53–58

SPCC 2019, Portland USA | 4

## METHODOLOGY
## FREE-$NH_3$ SET-UP DEVELOPMENT

Liquid Phase Extraction (LPE process) → Quantify $NH_3$ on Cr surface

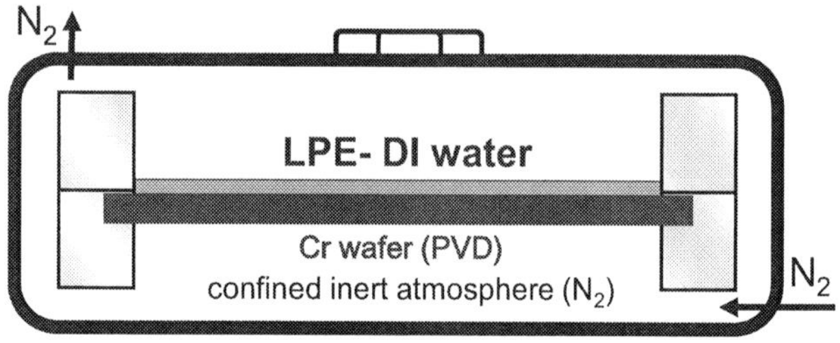


Liquid Phase Extraction (LPE process)
Collect $NH_3$ with DI water on Cr surface

$NH_3 + H_2O \rightarrow NH_4^+ + OH^-$

Quantify ion $NH_4^+$ by Ionic Chromatography (IC)

↓

$NH_4^+$ ion deposition on Cr surface

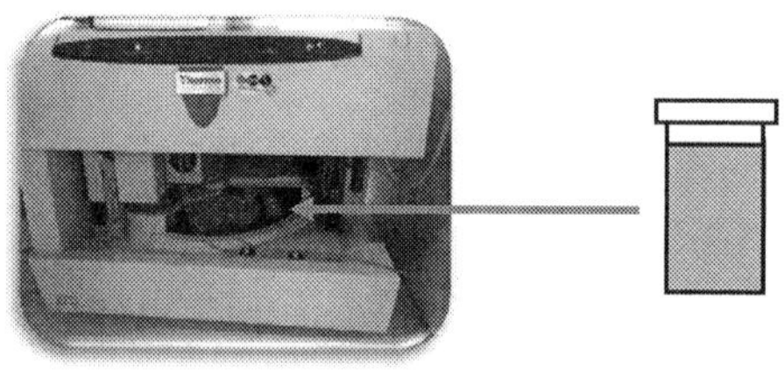

Efficiency of LPE-IC

- Collection by LPE is >95% at first extraction (LPE1)
- LLD for Ammonia: **<1.0 x $10^{11}$ ion/cm²**



## METHODOLOGY
## FREE-$NH_3$ SET-UP DEVELOPMENT

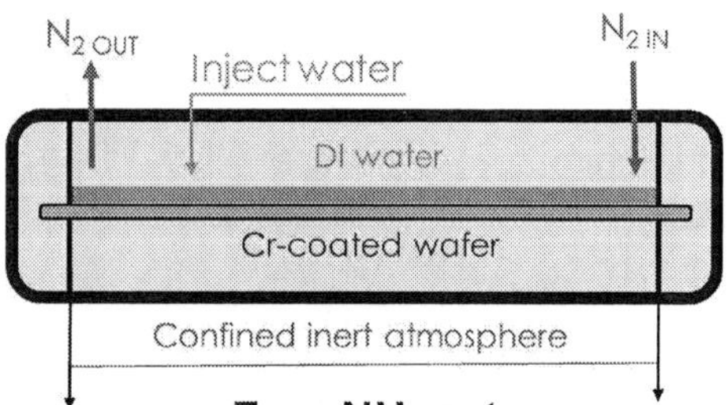


**Free-$NH_3$ set-up**

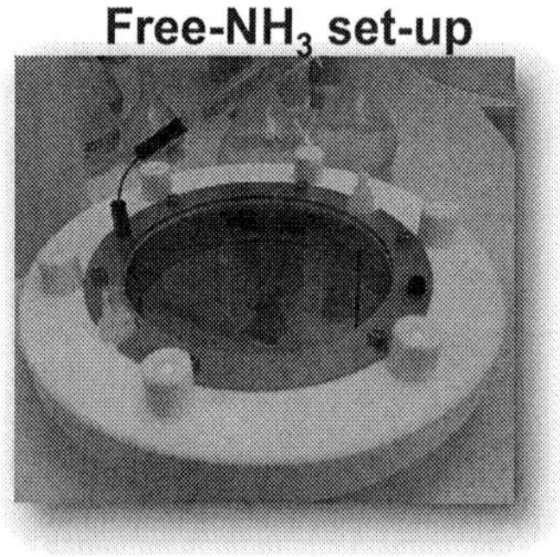

Free-$NH_3$ extract protocol
Water spreading & sampling
in free-$NH_3$ condition

**IC analysis of blank surfaces**

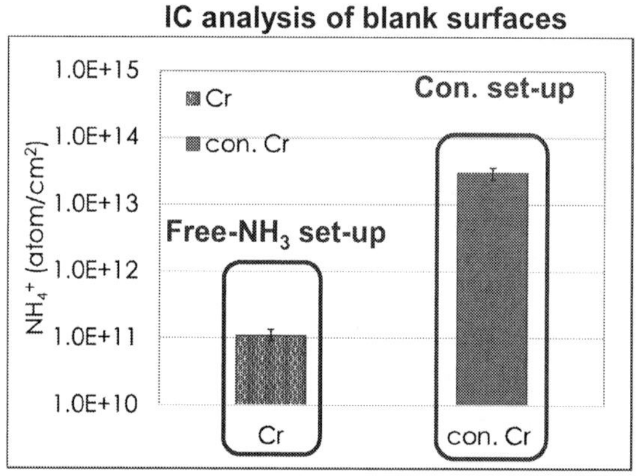


**Free-$NH_3$ set-up helps keeping proper surface from $NH_3$ atmosphere in clean room (<2E11 ion.cm$^{-2}$)**

## METHODOLOGY – SURFACE PREPARATION

- Water rinse (DIW)
- Thermal desorption (350°C – 10 min)

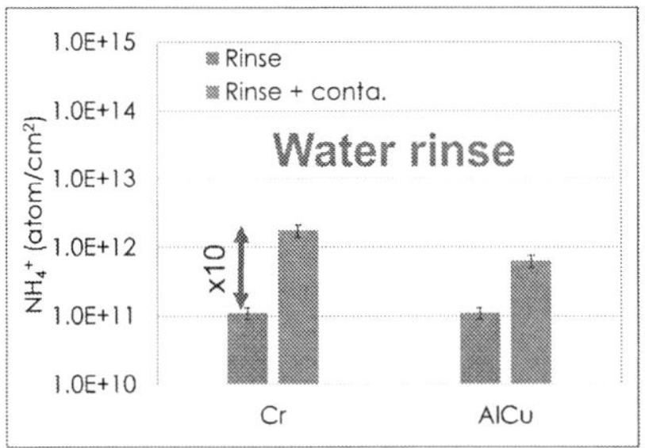


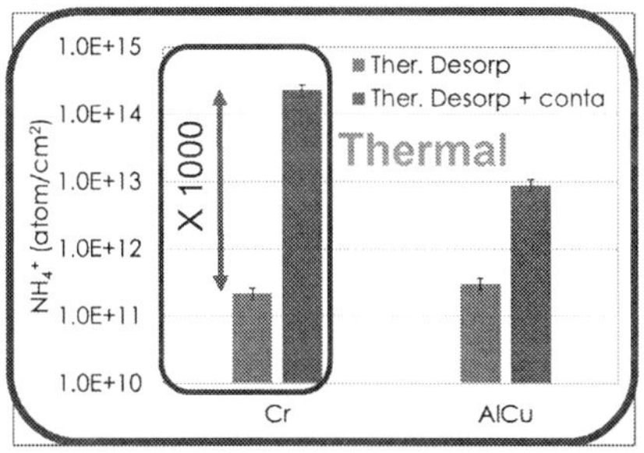


- Clean wafer contaminated in $NH_3$ (800ppbv – 2h)

Cr wafer (200mm)

| N° | Requirements Test-vehicle | Si | AlCu | Cr |
|---|---|---|---|---|
| 1 | Blank – Saturation level (orders of magnitude) | ☹ | ☹ | ☺ |
| 2 | High affinity to $NH_3$ | ☹ | ☹ | ☺ |
| 6 | No corrosion caused by $NH_3$ | ☺ | ☺ | ☺ |

**Cr-coated wafer - Best candidate as Test-vehicle → Establish deposition kinetics**



## EXPERIMENTAL SETUP

Cleaning surface by Thermal desorption (350°C – 10 min)

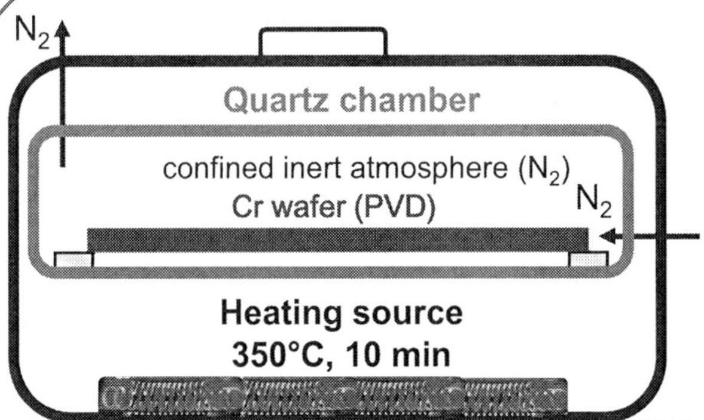


**Thermal Desorption Chamber**

Remove efficiently → Clean surface (as deposited) 2.1E11 ion.cm$^{-2}$

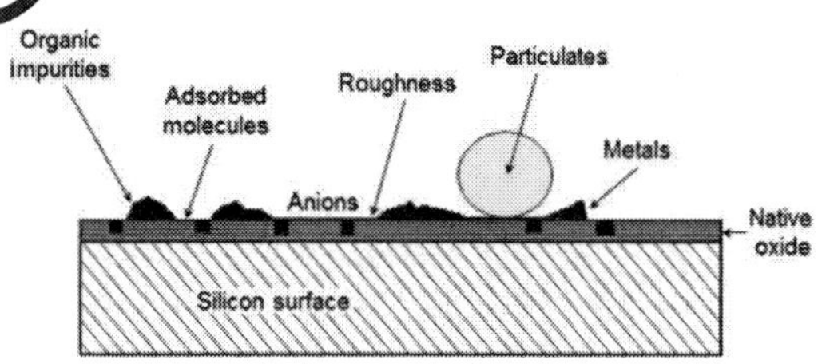


Cr wafer (PVD)

1st

leti ceatech

## EXPERIMENTAL SETUP

Entegris

**$NH_3$ generator**
**Constant $C_{NH3}$**

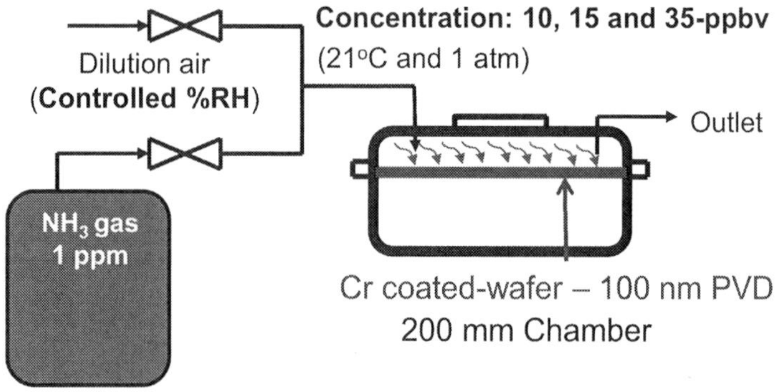


- Cr-coated wafers: Physical Vapor Deposition (PVD) - 100nm Cr
- Diluted $NH_3$ gas by clean air, controlled humidity

**2nd**



leti ceatech

## RESULTS – $NH_3$ DEPOSITION ON Cr WAFER

$NH_3$ Deposition Kinetics on Cr at 40%RH, varying $NH_3$ concentration

Entegris

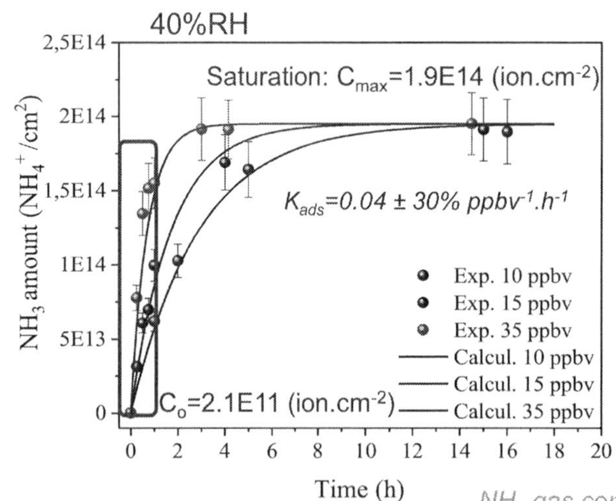


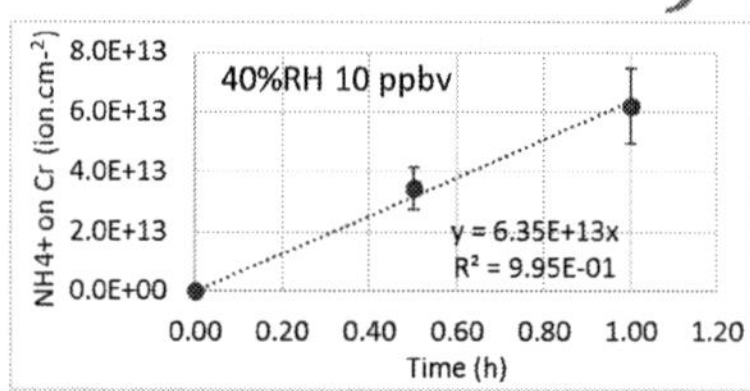


*$K_{ads}$ determined from slope (short time)*

$$C = (C_{max} - C_o)k_{ads}C_g t - C_o$$

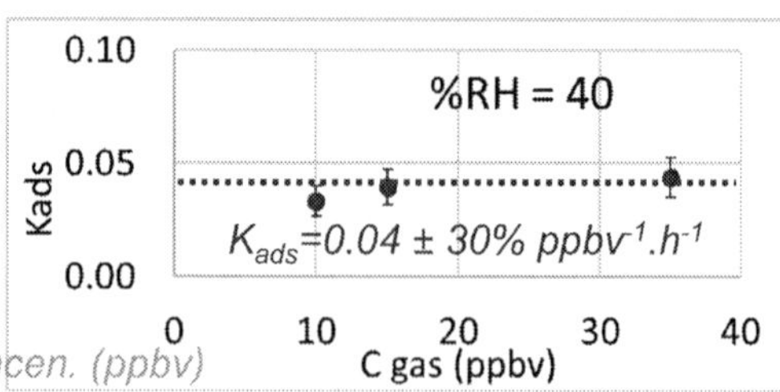


Type Langmuir deposition

$NH_3$ gas concen. (ppbv)

time

$$C = C_{max} - (C_{max} - C_o)\exp[-k_{ads}C_g t]$$

- Saturation at 1.9E14 (ion.cm$^{-2}$)
- $C_{max}$ and $K_{ads}$ independent on $C_{NH3\text{-}gas}$

## RESULTS – $NH_3$ DEPOSITION ON Cr WAFER

Deposition Kinetics on Cr at 15-ppbv, varying %RH (Humidity)

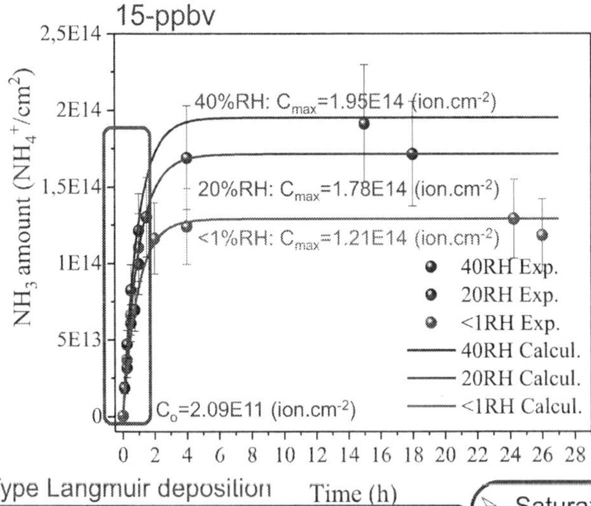


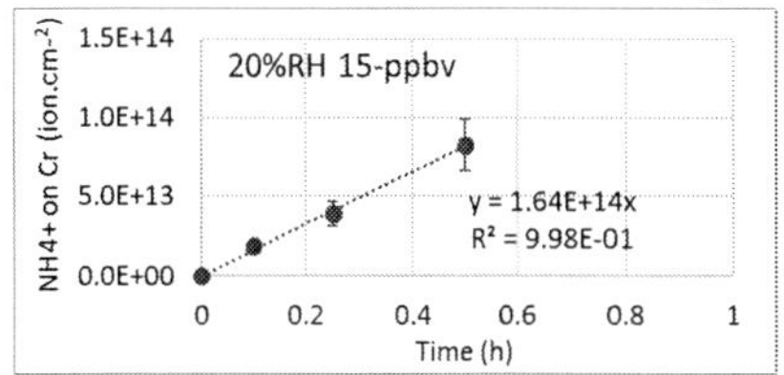


*$K_{ads}$ determined from slope (short time)*

$$C = (C_{max} - C_o)k_{ads}C_g t - C_o$$

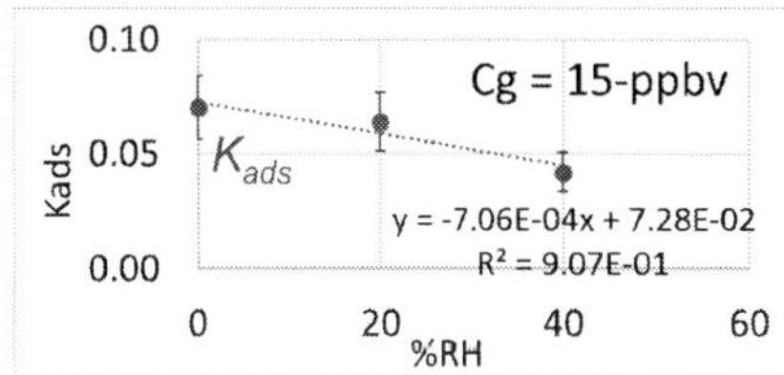


Type Langmuir deposition

$$C = C_{max} - (C_{max} - C_o)\exp[-k_{ads}C_g t]$$

- Saturation level, $C_{max}$ depending on %RH
- Increase %RH → increasing $C_{max}$ & decrease $K_{ads}$
- $H_2O_{vapor}$ favors $NH_3$ deposition
- Agreement with HF deposition on Cu-coated surface *[Ref]*

F. Herrán, H. Fontaine et al., Solid State Phenomena, 2016, vol. 255, pp. 323–328



## RESULTS – $NH_3$ DEPOSITION ON Cr WAFER
## MATHEMATICAL MODEL

**Langmuir modified model – Deposition**

$$C = C_{max} - (C_{max} - C_o)exp[-K_{ads}C_g t]$$

$$C_{surface} = f(RH, C_g, t)$$

$$[NH_3]_{surf} = f(RH, C_{NH3}, time)$$

Experimental results

$C_{max}$ independent on $C_g$ at %RH = 40

$C_{max}$ linearly dependent on %RH at $C_g$= 15 (ppbv)

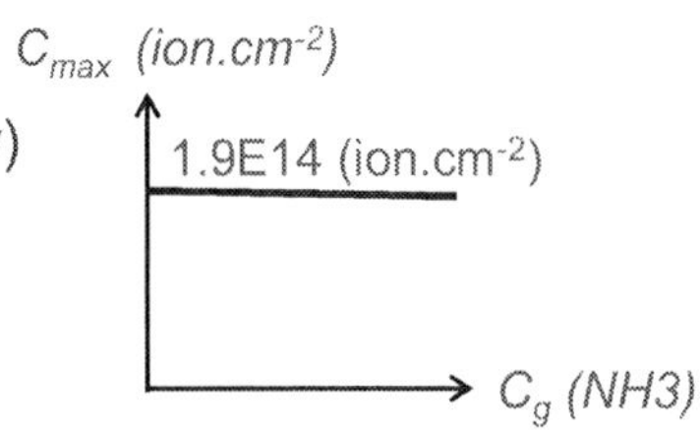


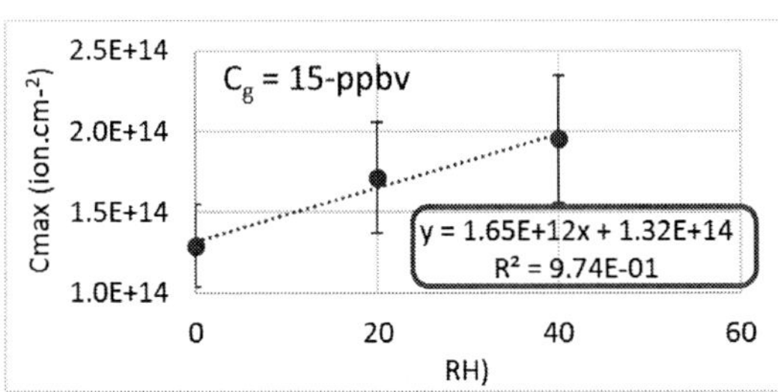


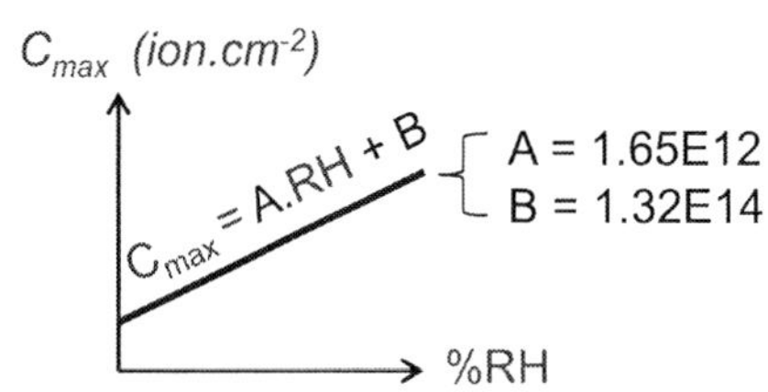


$$C_{max}\ (ion.cm^{-2}) = 1.65E12(RH) + 1.32E14 \quad (1)$$

## RESULTS – $NH_3$ DEPOSITION ON Cr WAFER
## MATHEMATICAL MODEL

**Langmuir modified model**

**– Deposition**

$$C = C_{max} - (C_{max} - C_o)exp[-K_{ads}C_g t]$$

**$C_{surface}$ = *f* (RH, $C_g$, t)**

$[NH_3]_{surf}$ = f( RH, $C_{NH3}$, time)

Experimental results

$K_{ads}$ independent on $C_g$ at %RH = 40

$K_{ads}$ linearly dependent on %RH at $C_g$= 15 (ppbv)

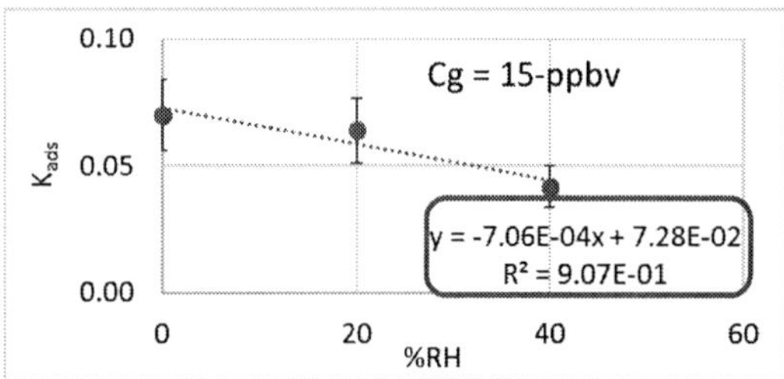


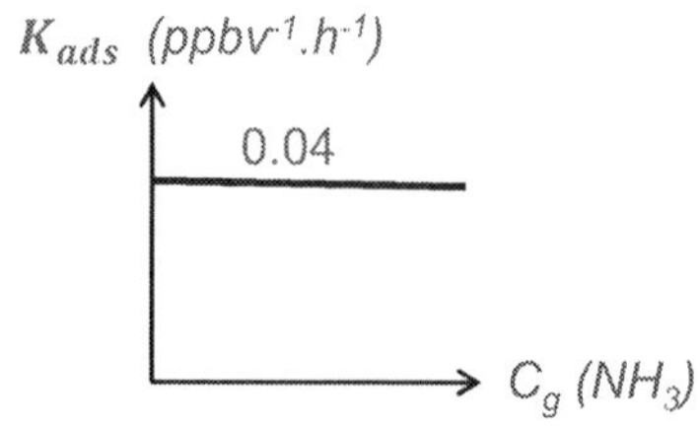


$K_{ads}$ ($ppbv^{-1}.h^{-1}$) = -7.06E-4(RH) + 7.28E-2 (2)

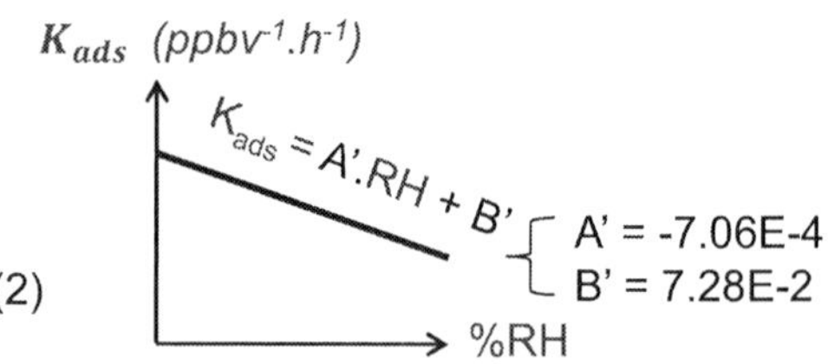




## RESULTS – $NH_3$ DEPOSITION ON Cr WAFER
## MATHEMATICAL MODEL

**Langmuir modified model**

**– Deposition**

**$C_{surface}$ = *f* (RH, $C_g$, t)**

$[NH_3]_{surf}$ = *f* ( RH, $C_{NH3}$, time)

$$C = C_{max} - (C_{max} - C_o)exp[-K_{ads}C_g t]$$

$C_{max}$ ($ion.cm^{-2}$) = 1.65E12(RH) + 1.32E14 (1)

$C_o$ = 2.1E11 $ion.cm^{-2}$

$K_{ads}$ ($ppbv^{-1}.h^{-1}$) = -7.06E-4(RH) + 7.28E-2 (2)

$$C = 1.65E12RH{+}1.32E14 - (1.65E12RH{+}1.32E14 - C_o)exp[-(-7.06E{-}4)RH + 7.28E{-}2)C_g t]$$

- Model validation: <1% – 40%RH (Humidity) & 1 – 35-ppbv ($NH_3$ airborne concentration)
- Initial Clean surface: $C_o$ = 2.1E11 $ion.cm^{-2}$

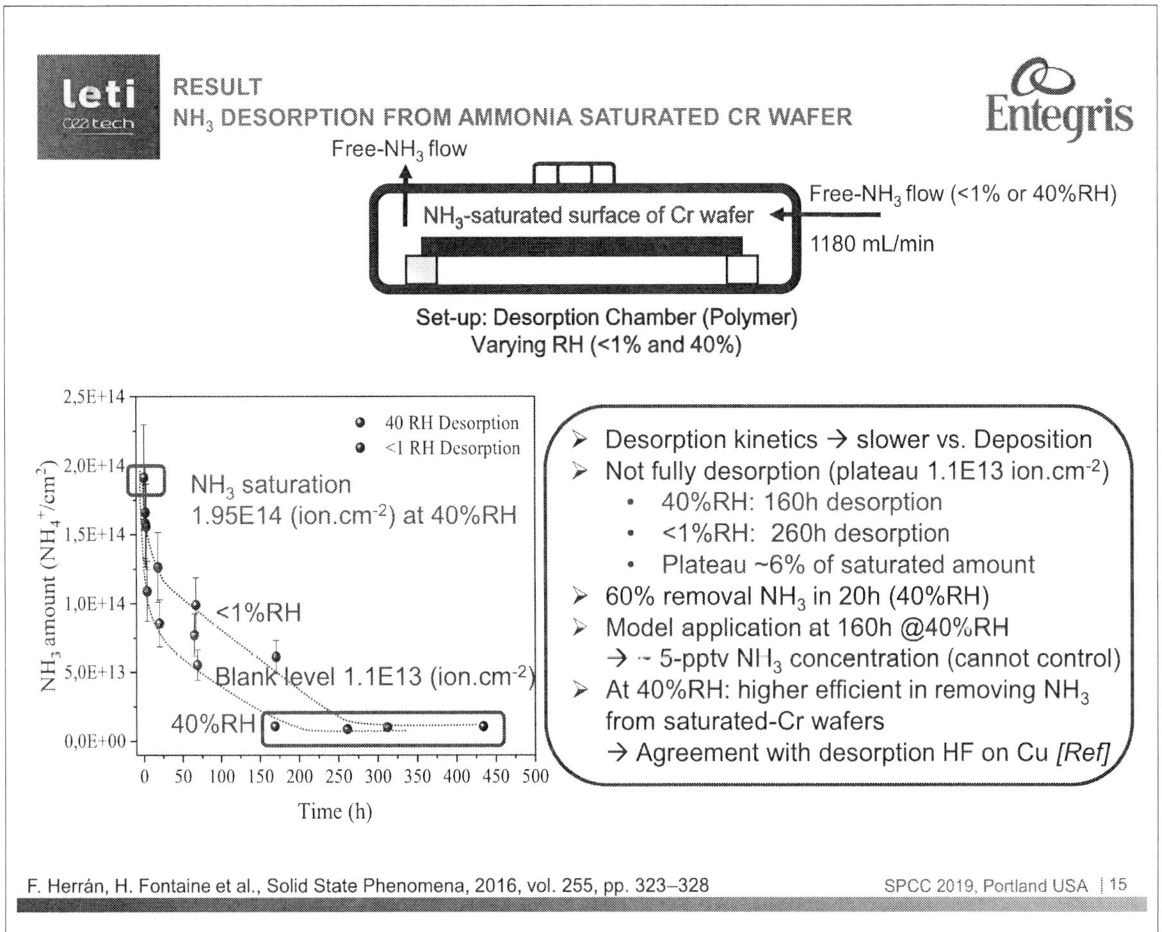
leti ceatech
RESULT
$NH_3$ DESORPTION FROM AMMONIA SATURATED CR WAFER
Entegris
Free-$NH_3$ flow
$NH_3$-saturated surface of Cr wafer
Free-$NH_3$ flow (<1% or 40%RH)
1180 mL/min
Set-up: Desorption Chamber (Polymer)
Varying RH (<1% and 40%)
40 RH Desorption
<1 RH Desorption
$NH_3$ saturation
1.95E14 (ion.cm$^{-2}$) at 40%RH
<1%RH
Blank level 1.1E13 (ion.cm$^{-2}$)
40%RH
$NH_3$ amount ($NH_4^+$/cm$^2$)
Time (h)
➢ Desorption kinetics → slower vs. Deposition
➢ Not fully desorption (plateau 1.1E13 ion.cm$^{-2}$)
• 40%RH: 160h desorption
• <1%RH: 260h desorption
• Plateau ~6% of saturated amount
➢ 60% removal $NH_3$ in 20h (40%RH)
➢ Model application at 160h @40%RH
→ ~ 5-pptv $NH_3$ concentration (cannot control)
➢ At 40%RH: higher efficient in removing $NH_3$ from saturated-Cr wafers
→ Agreement with desorption HF on Cu [Ref]
F. Herrán, H. Fontaine et al., Solid State Phenomena, 2016, vol. 255, pp. 323–328
SPCC 2019, Portland USA | 15

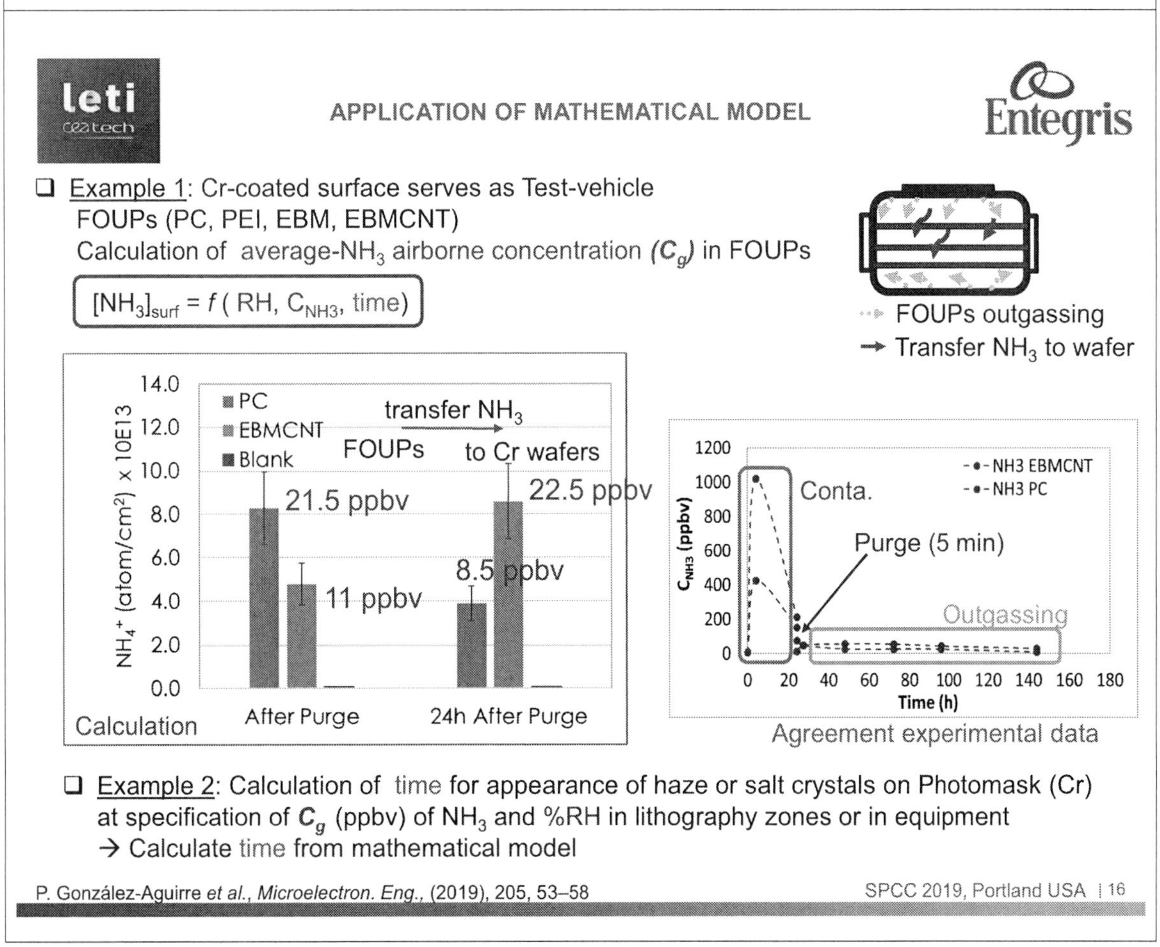
leti ceatech
APPLICATION OF MATHEMATICAL MODEL
Entegris
❑ Example 1: Cr-coated surface serves as Test-vehicle
FOUPs (PC, PEI, EBM, EBMCNT)
Calculation of average-$NH_3$ airborne concentration ($C_g$) in FOUPs
$[NH_3]_{surf} = f(RH, C_{NH3}, time)$
FOUPs outgassing
Transfer $NH_3$ to wafer
PC
EBMCNT
Blank
transfer $NH_3$
FOUPs
to Cr wafers
21.5 ppbv
22.5 ppbv
8.5 ppbv
11 ppbv
$NH_4^+$ (atom/cm$^2$) x 10E13
After Purge
24h After Purge
Calculation
Conta.
Purge (5 min)
Outgassing
NH3 EBMCNT
NH3 PC
$C_{NH3}$ (ppbv)
Time (h)
Agreement experimental data
❑ Example 2: Calculation of time for appearance of haze or salt crystals on Photomask (Cr) at specification of $C_g$ (ppbv) of $NH_3$ and %RH in lithography zones or in equipment
→ Calculate time from mathematical model
P. González-Aguirre et al., Microelectron. Eng., (2019), 205, 53–58
SPCC 2019, Portland USA | 16

leti
ceatech

## CONCLUSIONS

Entegris

- Development of free-$NH_3$ set-up for LPE-IC protocol → avoid cross conta. from clean room air
- Establishment of $NH_3$ deposition kinetics → knowledge of $NH_3$ contamination model
  $C_{max}$ (saturation level) and $K_{ads}$ (adsorption coefficient)
  - ✓ Dependent on humidity (%RH)
  - ✓ Independent on (**$C_g$**)
- Langmuir modified deposition model

$$C = C_{max} - (C_{max} - C_o)exp[-K_{ads}C_g t]$$

$[NH_3]_{surf} = f\,(RH, C_{g\text{-}NH3}, t)$

$K_{ads} = A'(RH) + B'$
$C_{max} = A(RH) + B$
$\boldsymbol{C_o} = 2.1E11\ ion.cm^{-2}$

*C=[NH$_3$]$_{surf}$*

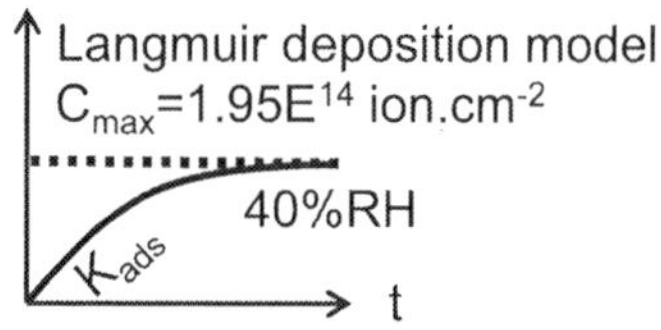


$C_g$ = 0 – 35 (ppbv)
%RH = 0 - 40

- Desorption
  - ✓ Humid condition favors desorption (vs. Dry)
  - ✓ Desorption kinetic: Slower than Deposition
  - ✓ No full desorption → presence of few *pptv* of $NH_3$ in desorption chamber

- Apply model to calculate average-**$C_g$** and time in mini-enviroment
- Dry condition is highly recommended in sensible zones to AMCs



leti
ceatech

Entegris

**Univ. Grenoble Alpes, CEA, LETI**

*Thanks for your attention*

# Improvement in post-Chemical Mechanical Planarization cleaning process for Ru interconnects

Ken Harada[1,2], Toshiaki Shibata[1], Yasuhiro Kawase[1],
Lieve Teugels[2], Nancy Heylen[2], Katia Devriendt[2], and Herbert Struyf[2]

[1]Kurosaki R&D Center, Mitsubishi Chemical Corporation, Kitakyushu, Fukuoka, 806-0004, Japan, E-mail; ken.harada.ext@imec.be
[2]imec, Kapeldreef 75, B-3001 Leuven, Belgium

## Introduction

### Motivation

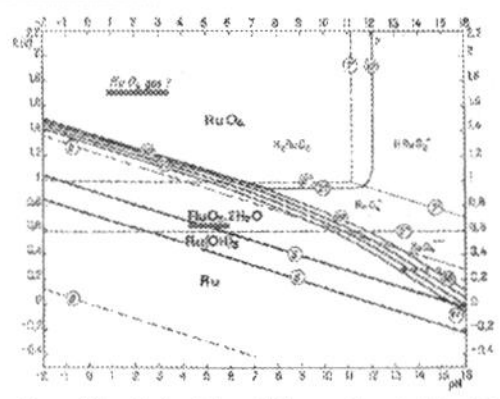

Marcel Pourbaix; *Atlas of Electrochemical Equilibria in Aqueous Solutions*

CMP in Acidic pH : toxic gas
CMP in Alkaline pH : RuOx (Oxide layer? Particles?)

In this investigation, we focused on removing small particles including RuOx to make an improvement in the post-CMP cleaning,

### DLS (Dynamic Light Scattering)

Tool. DLS-8000 (Otsuka Electronics)
Angle of the detector: 30°

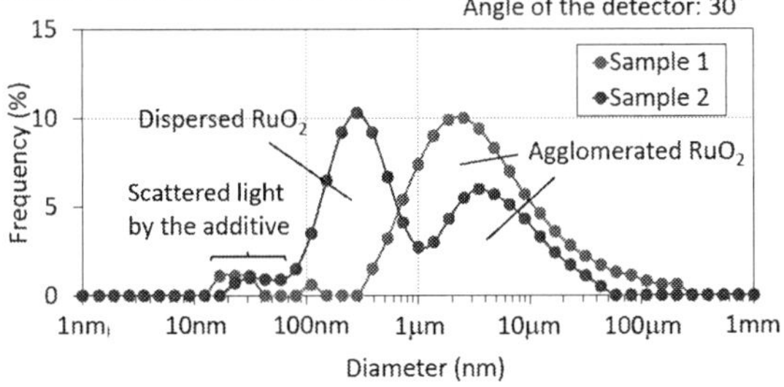


### Solutions for measurement

| | Sample 1 | Sample 2 |
|---|---|---|
| Alkali | | X |
| Additive | X | X |
| pH | ~7 | 10-11 |

## post-CMP Cleaning Test

### Defect count by SP2 (KLA-Tencor)

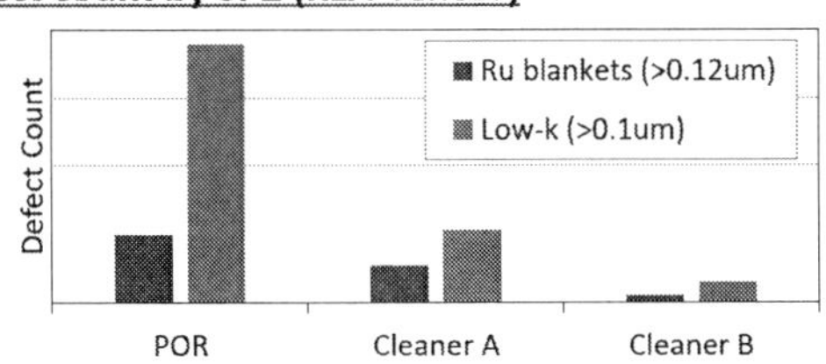


### CMP related materials

CMP tool: Reflexion LK (Applied Materials)
CMP Slurry: Alkaline
CMP Pad: Visionpad and Optivision
post-CMP cleaners: Alkaline

| | Cleaner A | Cleaner B |
|---|---|---|
| Alkali | X | X |
| Chelator | X | X |
| Additive | | X |

### Defect classification by EDR7100 (KLA-Tencor)

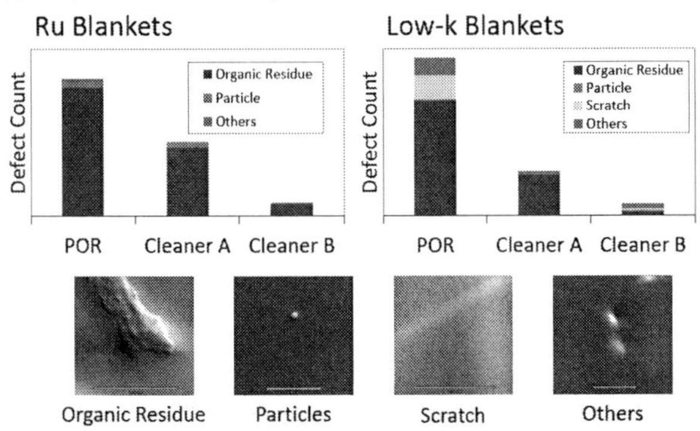


### Defects with Ru on Low-k

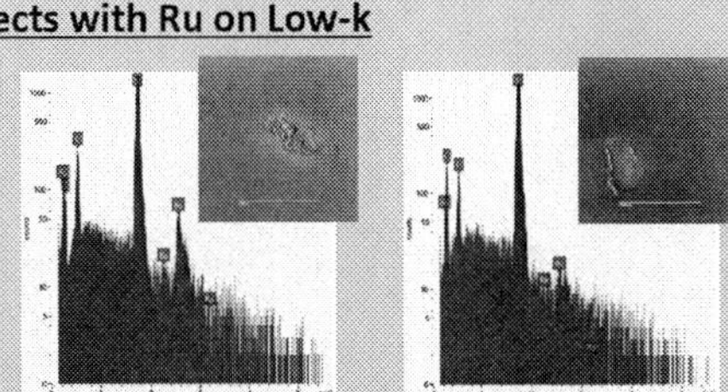

The defects containing Ru were detected on Low-k blankets (<5%). Although it might be a cross-contamination from the CMP pads or the brush, RuOx particle can be a target of post-CMP cleaning.

## Conclusion

**The novel additive in cleaner B drastically contributes to removing small particles and organic residue.**

MITSUBISHI CHEMICAL imec

# The Effects of Reductant on the Cleaning Performance for Ceria slurries with Post CMP cleaning

Areji Takanaka*, Yumiko Taniguchi, Takuo Owada and Hideki Takahashi
KANTO CHEMICAL CO., INC. *E-mail: takanaka-areji@gms.kanto.co.jp

## Introduction

Currently, various ceria-based slurries are applied for semiconductor manufacturing in two kinds of CMP processes: ILD-CMP and STI-CMP; especially in DRAM and NAND flash manufacturing. In these processes, high speed polishing is accomplished by using ceria-based slurries instead of conventional silica-based slurries. Positive-charged ceria slurries have characteristicly high adsorption rates on polished CMP film surfaces, but have more efficient polishing rates compared with silica-based slurries. This is due to effects of both chemical reactions between ceria and silicon dioxide ILD, and the superior mechanical power during CMP polish. However, ceria abrasives frequently remain on polished surface after CMP process due to the formation of chemical bonding "Si-O-Ce" during polish. This bonding makes the post-polish cleaning and removal process difficult. Typically, dHF (diluted HF), SC-1 ($NH_4OH+H_2O_2$), and SPM ($H_2SO_4+H_2O_2$) have been used to remove ceria abrasives from post CMP ILD surfaces. Recent technology node advances in semiconductor manufacturing have required the ILD surface post CMP process to be have a lower remaining ceria concentration.

## Strategy

Aim — Breaking " Si-O-Ce" chemical bonding.

Approach — Using Reducing agent.
Change Cerium ion valence from $Ce^{4+}$ to $Ce^{3+}$.
This makes "Si-O-Ce" chemical bonding weaker.

*Oxygen defects make chemical bonding weaker*

$CeO_2$: Cubic (Fluorite structure) $Ce_2O_3$: Hexagonal (hp5 structure)

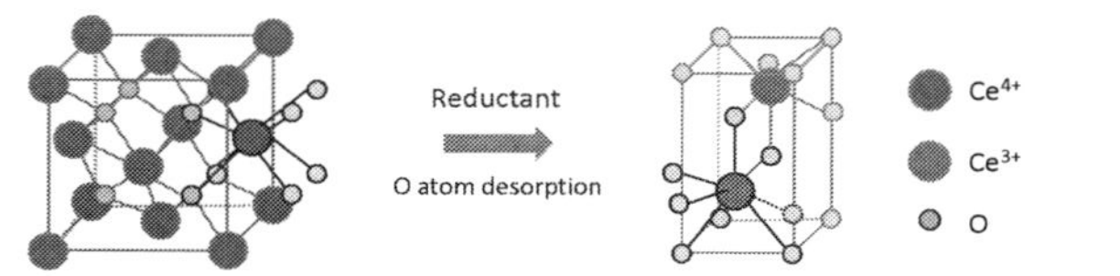


The number of O atoms connected in Ce atom. : 8 The number of O atoms connected in Ce atom. : 7

Surface structural change causes oxygen atom desorption on the outermost TEOS surface

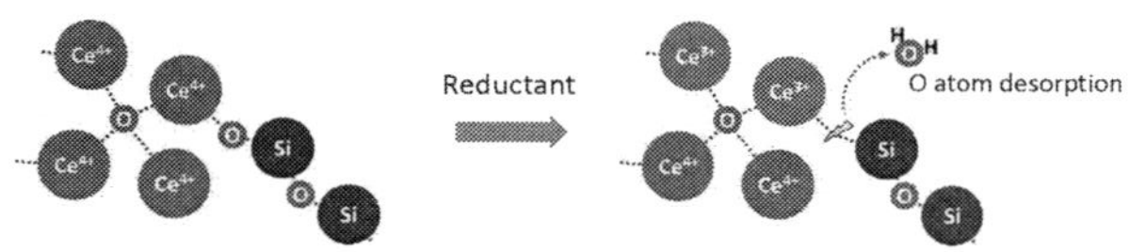


□Ce(IV)→Ce(III) suggests the local structural change from $CeO_2$ to $Ce_2O_3$.
This reaction causes an oxygen defect.

□This makes Ce-O-Si bonding on TEOS surface after CMP weak due to the crystal strain.

## Results and Discussion

*(a) Relationship between Reduction potential and Cerium surface concentration after TEOS cleaning*

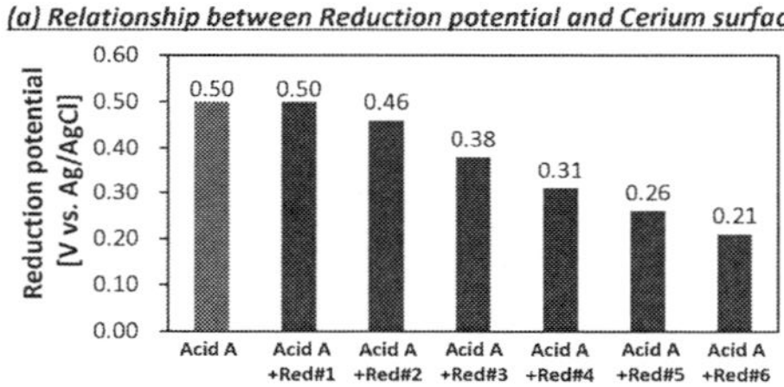


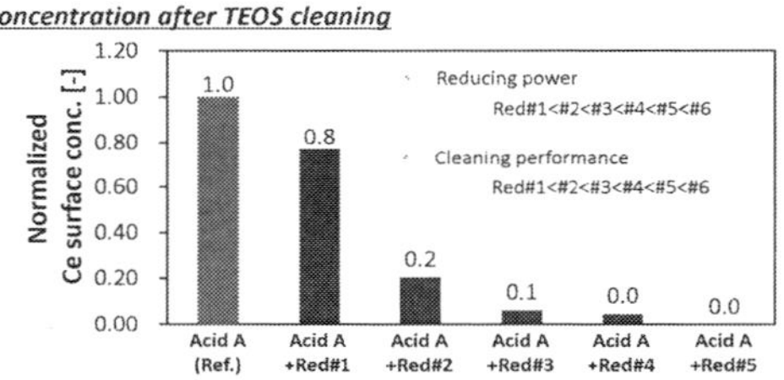


*(b) XPS results for $CeO_2$ films after Reductant dipping treatment*

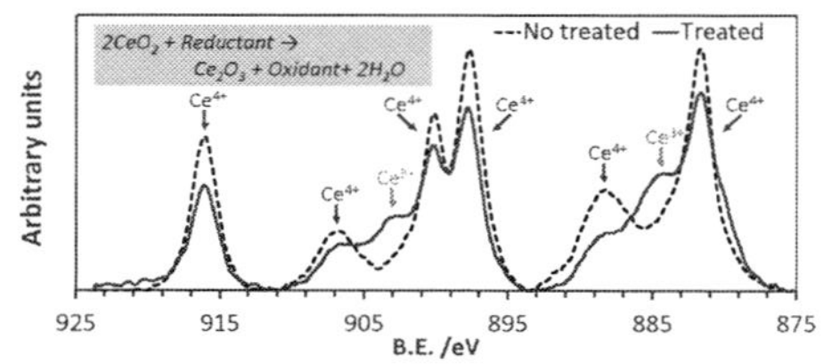


*(c) Cerium surface concentration after TEOS cleaning*

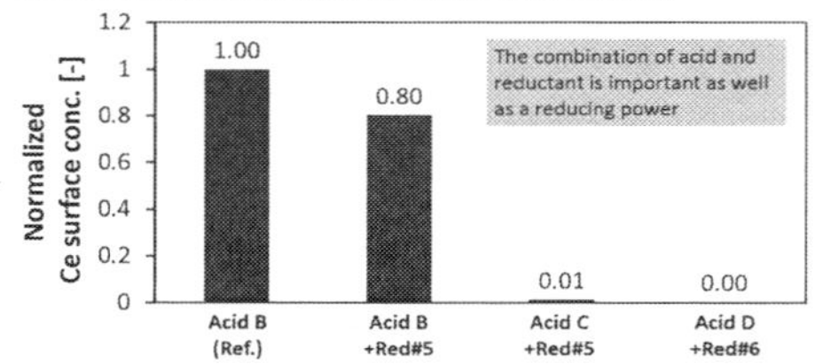


*(d) $CeO_2$ dissolution rate and TEOS E/Rs*

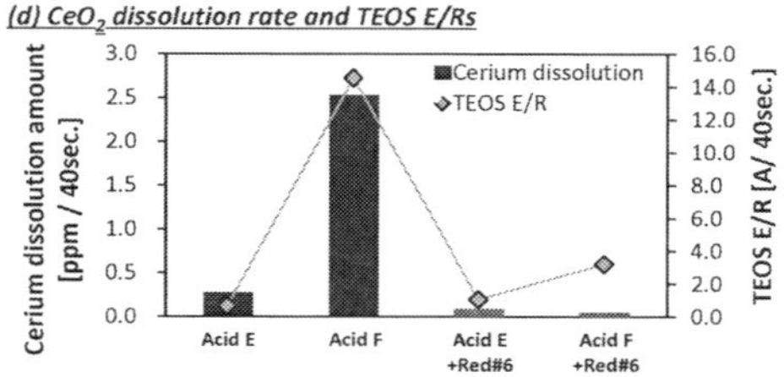


*(e) Cerium surface concentration after TEOS cleaning*

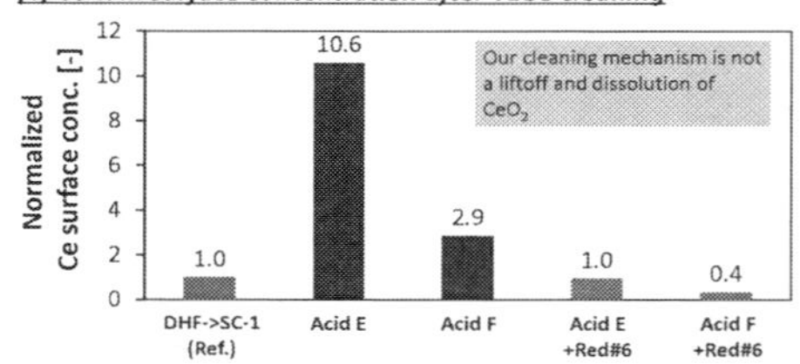


Figure (a) Relationship between Reduction potential and Cerium surface concentration on TEOS after CMP and cleaning.

Figure (b) XPS spectrum (Ce $3d_{3/2}$ and $3d_{5/2}$). We analyzed the surface state of ceria deposited film after a solution treatment including reductant by XPS.

Figure (d) Cerium dissolution rate and TEOS E/Rs of cleaners. We used cerium powder and ICP-MS when we evaluated cerium dissolution amount.

Figure (c) Cerium surface concentration analysis on TEOS by TXRF after CMP and cleaning. We used positive charged slurry made by Hitachi Chemical Company, Ltd.

Figure (e) The cerium surface concentration comparison on TEOS after CMP and cleaning between our cleaners and DHF-> SC-1 continuous process.

## Conclusion and Application

Combination of acid and reductant showed significant cleaning performance on TEOS.
We developed cleaner "CMP-B312" based on these techniques. CMP-B312 shows excellent cleaning performance not only on TEOS but also SiN.

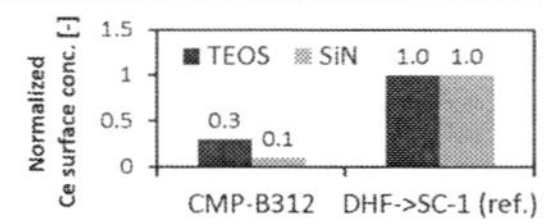

# Novel Method to Characterize the Retention of Polydisperse Particles by Advanced Membranes

*Suwen Liu*, David Hamilton, Ashutosh Bhabhe, and Anthony Ozzello – Entegris, Inc.*

## INTRODUCTION

Chemicals, movable and immovable parts in the fluid path, and auxiliary equipment all introduce contamination into semiconductor processes. Particulate contamination in semiconductor manufacturing processes is polydisperse in nature with the typical distribution skewed toward particle sizes <20 nm. Filters made from various polymer membranes are used by semiconductor fabs to reduce incoming contamination.

Historically, membrane retention efficiency has been characterized using Gold Nanoparticles (NP), Silica NP and Quantum dots, all of which are monodisperse in nature. Though a significant amount of information about membrane retention performance has been acquired using monodisperse, 2 nm to 1 μm particles, there is a substantial gap in understanding the outcome of a polydisperse particle challenge.

In this study, we present a novel method using atomic force microscopy (AFM) to characterize retention performance of advanced membranes when challenged with polydisperse particles whose size distribution better reflects contamination in process chemicals. Two modified polytetrafluoroethylene (PTFE) membranes with different retention ratings were tested using the new method. Results from the polydisperse retention study are compared between two filters to highlight the differences and illustrate the utility of this method to better understand retention in process chemistries.

## METHODS

- **Instruments:** Bruker FastScan AFM; Hitachi High-Technologies F-7000 Fluorescence Spectrophotometer
- **Samples:** Two modified polytetrafluoroethylene (PTFE) filters (Filter 1 and Filter 2)
- **Test particles:** Thermo Fisher Scientific Fluorescent polystyrene latex (G25 PSL) beads
- **Experiments:** 32 ppb G25 were used to challenge the filters, the filtrates and feed were collected and measured with spectrometer and AFM, respectively. Each sample was scanned as 300 images on a wafer surface, and the height of particles was collected as the particle size.

## RESULTS AND DISCUSSIONS

### G25 particle and its size distribution

G25 particles were chosen for testing because:

G25 PSL beads are usually measured by dynamic light scattering (DLS) methods, which indicate an average particle size of 25 nm. Figure 1, comparing different measurement methods for the same beads, suggests that the PSL beads are polydisperse and have an average size of 12 nm, as demonstrated by the number weighted distribution.

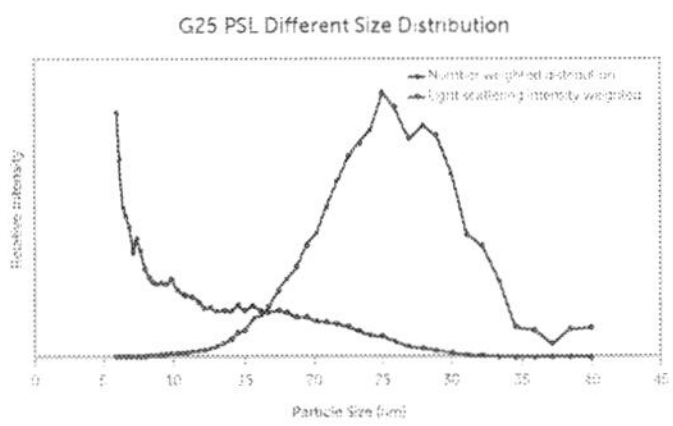


*Figure 1. G25 particle size distribution collected using Liquid Nanosizer as a number weighted size distribution (in red), the average size is about 12 nm. However, when converted to light intensity size distribution, the average size is about 25 nm which agrees with manufacturer claimed DLS test results.*

The PSL beads are spherical and are therefore easily measured by the AFM technique. G25 particle size distribution measured by AFM matches well with Cryo-EM (electron microscope) analysis results for the particle larger than 8 nm. Figure 2 shows the particle size distribution measured by both AFM and Cryo-EM.

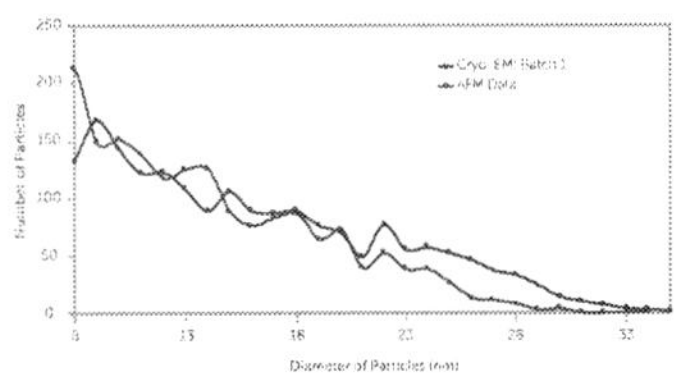


*Figure 2. AFM test results match well with Cryo-EM test results on G25 analysis when particle size is larger than 8 nm.*

### Particle size distribution analysis after filtration

Two modified PTFE membranes with different retention ratings were challenged by using 32 ppb G25 feed. Fluorescence intensity was used to calculate the retention of two filters at 1% Monolayer particle loading. The retention results of Filter 1 and Filter 2 are 75.6% and 68.5%, respectively. 300 AFM images from each sample were scanned. The particles ≥5 nm in the feed and filtrate were further analyzed and compared. Figure 3 shows the particle size distribution of the feed and filtrate. Clearly, after filtration, the particle counts in all size ranges decrease when compared to the feed. Also, in all size ranges, Filter 1 shows fewer particles than Filter 2, which aligns with retention ratings and fluorescence retention test results.

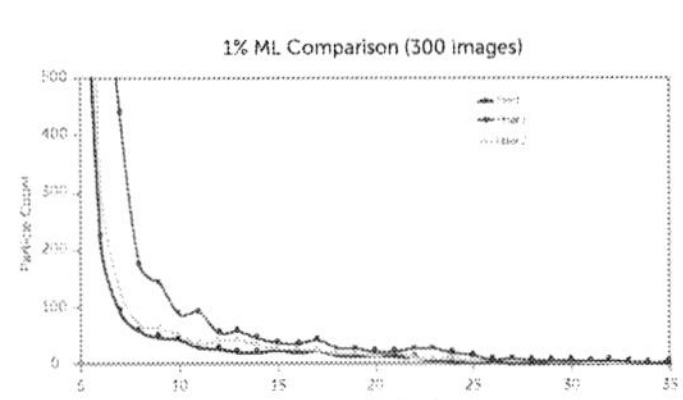


*Figure 3. Particle size distribution comparison before and after filtration.*

### An estimation of particle retention by different size bins

Two significant reasons limit this method as a quantitative method:

- AFM detects all particles and non-volatile residues on the wafer surface, which includes the mixture of G25 beads and any defects on the wafer surface.
- The uniformity of particles dispersed on the wafer

Considering these two reasons, the test system was flushed for >1 hour to avoid possible shedding of particles. We compared the particle size distribution (5 nm to 35 nm) of the two filters before and after filtration. In this case, we could estimate the retention performance in each single size bin (see Figure 4). Filter 1 has a better retention performance at almost every size bin from 5 nm to 35 nm. The results agree with our regular fluorescence retention results and the filter retention ratings.

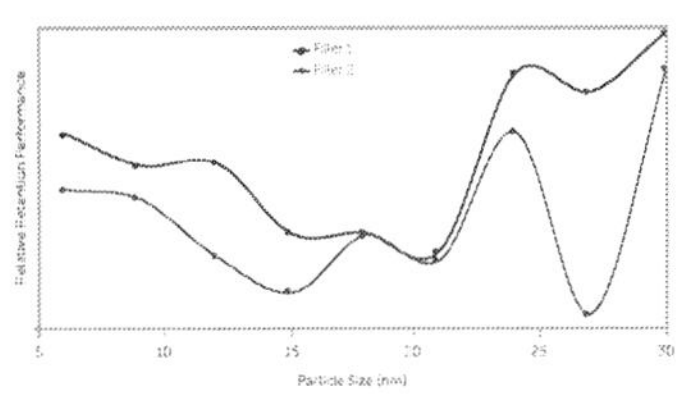


*Figure 4. Filter performance comparison: Filter 1 has a better performance than Filter 2 at almost every size bin from 5 nm to 35 nm.*

## SUMMARY

- Polydisperse, spherical G25 particles, ranging in size from 3 to 35 nm, are useful for evaluating relative filter performance
- G25 size distribution collected with AFM matches the results collected using other characterization techniques. The AFM method can provide particle information at particle sizes as low as 5 nm particles
- Filter retention performance can be observed by comparing the changes in the G25 size distribution post filtration
- While not quantitative, the method provides retention information in each size bin (from 5 to 35 nm)
- Filter 1 has a better performance than Filter 2 in almost all size ranges which aligns with fluorescence test results

## ACKNOWLEDGEMENT

The authors would like to thank Jennifer Braggin, Haizheng Zhang, and Thomas Phely-Bobin for their support and discussions.

**Corresponding author: suwen.liu@entegris.com*

BASF
We create chemistry

# TiN Hardmask Wet Etching and Post Etch Residue Removal in BEoL for advanced Technology nodes: Effects of material deposition and etch compositions

*Andreas Klipp[a*], Johan Hoogboom[a], [a]Electronic Materials, BASF SE, Ludwigshafen, 67056, Germany. (e-mail: * andreas.klipp@basf.com)*

In the semiconductor industry, titanium nitride is often used as an inorganic hard mask, enabling the robust creation of structures (vias and trenches) with widths down to at least 5 nm. After the vias and trenches are created, using plasma etching techniques, the hard mask has to be removed again by a wet chemical process, as its presence during final manufacturing steps would severely impact device performance or would increase process complexity in the subsequent process steps.

As the trench width is continuously decreasing with advanced technology nodes, the aspect ratio (depth to width) is increasing and can also lead to severe pattern collapse in BEoL. The selective removal of the titanium nitride hardmask after patterning can help to minimize pattern collapse and also reduce the aspect ratio in order to improve gap fill performance of the subsequent metallization process like electroplating copper or cobalt deposition.

In general, the TiN layer is removed using oxidative wet etching chemistry. With differentiation between chip manufacturers increasing over the past years, multiple types of TiN are used, mostly differing in deposition technique (Chemical Vapor Deposition (CVD), Physical Vapor Deposition (PVD) and Atomic Layer Deposition (ALD)) and material composition (ratio between Ti : N : O : H). These different types of TiN hard masks show vastly differing wet etch rates and underlying mechanisms.

In this paper, we present data on the oxidative wet etching of different TiN hard masks, in which the deposition techniques and conditions were varied (PVD, CVD, ALD), as well as the oxidative wet etching chemistry. We discuss the impact of the deposition technique and composition on the etch rates, film morphology, thermodynamic parameters and underlying mechanisms for various products, and show that tuning of both the TiN film composition and the applied etch product results in a broad range of applicable chemistries which can be used in cutting-edge chip manufacturing technologies.

# Development of Wet-etch Chemistries for Selective Silicon Nitride Removal in 3D NAND Structure

Jhih Kuei Ge[1,†], Yi Chia Lee[1], Rick Tianniu Chen[1], Aiping Wu[1],
Taehyeon Kim[2], Changjin Son[2], and Sangwoo Lim[2]

1. Versum Materials Technology LLC., 2F, No.25, Lane 62, Chung Ho Street, Chupei City, Hsinchu County 30267, Taiwan
2. Department of Chemical and Biomolecular Engineering, Yonsei University, 50 Yonsei Ro, Seodaemun Gu, Seoul, Korea

† Corresponding author: ck.ge@versummaterials.com

2019 Surface Preparation and Cleaning Conference
April 02-03, 2018 in Portland, Oregon

VERSUM MATERIALS

1

# Technical Background

## Sacrificial Nitride Strip for V-NAND

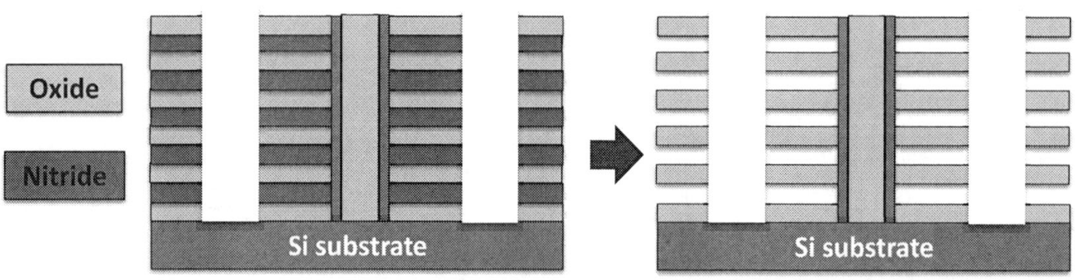


- To selectively remove the SiNx from the stack w/o damaging SiOx, high selective SiNx etch chemical is necessary.
- Hot phosphoric acid is a well-known wet etchant for SiNx removal, which can remove SiNx with minimal damage on SiOx.

$$3Si_3N_4 + 4H_3PO_4 + 27H_2O \rightarrow 4(NH_4)_3PO_4 + 9H_2SiO_3$$

- However, conventional $H_3PO_4$ chemistry faces challenges as the number of layers is >48. Therefore, the formulated $H_3PO_4$ chemistry is developed to further improve SiNx to SiOx selectivity.

VERSUM MATERIALS

2

# Design Concepts

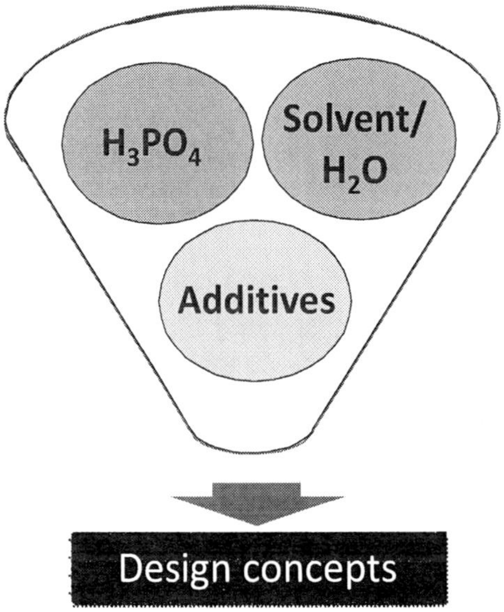


**$H_3PO_4$**

Etchant for SiNx

**Solvents/$H_2O$**

1. Improve SiNx to SiOx selectivity
2. Enhance the miscibility between $H_3PO_4$ and additives

**Additives**

1. Improve SiNx to SiOx selectivity
2. Oxide regrowth can be eliminated by adding proper additives

VERSUM MATERIALS

3

# Experimental Process & Measurement

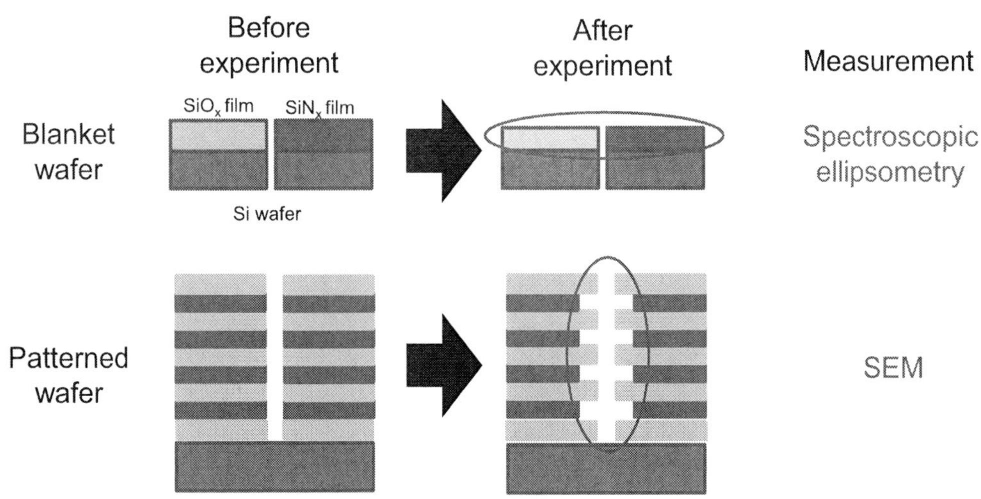


**DHF($H_2O$:HF=100:1) pretreatment was performed for 3mins on $SiN_x$ blanket wafer and patterned wafer before etching experiments were performed.**

VERSUM MATERIALS

4

# Effect from Solvent Addition

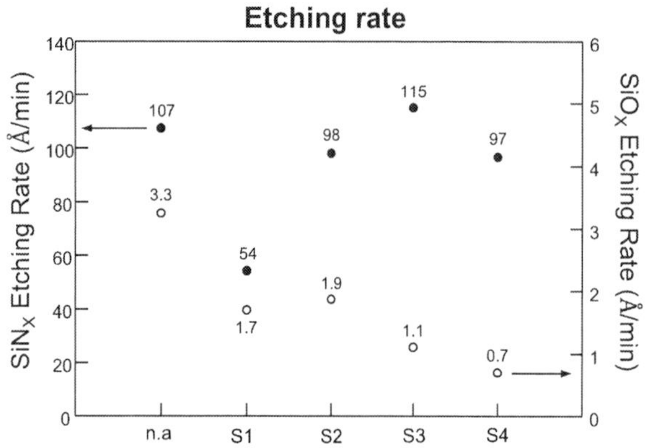


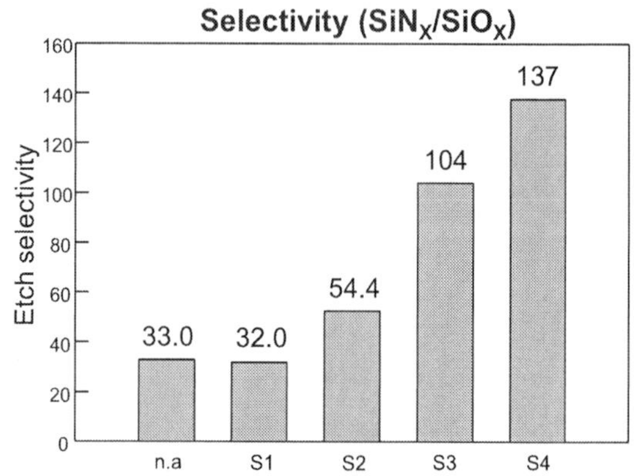


- With the addition of various solvents, etching of $SiO_X$ was suppressed.
- Addition of S1 decreased etch rates of both $SiN_X$ and $SiO_X$ without change of selectivity.
- Addition of S2 more decreased etch rate of $SiO_x$ with a selectivity increase to 54.
- Addition of S3 and S4 which have a specific functional group greatly suppressed the etching rate of $SiO_x$ with huge increases of selectivity to 104 and 137, respectively.

VERSUM MATERIALS

5

# Effect from Additive(Type A) Addition – A1

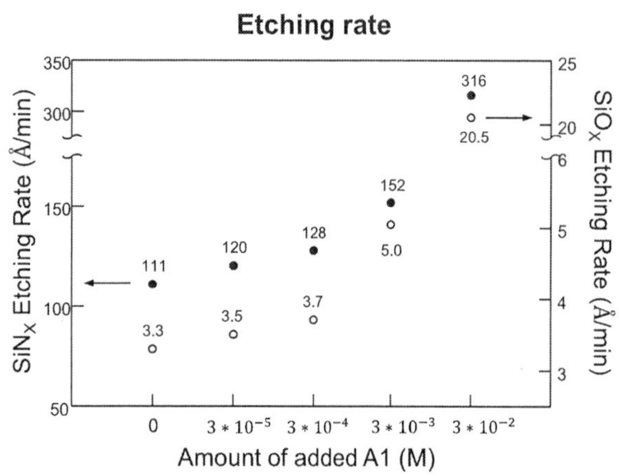


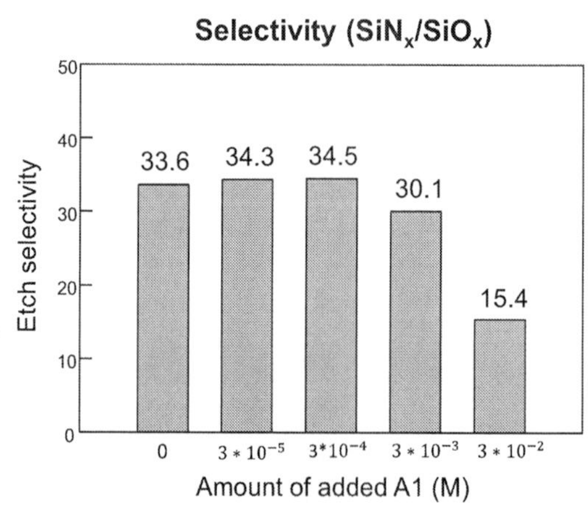


- As the amount of added A1 increases, the etching rates of $SiN_X$ and $SiO_X$ increase.
- When $3*10^{-2}$M of A1 is added, the etching rate of $SiN_x$ increases 2.8 times and that of $SiO_x$ increases 6.2 times.
- With the addition of A1, etching rate of $SiN_X$ increases but etching rate of $SiO_X$ increases more, thus selectivity decreases.

VERSUM MATERIALS

6

# Effect from Additive(Type A) Addition – A2

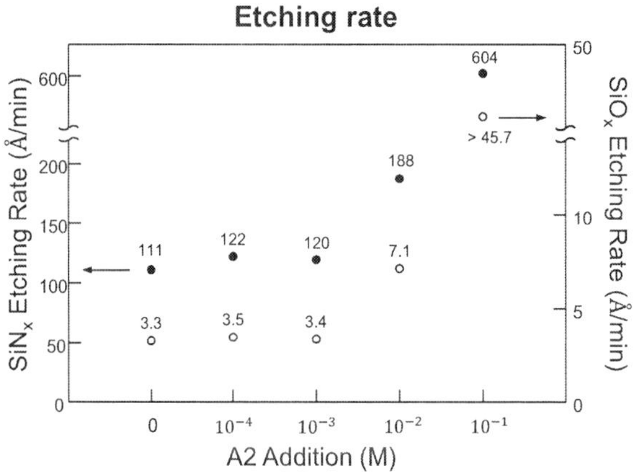


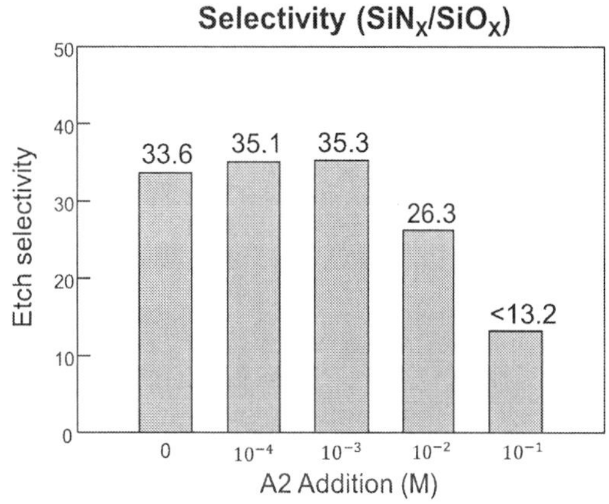


- Etching rates of $SiN_X$ and $SiO_X$ remarkably increased with addition of A2 over 0.01 M. With 0.1 M addition, etching rates increased by 5.4 and 13.8 times for $SiN_X$ and $SiO_X$, respectively.
- With the addition of A2, etching rate of $SiN_X$ increases but etching rate of $SiO_X$ increases more, thus selectivity decreases.

VERSUM MATERIALS



# Comparison of the Effects from A1 & A2

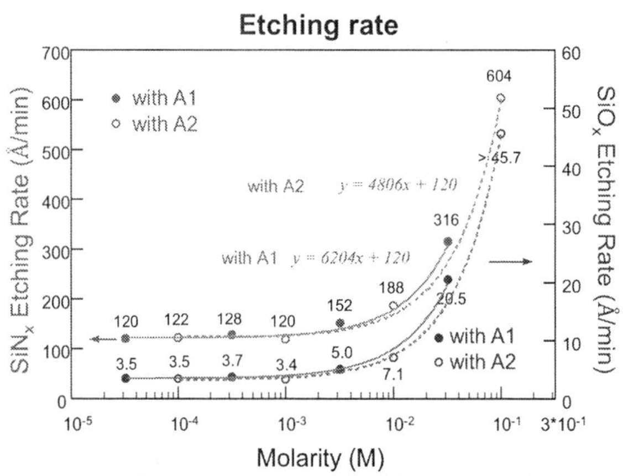


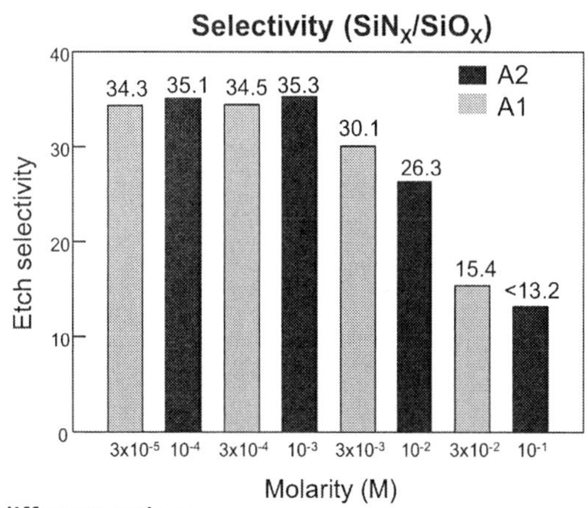


- A1 and A2 are composed of the same anion, but different cations.
- Etching rates of $SiN_x$ and $SiO_x$ were fitted by linear function of additive's molarity. At a given $ER_0$ (intercept) of 120 Å/min, the coefficient (slope) of the molarity function for the etching rate of $SiN_x$ was larger with addition of A1 (~6200) than A2 (~4800), and the coefficient of the molarity function for the etching rate of $SiO_x$ was also larger with addition of A1 (542) than A2 (425).
- Etching rates of $SiN_X$ and $SiO_X$ with addition of A1 were observed to be as high as 1.3 times of those with addition of A2. No significant effect from the cation on the increase in the etching rate is observed.

VERSUM MATERIALS

# Effect from Additive(Type B) Addition – B1

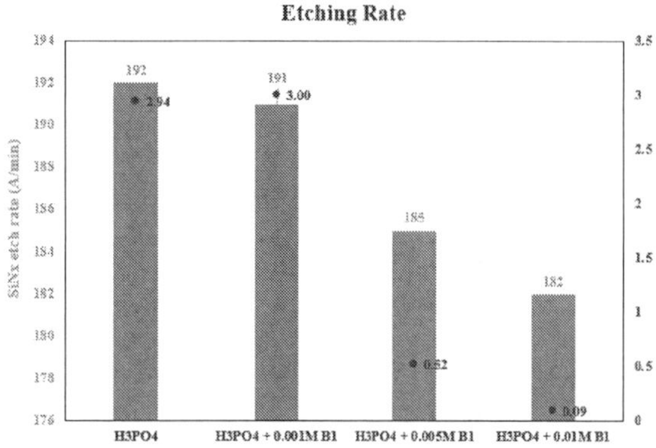


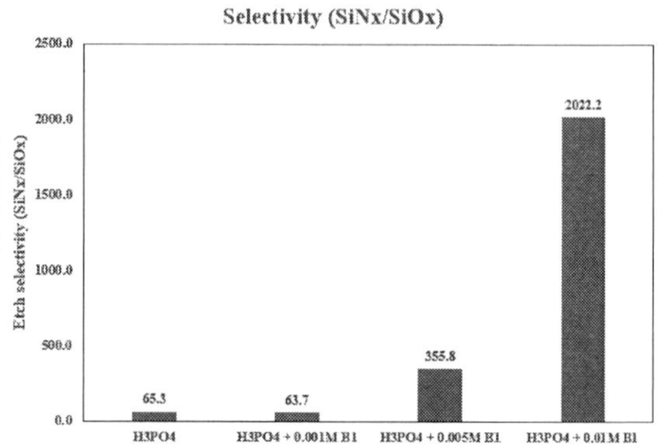


- With the decrease in B1 concentration from 0.01M to 0.001M, SiNx etch rate increased, but SiOx etch rate increase more, thus SiNx/SiOx selectivity decreased.
- The concentration of B1 at around 0.01M can achieve 2000:1 etch selectivity.
- However, it showed oxide regrowth and clogging problem on pattern structure by processing with B1 contained chemicals.

VERSUM MATERIALS

9

# Effect from Additive(Type B) Addition – B2

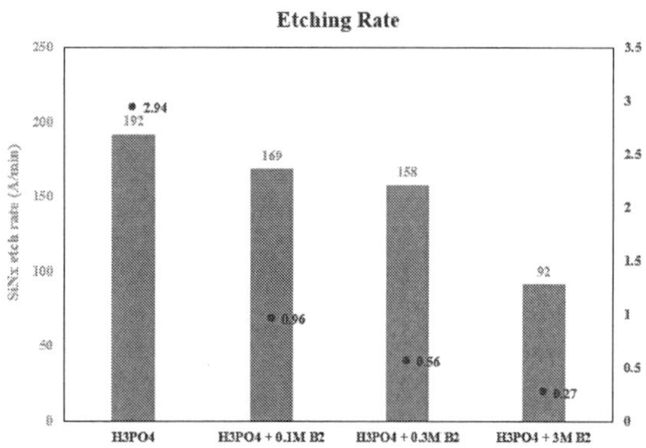


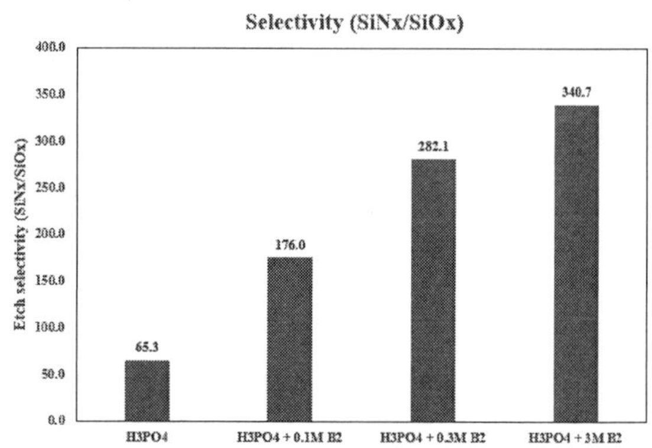


- With the increase in B2 concentration from 0.1M to 3M, both SiNx and SiOx etch rates decreased, and SiOx etch rate decrease more, thus SiNx/SiOx selectivity increased.
- However, with such high concentration(3M) of B2 addition, it still can't achieve 2000:1 etch selectivity, and SiNx etch rate became too low that would lead to long process time.
- By processing with B2 contained chemicals on pattern structure, it showed uniform etching from top to bottom. In addition, no clogging and no thinning of SiOx trench layer were observed.

VERSUM MATERIALS

10

# Effect from Additive(Type B) Addition – B3

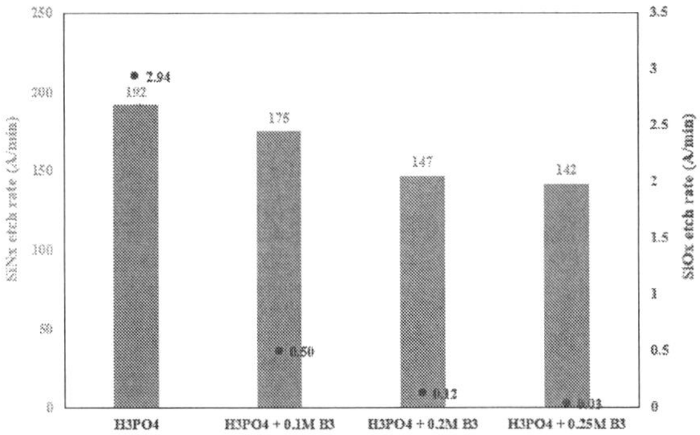


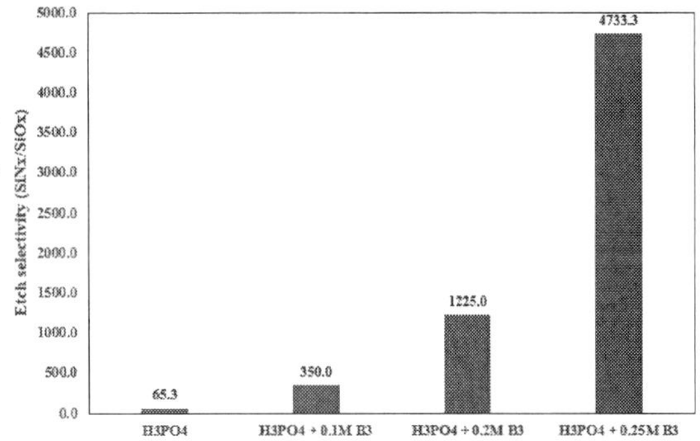


- With the increase in B3 concentration from 0.1M to 0.25M, both SiNx and SiOx etch rates decreased, and SiOx etch rate decrease more, thus SiNx/SiOx selectivity increased.
- The concentration of B3 at 0.25M can achieve the etch selectivity higher than 2000:1.
- By processing with B3 contained chemicals on pattern structure, it showed uniform etching from top to bottom. In addition, no clogging and no thinning of SiOx trench layer were observed.
- However, the miscibility between $H_3PO_4$ and B3 is not good, and it needs to add other components to improve miscibility.

VERSUM MATERIALS

11

# Comparison of B1, B2 & B3

| Chemical | $SiN_X$ E/R (A/min) | $SiO_X$ E/R (A/min) | Selectivity ($SiN_x/SiO_x$) | Miscibility | Pattern Performance |
|---|---|---|---|---|---|
| $H_3PO_4$ | 192 | 2.94 | 65.3 | - | - |
| $H_3PO_4$ + 0.01 M B1 | 182 | 0.09 | 2022.2 | Bad | Clogging |
| $H_3PO_4$ + 3.0 M B2 | 92 | 0.27 | 340.7 | Good | Uniform etching<br>No clogging<br>No thinning |
| $H_3PO_4$ + 0.25 M B3 | 142 | 0.03 | 4733.3 | Not good | Uniform etching<br>No clogging<br>No thinning |

VERSUM MATERIALS

12

# Miscibility Improvement by Solvent Addition

| Chemical | $SiN_X$ E/R (A/min) | $SiO_X$ E/R (A/min) | Selectivity ($SiN_x/SiO_x$) | Miscibility |
|---|---|---|---|---|
| $H_3PO_4$ + 0.25 M B3 | 142 | 0.03 | 4733.3 | Cloudy |
| $H_3PO_4$ + 0.25 M B3 + 3wt% S6 | 148 | 0.10 | 1480.0 | Cloudy |
| $H_3PO_4$ + 0.25 M B3 + 3wt% S7 | 146 | 0.05 | 2920.0 | Opaque |
| $H_3PO_4$ + 0.25 M B3 + 3wt% S8 | 145 | 0.07 | 2071.4 | Slightly Opaque |

- S6, S7 and S8 are solvents with the same carbon number but different functional groups.
- Polarity: S6 > S7 > S8
- By adding S8, the solvent with lower polarity, in $H_3PO_4$ + 0.25 M B3, the miscibility is improved most.
- S7 contains the same functional group as S3, and it showed the best SiNx/SiOx selectivity.

VERSUM MATERIALS

13

# Miscibility Improvement by Solvent Addition

| Chemical | $SiN_X$ E/R (A/min) | $SiO_X$ E/R (A/min) | Selectivity ($SiN_x/SiO_x$) | Miscibility |
|---|---|---|---|---|
| $H_3PO_4$ + 0.25 M B3 | 142 | 0.03 | 4733.3 | Cloudy |
| $H_3PO_4$ + 0.25 M B3 + 3wt% S7 | 146 | 0.05 | 2920.0 | Opaque |
| $H_3PO_4$ + 0.25 M B3 + 3wt% S9 | 148 | 0.07 | 2114.3 | Slightly Opaque |
| $H_3PO_4$ + 0.25 M B3 + 3wt% S10 | 153 | 0.06 | 2550.0 | Transparent |
| $H_3PO_4$ + 0.25 M B3 + 3wt% S11 | 158 | 0.14 | 1128.6 | Slightly Opaque |

- S7, S9, S10 and S11 are solvents with the same functional group but different carbon numbers.
- Carbon number: S11 > S10 > S9 > S7; Polarity: S7 > S9 > S10 > S11
- By adding S10, the solvent with proper polarity, in $H_3PO_4$ + 0.25 M B3, the miscibility is improved most and the appearance is transparent. Meanwhile, SiNx/SiOx selectivity is still kept higher than 2000:1.

VERSUM MATERIALS

14

# Effect from S10 Addition

| Chemical | $SiN_X$ E/R (A/min) | $SiO_X$ E/R (A/min) | Selectivity ($SiN_x/SiO_x$) | Miscibility |
|---|---|---|---|---|
| $H_3PO_4$ + 0.25 M B3 | 142 | 0.03 | 4733.3 | Cloudy |
| $H_3PO_4$ + 0.25 M B3 + 1wt% S10 | 149 | 0.05 | 2980.0 | Slightly Opaque |
| $H_3PO_4$ + 0.25 M B3 + 3wt% S10 | 153 | 0.06 | 2550.0 | Transparent |
| $H_3PO_4$ + 0.25 M B3 + 5wt% S10 | 162 | 0.10 | 1620.0 | Transparent |

- With higher concentration of S10 addition, the miscibility of the mixtures became better.
- With the increase in S10 concentration from 1wt% to 5wt%, SiNx etch rate increased, but SiOx etch rate increase more, thus SiNx/SiOx selectivity decreased.
- From both miscibility and selectivity aspects, $H_3PO_4$ + 0.25 M B3 + 3wt% S10 is the best combination.

VERSUM MATERIALS

15

# Performance of $H_3PO_4$+0.25M B3+3wt% S10

| Chemical | $SiN_X$ E/R (A/min) | $SiO_X$ E/R (A/min) | Selectivity ($SiN_x/SiO_x$) | Miscibility | Pattern Performance |
|---|---|---|---|---|---|
| $H_3PO_4$ + 0.25 M B3 + 3wt% S10 | 153 | 0.06 | 2550.0 | Transparent | Uniform etching<br>No clogging<br>No thinning |

- By combination of $H_3PO_4$ with 0.25M B3 and 3wt% S10, it showed good miscibility and the appearance is transparent.
- $H_3PO_4$ + 0.25 M B3 + 3wt% S10 can achieve the SiNx/SiOx selectivity higher than 2000:1
- Additionally, it also performs uniform etching on patterned structure. Besides, no thinning of SiOx layer and no oxide regrowth while SiNx layer is totally removed.

VERSUM MATERIALS

16

# Summary

- Conventional $H_3PO_4$ chemistry faces challenges as the number of layers is >48. Therefore, we are developing the formulated $H_3PO_4$ chemistry to further improve SiNx/SiOx selectivity.
- Addition of the solvents with a specific functional group can greatly increase SiNx/SiOx selectivity.
- Addition of type A additives can promote $SiN_X$ etch rate, but $SiO_X$ etch rate would increase more at the same time, thus SiNx/SiOx selectivity decreases.
- By adding 0.25M B3 in $H_3PO_4$ can achieve the etch selectivity higher than 2000:1. In addition, it showed uniform etching from top to bottom on pattern structure, meanwhile no clogging and no thinning of SiOx trench layer were observed. However, the miscibility between $H_3PO_4$ and B3 is not good, and it needs to add other components to improve miscibility.
- By combination of $H_3PO_4$ with 0.25M B3 and 3wt% S10, it showed good miscibility and can achieve the SiNx/SiOx selectivity higher than 2000:1. Additionally, it also performs uniform etching on patterned structure. Besides, no thinning of SiOx layer and no oxide regrowth while SiNx layer is totally removed.

VERSUM MATERIALS

17

# Acknowledgments

✓ I would like to express my appreciation to all of the group members for supporting this research.

- *Yi-Chia Lee*
- *Aiping Wu*
- *Hoon Song*
- *Wen Dar Liu*
- *Chuck Lhota*
- *Steve Ryu*

✓ This work was also supported by Professor Sangwoo Lim's group from Department of Chemical and Biomolecular Engineering, Yonsei University

VERSUM MATERIALS

18

# THANK YOU

VERSUM
MATERIALS

19

# Post-Etch Residue Removal for Cobalt Contact and Interconnect

Wilson Yeh*, Wade Lee, Carol Cheng, Alan Chen, Cindy Chiang, Lester Lee, Paul Bernatis, Sang-Min Wong, and Chi Yen

*EKC Technology, DuPont Electronics & Imaging*

Contact info: Wilson.Yeh@dupont.com; Sang-Min.Wong@dupont.com

## Introduction

As the critical dimension of transistor scaling down to 7nm and below, cobalt has been introduced for contacts and lower level of interconnects in the middle-of-the-line (MOL). Co can bring several benefits for chip making, including greater volume of contact metal with thin TiN barrier layer, and enabling high-aspect ratio gap filling at very fine dimensions. When the linewidth continues to shrink, Co will be a better option for offering lower resistance and improving reliability. Therefore, many device makers have already used Co instead of W and Cu for contacts and MOL interconnects. For customers who want to develop the advanced technology nodes, they must need new chemistries which able to clean Co contacts/interconnects after dry etching. One key attribute of the chemical is the ability to clean dry etching residues on Co contact/interconnects without damaging the underlying materials. In response to the industry need, we have developed an advanced Co post-etch cleaning solution with superior Co compatibility.

## Cleaning Approaches

- Effective cleaning reagents to remove the post-etching and post-ashing Co residues.
- Introduce Co inhibitor with high passivation efficiency.
- Adopt alkaline pH regime for better Co compatibility and clean capability.

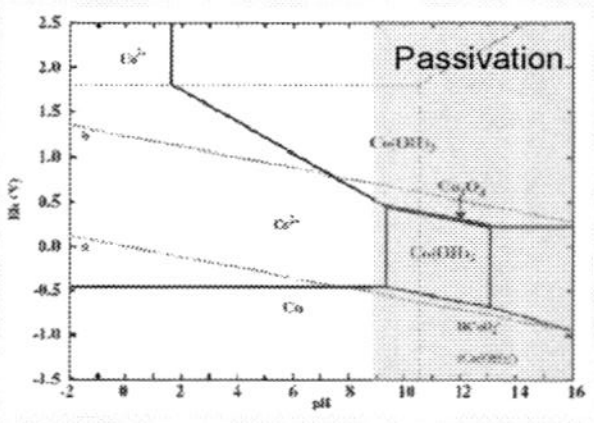


J. Chivot et al. / Corrosion Science 50 (2008) 62-69

## Results and Discussion

(a)

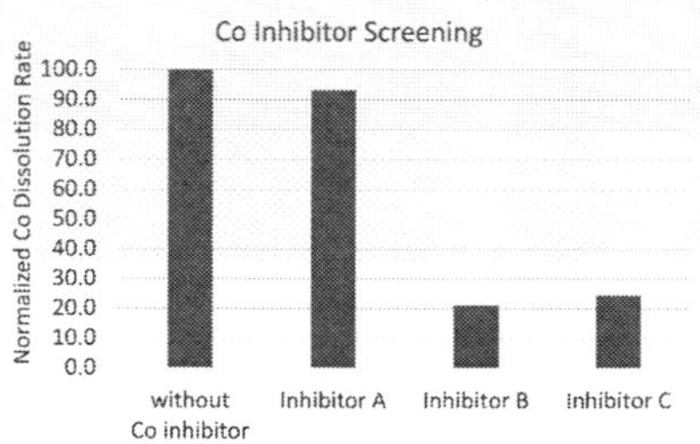


(b)

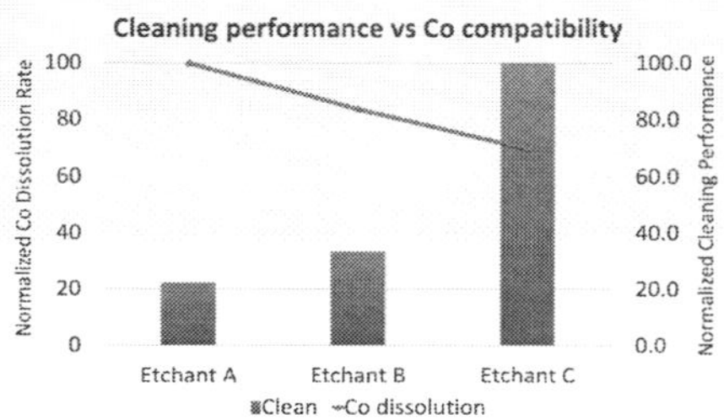


**Figure 1.** (a) Inhibitor screening for excellent cobalt protection; (b) Cleaning performance versus Co compatibility.

(a)

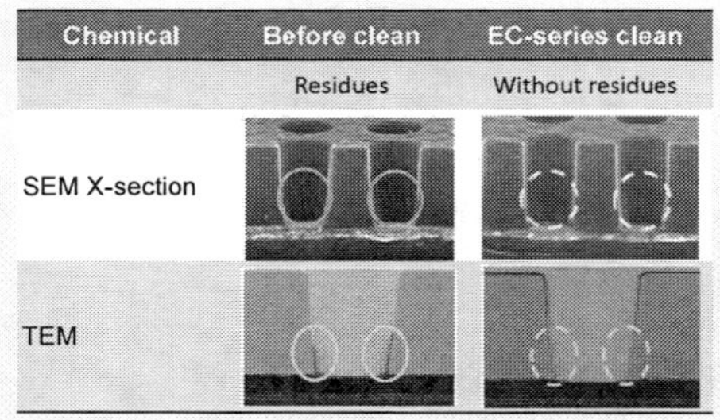


(b)

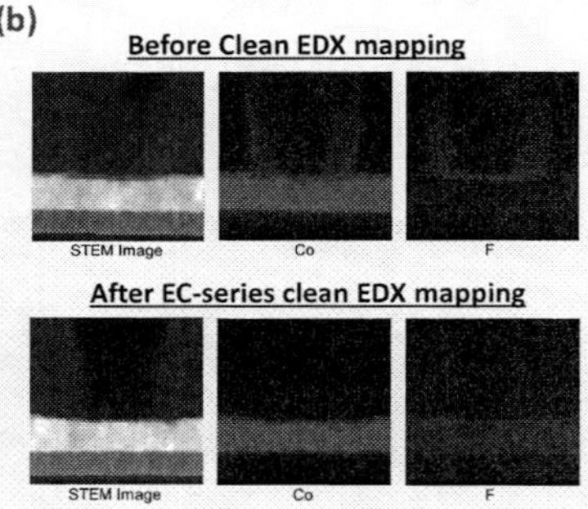


(c)

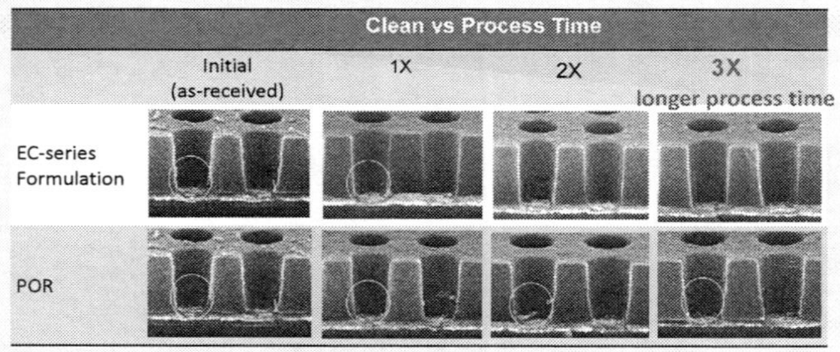


(d)

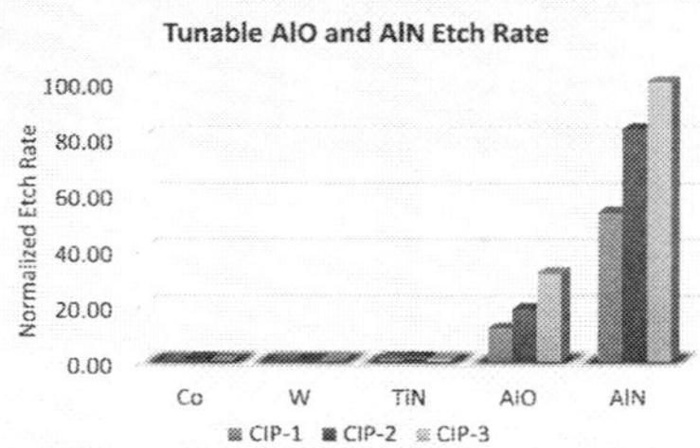


**Figure 2.** (a)-(b) SEM, TEM and EDX mapping of the Co contacts cleaned by EC-series formulation; (c) EC-series formulations can provide shorter process time; (d) EC-series formulations have tunable AlO and AlN etch rate.

## Conclusions

- EC-series formulations possess excellent cleaning capability for removing post-etch residues on Co contact and keep excellent Co compatibility.
- EC-series formulations are compatible with Co, W, TiN and dielectric, and also have tunable AlO and AlN etching rate.
- EC can provide better cost-of-ownership through upgrading clean performance and reducing process time.

HORIBA Advanced Techno

Beyond Water with You

# Development of Monitoring Technique for High Temperature Phosphoric Acid by Spectroscopy

Yutaro Tsuchisaka[1*], Takuya Onoda[2], Takashi Ohyama[2], Yohei Takahashi[2], Issei Yokoyama[2], and Kentaro Inoue[2]

*[1]HORIBA Instruments Incorporated, [2] HORIBA Advanced Techno Co., Ltd.*

**Address: 7007 SW Cardinal Ln., #185, Portland, OR 97224, *E-mail: yutaro.tsuchisaka@horiba.com*

## Introduction

### ■ Selective removal of silicon nitride in 3D NAND process

Selective removal of silicon nitride ($Si_3N_4$) in multi-layers of silicon oxide and silicon nitride is a crucial step in the production of 3D-NAND devices[1]. Phosphoric acid is a typical etchant used for the removal of $Si_3N_4$ due to its high selectivity of $Si_3N_4$ to $SiO_2$[2].
In this process, concentration of $H_3PO_4$ and the temperature are the key parameters for process control because the etching rate depends on both of them [2].

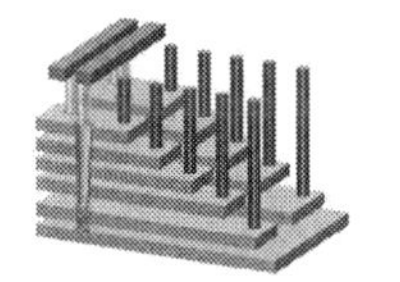

Fig.1 Schematic Diagram of Multi Stuck Layer in 3D-NAND[3]

### ■ Measurement metrology for hot phosphoric acid

For the measurement of $H_3PO_4$ concentration, conductivity measurement and optical measurement are strong candidate for this purpose[4]. Table1 shows the comparison of advantage and disadvantage for each metrology. Usually, the process temperature is generally over 140 deg.C, making it difficult to measure the concentration with conductivity.

Table.1 Comparison of advantage for metrology of hot phosphoric acid measurement

| Metrology | Conductivity | Monochromatic | Spectroscopy |
|---|---|---|---|
| Picture |  |  |  |
| High temp. (>100 deg.C) | × | × (PFA cell) | ✔ |
| Effect of temp. change | ✔ | △ | ✔ |
| Repeatability | ✔(<100 deg.C) | △ | ✔ |
| Cost | ✔ | ✔ | △ |

### ■ Objective

1. To develop spectroscopy that can directly measure hot phosphoric acid over 140 deg.C
2. To show the results of feasibility study for the performance of newly developing monitor and compared with the performance of conductivity measurement

## Materials & Method

### ■ Materials

- **Sample preparation**

Aqueous solutions of $H_3PO_4$ were prepared by diluting 85 wt% phosphoric acid solution with deionized water.

$H_3PO_4$ conc. range : 70 – 85 wt%
Temp. range : 25 – 160 deg.C

- **Experiment system**

Sample temperature was controlled between 25 and 160 deg.C using an oil bath. The image of experiment system is showed in Fig.2.

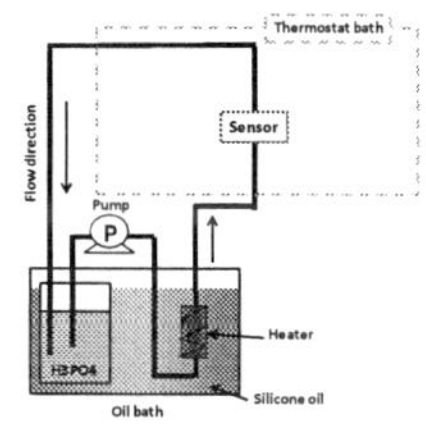


Fig.2 Experiment system

- **Apertures**

***NIR Spectrometer***

Type : Not decided (HORIBA)
Principal : Near Infrared Spectroscopy
Wavelength region : 1400 – 1600 nm
Sample temp. : 5 – 180 deg.C (Tentative)
Measurem. Range : 80 – 90 wt% (H3PO4)

***Conductivity meter***

Type : FES-510-1/4 (HORIBA)
Principal : 4 electrode method
Cell constant : About 1.4/cm
Sample temp. : 5 – 100 deg.C
Measurem. Range : 80 – 90 wt% ($H_3PO_4$)

Fig.3 NIR Spectrometer (Prototype)

Fig.4 Conductivity meter FES-510S-1/4S

### ■ Methods

- **Data aquisition**

Investigation of the major performance parameters (ex. linearity between $H_3PO_4$ concentration & conductivity,absorbance, long term measurement stability etc.) was performed as basic investigations.

## Results & Discussion

### ■ Performance test for $H_3PO_4$ sample with low temp.

#### 1. Row spectra for $H_3PO_4$ with NIR spectroscopy

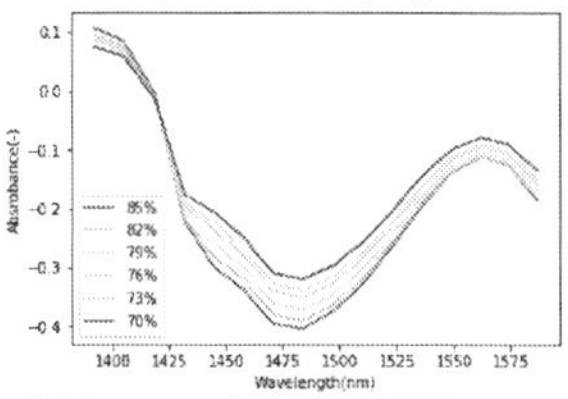


Raw spectra for $H_3PO_4$ were acquired with developing spectroscopy which can measure the region 1400 – 1600 nm called 1st overtone region of OH stretching vibration. The peak absorbance around 1480 nm decreased with increasing concentration of $H_3PO_4$ since this peak derived from OH absorbance for water.

Fig.5 Raw spectra for $H_3PO_4$ 70 – 85 % aqueous solution sample with RT sample

#### 2. Linearity between single absorbance value and $H_3PO_4$ conc.

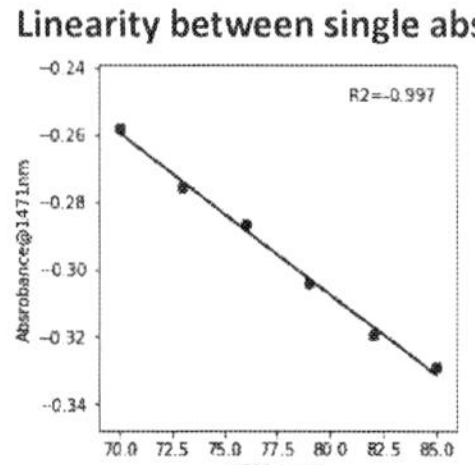


Linearity between the absorbance of 1471 nm and $H_3PO_4$ concentration was investigated. The result showed strong correlation between the absorbance and concentration since the absorbance of OH stretching vibration was changed by increasing $H_3PO_4$ concentration.

Fig.6 Linearity between the absorbance of 1471 nm and $H_3PO_4$ concentration.

### ■ Performance test for $H_3PO_4$ sample with high temp.

#### 1. Comparison of metrology for $H_3PO_4$ 80wt% sample over 100 deg.C

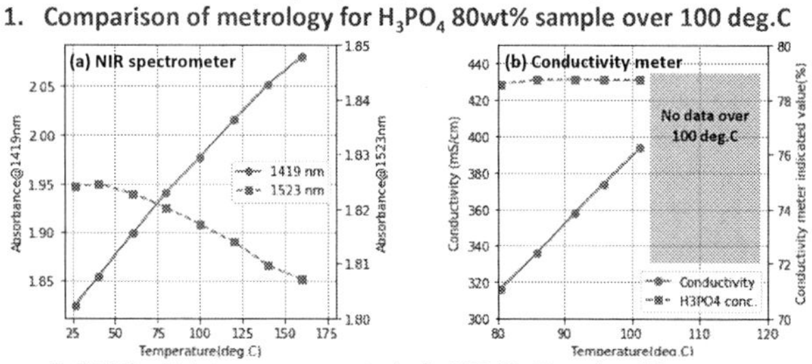


Fig.7 Relationship between measurement value for $H_3PO_4$ 80wt% sample and Temperature
(a) NIR Spectroscopy (b) Conductivity meter

#### 2. Monitoring with NIR spectrometer for $H_3PO_4$ 80wt% 25 - 160 deg.C

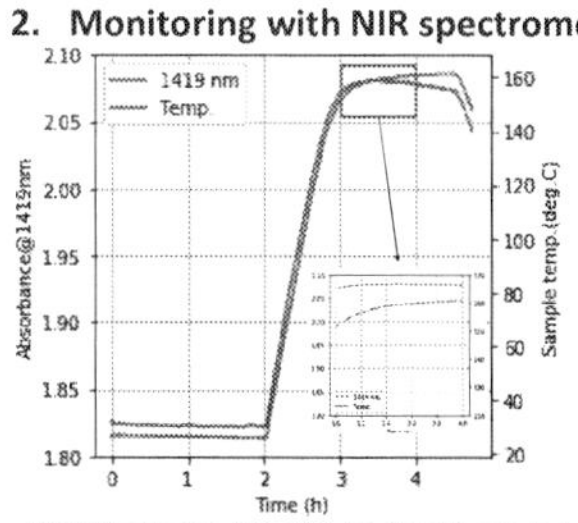


**1.** NIR spectra for $H_3PO_4$ 80wt% sample with temperature 25 - 160 deg.C were acquired. Though the absorbance value for 1419 nm and 1523 nm changed with temperature since these wavelengths related to condition of hydrogen bonding, the results showed the potential for measurement of the sample with 160 deg.C or higher. Conductivity meter could measure the sample under 100.degC but couldn't measure it over 100 deg.C since the sample cell deformed.
**2.** Monitoring data showed the possibility to measure 160.degC sample for long time measurement.

Fig.8 Monitoring data of $H_3PO_4$ 80wt% sample by spectrometer with temperature changing 25 – 160 deg.C

## Conclusion

■ This results showed NIR spectroscopy could be applied to measurement of hot $H_3PO_4$ aqueous solution over 100 deg.C and continuous monitoring for 160 deg.C.

### ■ Reference

[1] DW Bassett *et al.*, *Solid State Phenomena.*, **255**, pp. 285-290 (2016)
[2] W. Van Gelder *et al.*, *J. Electrochem. Soc.*, **114-8**, pp.869-872 (1967)
[3] Hiromichi Enami, *HORIBA Readout*, **47**, pp.10 – 15 (2016)
[4] D. Chin *et al.*, *J. Applied Electro. Chem.*, **19**, pp.95-99 (1989)

Explore the future | Automotive Test Systems | Process & Environmental | Medical | Semiconductor | Scientific | HORIBA

# Selective Etchant to Remove Metal Gate Electrodes and Protect High-k Dielectrics

Wei-Shao Tung*, Wade Lee, Carol Cheng, Alan Chen, Cindy Chiang, Charlene Chuang, Lester Lee, Sang-Min Wong, and Chi Yen
*EKC Technology, Semiconductor Technologies, DuPont Electronics & Imaging*
Contact info: wei-shao.tung@dupont.com; Sang-Min.Wong@dupont.com

## Introduction

High-k metal gate (HKMG) technology has enabled improved device performance in next-generation CMOS devices, by reducing gate leakage and threshold voltages. As the critical dimension of transistor is scaling down to 7nm and below, multiple WFMs (Work Function Metals) are used to tune threshold voltages with no channel doping in a replacement metal gate (RMG) integration scheme. In order to form different NMOS WFM and PMOS WFM stacks, it is necessary to deposit, pattern and etch one WFM followed by deposition of another WFM. One such process step involves the removal of TiN WFM with good compatibility to high-k dielectrics such as lanthanum oxide, zirconium dioxide, hafnium oxide, hafnium silicon oxynitride, and aluminum oxide. The objective of this study is to develop a formulation with high TiN to high-k dielectric selectivity.

## Results and Discussion

### Hydrogen Peroxide and Chelators to remove TiN

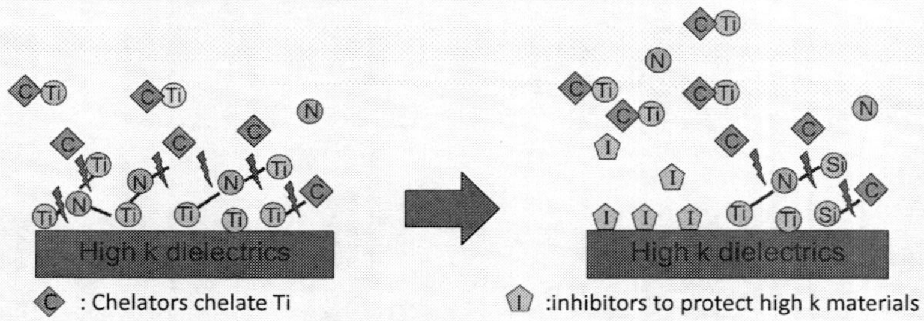


- $H_2O_2$ oxidize TiN to $TiO^{2+}$,and the TiN etch rate increases with increasing $H_2O_2$

$$TiN + H_2O_2 + 3\,H_2O \longrightarrow TiO^{2+} + 3\,OH^- + NH_4OH$$

- TiN etch Enhancer to accelerate TiN etch by forming complex with $TiO^{2+}$

$$TiO^{2+} + E \longrightarrow TiO^{2+}.E \text{ (Soluble)}$$

[1] Mat. Res.Soc.Symp.Proc.Vol. 477, 1997

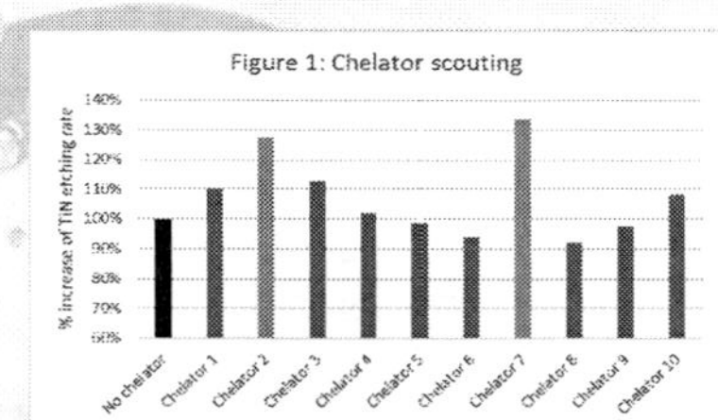

Figure 1: Chelator scouting

- Chelator 2 and chelator 7 show significant increase of TiN removal rate.

### Effective Inhibitor to passivate high-k materials

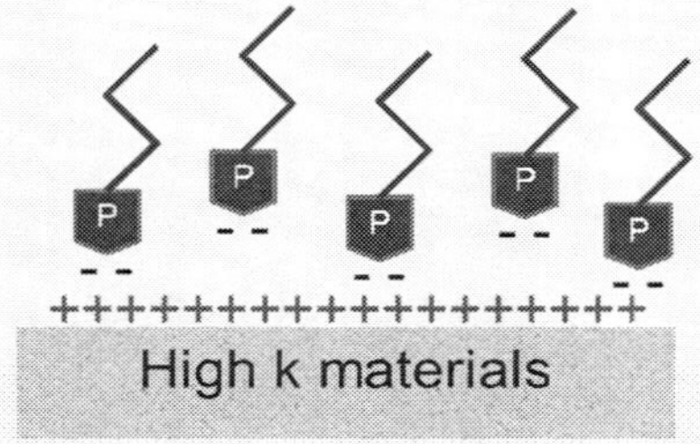


Surfactant type inhibitor contains P in WTX-series provides effective passivation to the high k surface
→ high-k etching rate reduced by 50%

### Inhibitor Scouting

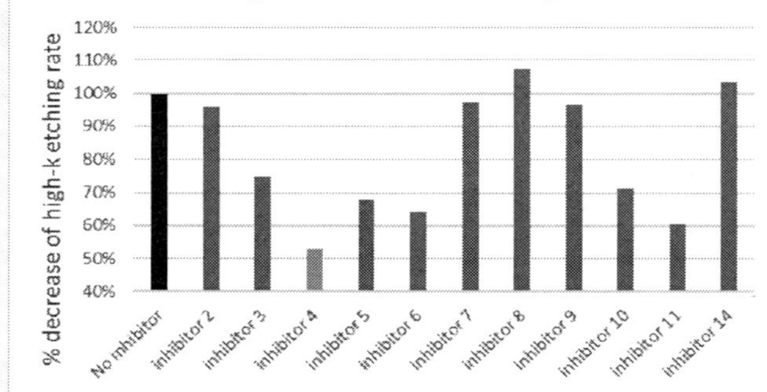


Figure 2: Inhibitor Scouting

- Couple of inhibitors show effective protection for high-k materials.
- Inhibitor 4 shows outstanding performance regarding to lowering the high-k etching rate.

## Conclusions

In this study, we report series of novel selective etchant to remove metal gate electrodes and compatible to high-k insulators. Our formulation can achieve high selectivity (metal gate etching rate/high-k etching rate > 10) to remove gate electrodes. Effective chelators and inhibitors were found to achieve the selectivity between the metal electrode materials and high-k dielectrics.

# Chemically Controlled Megasonic Cleaning Technology for Higher PRE and Lower Pattern Damage

Bongkyun Kang[c]*, Nagendra Prasad Yerriboina[b], Hyuntae Kim[a], Yujin Lee[b], Samrina Sahid[a], Andreas Klipp[d], Berthold Ferstl[d], Won-Seob Cho[c] and Jin-Goo Park[a,b]†.

[a] Department of Bio-Nano Technology and bMaterials Science and Chemical Engineering Hanyang University, Ansan, 15588, Korea.

[c]Electronic Material R&D Center Asia, BASF Company Ltd. , Suwon,16419, Korea

[d]BASF SE, Ludwigshafen, 67056, Germany

(E-mail: * brown.kang@basf.com, †jgpark@hanyang.ac.kr)

## Abstract

Megasonic cleaning has been know as the most effective way to clean the semiconductor wafers and masks. Controlling of transient cavitation is necessary to meet the stringent requirements of achieving high PRE and low pattern damage. Addition of a suitable surfactant would play a key role in achieving the controlled cavitation by means of the surfactant adsorption on acoustic bubble interfaces effectively. In this work, the effect of commercial surfactant cleaning solution (SELECTIPUR C-2106) concentration on PRE and pattern damage was evaluated. It was observed that higher PRE can be achieved at lower concentrations and it becomes negligible at higher concentrations. However, pattern damage was reduced gradually with increase of surfactant concentration. It was observed that change of zeta potential or static surface tension are not the key factors in affecting the PRE values. From the results, it is understood that the dynamic surface tension of the surfactant would be the critical factor in controlling the transient cavitation and translating it into stable cavitation to influence both PRE and pattern damage. It is hypothesized that dynamic surface tension of the surfactant affects the cavitation bubble size differently with increase of concentration and as a result influences the PRE and pattern damages differently. It was found that bubble size increases at lower concentration and it decreases to a greater extent at higher concentration by producing different microstreaming to affect PRE and pattern damage differently. From these results it can be concluded that suitable concentration of surfactant should be used to achieve better cleaning performance.

# Study on Wet Etching of Dummy Poly-Si

Dongjoo Shin[1], Yerim Yeon[2], Kwangsu Kim[3], Younggi Ahn[3], Kubra Aydin[4], Taesung Kim[1,4,*]
[1]School of Mechanical Engineering, Sungkyunkwan University, Suwon, Republic of Korea
[2]Semiconductor R&D Center, SEMES, Hwaseong, Republic of Korea
[3]Department of Mechatronics, Daelim University College, Anyang, Republic of Korea
[4]SKKU Advanced Institute of Nanotechnology (SAINT), Sungkyunkwan University, Suwon, Republic of Korea
* Email: tkim@skku.edu

## Introduction

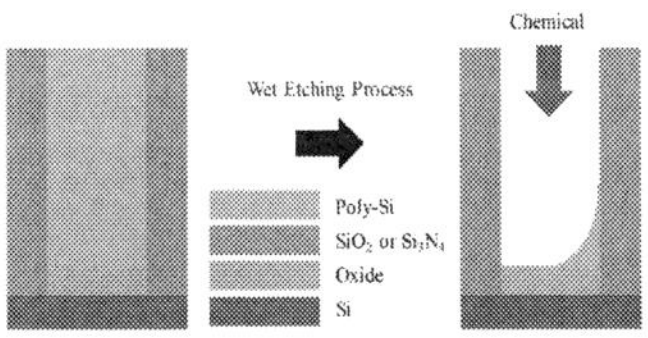


- Before making High-K/Metal gate, polysilicon was filled at gate space for building source and drain
- The complete removal of polysilicon is necessary for transistor

## Experiment

### Experiment Parameters

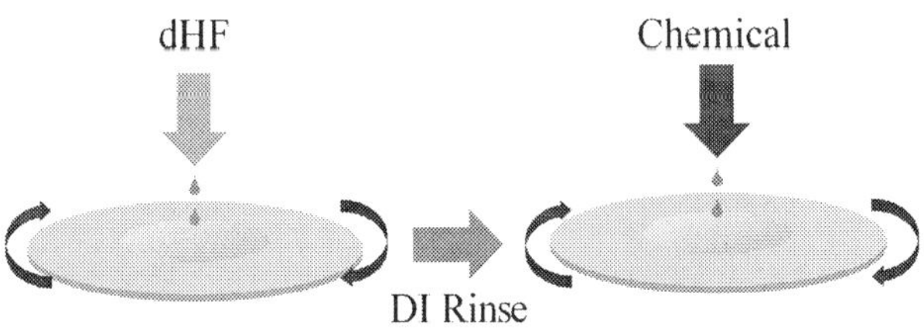


| Parameter | Method |
|---|---|
| Recipe Condition | Temperature |
| | RPM |
| Chemical Condition | Chemical Concentration |
| | Chemical Type |
| Removal of Native Oxide | 0.5wt% of dHF, 1 min |

### Experiment Process

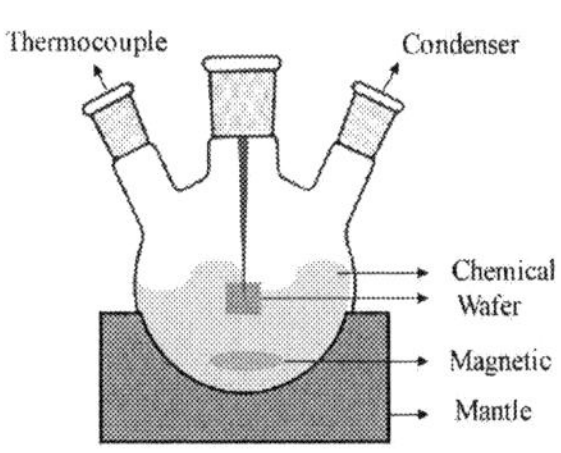


| | Parameters |
|---|---|
| Temperature | 70, 80, 90 °C |
| Chemical Conc. | 1, 6, 10 wt% |
| Chemical Type | Chemical-A |
| | Chemical-B |
| RPM | 0, 100, 200 rpm |

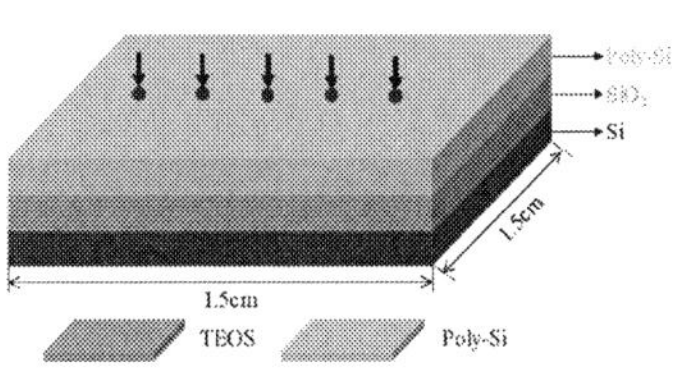


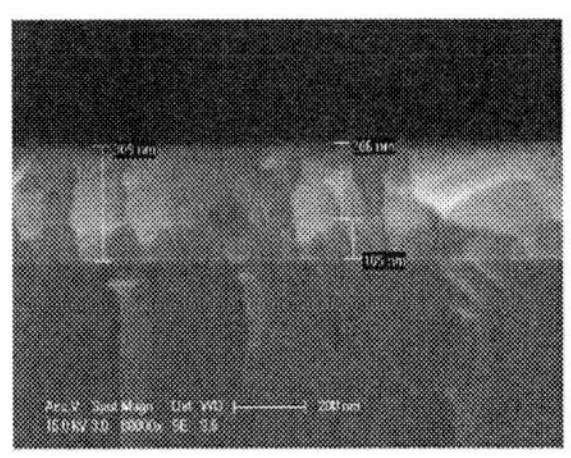

- 2000 Å of Poly-Si and 1000 Å of $SiO_2$ in coupon wafer
- Coupon wafer was etched by dipping in chemical using mantle and magnetic stirring in flask

**Main Goal → Suggest unremoved factors including improvement of etchant chemical**

## Results and Discussions

### Etch Rate Results

**Chemical-A**

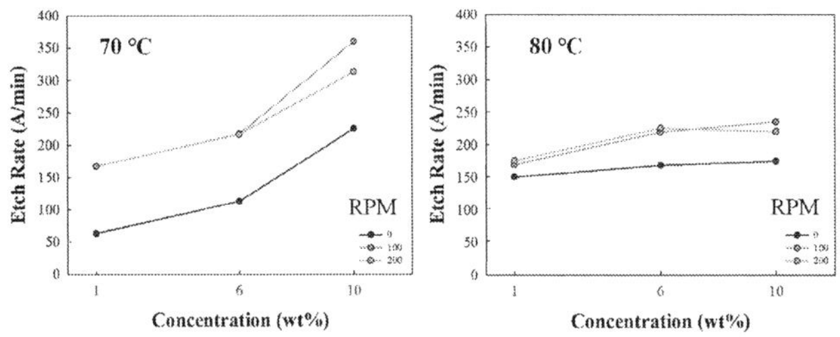


**Chemical-B**

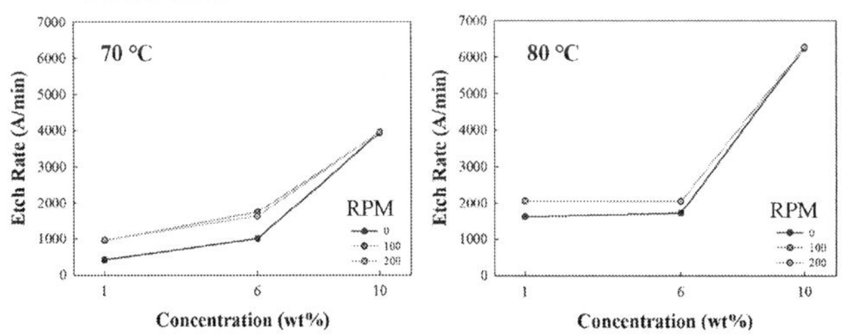


**Chemical-A**

| Unit: Å/min | 1wt% | | 6wt% | | 10wt% | |
|---|---|---|---|---|---|---|
| Temp / RPM | 70 °C | 80 °C | 70 °C | 80 °C | 70 °C | 80 °C |
| 0 rpm | 62 | 150 | 167 | 169 | 167 | 175 |
| 100 rpm | 167 | 169 | 217 | 219 | 359 | 235 |
| 200 rpm | 313 | 226 | 359 | 235 | 313 | 220 |

**Chemical-B**

| Unit: Å/min | 1wt% | | 6wt% | | 10wt% | |
|---|---|---|---|---|---|---|
| Temp / RPM | 70 °C | 80 °C | 70 °C | 80 °C | 70 °C | 80 °C |
| 0 rpm | 417 | 1615 | 1064 | 1717 | 3934 | 6242 |
| 100 rpm | 952 | 2059 | 1744 | 2054 | 3929 | 6244 |
| 200 rpm | 967 | 2059 | 1631 | 2046 | 3954 | 6270 |

- Etch rate of Chemical-B is much higher than Chemical-B at each conditions such as temperature, rpm and chemical concentration
- Chemical type and concentration are main factor of polysilicon etching process

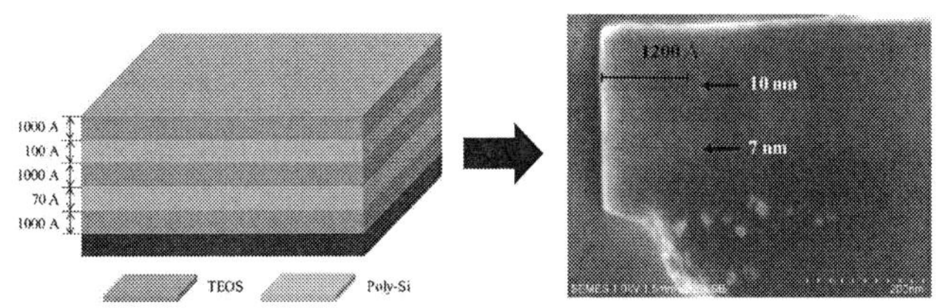


- About 1200 Å of polysilicon was successfully etched at gap of 7 nm and 10 nm in SEM image using Chemical-B

## Summary

- The poly-Si was etched by Chemical-A and Chemical-B
- Temperature, agitation, chemical concentration and chemical type are factors in wet etching process
- Chemical type and concentration are main factor of polysilicon etching process
- About 1200 Å of polysilicon was successfully etched at gap of 7 nm and 10 nm in SEM image at Chemical-B

Nano Particle Technology Laboratory (NPTL), +82-31-299-4194
Room# 83502 Sungkyunkwan Univsersity
Sungkyunkwan University, Seobu-Ro, Jangan-Gu, Suwon, Republic of Korea
http://nptl.skku.edu/

# Optimization of Water Polishing for and Improved Abrasive Removal

Seokjun Hong[1], Juhwan Kim[1], Vinit K. Kanade[2], Taesung Kim[1,2,*]
Wada Yutaka[3], Hiyama Hirokuni[4] and Satomi Hamada[3]

[1]School of Mechanical Engineering, Sungkyunkwan University, Suwon, South Korea
[2]SKKU Advanced Institute of Nanotechnology, Sungkyunkwan University, Suwon, South Korea
[3]Ebara Corp., 4-2-1 Honfujisawa, Kanagawa 251-8502 Japan
[4]Ebara Corp., 11-1 Haneda Asahi-cho, Ohta-ku, Tokyo 144-8510, Japan
*tkim@skku.edu

SPCC2019
April 2-3, 2019
Portland, USA

## Introduction

### BACKGROUND AND OBJECTIVE

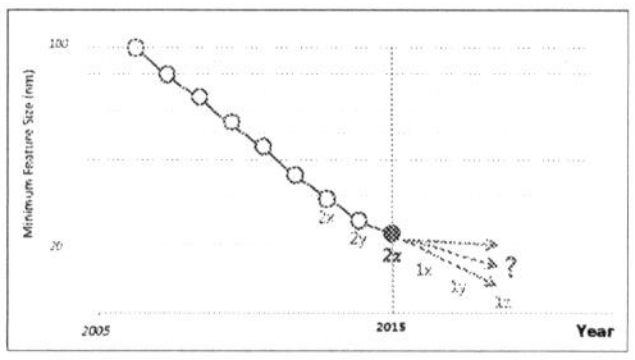


**Fig. 1. DRAM Technology Trend & Future (ITRS2015)**

- Design rule ↓
- Size of pattern ↓
- Required level of scratch & defect ↓ .

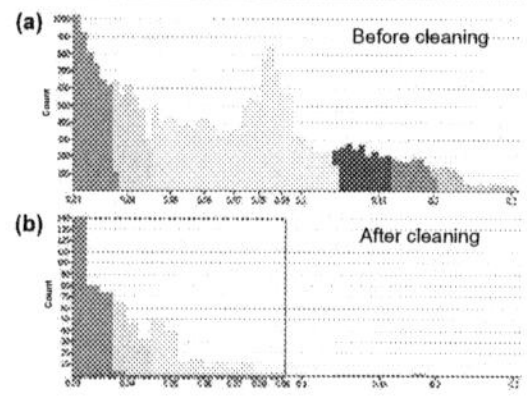


**Fig. 2. Defect counts of SiO2 wafer (a) before and (b) after cleaning**

- CMP → Post CMP Cleaning
- However, small particles still remain on the wafer surface.

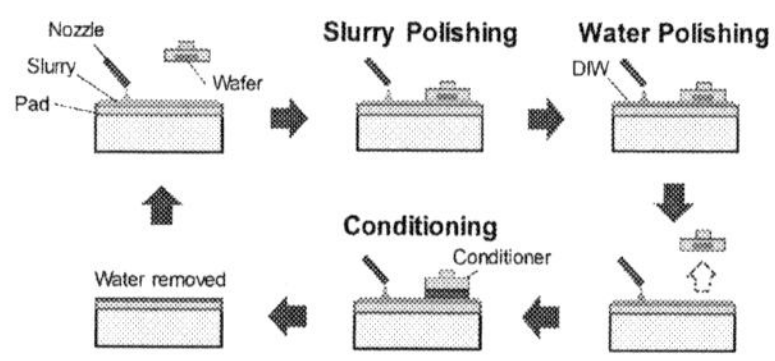


**Fig. 3. Schematic of polishing step**

- Slurry polishing : main process
- Water polishing : cleaning & surface treatment
- Post CMP Cleaning : removal of defects

→ **Optimization of water polishing step for more effective post-CMP cleaning**

## Experimental

### EXPERIMENTAL METHOD

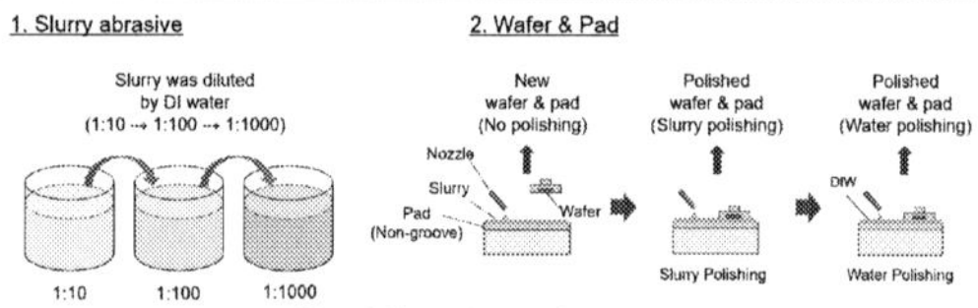


**Fig. 4. Experimental process**

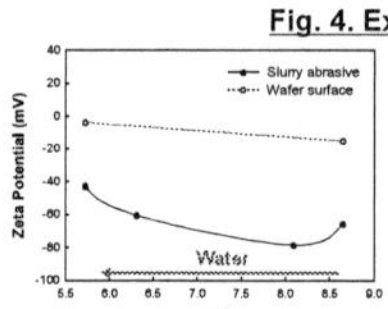


**Fig. 5. Zeta potential changes of slurry abrasive & wafer surface**

- Repulsion between abrasive & wafer ↓
- We need more mechanical force or other chemical.

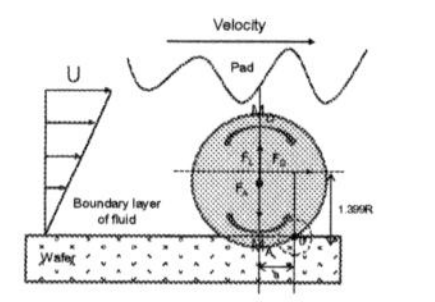


**Fig. 6. Forces acting on abrasives**

$$RM = \frac{F_D \times 1.399R}{(F_A - F_L)\times a} \text{ (non-contact)}$$

$$RM = \frac{I_p/\omega_p}{(F_A - F_L)\times a} \text{ (contact)}$$

→ Increasing of $F_D$, $F_L$, $F_C$ can Induce abrasive removal

**Table 1. Process conditions of water polishing**

| | |
|---|---|
| Head pressure (hPa) | 50, 100, 200 |
| Head speed (RPM) | 60, 90, 120 |
| Platen speed (RPM) | 61, 91, 121 |
| Flow rate (ml/min) | 100, 500, 1000 |
| Polishing time (sec) | 5, 10, 30 |
| Water temperature (℃) | 4, 20, 50 |

Contact area(%) between the pad & wafer is very low (<1%)

**Drag force** > contact force → more dominant

$$F_{drag} = 3\pi\mu d_p V_{rel}$$

$\mu\uparrow, V_{rel}\uparrow \rightarrow F_{drag}\uparrow$

## Results

### RESULTS AND DISSCUSSION

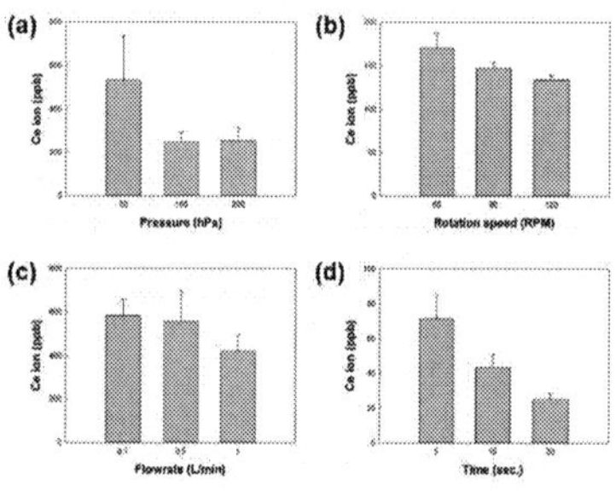


**Fig. 7. Residual Ce-ion change on wafer surface according to (a) pressure, (b) rotation speed, (c) flowrate and (d) processing time using ICP-MS**

$F_{drag}\uparrow = 3\pi\mu d_p V_{rel}\uparrow$

Pressure (P) ↑
Rotation velocity (V) ↑
Flowrate (Q) ↑
Processing time (t) ↑

⇩

Abrasive removal ↑

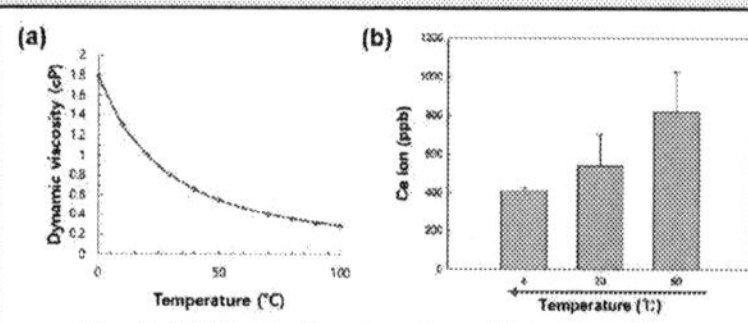


**Fig. 8. (a)Viscosity of water with temperature, (b)residual Ce-ion with water temperature**

$F_{drag}\uparrow = 3\pi\mu\uparrow d_p V_{rel}$

Temperature (T) ↓
↓
Abrasive removal ↑

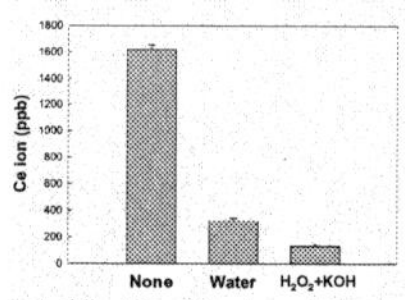


**Fig. 9. Residual Ce-ion with chemical**

Chemical → Repulsion ↑ → $F_L$ ↑

Water with chemical can increase abrasive removal.

## Conclusion

### SUMMARY AND CONCLUSION

❖ **Suggested mechanism of abrasive removal**

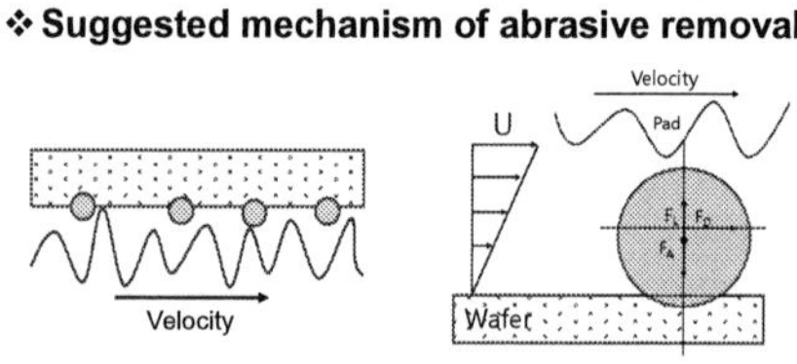


$$\text{Abrasive removal} = f(C, F_{drag}) = f(C, P, V, Q, \frac{1}{T})$$

C : **Chemical enhancement** (Electrical repulsion, hydrophilicity…)

$F_{drag}$ = **f(Pressure, Rotation speed, Temperature, Flowrate)**

#### Conclusion

- Abrasive removal increased with pressure, rotation speed, flowrate and processing time.
- Lower temperature increases abrasive removal with increasing drag force.
- Water with chemical increases abrasive removal.
- Increasing drag force and chemical enhancement can help removing abrasives during water polishing.

#### References

1. Ahmed A. Busnaina et al., IEEE TRANSACTIONS ON SEMICONDUCTOR MANUFACTURING, VOL. 15, NO. 4, (2002)
2. A. Philipossian and S. Olsen, Japanese journal of applied physics, 42(10R), 6371 (2003).
3. Ting Sun et al., Japanese Journal of Applied Physics 49 (2010)

# Removal of Nano Particles on the Hard Mask Wafer Backside using Brush Cleaning


한양대학교
HANYANG UNIVERSITY

**Hyun-Tae Kim**[a*], Yeo-Ho Kim[a], Nagendra Prasad Yerriboina[b], Tae-Gon Kim[c], KyuHwang Won[d] and Jin-Goo Park[a, b, †]

[a] Department of Bio-Nano Technology, [b] Materials Science and Chemical Engineering,
[c] Smart Convergence Engineering, Hanyang University, Ansan, 15588, Korea
[d] Samsung Electronics, Hwaseong, 18448, Korea
(E-mail: *hhyun126@hanyang.ac.kr, †jgpark@hanyang.ac.kr)

NEMPL Hanyang University
Nano-Bio Electronic Material & Processing Lab



## Abstract

As pattern nodes continue to scale down and become more complex, lithography technologies have required many solutions to create extremely small patterns. For a certain process such as, high aspect ratio trench patterning, the hard mask has been chosen for a high etch resistance and selectivity. Amorphous carbon and siloxane based materials were commonly used as hard mask materials which can be prepared by CVD or spin-on processes. During deposition process, the physical contacts of the wafer backside to pedestals such as chuck or lift pins are necessary to support and transfer the wafer. The electrostatic chuck (ESC) is widely utilized in deposition process because of its advantages in nonedge exclusion, wafer temperature control and particle defect reduction. However, the pin contact is the protrusion part on ESC that directly contacts wafer backside can make exfoliation of pre-existed backside film which results in the generation of particle defects. These backside particles may travel to the front side. These backside particles vary in size and materials because of backside particles which were crushed by the chucking pressure. It also greatly affects ESC chucking performance, adhesion uniformity and flatness stability. Especially, these backside particles were exposed to the thermal energy during the CVD process. It induces deformation of the particles, leading to an increase in adhesion to the substrate. This aging effect results in the reduction of particle removal efficiency. To solve these particle issues, an effective cleaning process is necessary to remove the contaminated particles on the backside. The adhesion and removal of these particles were first analyzed on backside surfaces. PVA brush cleaning, generally used for post CMP cleaning process, was applied to effectively remove the particles by physical cleaning in this study.


## Introduction

### ❖ Back side defect

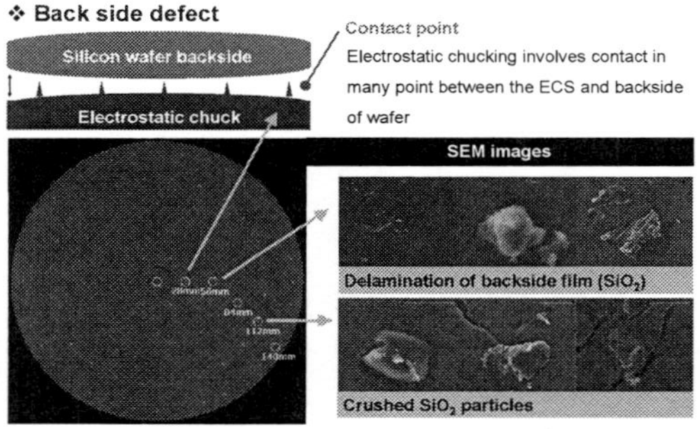


### ❖ Research Objective

**Backside Particle**

Backside particles which were crushed by the chucking pressure

- particle size : nm - 10 μm
- composition : $SiO_2$
- process temp : - 550°C, hard mask coating process

**Brush Cleaning**

To remove these particles, Brush cleaning was applied with DIW

Pin-type brush — Top view

Parameters
- Only DIW use
- Pressure
- Spin speed of brush
- Spin speed of wafer

Process was optimized by experiential approaching

How much force do we need to remove particles?

Quantification of Cleaning force + Quantification of Adhesion force → Optimization Backside cleaning condition

### ❖ Quantification of adhesion force using AFM

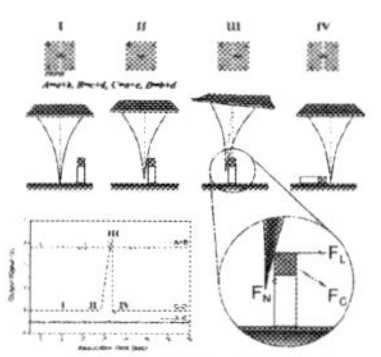

- C-D signal (a+c)-(b+d)

- Torsional spring constant ($k_t$) = nN/rad

$$k_t = \frac{Gwt^3}{3l} = \frac{75GPa \cdot 48\mu m \cdot 1\mu m^3}{3 \cdot 225\mu m} = 5.333nNm$$

- Angle conversion factor ($\eta$)

$$\kappa = \frac{\theta}{V} \cdot \frac{k_i}{H} = \eta \cdot \frac{k_i}{H} = \frac{0.01573\frac{\mu rad}{nN}}{0.441\frac{mV}{nN}} = 35.7\,\mu rad/V$$

- Force conversion factor (κ)

$$\kappa = \frac{\theta}{V} \cdot \frac{k_i}{H} = \eta \cdot \frac{k_i}{H} = 35.7\frac{\mu rad}{V} \cdot \frac{5.333nNm}{10\mu m} = 190.3nN/V$$

- Lateral force($F_L$)

$$F_L = \kappa \cdot V = 190.3 \cdot V\,nN$$

## Experimental

### ❖ Lateral force analysis

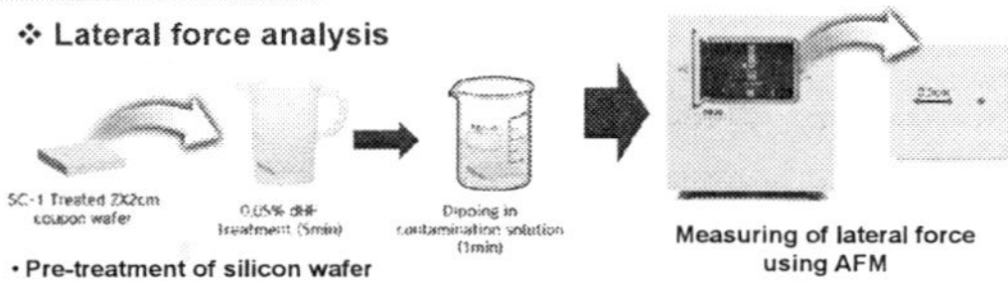


Measuring of lateral force using AFM

- Pre-treatment of silicon wafer
  1. SC1 ($NH_4OH$:$H_2O_2$:DIW=1:2:50) 10min. (dipping) – particle removal
  2. Rinse 5min. + Dry $N_2$ blowing
  3. DHF (1:100) 3min. (dipping) – native oxide removal
  4. Rinse 5min. + Dry $N_2$ blowing
- Deposition of PSL particle on silicon surface (~3μm)
  Colloidal PSL solution (100-500ppm)+ DIW 1min. (dipping)

### ❖ Brush cleaning

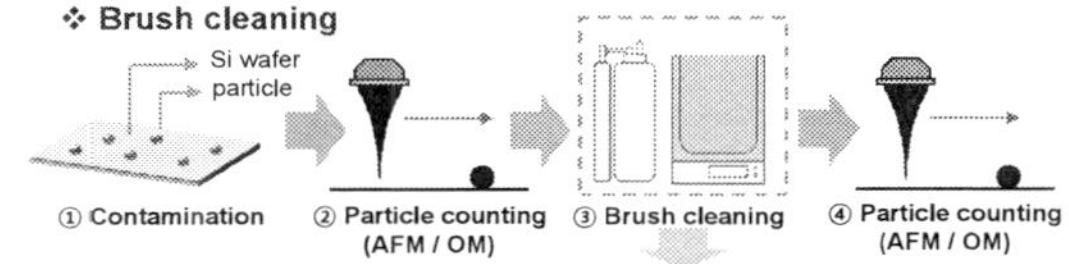


- Lab scale brush cleaning (Pin type)

Parameters
- z axis (pressure) ±30mm
- brush rotation speed (500rpm)
- wafer rotation speed (1000rpm)
- DIW flow
- brush type

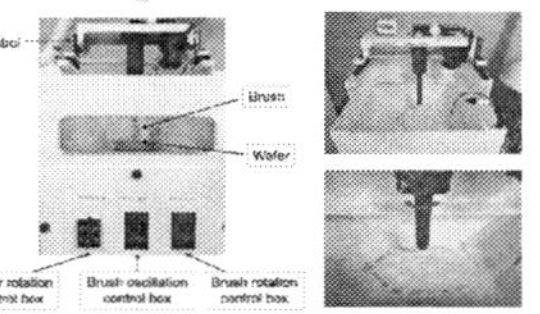


## Results

- Lateral force (1.0-3.0μm)

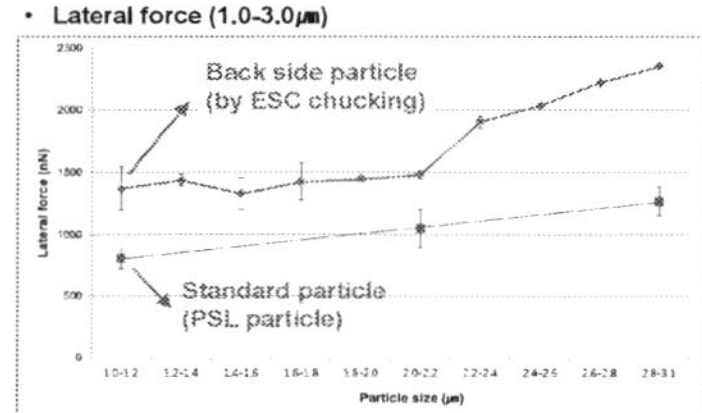


- Lateral force (20, 50, 100nm)

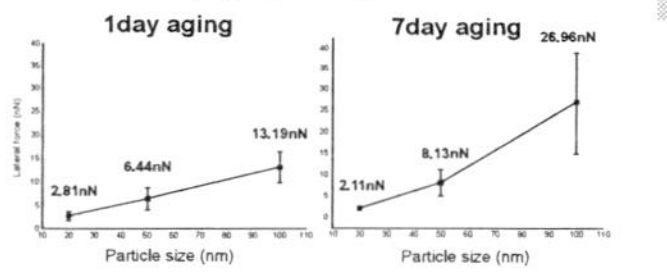


A smaller than 100nm size of particle was sensitive to aging time
Back side particle shows higher lateral force than standard particle

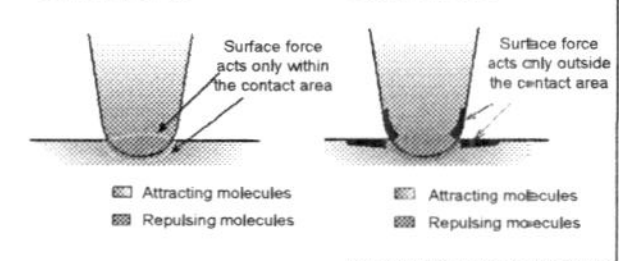


- Lateral force (0.1-3.0μm)

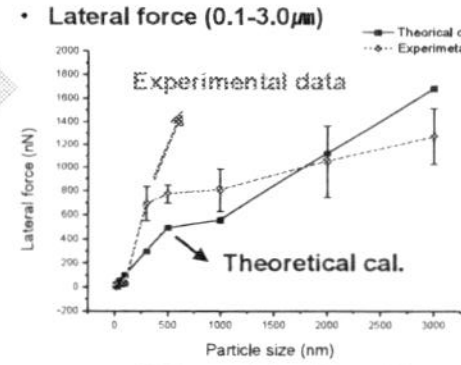


<1000nm particle : JKR model
>2000nm particle : DMT model

- Brush cleaning optimization

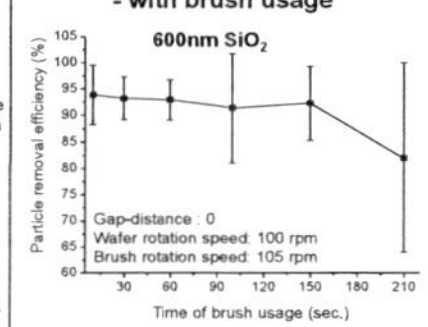


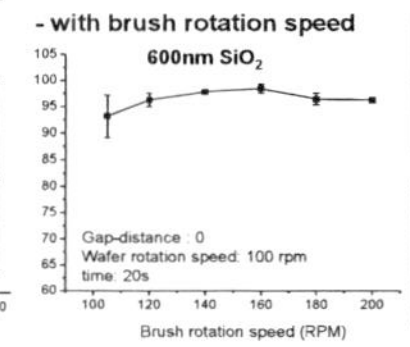


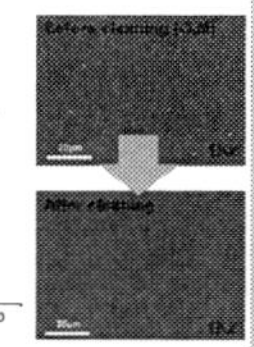


Optimized condition : >90% removal efficiency (over 600nm $SiO_2$ particle)

- Particle removal evaluation

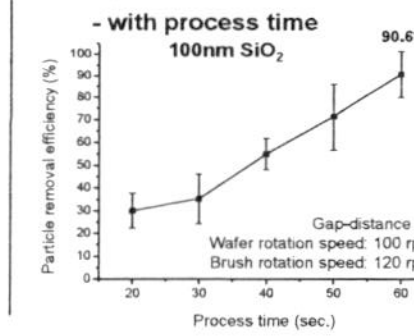


- Aging effect

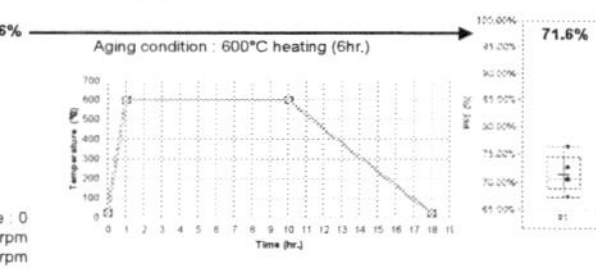


100nm $SiO_2$ removal efficiency : 90.6% (60sec.)
After aging process, removal efficiency is a 71.6%

## Conclusion

- To quantify the adhesion force of back side particle attached to the substrate, the lateral force was measured using the AFM
- Because of high temperature aging, back side particles (25.96nN) are harder to remove compared to the standard particle(13.19nN)
- Pin-type brush shows high removal efficiency of 100nm $SiO_2$ particle over 90%. However, after aging process, removal efficiency reduced to 19% decrease. Hence chemical forces are required to remove heat treated particle on backside


## Acknowledgment

This research was supported by a Semiconductor Industry Collaborative Project between Hanyang University and Samsung Electronics Co. Ltd.

SPCC 2019 | Portland, Oregon, USA | April 2-3, 2019

BALAZS™ nanoanalysis

Air Liquide creative oxygen ELECTRONIC

# Surface Analysis for EUV Lithography Thermal Outgassing Studies


Xiaoyu Zou, Hugh Gotts
Air Liquide Electronics – Balazs NanoAnalysis
46409 Landing Parkway, Fremont, CA 94538

*Contact:*
*xiaoyu.zou@airliquide.com*
*510-624-4000*




BALAZS™ nanoanalysis

# Introduction

- Extreme ultraviolet (EUV) lithography is a key enabling technology for the sub-10 nm process technology node.
- A major challenge for EUV lithography components and materials is the requirement for high cleanliness, particularly with regard to outgassed organic contamination.
- Volatile organic contaminants can foul EUV optics.
- A common source of organic outgassing is the photoresist. Due to their high absorptivity in the EUV, component heating may occur, resulting in thermal outgassing (*Proc. SPIE*, Vol. 8322, 83222X, 2012)
- To study temperature-dependent outgassing phenomena, we investigate the surface chemistry of a model photoresist formulation containing poly(methyl methacrylate) (PMMA) and the photosensitizer Irgacure 907 (IG907) under thermal vacuum conditions.

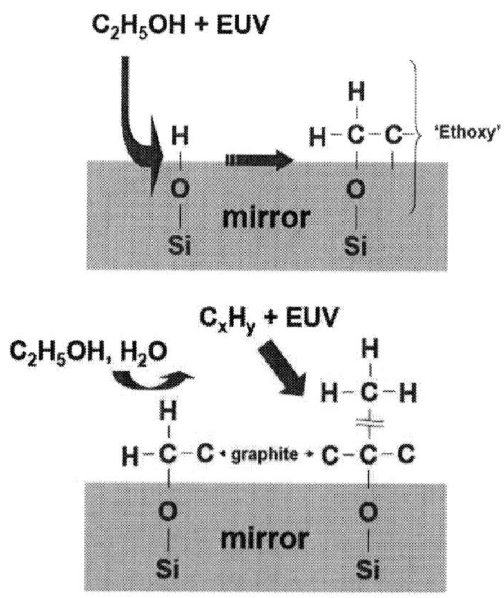

BALAZS nanoanalysis

# Experimental Design

- PMMA is a commonly used model photoresist for DUV and EUV lithography.
  - First EUV use demonstrated in the 80's (Namioka, *Revue Phys. Appl.* 1998, 23, 1711-1726)
  - Frequently used to demonstrate ultimate resolution in lithography research
- IG907 is a commonly used organic UV photoinitiator with high solubility in common solvents
- Mixture of PMMA/IG907 as a model photoresist formulation to investigate thermal outgassing of small molecule photoinitiators in polymer matrix

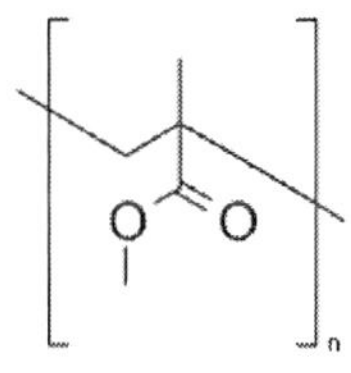

PMMA

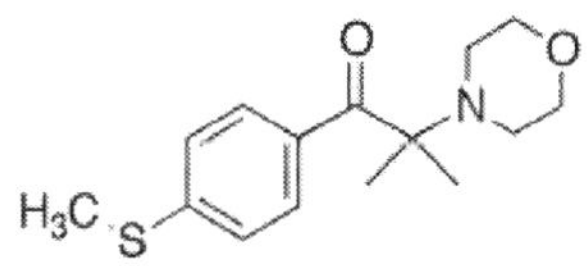

Irgacure 907



BALAZS nanoanalysis

# Analytical Methods

- X-ray photoelectron spectroscopy (XPS) uses photoelectric effect to obtain elemental and chemical information of a surface
  - Ejects core electrons with elemental-unique binding energy
  - Only top 1-12 nm electrons escape without scatter
- Residual gas analyzers (RGA) measures the vacuum outgassing of solid materials
  - Quadrupole mass spectrometry in UHV
  - Electron impact ionization
- Thermal desorption spectroscopy (TDS) uses temperature to separate gases for RGA mass spectrometry analysis

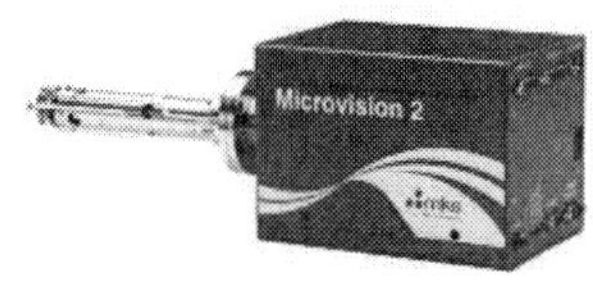


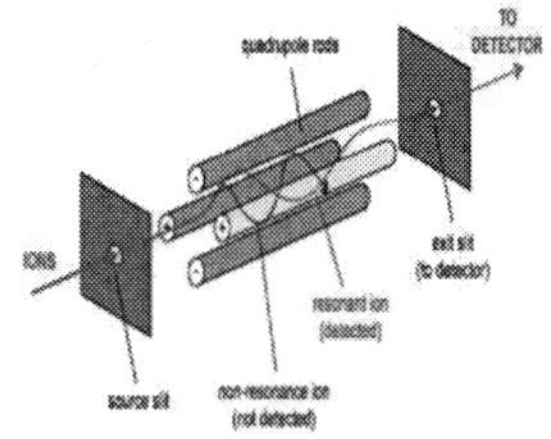


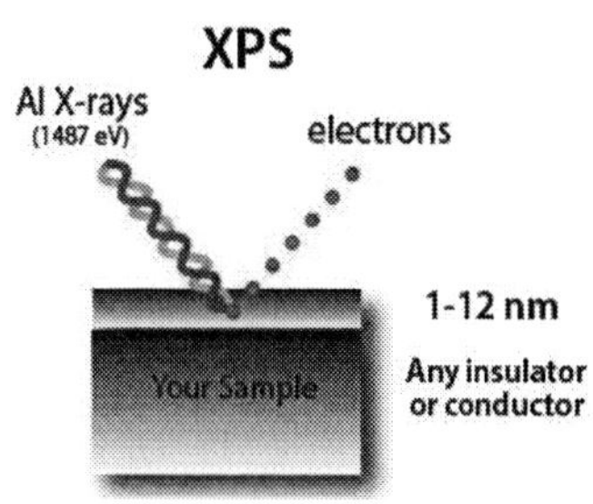

BALAZS™ nanoanalysis

# Analytical Methods

- RGA/TDS used to identify IG907 and PMMA chain scission product partial pressure as a function of temperature
  - Fragmentation patterns used to identify gas phase composition
  - Desorption temperature depends on strength of **chemical** forces between the gas molecule and the substrate
- XPS produces surface chemistry information
  - Ratio of IG907:PMMA on surface
  - Chemical environment of each atom
  - Pre- and post- heating measurements

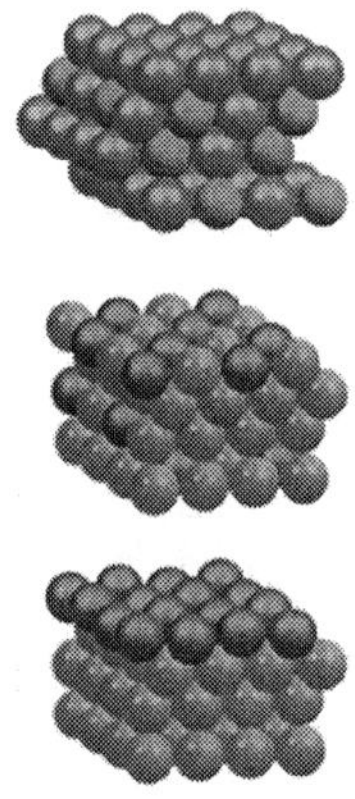

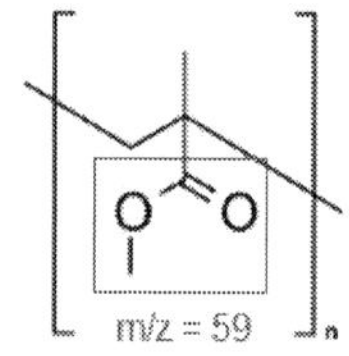


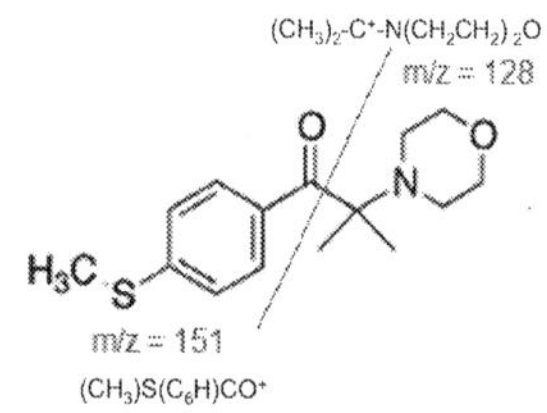




BALAZS™ nanoanalysis

# Sample Preparation

- Poly(methyl methacrylate) (PMMA, 99%, average $M_w$ = 120 kDa) and Irgacure 907 (98%) were obtained from Sigma-Aldrich and used without further purification.
- 0.5% PMMA was dissolved in analytical grade dichloromethane at room temperature. Equal weight percent IG907 was separately dissolved in equal volume of dichloromethane at room temperature.
- 25 ml of each solution was then mixed in a third container to form a 1:1 mixed PMMA/IG907 solution.
- 1x1 cm Si (100) wafer fragments were immersed in the mixed solution to form a PMMA-IG907 photoresist formulation film.

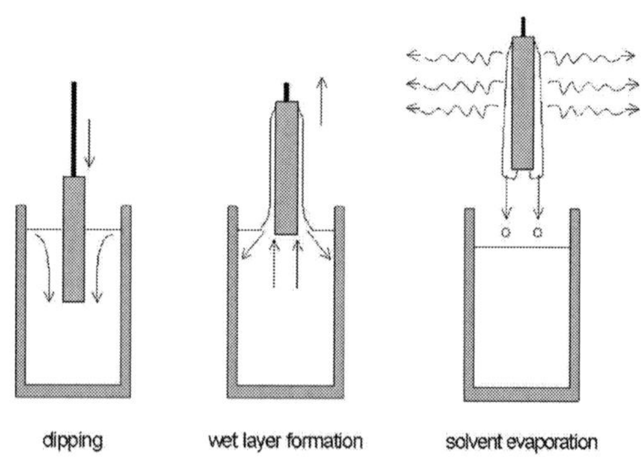

BALAZS™ nanoanalysis

# Results: RGA/TDS

25 C Isothermal Outgassing

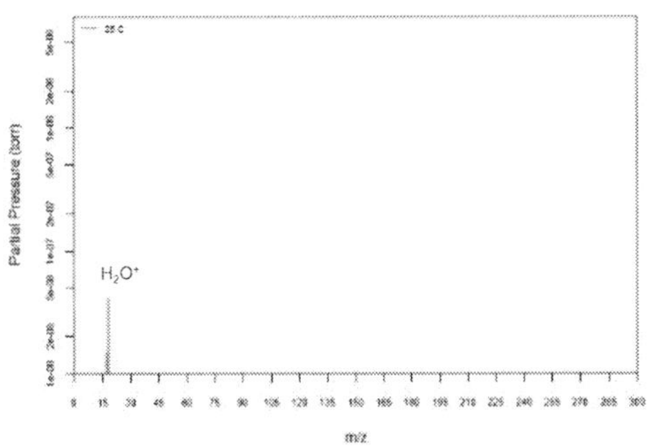


50 C Isothermal Outgassing

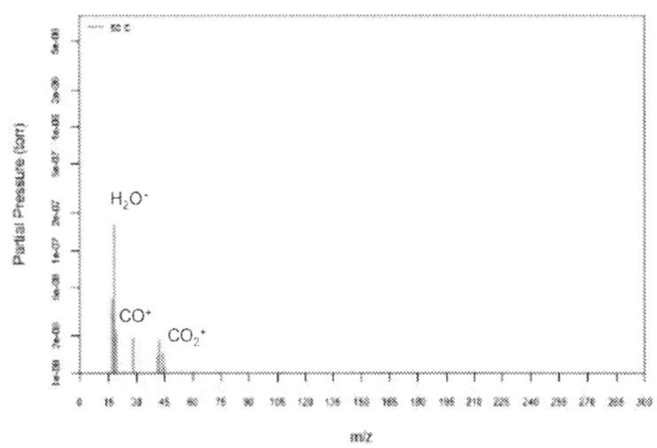


75 C Isothermal Outgassing

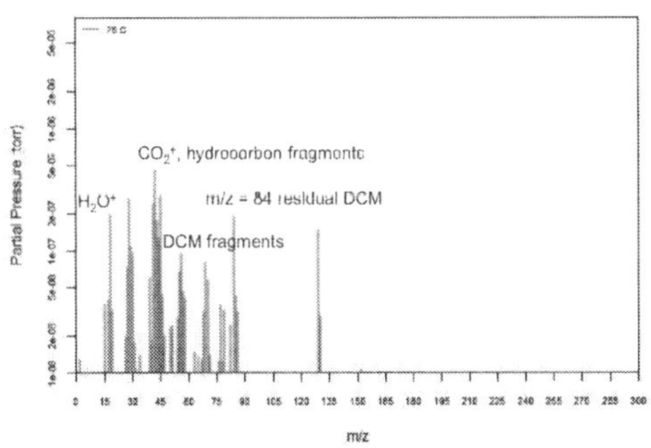


100 C Isothermal Outgassing

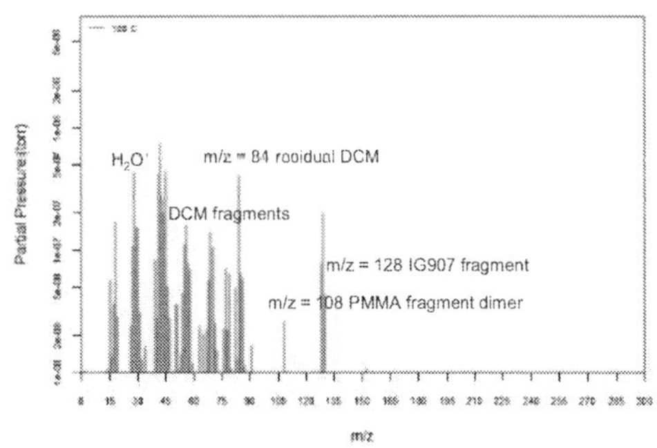




BALAZS™ nanoanalysis

# Results: RGA/TDS

120 C Isothermal Outgassing

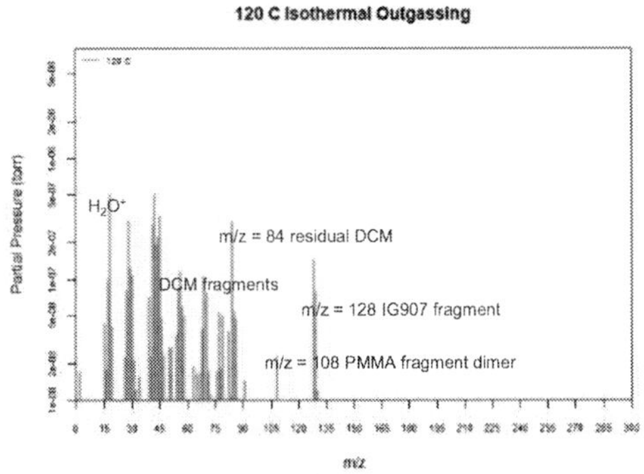


145 C Isothermal Outgassing

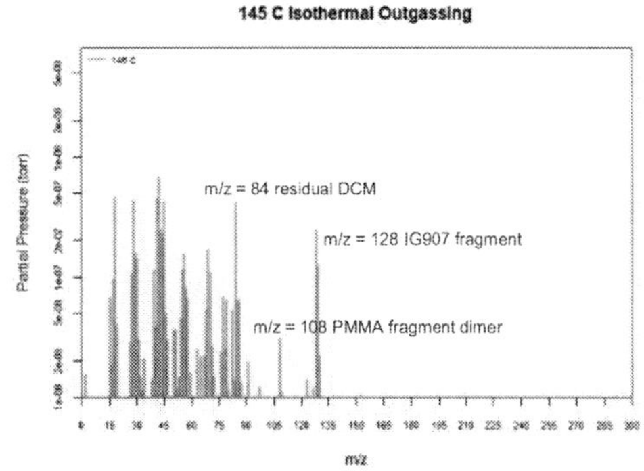


170 C Isothermal Outgassing

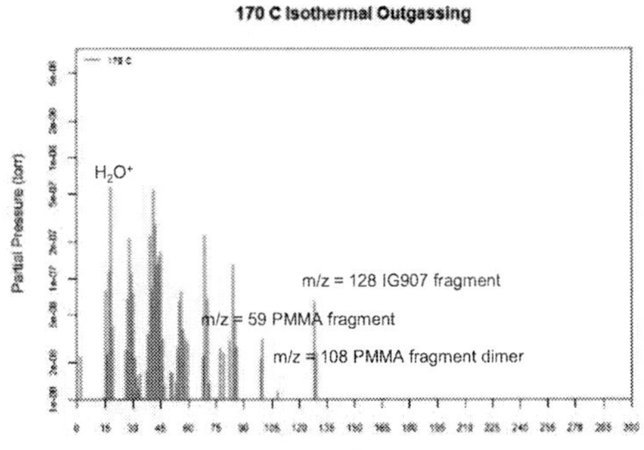


180 C Isothermal Outgassing

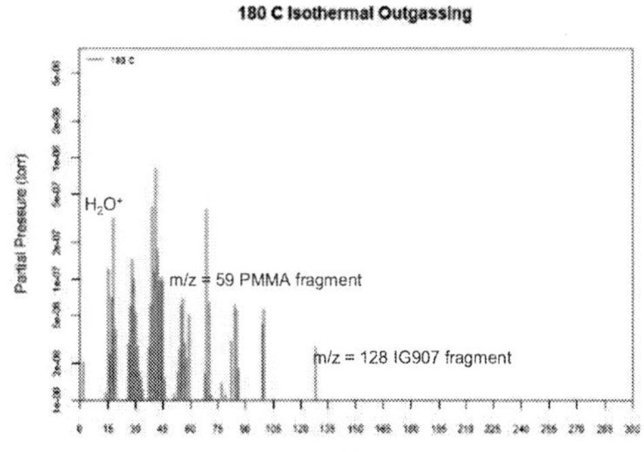

BALAZS™ nanoanalysis

# Results: RGA/TDS

- PMMA decomposition begins as low as 75°C
  - Decomposition starts dominated by sidechain dimer outgassing
  - At high temperatures (>160°C) outgassing dominated by monomer
  - Monomer outgassing continues to increase with temperature while dimer outgassing declines
- IG907 outgassing began at 75°C and peaked at ~140°C
  - IG907 is a tertiary amine and may form stable salt with addition of water in ambient atmosphere
  - Loss of hydrated salt form and conversion to neutral form at high temperature

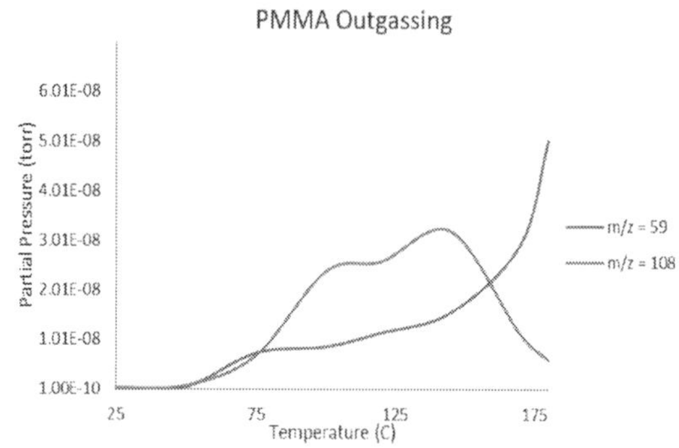


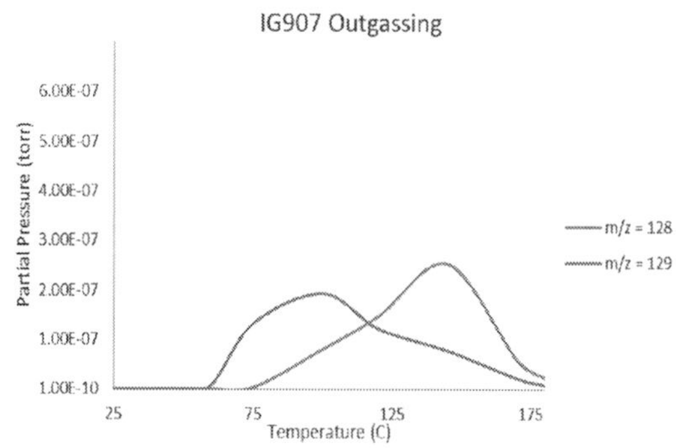




BALAZS™ nanoanalysis

# Results: XPS

- XPS detects surface chemistry
  - N:S ratio in IG907 maintained at 1:1 for both pre-heat and post-heated films
  - No thermal decomposition
- IG907 highly diluted at surface of PMMA formulation
  - N:C ratio of the PMMA-IG907 mixture is 7x lower than N:C ratio of IG907
  - After vacuum heating, IG907 is completely removed from mixture
- N 1s spectra differentiates between basic and neutral forms of IG907
  - Pure IG907 is in a mixed neutral and ionic form
  - After processing into a mixed PMMA-IG907 film, IG907 is in neutral form
  - Ionic form outgasses first from RGA

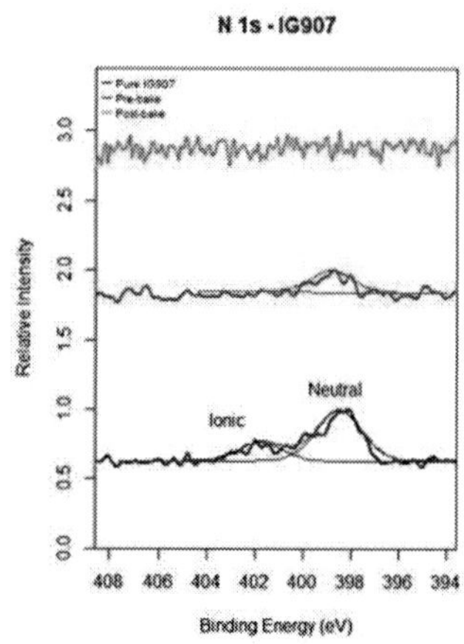


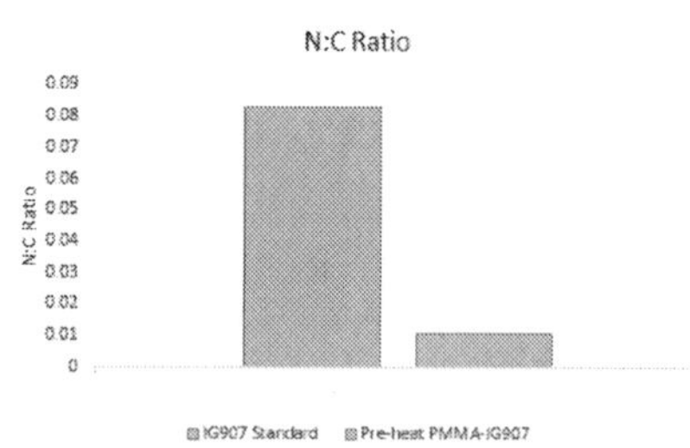

BALAZS™ nanoanalysis

# Results: XPS

- PMMA/IG907 films are visually smooth and free of macroscopic porosity
  - Unaffected by thermal processing
  - No Si signal observed
- Post-heating films show signs of thermal degradation near melting point
  - C:O ratio in PMMA-IG907 increases after heating, consistent with side chain loss
  - C 1s spectra shows little compositional change before and after heating

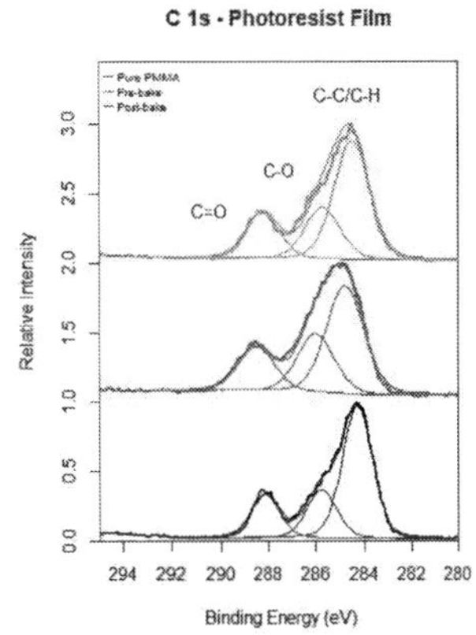


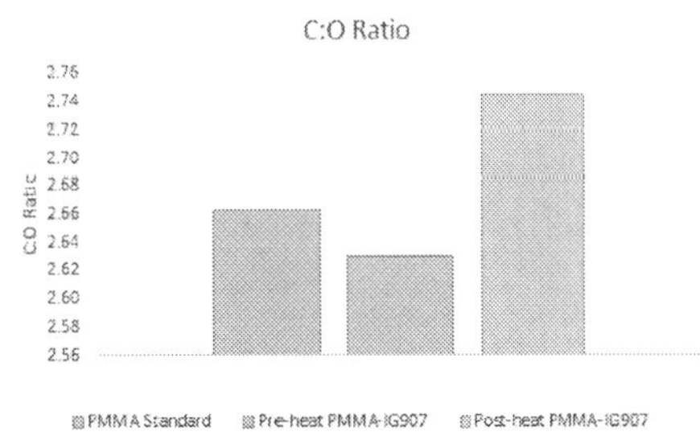




BALAZS™ nanoanalysis

# Conclusions

- Room temperature outgassing measurements may not accurately represent process conditions
  - Outgassing easily detected using mass spectrometry but only at elevated temperatures
  - No outgassing at room temperature – current standardized tests insufficient for detection of real outgassing during processing!
  - Complete loss of photoinitiator with only slight polymer degradation
- Alternative reaction paths of photoresist formulation components are important to consider
  - Hydration and moisture uptake during processing creates new acid/base chemistry
  - Different volatility of ionic and neutral forms
  - Implications for resist formulations to ensure correct function and reduce outgassing
- **Complex surface and gas phase chemistry occurs during any real process!**

# Surface Analysis for EUV Lithography Thermal Outgassing Studies


BALAZS™ nanoanalysis

Xiaoyu Zou, Hugh Gotts

Air Liquide Electronics – Balazs NanoAnalysis

46409 Landing Parkway, Fremont, CA 94538

Air Liquide creative oxygen ELECTRONICS


## Introduction

- Extreme ultraviolet (EUV) lithography is a key enabling technology for the sub-10 nm process technology node.
- A major challenge for EUV lithography components and materials is the requirement for high cleanliness, particularly with regard to outgassed organic contamination.
- Volatile organic contaminants can foul EUV optics.
- A common source of organic outgassing is the photoresist. Due to their high absorptivity in the EUV, component heating is likely to occur, resulting in thermal outgassing (*Proc. SPIE*, Vol. 8322, 83222X, 2012)
- To study temperature-dependent outgassing phenomena in lithography processes, we investigate the surface chemistry of a model photoresist formulation containing poly(methyl methacrylate) (PMMA) and the photosensitizer Irgacure 907

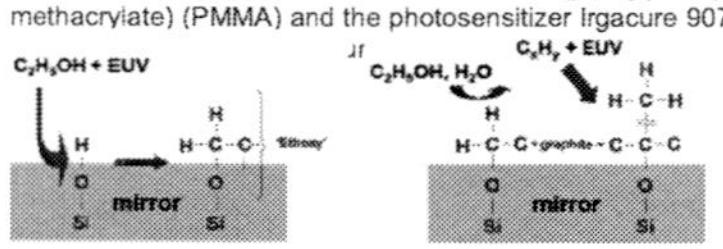


## Experimental Design

- PMMA is a commonly used model photoresist for DUV and EUV lithography.
  - First EUV use demonstrated in the 80's (Namioka, *Revue Phys. Appl.* 1998, 23, 1711-1726)
  - Frequently used to demonstrate ultimate resolution in lithography research
- IG907 is a commonly used organic UV photoinitiator with high solubility in common solvents
- Mixture of PMMA/IG907 as a model photoresist formulation to investigate thermal outgassing of small molecule photoinitiators in polymer matrix
- X-ray photoelectron spectroscopy (XPS) and residual gas analysis/thermal desorption spectroscopy (RGA/TDS) measures surface chemistry changes

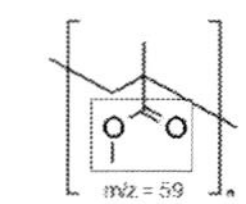


PMMA

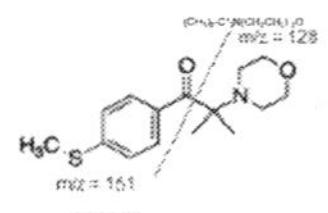


Irgacure 907

## Sample Treatment

- Poly(methyl methacrylate) (PMMA, 99%, average $M_w$ = 120 kDa) and Irgacure 907 (98%) obtained from Sigma-Aldrich and used without further purification.
- 0.5% PMMA was dissolved in analytical grade dichloromethane at room temperature. Equal weight percent IG907 was separately dissolved in equal volume of dichloromethane at room temperature.
- 25 ml of each solution was mixed in a third container to form a 1:1 mixed PMMA/IG907 solution.
- 1x1 cm Si (100) wafer fragments were immersed in the mixed solution to form a PMMA-IG907 photoresist formulation film.
- Samples heated in vacuum chamber for thermal outgassing

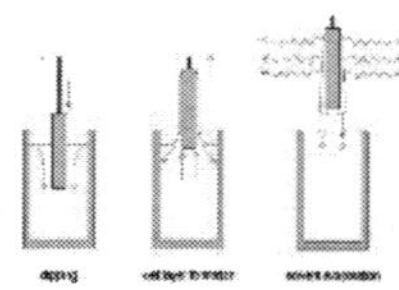

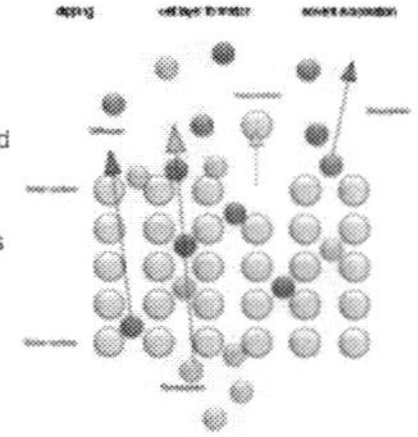

## Analytical Methods

- XPS uses photoelectric effect to obtain elemental and chemical information of a surface
  - Ejects core electrons with elemental-unique binding energy
  - Only top 1-12 nm electrons escape without scatter
- XPS produces information about the sample surface including:
  - Ratio of IG907:PMMA on surface
  - Pre/post-heating surface chemistry
- Residual gas analyzers (RGA) measures the vacuum outgassing of solid materials
  - Quadrupole mass spectrometry in UHV
  - Electron impact ionization
- TDS uses temperature to separate gases for RGA mass spectrometry analysis
  - Desorption temperature depends on **chemical** forces between gas molecule and substrate
- RGA/TDS used to identify IG907 and PMMA outgassing product partial pressure as a function of temperature

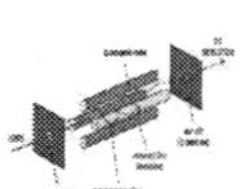

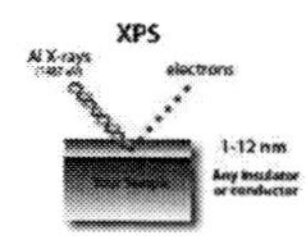


## Results

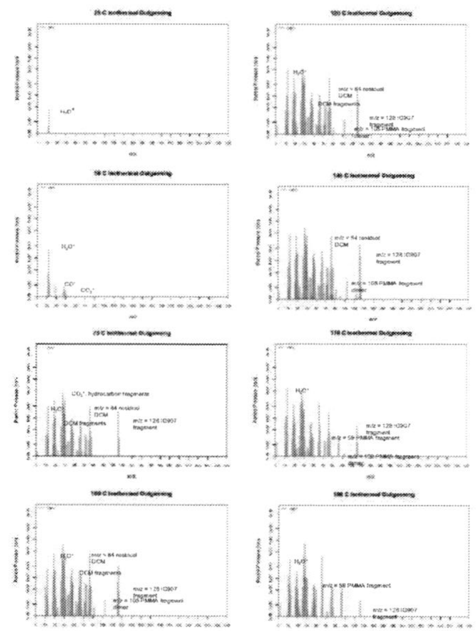

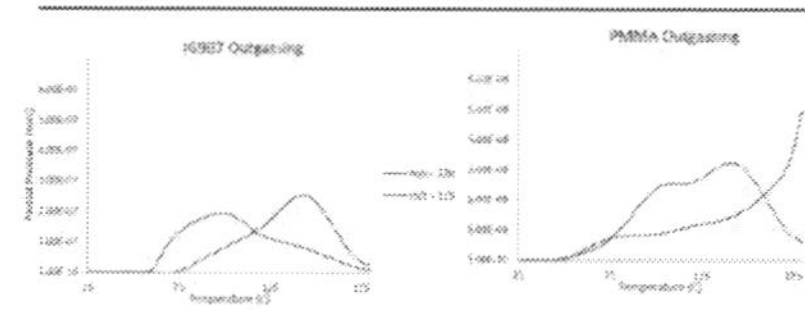


- IG907 outgassing begins at 75°C and peaks at ~140°C
  - IG907 is a tertiary amine and may form stable salt with ambient water
  - Loss of hydration and conversion to neutral form at high temperature
- PMMA decomposition begins as low as 75°C
  - Decomposition starts dominated by sidechain dimer outgassing
  - At high temperatures (>160°C) outgassing dominated by monomer

## Results (cont.)

- XPS detects surface chemistry
  - N:S ratio in IG907 maintained at 1:1
- IG907 highly diluted at surface of PMMA formulation
  - N:C ratio of PMMA-IG907 mixture is 7x lower than N:C ratio of IG907
- N 1s spectra differentiates between basic and neutral forms of IG907
  - Pure IG907 is in a mixed neutral and ionic form
  - After processing into a mixed PMMA-IG907 film, IG907 is in neutral form
  - Ionic form outgasses first
- Post-heating films show signs of thermal degradation near melting point
  - C:O ratio in PMMA-IG907 increases after heating, consistent ... change

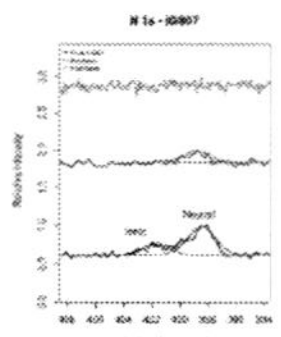


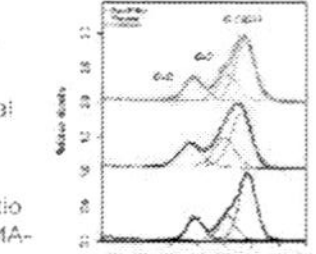


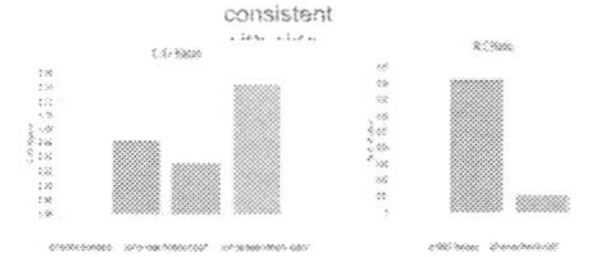

## Conclusion

- Room temperature outgassing measurements may not accurately represent all conditions
  - Outgassing detected only at elevated temperatures; only water at room temperature
- Alternative reaction paths of photoresist formulation components are important to consider
  - Hydration and moisture uptake during processing creates new acid/base chemistry
  - Implications for resist formulations to ensure correct function and reduce outgassing
- Complex surface and gas phase chemistry occurs during real process!

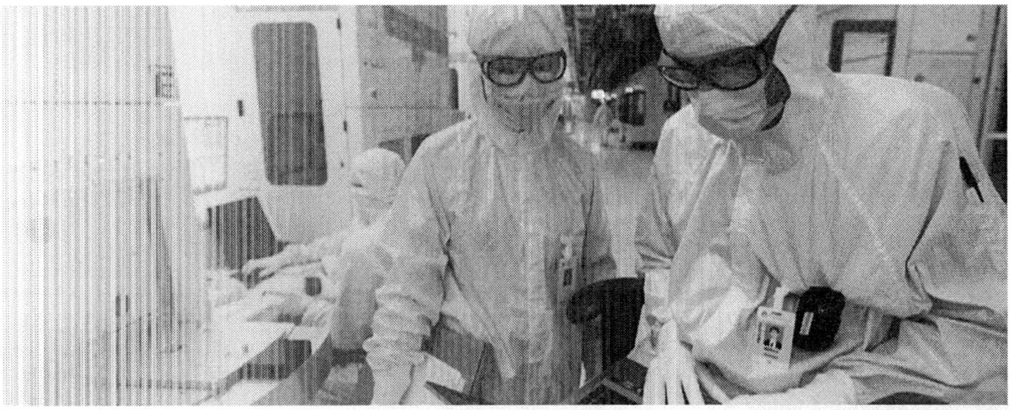

# Elimination of Surface Particles with Lobe Signature by Optimizing Post-Ash Cleans

**Dhiman Bhattacharyya*, Fauzia Khatkhatay, Sneha Gaddam & Crystal Kenney**
GLOBALFOUNDRIES
400 Stonebreak Rd. Ext., Malta, NY - 12020
Mobile: 518-948-3123
Email: dhiman.bhattacharyya@globalfoundries.com

GLOBALFOUNDRIES

**April 2, 2019**

## Background

- Post-ash wet cleaning processes are there to remove any photo resists and particles left on the wafer after ashing process.
- Commonly used processes involve sulfuric peroxide mixture[1] (SPM) which is a blend of sulfuric acid and hydrogen peroxide in a predetermined ratio and/or standard clean-1[2] (SC1) which is a mixture of ammonium hydroxide, hydrogen peroxide and deionized water in a specific ratio, or ozonized deionized water[3].
- Any particulate defect left behind after these processes may be a yield killer defect, particularly when the feature size of the semiconductor device is shrinking and a single large particle may kill a whole die.

GLOBALFOUNDRIES 2

## Background

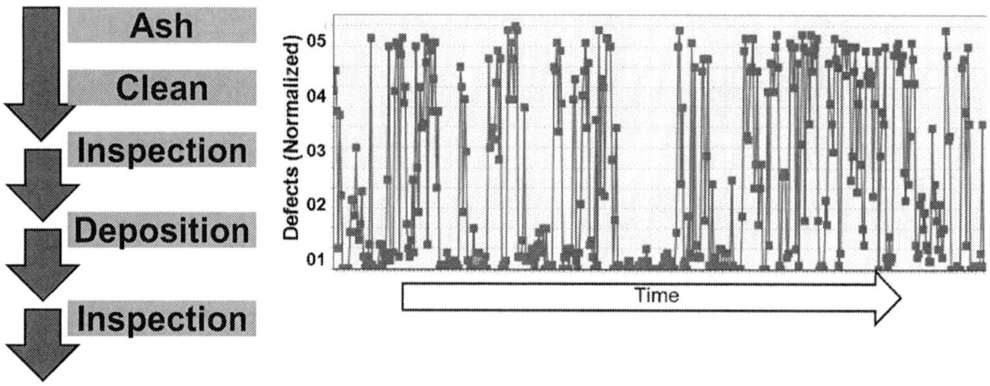


- POR Post-Ash clean was directly correlated to the defects.
- Pre and post cleans wafer scans confirmed the source of the particles to be the associated POR cleans process.
- Frequent high fliers with these defects observed at inspection steps.



## Background

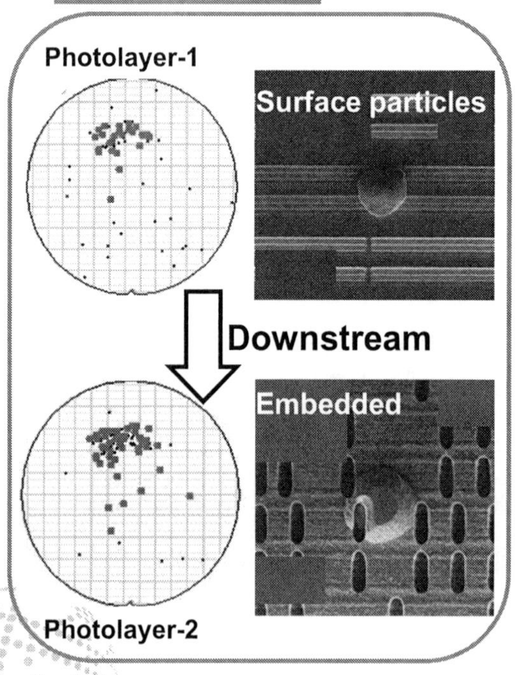


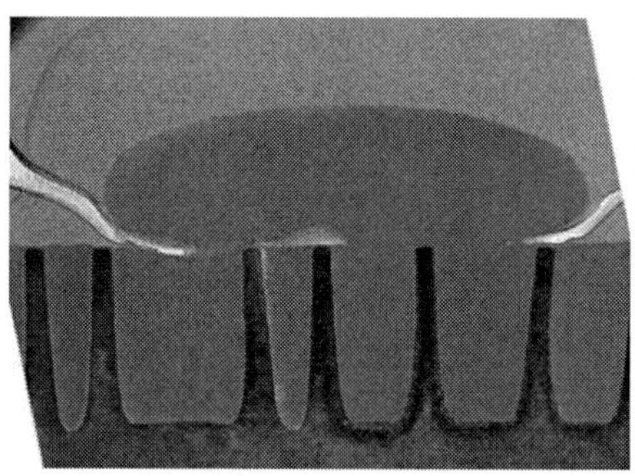

Particles were made of Si & O and sitting on top of fins.

## Approach

- To remove the defects, or the source of the defects
- A GoTo recipe was evaluated on blanket wafers before product.
- GoTo recipe proved to be much cleaner, i.e. less particle adders.
- GoTo has more particle removal efficiency.
- GoTo recipe expected to show more inline loss than POR because of the nature of the chemistry.

GLOBALFOUNDRIES 5

## Particle Removal Delta on Blanket Si

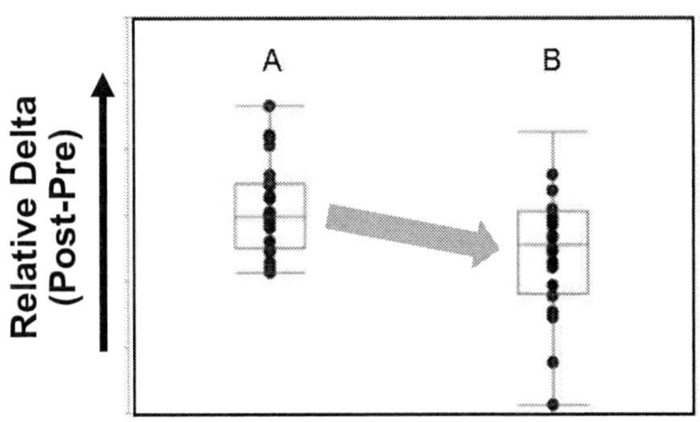


**Improvement of Particle Removal by 10 counts**

Comparison of 22nm+ particle delta performances between the POR (A) and GoTo (B) processes on blanket silicon wafers.

GLOBALFOUNDRIES 6

## Particle Adders Performance on Blanket Si

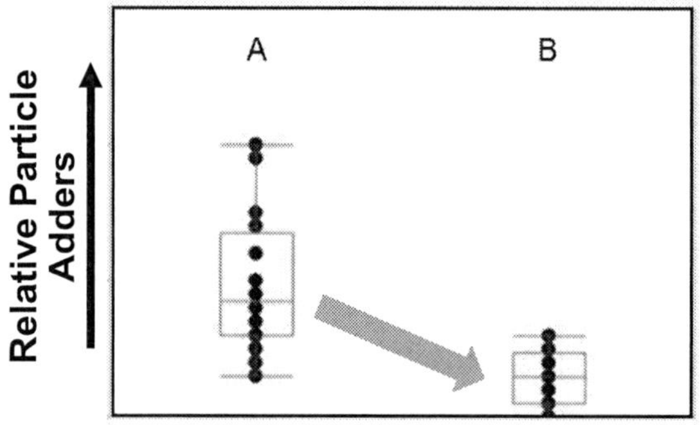


**65% Reduction in Particle Adders**

Comparison of 22nm+ particle adders performances between the POR (A) and GoTo (B) processes on blanket silicon wafers.

GLOBALFOUNDRIES 7

## Defect Improvement on Product Wafers

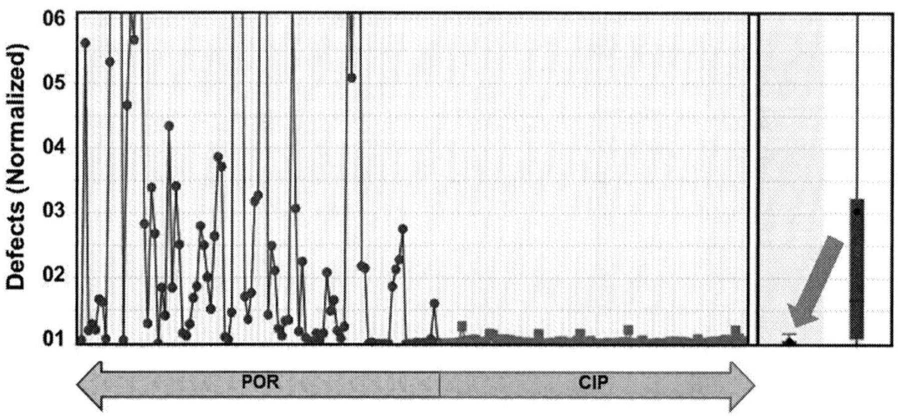


- Implementation of the CIP reduced the defects by >3x.
- Eliminated the frequent high fliers.

GLOBALFOUNDRIES 8

# Inline Trend on Product Wafers

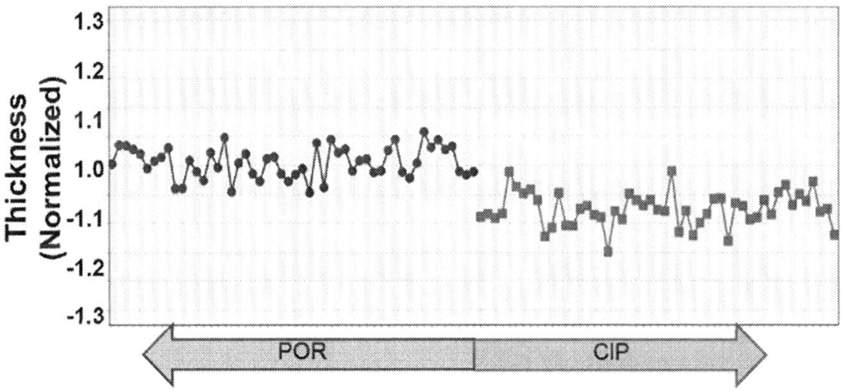


- Implementation of the CIP reduced the thickness of the exposed layer without impacting the device.



# Yield Analysis on Product Wafers

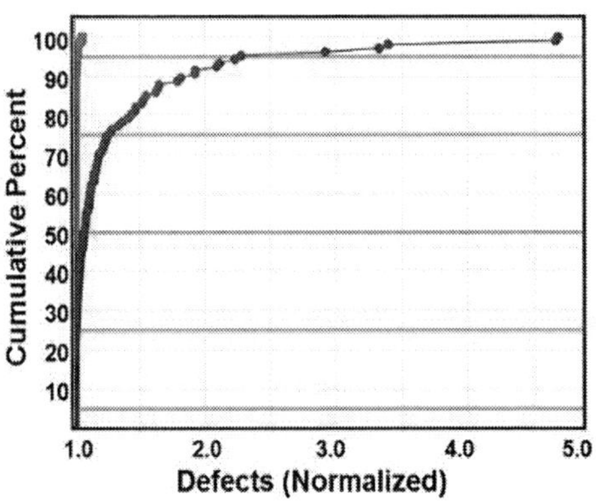


The CDF (the cumulative distribution function) plot of defect population at inspection step vs. the **POR** & **CIP.**

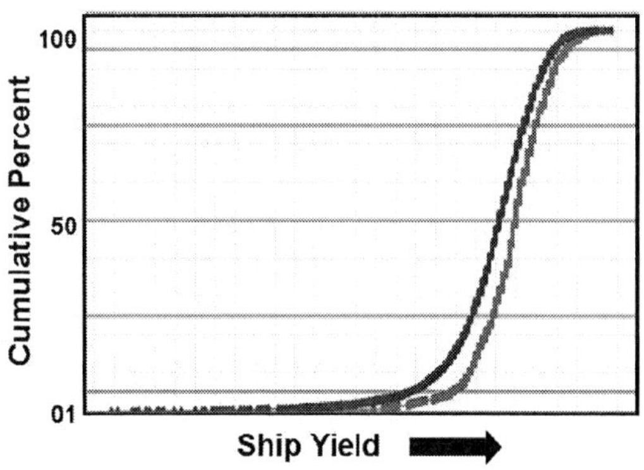


The CDF (the cumulative distribution function) plot of ship yield vs. the **POR** & **CIP.**

Elimination in frequent high fliers of surface particles, as observed at the defect inspection step, improved the ship yield of the low yielding wafers.

## Summary

- POR Post-Ash clean was directly correlated to the defects.
- Pre and post cleans wafer scans confirmed the source of the particles to be the associated POR cleans process.
- Frequent high fliers with these defects observed at inspection steps.
- Particles were made of Si & O and sitting on top of fins.
- Implementation of the CIP reduced the defects by >3x.
- Eliminated the frequent high fliers.
- Implementation of the CIP reduced the thickness of the exposed layer without impacting the device.
- Elimination in frequent high fliers of surface particles, as observed at the defect inspection step, improved the ship yield of the low yielding wafers.

GLOBALFOUNDRIES 11


## Acknowledgement:

*All colleagues in GLOBALFOUNDRIES*


GLOBALFOUNDRIES 12

# Wet Etchant for DRAM Word-line Titanium Nitride Recess with Selectivity to Tungsten

Wilson Yeh*, Wade Lee, Carol Cheng, Alan Chen, Lester Lee, Cami Shih, Paul Bernatis, Sang-Min Wong, and Chi Yen

*EKC Technology, DuPont Electronics & Imaging*

Contact info: Wilson.Yeh@dupont.com; Sang-Min.Wong@dupont.com

## Introduction

As dynamic random access memory (DRAM) scales to 20 nm and sub-20 nm, titanium nitride (TiN) recess with selectivity to tungsten (W) becomes one of the key process for fabricating the word-line structure in DRAM. Conventional approach of TiN recess was done by dry etching to etch back the TiN, but in advanced nodes it would damage the underneath gate oxide and induce the time-dependent dielectric breakdown leading to reliability issues. Therefore, many DRAM makers turn to achieve this process by using some commodity cleaning solutions, such as SC1 (mixture of $NH_4OH$ and $H_2O_2$), SC2 (mixture of HCl and $H_2O_2$) and SPM (mixture of $H_2SO_4$ and $H_2O_2$). However, the TiN to W etch rate selectivity of these commodity cleaning solutions are low, and most of these chemicals have high W corrosion rate. In this study, we present a novel selective etchant having high TiN etching with selectivity to W, and compatible with thermal oxide (ThOx), SiN and high-K material.

## Technical Background

TiN recess with selectivity to W and ThOx becomes one of the key process for fabricating the word-line structure in DRAM.

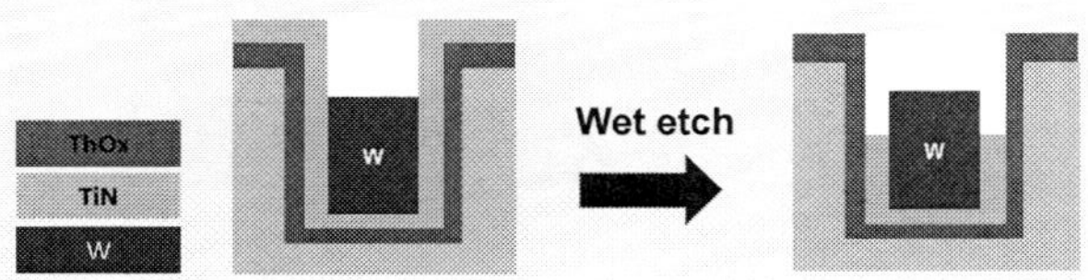


## Design Concept

- Oxidize TiN to soluble $TiO^{2+}$.
- Introduce effective W inhibitor with good W protection.
- Employ TiN etching enhancer to accelerate TiN recess.

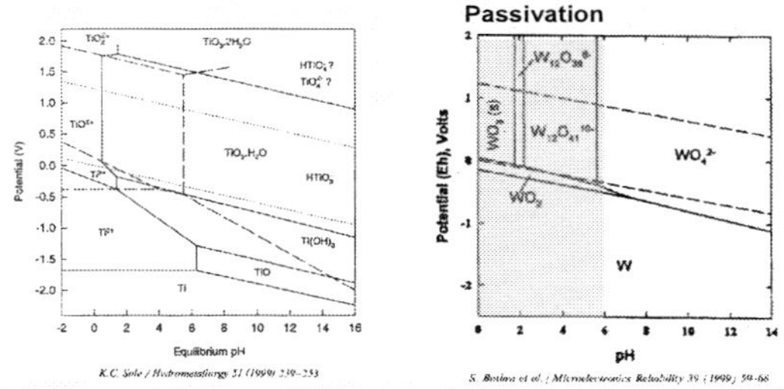


## Results and Discussion

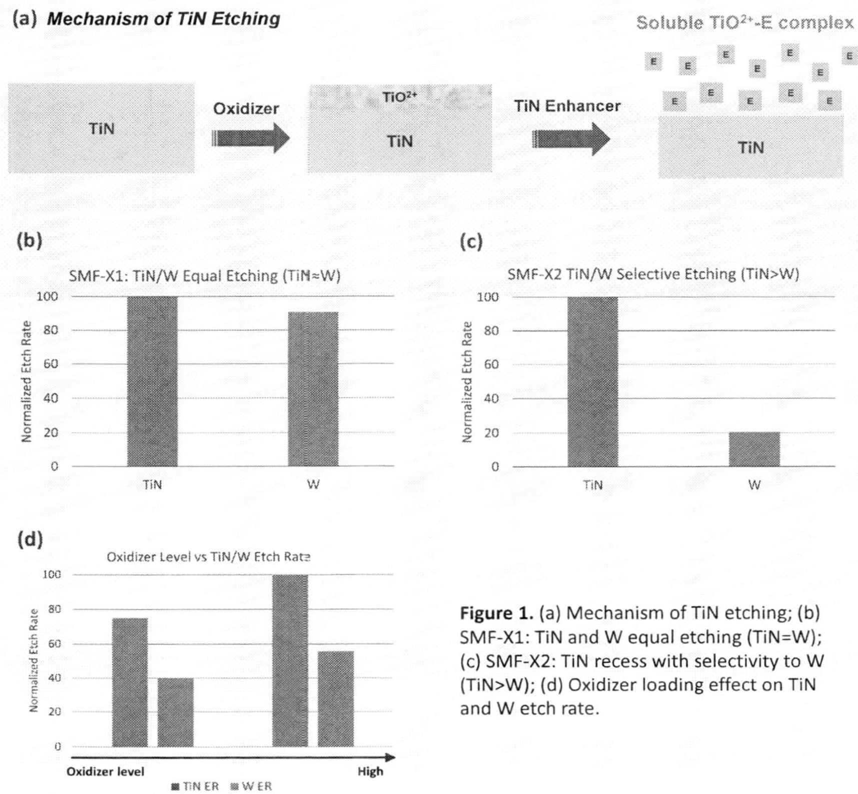


**Figure 1.** (a) Mechanism of TiN etching; (b) SMF-X1: TiN and W equal etching (TiN=W); (c) SMF-X2: TiN recess with selectivity to W (TiN>W); (d) Oxidizer loading effect on TiN and W etch rate.

## Conclusions

- We have developed SMF-series formulations for DRAM word-line TiN recess with selectivity to W and ThOx; the effective W protection allowing higher TiN/W selectivity.
- SMF-series formulations are tunable on TiN and W etch rate according to customer's requirements, and have good chemical stability.

# Humidity Control in a FOUP by Laminar Air Curtain (LAC) When FOUP Door is in Open Condition

***Shih-Cheng Hu** (schu.ntut@gmail.com)*
***Ti Lin** (whitelin1112@gmail.com)*
***Oscar Juina** (oscar1605ju@gmail.com)*
**David Benalcazar** (dfbenalcazarc@gmail.com)

National Taipei University of Technology
Department of Energy, Refrigeration and Air Conditioning.
Center for Cleaning Technology Research

The Surface Preparation and Cleaning Conference (SPCC)
April 2, 2019. Portland OR USA



## 1. Introduction

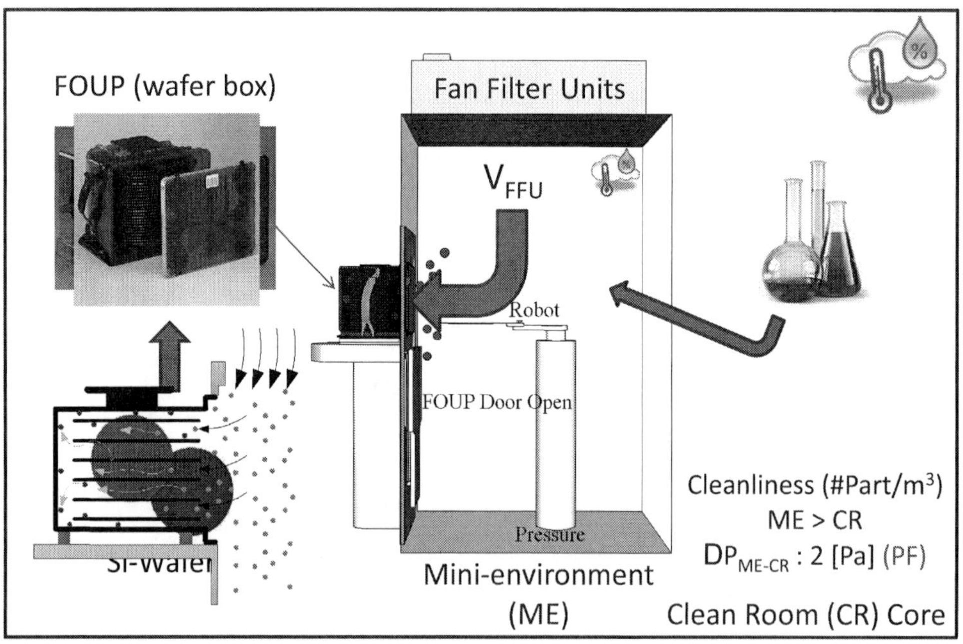

TAIPEI TECH

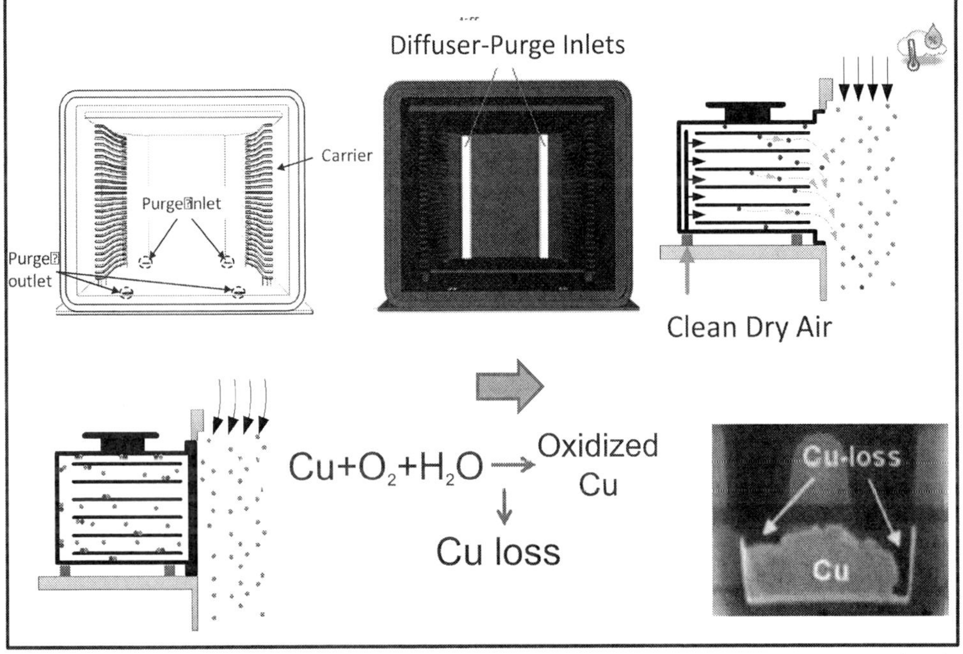


Mini-environment (ME)

CCTR

3

## 2. The Problems

TAIPEI TECH

- Tool arrangement and ME configuration cases

**Case 1**

Dead Zone (Blank Panel)

Lack of FFU coverage. A solid plate covers the FFU surface.

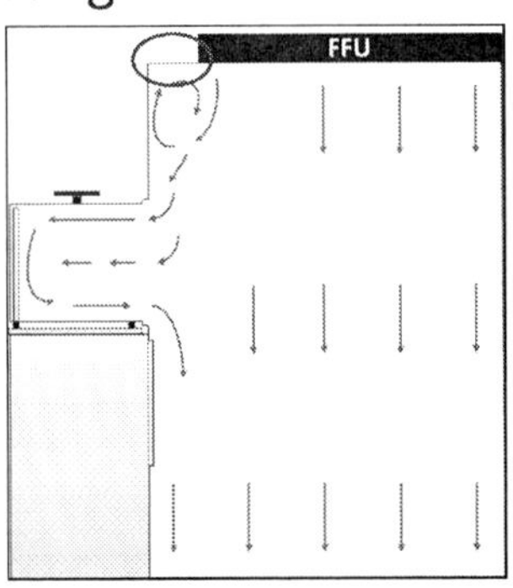


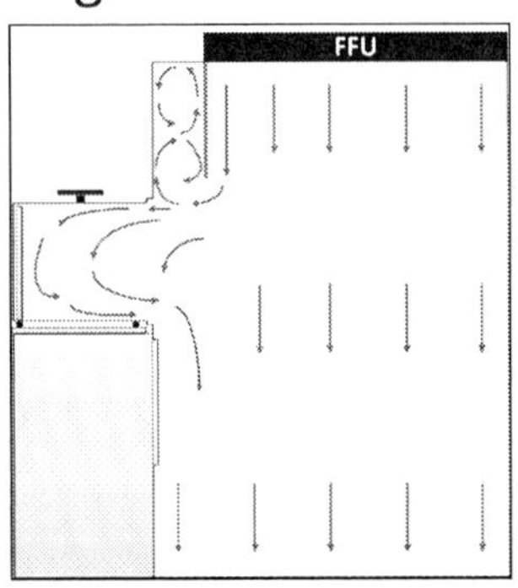


**Case 2**

Dead Zone & Eyelid

Lack of FFU coverage and vertical plate between FFU and Dead zone

**Case 3**

Objects in front of the FOUP

Perforated plate that reduces the cross section bellow the FOUP's interface.

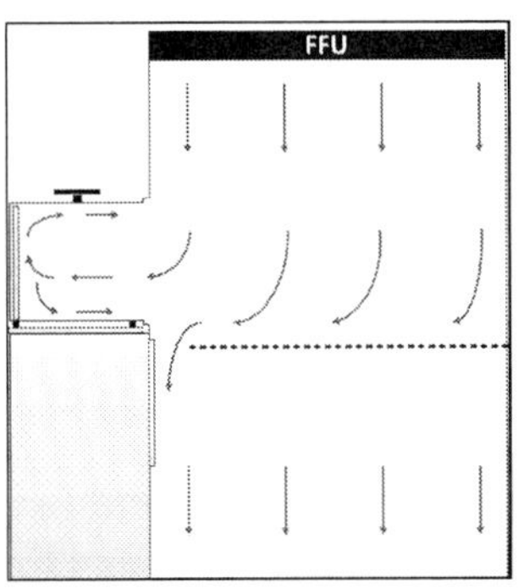


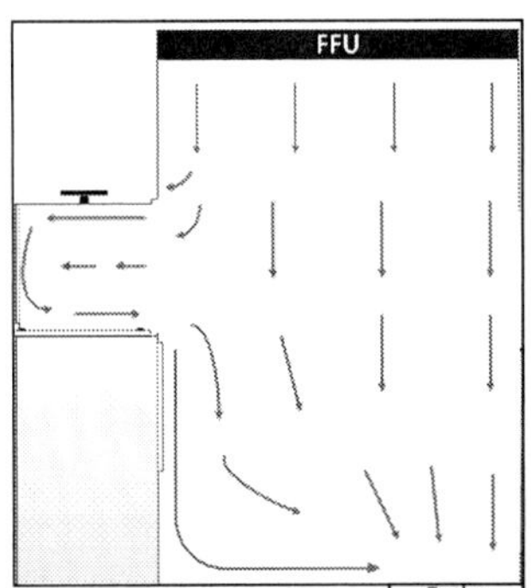


**Case 4**

Tool Exhaust suction (Reduced Outlet)

Outlet cross section is reduced due to tool's exhaust that forces air to escape from one specific zone of the floor.

CCTR

4

## Door open without purge

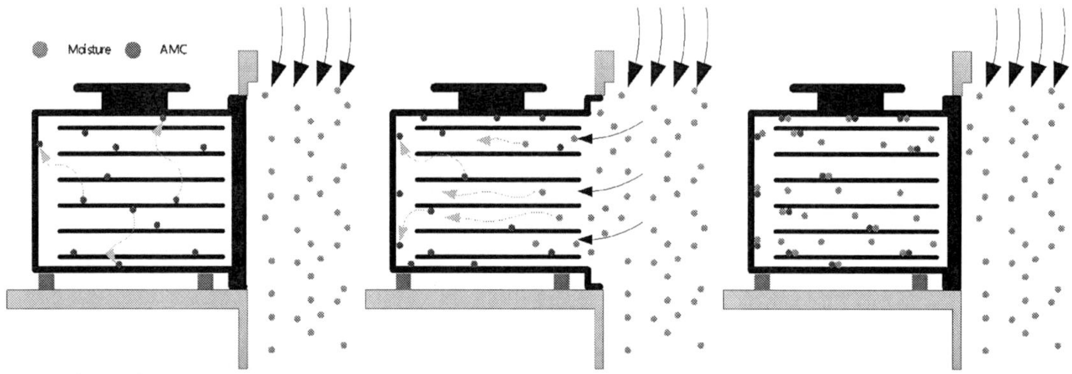


Door Closed. AMC on wafer evaporated and deposited on FOUP surfaces.

Door Open. Chemical reaction takes place by AMC and moisture.

Door Closed. Chemical reaction takes place by AMC and moisture

1 2 3

5 2019/3/25

## Perfect Scenario: No down-flow air invades to the FOUP when door is open (by diffuser only)

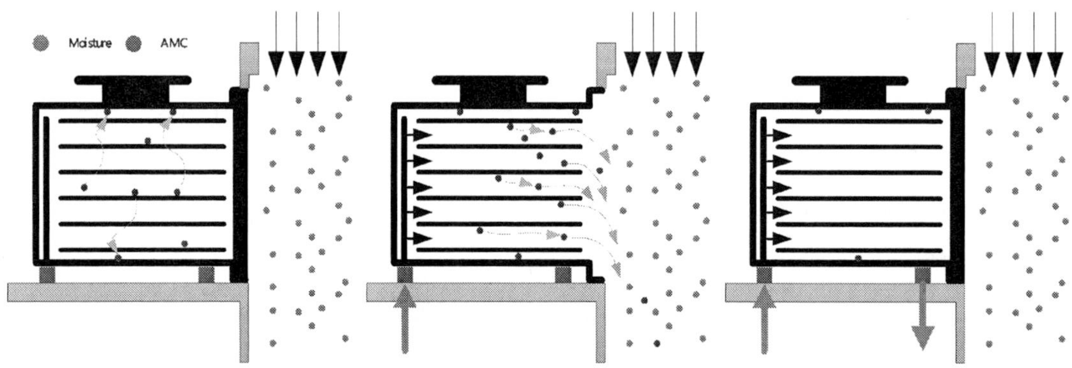


Door Closed. AMCs on wafer evaporated and deposited on FOUP surfaces.

Door Open condition. AMCs are expelled by purged CDA via diffuser.

(Flow rate is limited by possible particle suspension on wafer surfaces and vibration of diffuser holder)

FOUP is purged with CDA

6 2019/3/25

## Reality: down-flow air invades to the FOUP when door is open (by diffuser only)

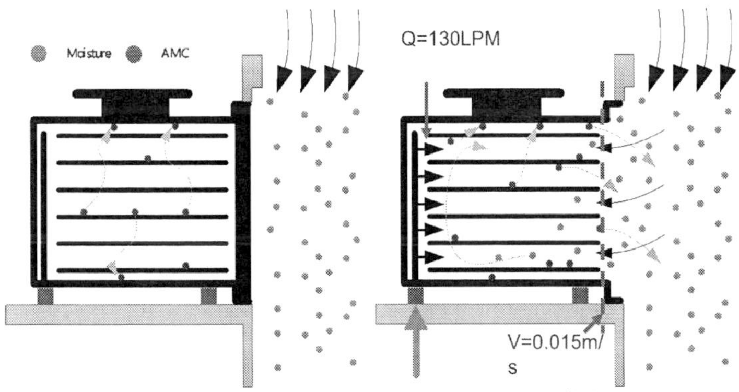


Door Closed. AMCs on wafer evaporated and deposited on FOUP surfaces.

**Door Open condition. AMCs can not be expelled by purge air and moisture invades, as the sectional purge air velocity is less than 0.015m/s. The purge flow rate of diffuser is limited to 130 L/min.**

Door closed. AMCs still reside in the FOUP.

7 2019/3/25

## Combine diffuser & air curtain

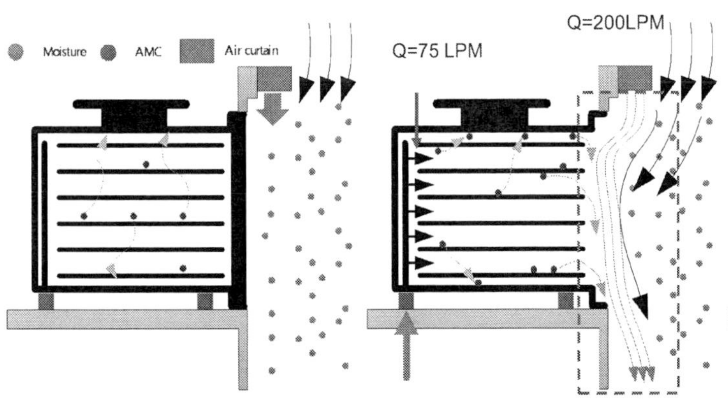


Door Closed. AMCs on wafer evaporated and deposited on FOUP surfaces.

Air curtain with purge air acts 3 seconds before door open.

**Door Open. Combined purge air by diffuser and purge air & air curtain can effectively keep the invasion of down-flow moisture and expel the residual AMCs.**

**Air curtain with purge air de-acts 3 seconds after door close.**

8 2019/3/25

# 3.Methods:

Equipment: FOUP & Purging Devices

FOUP (wafer box) & Porous Diffuser

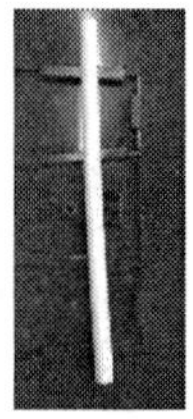

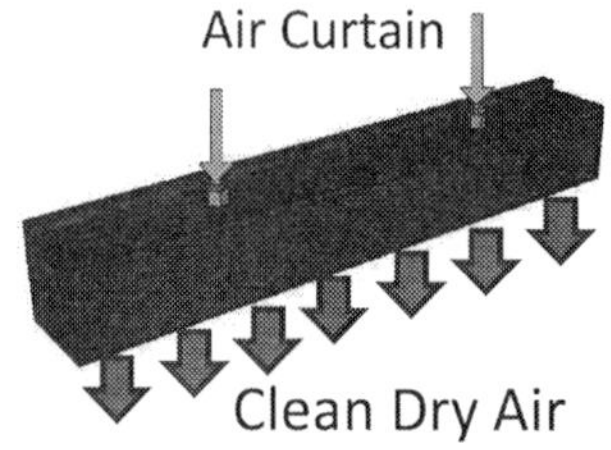


RH sensors location (Top View)

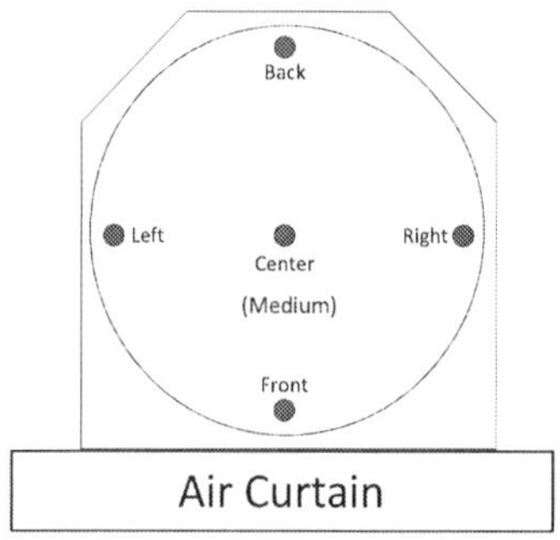


Al-Wafer & RH sensors (Top View)

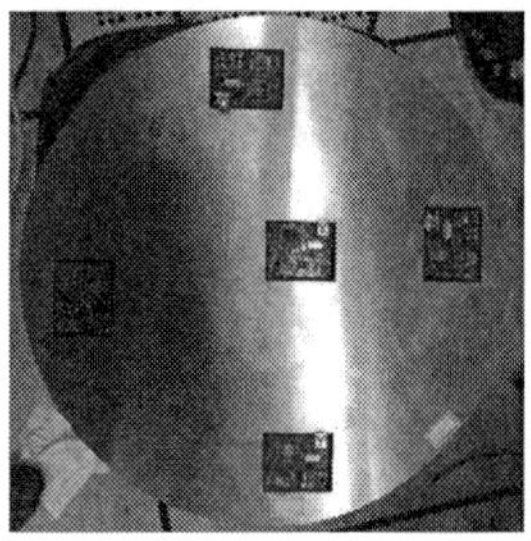

Slots Location:
25
19
13
7
1



## Purge by diffusers only with $Q_{DP}$=130 L/min.
## Slots 1 ~13 are influenced

| Left diffuser | center | Right diffuser |
|---|---|---|
| empty | | |
| 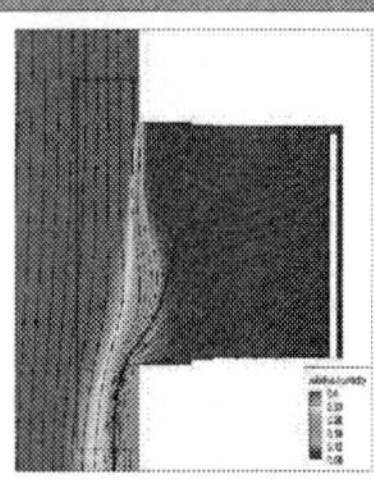 | 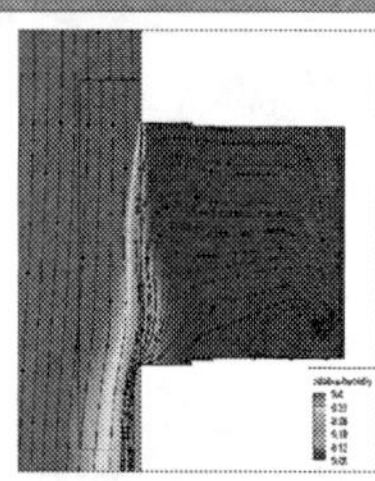 | 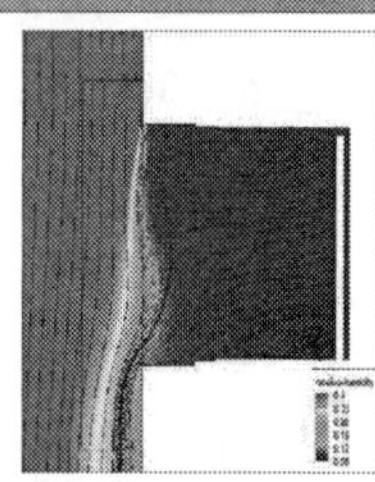 |
| Full slot | | |
| 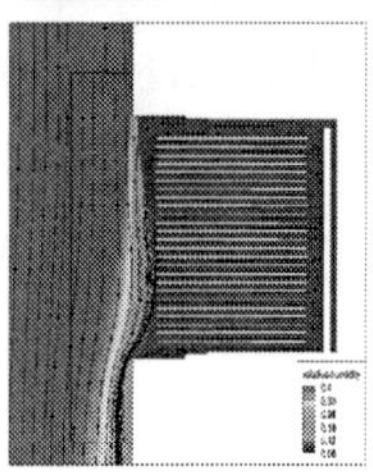 | 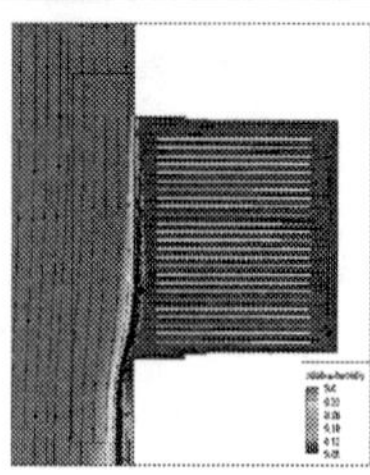 | 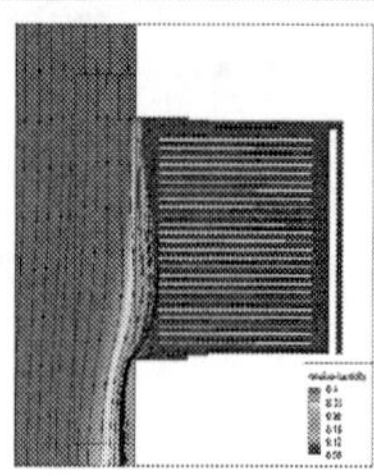 |

## Purge by diffusers only with $Q_{DP}$=130 L/min. Slots 1 ~13 are influenced

| slot01 | slot13 | slot25 |
|---|---|---|

empty

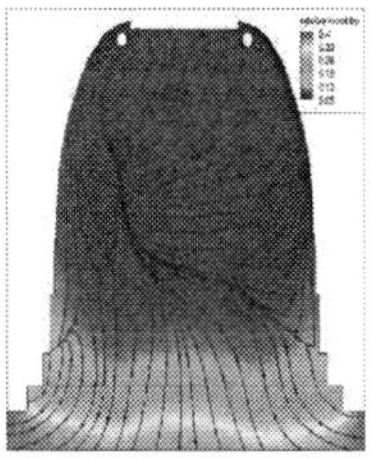
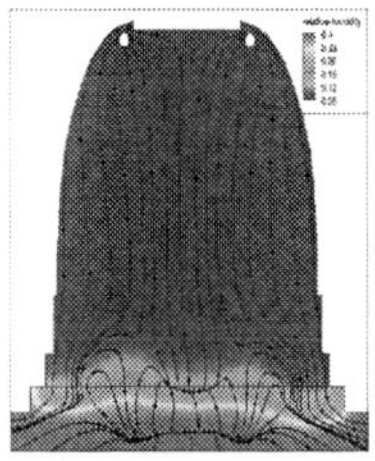
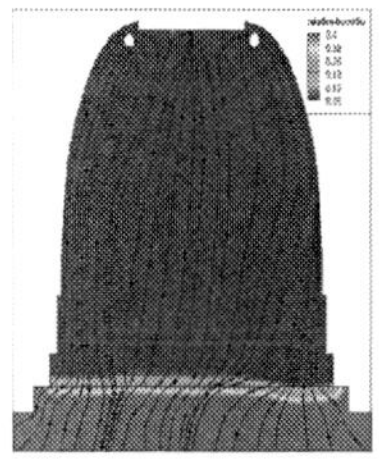

Full loaded

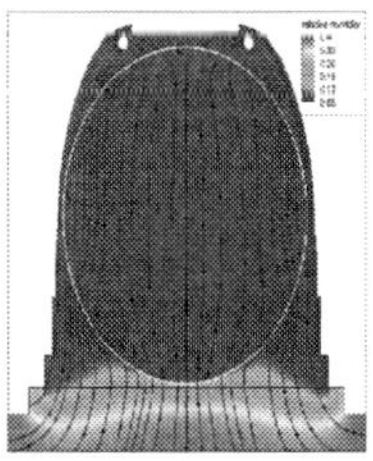
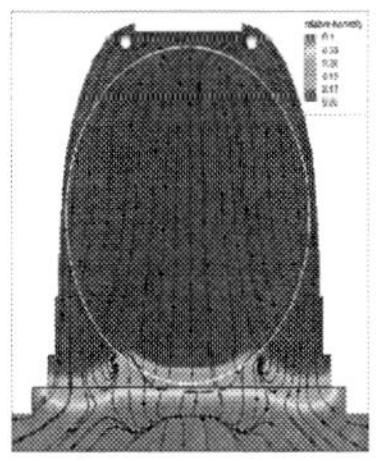
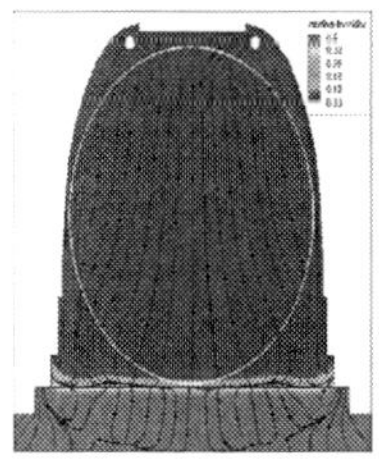

CCTR

11

## Purge by air-curtain. No slot is influenced

| Left diffuser | center | Right diffuser |
|---|---|---|

empty

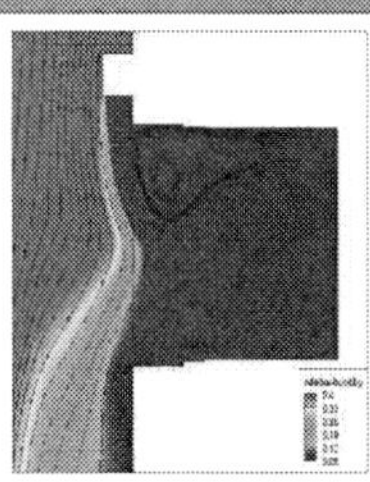
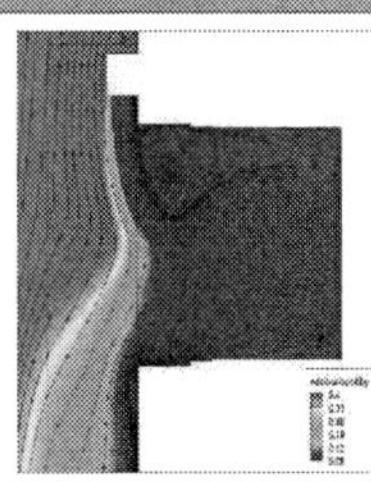
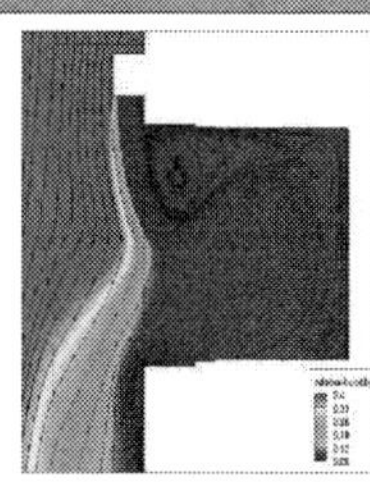

Full slot

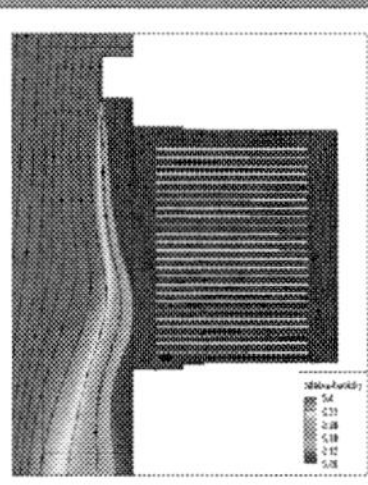
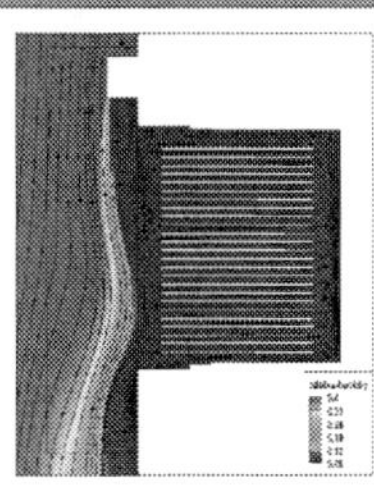
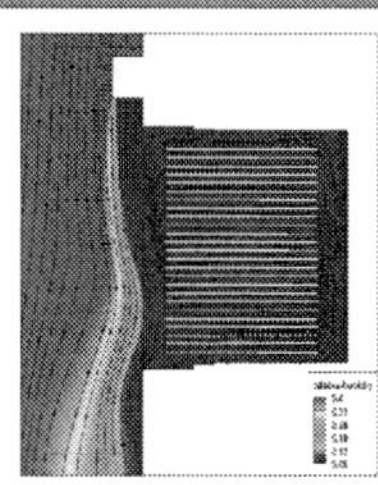

CCTR

12

## Purge by air-curtain only. No slot is influenced

| slot01 | slot13 | slot25 |
|---|---|---|

empty

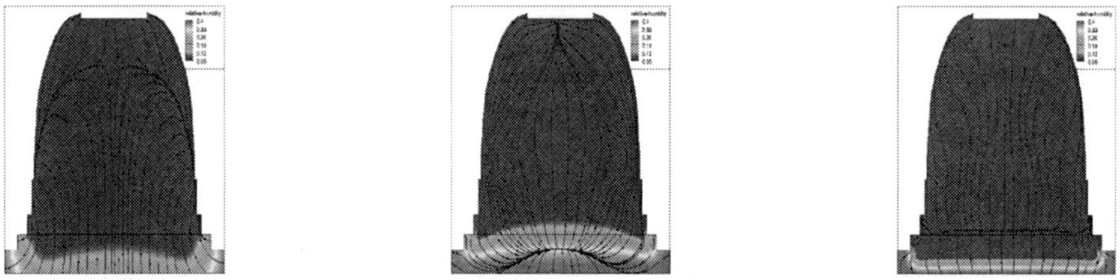

Full slot

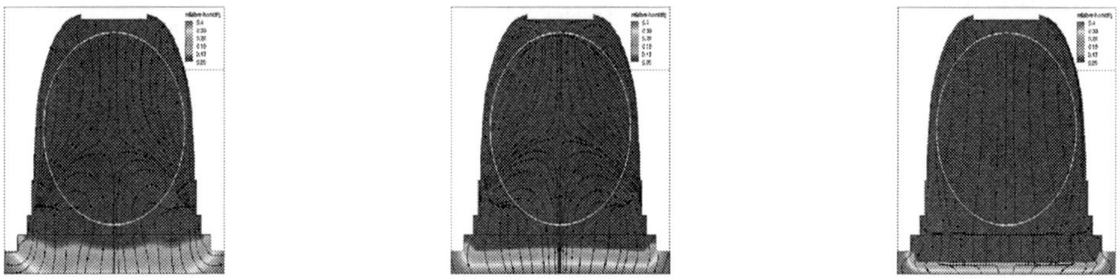

CCTR 13

RH trend chart at center of wafers of slot 1, 13, and 25
(Flow rate of XCDA : $Q_{Air\ Curtain}$:300L/min, $Q_{diffuesrs}$:0 L/min)
Total Wafer Q'ty : 24 pcs of wafers inside the FOUP

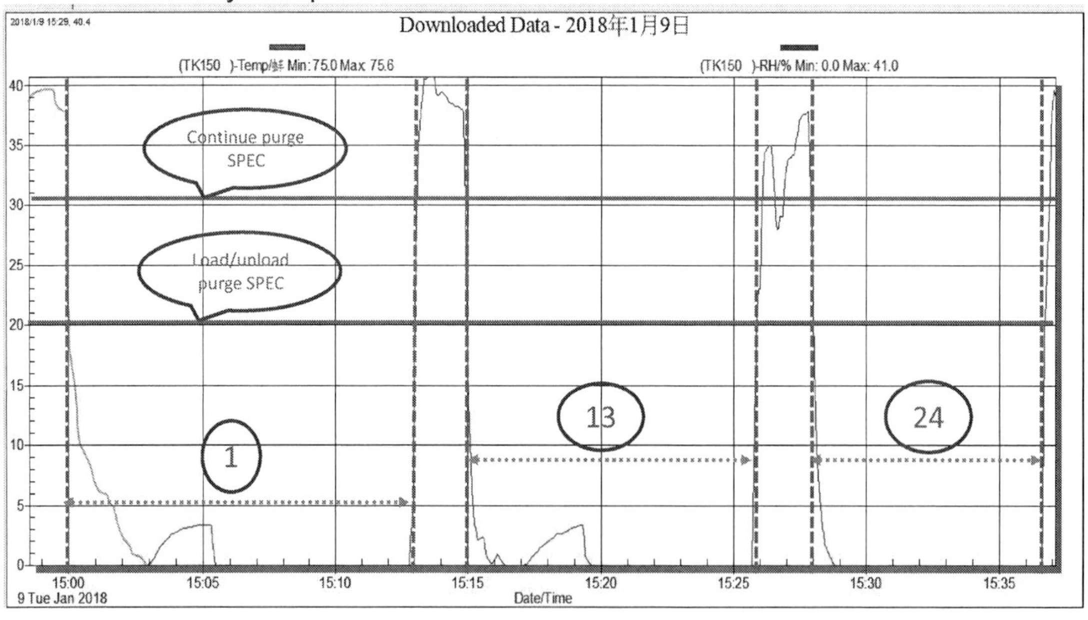


CCTR

# Conclusions

- Air-Curtain can effectively keep the invasion of down-flow moisture based on operation conditions of down-flow.
- Lower ownership: mount on each LPU vs. install two diffusers in each FOUP.
- Purge by diffusers only can not provide very low RH%, it reaches to about 30% even in good condition of EFEM. If EFEM is in a bad condition (i.e. the 5 scenarios), it won't reach to 20%.
- Air Curtain speeds up the decreasing of RH% and keeps the EFEM humid air from invading into the FOUP.
- Combine air curtain with conventional FOUP purge approach, the purge flow rate can be optimized to reduced the risk of particle re-suspension in the FOUP, while keep the FOUP in low RH levels.

CCTR 2019/3/25 15

## Thank you for your participation!

Dr. Shih-Cheng Hu
tel: 886-27712171 ext 3588 or 3512
emails: schu.ntut@gmail.com and f10870@ntut.edu.tw
*website: www.ntut.edu.tw/~f10870

## Thanks to CCTR projects sponsors :

AIRREX CO., LTD.
YESIANG ENTERPRISE CO., LTD.
AERO PRO CO., LTD.
AAF Taiwan Co., Ltd.
Gudeng Co.
Yi-Shen Science and Tech.
Wholetch Co.
Weltall Science and Tech
MayAir Technology Co Ltd
AEROPROFILTER IND (SHANGHAI) CO., LTD.
Hollingsworth & Vose (Suzhou) CO., LTD.
TSMC
Brillian Network & Automation Integrated System
Springfield Technologies.

## Thanks to ALL CCTR lab members

CCTR 16

The Surface Preparation and Cleaning Conference
April, 2-3, 2019
Portland, OR, United States of America

# Investigation of Ceria Abrasive Removal during Post Chemical Mechanical Polishing Cleaning

Juhwan Kim[1], Seokjun Hong[1] Vinit K. Kanade[2], Taesung Kim[1,2,*]
[1]School of Mechanical Engineering, Sungkyunkwan University, Suwon, South Korea
[2]SKKU Advanced Institute of Nanotechnology, Sungkyunkwan University, Suwon, South Korea
+82-31-299-4751 / 85277, Seoburo 2066, Jangangu, Suwonsi, Korea / kjh0045@skku.edu

## Introduction

### BACKGROUND AND OBJECTIVES

**BACKGROUND**

| Contaminants | Nonmetal CMP | Metal CMP | Effects |
|---|---|---|---|
| Particulate | Silica or ceria, fine fragments of film or pad | Alumina, or silica, metal hydroxide precipitates. Fine fragments of film or pad | Shorts by conductive particles Cause local roughness |
| Metallic | $K^+$,$Ca^{2+}$ | $Cu^{2+}$,$Al^{3+}$,$Fe^{3+}$ etc | Many metals can form silicide and/or affect the oxidation |
| Organic | Tetramethyl ammonium salts, buffers, surfactants | Buffers, surfactants, inhibitors | Affect wettability and cleanability |
| Other defects | Scratches, stress | Scratches, dishing and erosion, stress | Non planar surface Weak sites |

Table 1. CMP contamination

- Device shrinkage → Requirements smooth surface
- Global surface planarization → CMP
- After CMP process → CMP contamination
- Post CMP cleaning is needed

➡ **High cleaning performance is needed**

## Experiment

### EXPERIMENTAL METHOD

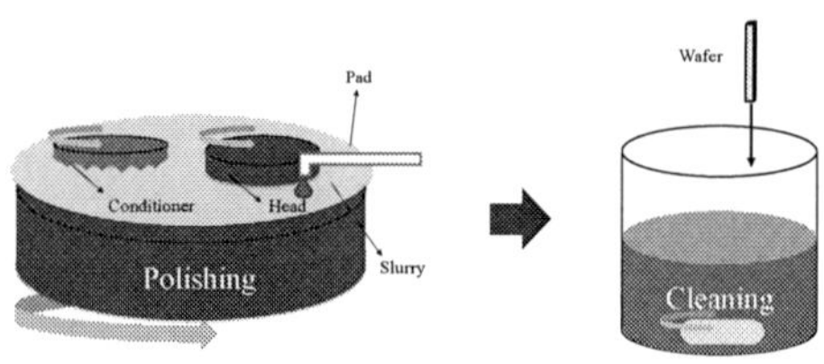


Fig 1. Experimental Setup : Polishing and Cleaning

| Equipment | 4 in wafer polisher |
|---|---|
| Wafer | PE-TEOS, Nitride coupon |
| Pad | IC1010 |
| Slurry flow (ml/min) | 80 |
| Rotation speed - pad/head (RPM) | 93/87 |
| Pressure (psi) | 3 |
| Polishing/Conditioning time (s) | 60/60 |

Table 2. Polishing Condition

| Step | 1 | 2 | 3 | 4 | 5 |
|---|---|---|---|---|---|
| Solution | DIW | Cleaner | DIW | Cleaner | DIW |
| Stirrer rotation speed (RPM) | 0 | 500 | 500 | 500 | 500 |
| Time (s) | 30 | 60 | 20 | 60 | 60 |

Table 3. Post CMP Cleaning Condition

## Results

### EXPERIMENTAL RESULTS

**Zeta potential**

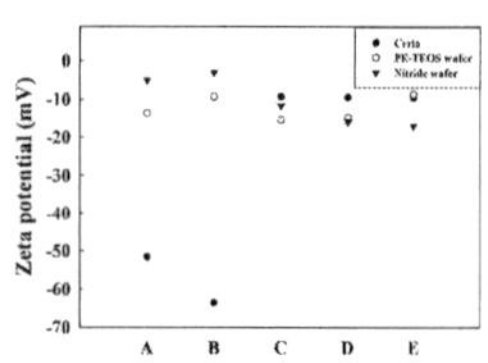


Fig 2. Zeta potential : Ceria particle, PE-TEOS and nitride wafer

※ Repulsion : $\zeta_1 \times \zeta_2$ (zeta potential : $\zeta$)

- Cleaner A, B : EDTA, TMAH base
- Cleaner C, D, E : Acetic acid, TMAH base
- EDTA : Chelating agent (prevents re-adsorption)
- TMAH : Make hydrophilic surface (prevents re-adsorption)
- Fig 2 shows that the zeta potential differs depending on the cleaning solution
- Cleaner A and B show better cleaning performance than others
- As increase of repulsion, amounts of Ce ions decrease

**ICP-MS (PE-TEOS wafer)**

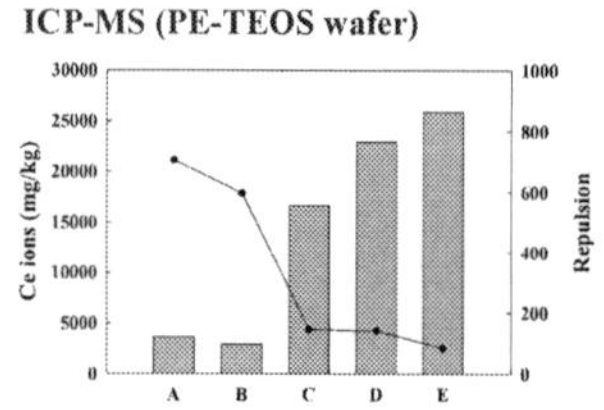


Fig 3. Amount of Ce ions after cleaning and repulsion

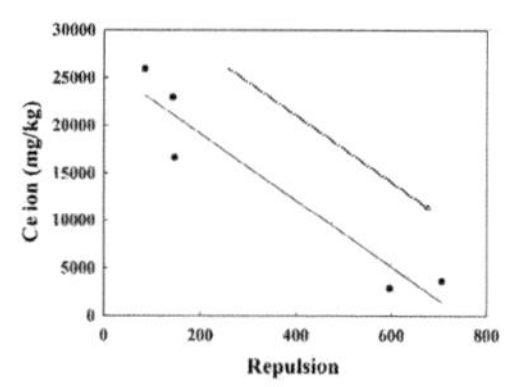


Fig 4. Relation between repulsion and Ce ions

**ICP-MS (Nitride wafer)**

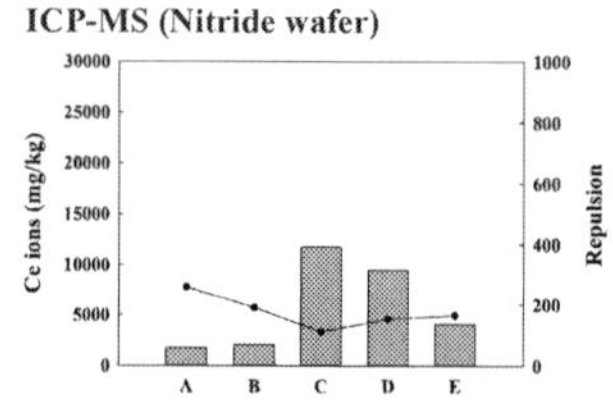


Fig 5. Amount of Ce ions after cleaning and repulsion

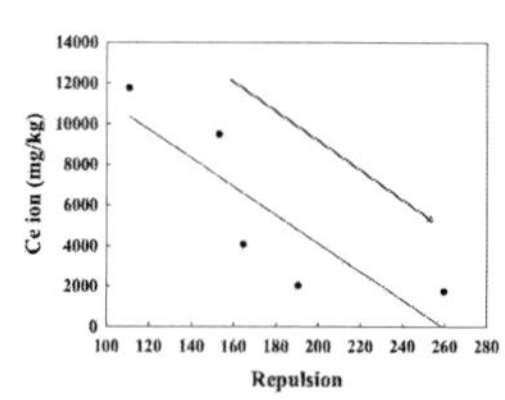


Fig 6. Relation between repulsion and Ce ions

## Conclusion

### SUMMARY

**Conclusion**

- 5 cleaners are developed for post CMP cleaning
- Zeta potential plays an important role in cleaning
- Zeta potential is related to repulsive force
- The larger repulsion, the better cleaning performance

**References**

1. P. B. Zantye, A. Kumar, A.K. Sikder, "Chemical mechanical planarization for micro electronics applications," Mater Sci Eng. R 2004; 45: 89-220.
2. N. Dedy, K. Subrata, K. Milind, L. Hong, "Role of surfactant molecules in post CMP cleaning," Electrochemical Society, 155 (2) H64-H68, 2008.
3. A.T. Jeremiah, C. Wonseop K. Rajiv "Effect of pH on ceria-silica interactions during chemical mechanical polishing," Materials Research Society vol. 20, No. 5 p. 1139-1145, 2005.
4. P. M. Tung, L. F Tan, C. C. Chao, C. S. Tien, L. C. Ming, Y. L. Wen, T. S. Ming, C. P. Lu, W. H. Chang, "Novel cleaning solutions for polysilicon film post chemical mechanical polishing," IEEE Electron Device Letters, vol. 21, No. 7 p. 338-340, 2000.

# ALUMINUM NITRIDE REMOVAL FORMULATION WITH MODIFIED ANISOTROPIC SILICON ETCHING PROPERTY

Chung Yi Chang[a], Wen Dar Liu[a] Yi Chia Lee[a] ,Tianniu Rick Chen[b], Aiping Wu[b]

a. Versum Materials Taiwan LLC., 2F, No.25, Lane 62, Chung Ho Street, Chupei City, Hsinchu County 30267, Taiwan
b. Versum Materials, 8555 South River Parkway, Tempe, AZ 85284, USA

E-mail address: joseph.chang@versummaterials.com

Phone: (886) 921439016

VERSUM MATERIALS

1

## Outline

1. **Background introduction**
   - Technical challenge of conventional AlN etchant
   - Anisotropic Si wet etching
2. **Design concept of novel AlN etchant**
   - Etch rate requirement
   - Formulation design
   - Experimental trails: solvent effect and additive effect
3. **Summary**

VERSUM MATERIALS

2

# Technical challenge of Alkaline solution of AlN wet etching

➢ Issue of conventional AlN wet etchant on silicon substrate.

High alkalinity solution was used in AlN etching widely but with an undesirable anisotropic etching properties and high etching rate of silicon, leading to roughness at back side of substrate by forming pyramidal defect.

**AlN removal application in BEoL** **AlN removal application in MEMs**

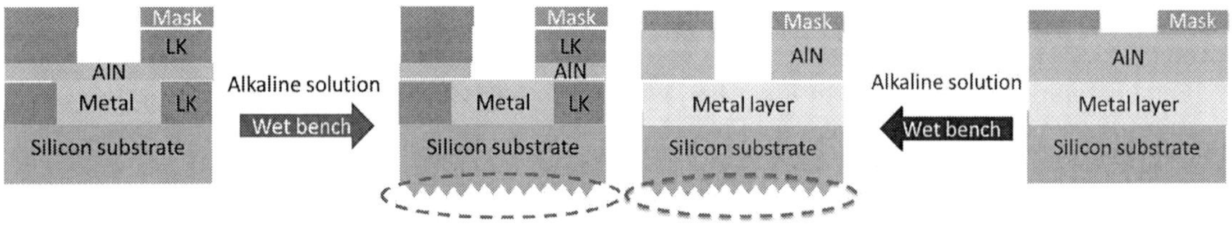


**roughness**

1. Silicon etching rate suppression
2. Adjusting the Si(110) to Si(100) etching rate ratio

VERSUM MATERIALS

3

# Aluminum nitride and Silicon etching by alkaline solution

➢ Etch rate of conventional AlN wet etchant

| | Silicon E/R(A/min) | | | | Etching selectivity | | |
|---|---|---|---|---|---|---|---|
| | AlN | Si(100) | Si(110) | Si(111) | AlN/Si(100) | Si(110)/Si(100) | Si(111)/Si(100) |
| Conventional etchant | 18241 | 12074 | 6367 | 1989 | 1.51 | 0.53 | 0.16 |

For conventional alkaline wet etchant, the etch rate in Si(100) direction is much higher than in (111) which results in pyramid shape etching profile.

Bare Si (100) → Conventional Si etchant

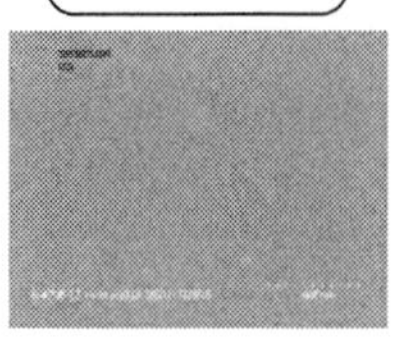

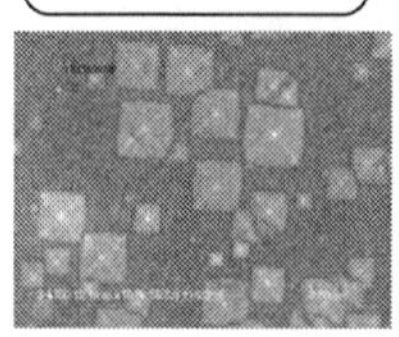

(Top view SEM)

pyramid shape etching profile

VERSUM MATERIALS

4

# Etch rate requirements

➢ Etch rate requirements of new AlN wet etchant

| | Silicon E/R(A/min) | | | | Etching selectivity | | |
|---|---|---|---|---|---|---|---|
| | AlN | Si(100) | Si(110) | Si(111) | AlN/Si(100) | Si(110)/Si(100) | Si(111)/Si(100) |
| Requirements | >15000 | <3000 | <3000 | <1000 | >5 | >1 | >0.3 |

To address the etch residue issue, both silicon etching rate and selectivity of Si(110)/Si(100) and Si(111)/Si(100) should be increased.

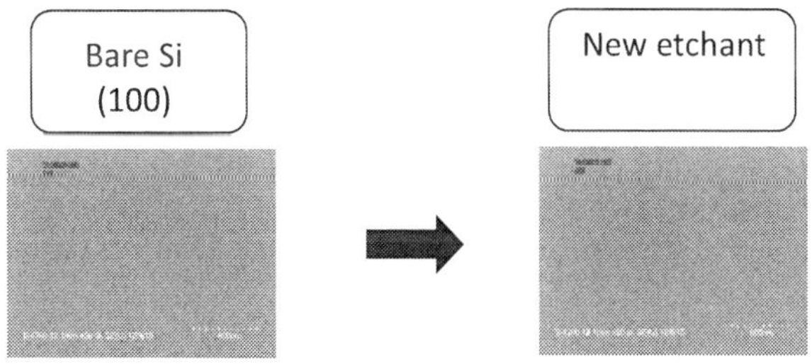


(Top view SEM)
No pyramid shape etching profile

VERSUM MATERIALS

5

# Formulation design

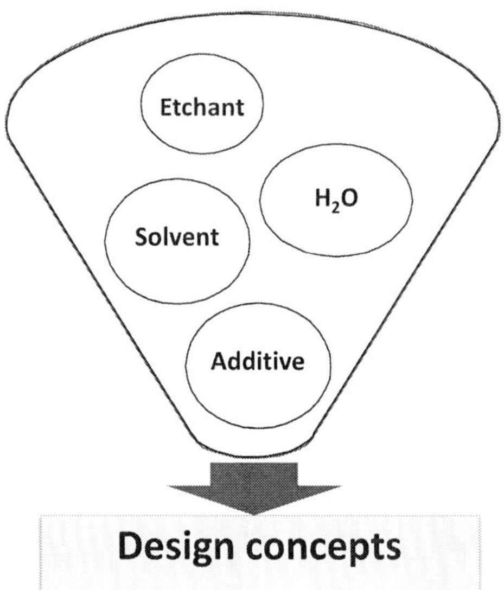


Etchant:
To dissolve AlN

Additive :
To suppress silicon E/R

H2O:
Hydrolysis medium

Solvent:
To adjust solution polarity

VERSUM MATERIALS

6

# Solvent effect in silicon etching rate suppression

- An effective solvent decrease all different orientation silicon e/r but cannot increase the Si(110)/Si(100) and Si(111)/Si(100) ratio

| Solvent effect | E/R(A/min) | | | Etching selectivity | |
|---|---|---|---|---|---|
| | Si(100) | Si(110) | Si(111) | Si(110)/Si(100) | Si(111)/Si(100) |
| Requirements | <3000 | <3000 | <1000 | >1 | >0.3 |
| Conventional etchant | 12074 | 6367 | 1989 | 0.53 | 0.16 |
| Solvent 1 | 7867 | 5107 | 1597 | 0.65 | 0.20 |
| Solvent 2 | 7529 | 3615 | 1349 | 0.48 | 0.18 |
| Solvent 3 | 4331 | 2920 | 941 | 0.67 | 0.22 |
| Solvent 4 | 2298 | 1245 | 661 | 0.54 | 0.27 |

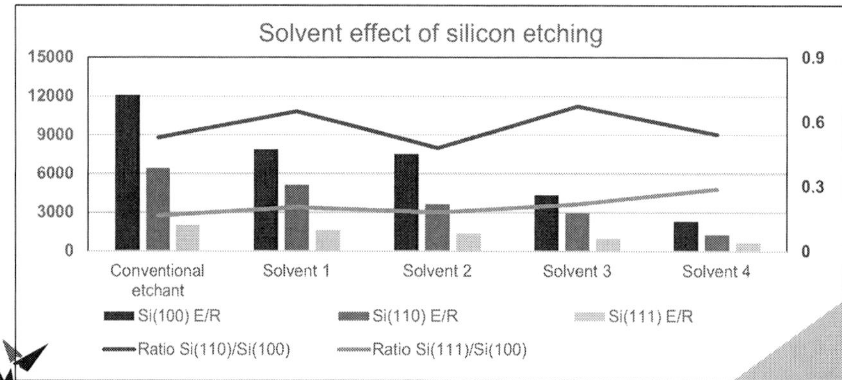


VERSUM MATERIALS

# SEM result of solvent effect in silicon etching

- Decreasing silicon e/r by adding solvent can reduce amounts and scale of pyramidal defect but cannot eliminate it.

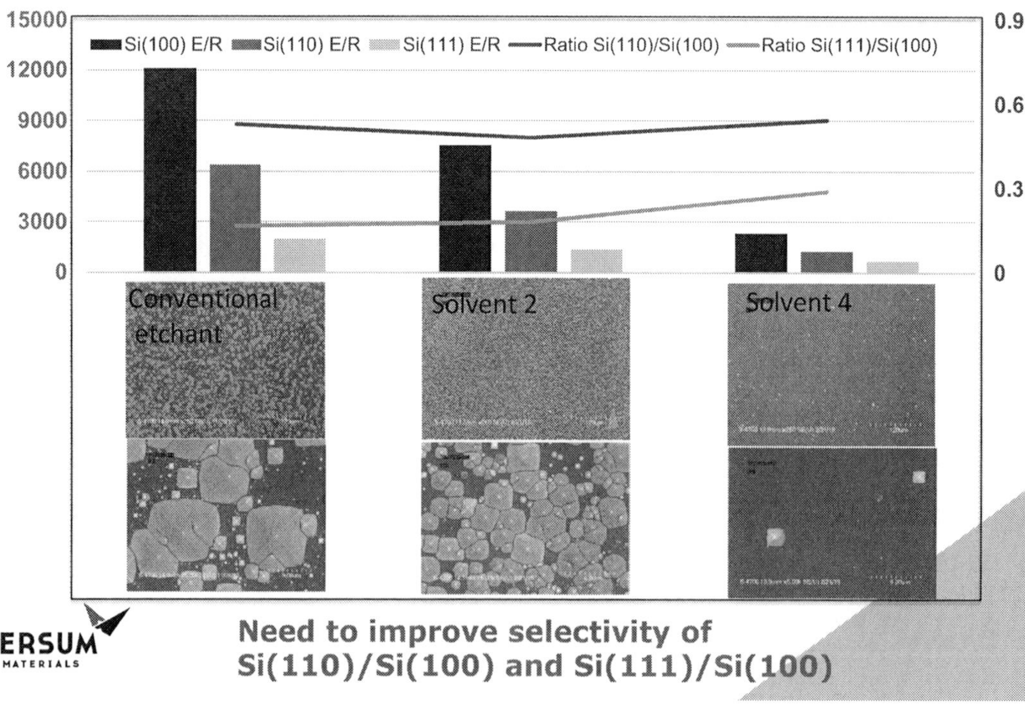


VERSUM MATERIALS

**Need to improve selectivity of Si(110)/Si(100) and Si(111)/Si(100)**

## Solvent effect in AlN etching and Si etching

➢ AlN etching rate would be undesirably decreased by changing solvent.

| Solvent effect | E/R(A/min) | | | |
|---|---|---|---|---|
| | AlN | Si(100) | Si(110) | Si(111) |
| Requirements | >15000 | <3000 | <3000 | <1000 |
| Conventional etchant | **18241** | 12074 | 6367 | 1989 |
| Solvent 1 | **18441** | 7867 | 5107 | 1597 |
| Solvent 2 | **15511** | 7529 | 3615 | 1349 |
| Solvent 3 | 13598 | 4331 | **2920** | **941** |
| Solvent 4 | 11430 | **2298** | **1245** | **661** |

AlN E/R requirement

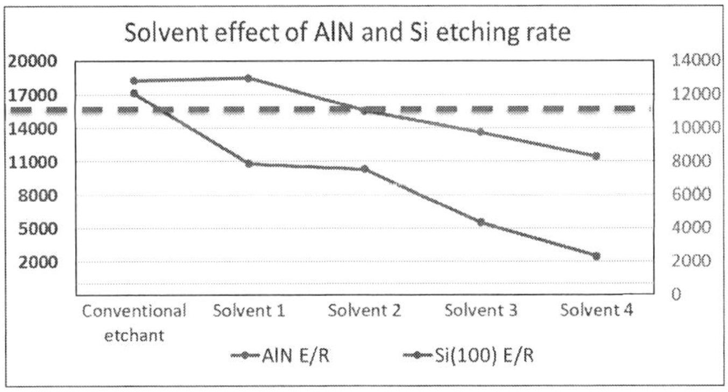


VERSUM MATERIALS

## Additive effect: Silicon E/R study

➢ Si E/R can be suppressed by additive, beside the selectivity of Si(110)/ Si(100) and Si(111)/ Si(100) can be also increased.

| Addtive effect | E/R(A/min) | | | Etching selectivity | |
|---|---|---|---|---|---|
| | Si(100) | Si(110) | Si(111) | Si(110)/Si(100) | Si(111)/Si(100) |
| Requirements | <3000 | <3000 | <1000 | >1 | >0.3 |
| Conventional etchant | 12074 | 6367 | 1989 | 0.53 | 0.16 |
| Conventional etchant +addtive | **2398** | **2772** | **768** | **1.16** | **0.32** |

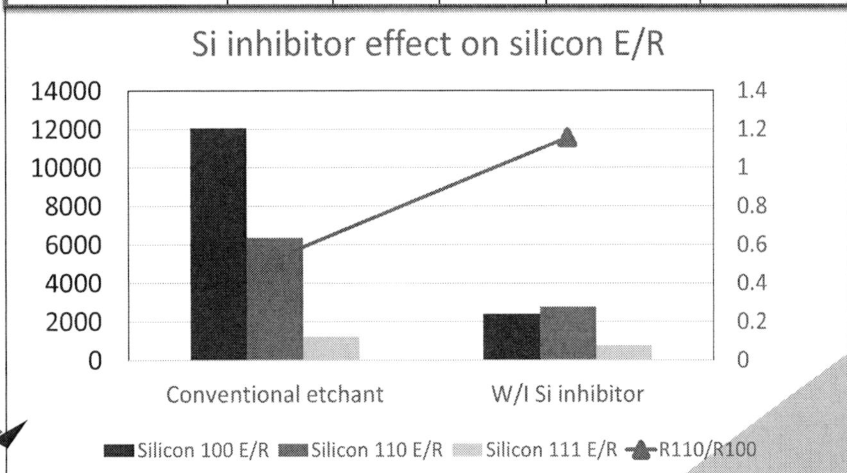


VERSUM MATERIALS

12

## Additive effect : SEM result in silicon surface

- SEM result of additive effect in silicon etching

Conventional etchant

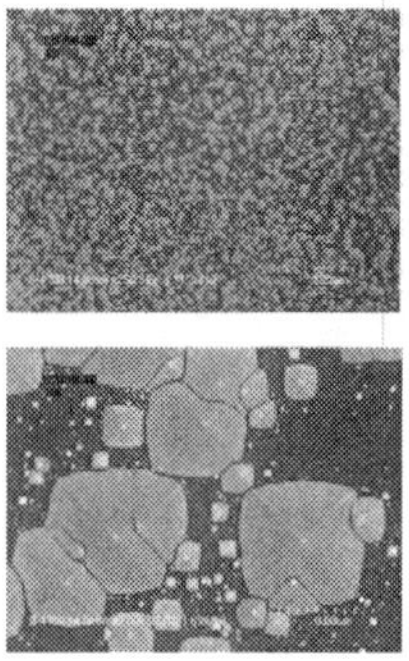

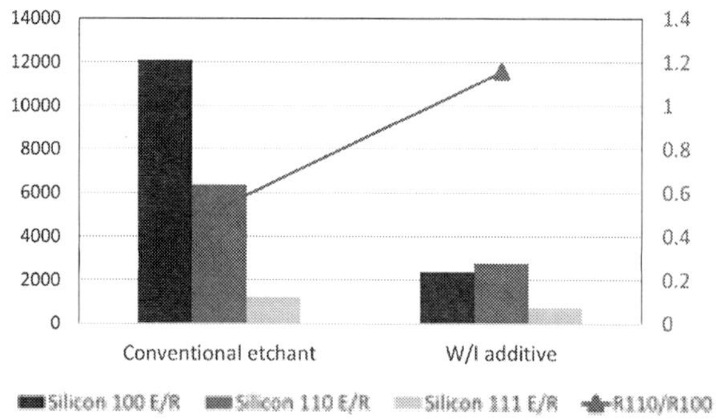


W/I Additive

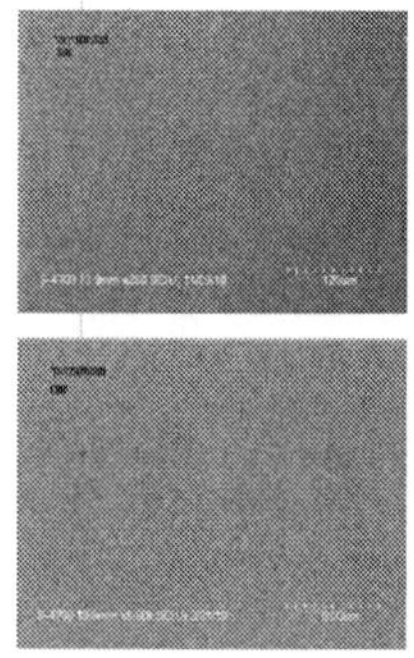

- Silicon roughness and pyramidal defect can be eliminated by suppression silicon etching rate and increasing Si(110) to Si(100) ratio.

VERSUM MATERIALS

## Additive effect : AlN etching performance

- Additive effect on AlN etching performance

| Solvent effect | E/R(A/min) | | | | | | Etching selectivity |
|---|---|---|---|---|---|---|---|
| | AlN | Si(100) | Si(110) | Si(111) | SiO2 | SiN | AlN/Si(100) |
| Requirements | >15000 | <3000 | <3000 | <1000 | <30 | <1 | >5 |
| Conventional etchant | **18241** | 12074 | 6367 | 1989 | 23.8 | 0.43 | 1.5 |
| Conventional etchant +additive | **18310** | **2397** | **2771** | **768** | 19.0 | 0.15 | **7.7** |

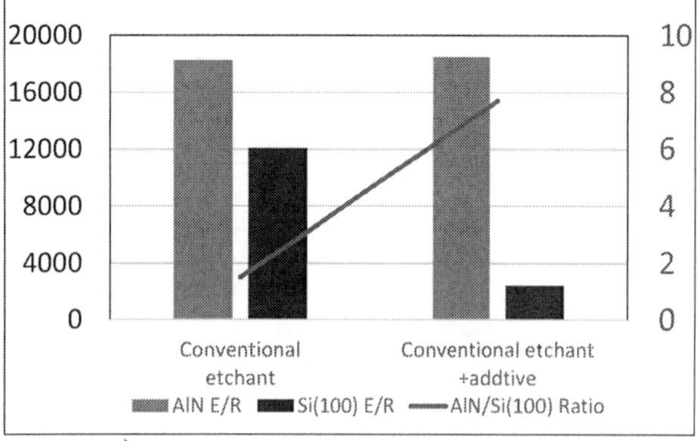


- AlN e/r is kept at high level and Si e/r is suppressed resulting in high AlN/Si etching selectivity by adding additive.
- Better compatibility with silicon oxide and silicon nitride

VERSUM MATERIALS

# Summary

➢ For the studies of Versum Material's novel AlN etchant, we demonstrate the following achievements:

✓ High AlN etching rate and high etching selectivity of AlN/Si

✓ Etching selectivity of Si (110)/ (100) and Si (111)/ (100) reach 1.5 and 0.3 respectively, which results in roughness free on silicon surface

✓ Better compatibility with $SiO_2$ and SiN

✓ Applicable on wet bench tool

VERSUM MATERIALS

14

DuPont Electronics & Imaging

# Material Solution toward Efficient Polysilicon Post-CMP Clean

*Jhih-Fong Lin*, Kathryn Gramigna, Paul Bernatis, Ling Chang, Adam Tang, and Chi Yen*

** jhih-fong.lin@dupont.com*

## 1. Introduction

Polycrystalline silicon is a material important in various applications, such as polysilicon contacts, capacitors, gates and floating gates. In these applications, chemical mechanical planarization (CMP) is utilized to achieve surface smoothing and/or global planarization for subsequent manufacturing. In general, the surface of polysilicon is predominately hydrogen-terminated and hydrophobic after CMP processing. Owing to the incompatibility with the surface, it is difficult to remove abrasive particles from the polished surface in the post-CMP (PCMP) cleaning process using a standard aqueous-based cleaner.

## 2. Fundamentals

### (i) Excellent wetting capability enables characteristics:

- "Flowability" to penetrate the interface between polySi surface and abrasive
- Improved organic residue removal

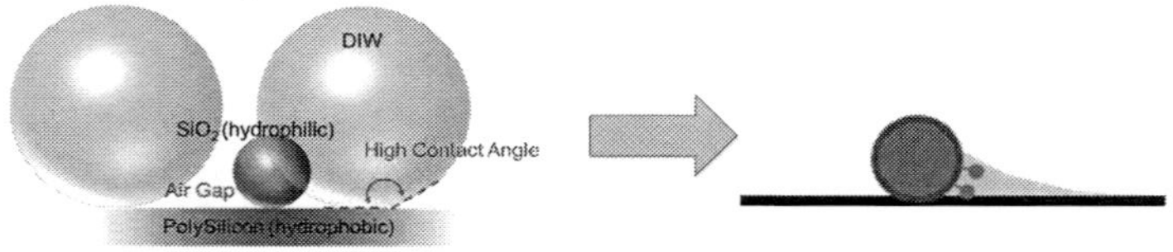


### (ii) Impacts of particle-polySi interaction can be derived from:

- Adhesion force of abrasive to polySi surface
- Electrostatic passivation/attraction (surface charge)

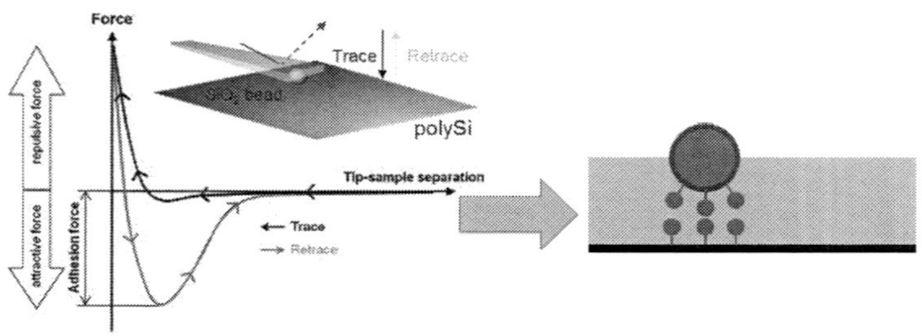


## 3. Results

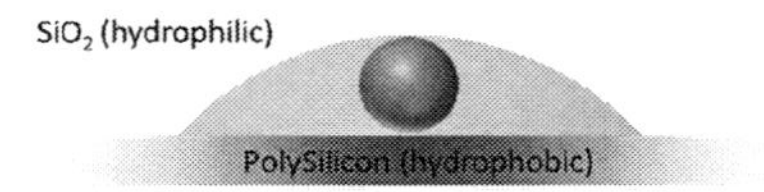


| Static contact angle on polySi | |
|---|---|
| DIW | EKC-477 |
| 75° | 15° |

**Figure 1. Enhanced wetting capability of EKC PCMP477 over DIW.**

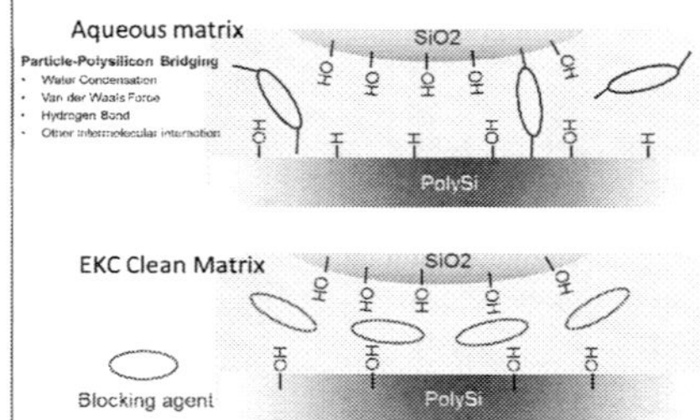


| Adhesion force | | |
|---|---|---|
| DIW | EKC-477 (w/o blocking agent) | EKC-477 |
| 11.6 nN | 9.8 nN | 0.5 nN |

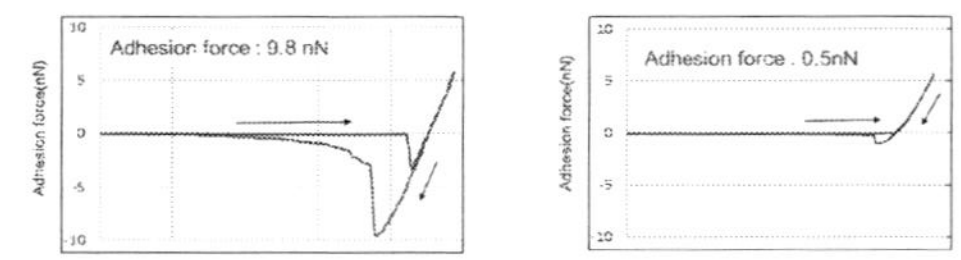


**Figure 2. EKC PCMP477 with blocking agent gives better particle inhibition capability during post-CMP cleaning process.**

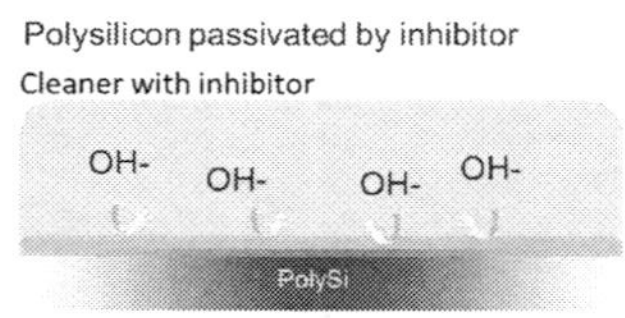


| Polysilicon loss in cleaner | |
|---|---|
| EKC-477 (mild alkaline) | $NH_4OH$ |
| <1 Å/min | 42 Å/min |

**Figure 3. EKC PCMP477 provides comprehensive polysilicon protection during post-CMP cleaning process.**

## 4. Conclusions and Future Applications

Currently, the cleaning process for polysilicon CMP is predominantly done by alkaline commodity chemicals such as $NH_4OH$ or SC1. As demonstrated by our in-house study, these chemicals can only enable electrostatic passivation (repulsion) between the abrasive particles and the polysilicon surface. Alternatively, through formulation design, our EKC formulated cleaners can easily reach the criteria listed above and outperform the commodity chemistry in practical PCMP cleaning.

DuPont Electronics & Imaging

# Advance of Oxide Post-CMP Clean Process: A Total Solution toward Superior Clean Performance

Jhih-Fong Lin*, Kathryn Gramigna, Paul Bernatis, Ling Chang, Adam Tang, and Chi Yen

* jhih-fong.lin@dupont.com

## 1. Introduction

Oxide (silica) removal is one of the most time- and cost-consuming applications in FEOL chemical mechanical planarization (CMP) processing. Recently, positively-charged silica has emerged as an advantageous direction in CMP processing with a lower cost of ownership. With tailored surface engineering on the abrasives, this kind of slurry can greatly enhance the interaction between the abrasive and the surface to be polished, eliminating solid content at point of use. However, the enhanced affinity (and thus removal rate) of positively-charged silica for the polished surface makes post-CMP cleaning challenging, and commodity chemistry-based cleaners are ineffective at addressing this issue.

## 2. Technical Challenge

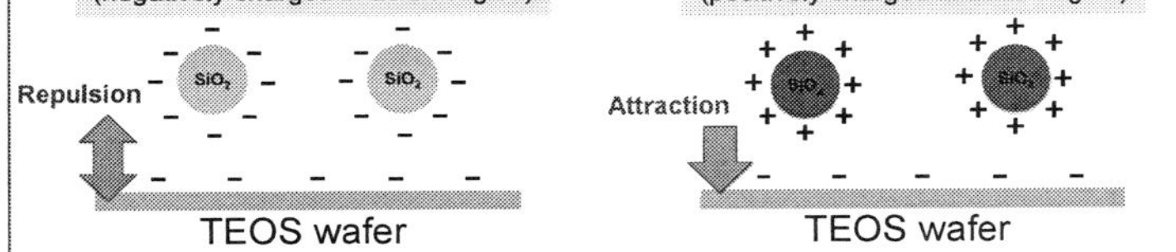


**Figure 1.** Interaction of different silica abrasives with TEOS surface.

## 3. Cleaning Principle

**Strong repulsive force results in excellent performance**

EKC-5710 induces a strong electrostatic repulsion between the TEOS/SiN surface and the slurry particle.

Surface charge Z.P. on various substrates [mV]

| Substrate | TEOS | SiN |
|---|---|---|
| Surface charge Z.P. [mV] | +16 | +34 |

Surface charge Z.P. on abrasive [mV]

| Particle | (Positively-charged colloidal SiO2 abrasive)* |
|---|---|
| Surface charge Z.P. [mV] | +3.5 |

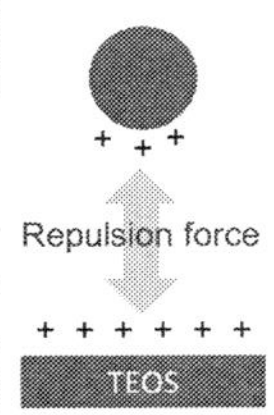


## 4. PCMP Cleaning Evaluation

(i) **Great material compatibility with dielectric surface**
(ii) **No organic residue/contamination**
(iii) **Broadened cleaning window versus commodity chemistry**

EKC PCMP5710 provides a total solution to PCMP processing of acidic/positively-charged silica slurries.

(a)

| | Cleaner | TEOS | SiN | PolySilicon |
|---|---|---|---|---|
| Compatibility (Å/min) | dHF (0.5%) | 63 | 3 | <1 |
| | EKC PCMP5710 | <1 | <1 | <1 |

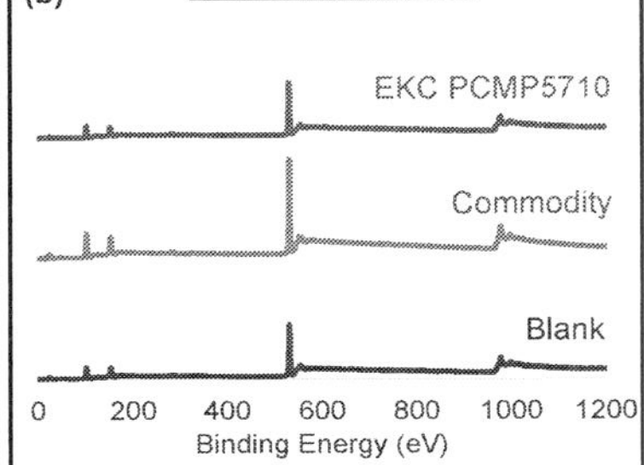


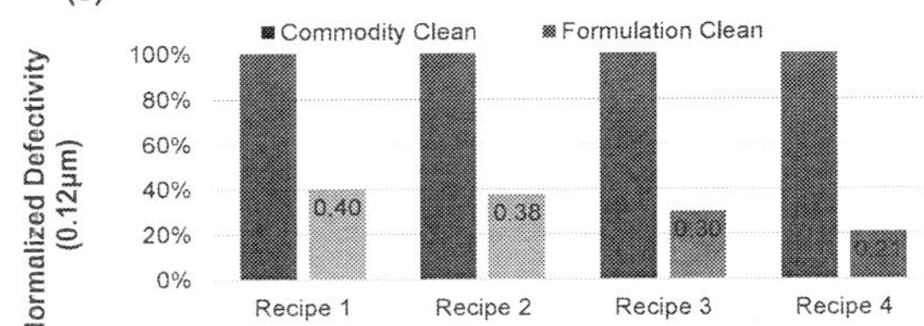


| Clean Configuration* | Mega | Brush 1 | Brush 2 |
|---|---|---|---|
| Commodity Clean | Alkaline | dHF | Alkaline |
| Formulation Clean | Alkaline | EKC PCMP5710 | Alkaline |

(i)Tool: AMAT; (ii) Inspection: KLA-SP2 (iii) Target: TEOS/HDP; (iv) Commerical competitive slurry

**Figure 2. (a)** Material compatibility of commodity chemistry and EKC PCMP5710; **(b)** Surface analysis of wet-cleaned TEOS wafer; **(c)** PCMP cleaning performance on positively-charged silica CMP.

## 5. Conclusions

Positively-charged silica CMP has been widely applied for interlayer dielectric (ILD) and shallow trench isolation (STI) processing, as the acidic silica abrasive provides higher TEOS removal efficiency. However, the electrostatic attraction poses a technical challenge in PCMP cleaning. Incorporating tailored surface engineering on the positively-charged silica and TEOS surface, EKC PCMP5710 enables a strong repulsive force and prevents particle redeposition during the PCMP cleaning process. In addition, EKC PCMP5710 provides both promising dielectric compatibility without organic residue/contamination left on the cleaned surface and a significantly broadened working window compared with commodity chemistry. EKC PCMP5710 substantially reduces defectivity, demonstrating a significant development toward the low cost of ownership for the FEOL CMP process.

DuPont Electronics & Imaging

# Approach toward Enhanced Cerium Residue Control on Post-CMP-Cleaned Surface

Jhih-Fong Lin*, Kathryn Gramigna, Paul Bernatis, Ling Chang, Adam Tang and Chi Yen

* jhih-fong.lin@dupont.com

## 1. Introduction

Owing to the push of the semiconductor industry toward advanced technology nodes, ceria-based slurries have been widely applied to achieve rapid removal rates of silicon oxide (TEOS) in front-end-of-line (FEOL) chemical mechanical planarization (CMP) processing. In addition to the relatively strong chemical bonding with the oxide surface, the opposite charge and smaller size of ceria abrasives make the post-CMP cleaning of ceria-based slurries a greater challenge than that of traditional silica-based CMP slurries. As a result, formulation chemistry has emerged in the market, gradually replacing commodity chemistry, such as dilute HF (dHF), SPM, and SC1, in both logic and memory applications.

## 2. Formulation vs. Commodity

Bulk chemicals are used to remove ceria residue by breaking the bonds between ceria and dielectrics. However, most commodity chemical-based cleaners have limitations such as compatibility issues with dielectrics and tungsten as well as safety concerns. In addition, ceria post-CMP cleaning relies on some aggressive matrixes such as SPM, which is not suitable for an in-line clean system and consequently will lower the overall production throughput. Alternatively, the use of EKC PCMP2110 (formulating chemistry) provides great materials compatibility and is also suitable for an in-line post-CMP clean system without safety concerns.

| Cleaner | Commodity | | | | | Formulation |
|---|---|---|---|---|---|---|
| | Ammonia | TMAH | SC1 | SPM | HF | EKC PCMP2110 |
| Dielectric Compatibility | Poly-Si loss | Poly-Si loss | Poly-Si loss | - | TEOS loss | O |
| Tungsten Compatibility | X | X | X | X | X | O |
| Safety | Hazard | Hazard | Hazard | Hazard | Hazard | O |
| Direct Clean | O | O | O | EX-Situ | O | O |

## 3. Inspection Methods

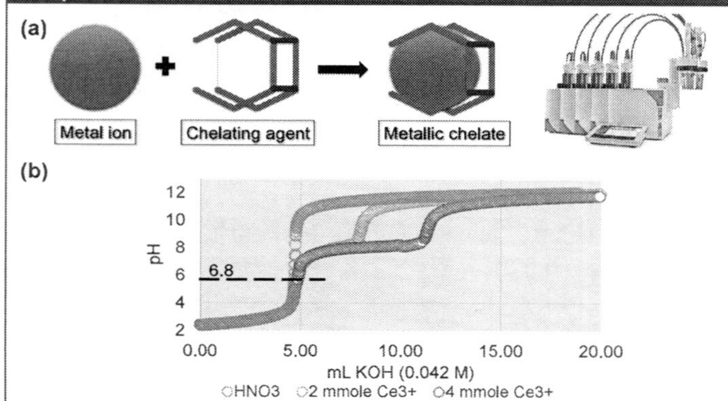


**Figure 1. (a)** Titration-based technique as the chelator screening methodology and **(b)** hydrolysis of $Ce^{3+}$, which explains the decreasing complexation for some chelators at pH above 7.

## 4. Results

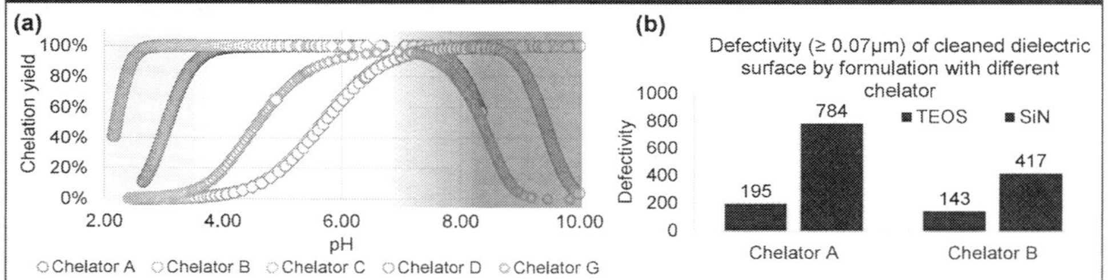


**Figure 2. (a)** Complexation yield of chelator upon $Ce^{3+}$ ion in aqueous matrix, chelator B with broaden working window outperforms other complexation agents; **(b)** Template with chelator B gives better cleaning performance than that of chelator A.

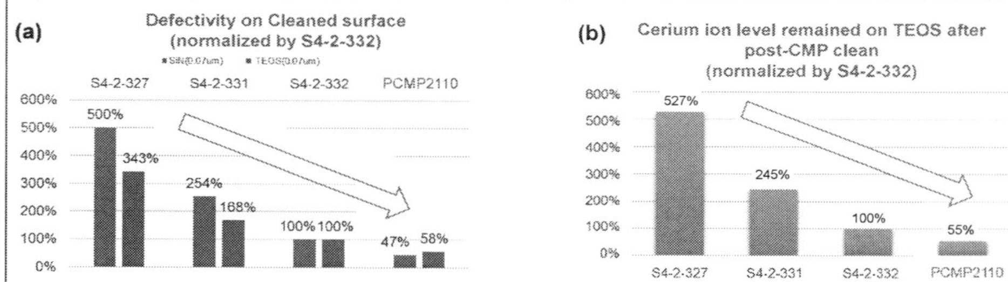


| Template modification | S4-2-327 | S4-2-331 | S4-2-332 | PCMP2110 |
|---|---|---|---|---|
| Chelator Template #1 | 1X | | | |
| Chelator Template #2 | | 1X | | |
| Chelator Template #3 | | | 1X | |
| Chelator Template #4 | | | | 1X |

**Figure 3.** Ceria residue control after post-CMP clean in brush box: (a) defectivity and (b) cerium ion level.

## 5. Conclusions and Future Applications

- EKC PCMP2110 (formulating chemistry) provides great materials compatibility and is also suitable for an in-line post-CMP clean system without safety concerns, making it a viable replacement for commodity chemicals in ceria post-CMP cleaning.
- Voltammetry titration provides valuable information for PCMP processing, such as the working window of complexation in cleaner matrix.
- Chelator B has the most broadened complexation window in the tested pH rage and shows the greatest cleaning capability.
- By optimizing the chelant, EKC cleaners can achieve improved cerium residue removal as manifested in both defectivity and cerium ion level.
- EKC PCMP2110 performs state-of-the-art ceria post-CMP cleaning, which has been validated by a major chip maker.

Entegris

# Understanding Interfacial Surface Interactions

Daniela White, Thomas Parson, Ruben Lieten, Bianzhu Fu, and Michael White

SPCC April 2019



## Challenges for W Post-CMP Cleaners

Slurry particles and organic residue removal from W and dielectric surfaces (PETEOS, Silicon Nitride, Polysilicon);

Metal residue in any form (Ions, Salts, Metal Oxide)

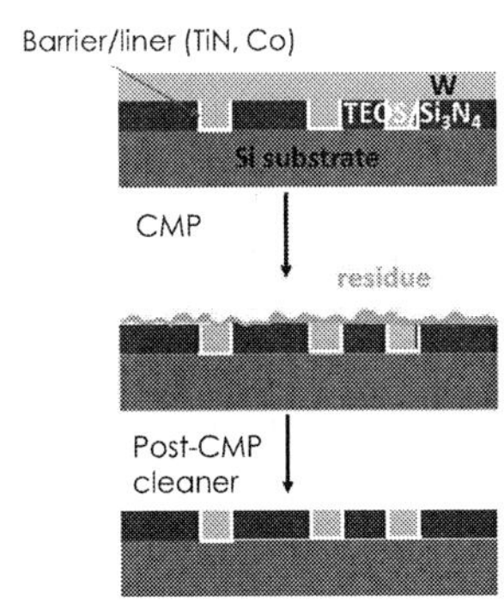


Cleaning Requirements:
- W ER < 1 Å/min
- TiN ER < 1 Å/min
- Dielectrics ER < 1 Å/min
- Dielectrics: $Si_3N_4$, TEOS, SiC, etc.
- Defect counts DDC ≥ 0.065 mm lower than commodities: dAmmonia, SC-1
- Low W/TiN galvanic corrosion
- Mt atoms < $10^{10}$ Mt/cm$^2$

- No increased roughness
- Market increasingly challenged by W recess
  - High pH commodities (SC1, dil $NH_3$)
  - Traditional low pH cleaners
- Low W etch rates (<2 Å/min) cannot be achieved with commodity cleaners
- No organic Residue
  Nitride cleaning is particularly problematic
- No silica particles or clusters
- Green chemistry (TMAH free)

Entegris

2

# Post-CMP W Cleaning Mechanisms vs. pH

Entegris

**W CMP residue: silica, Mt oxide, organics**

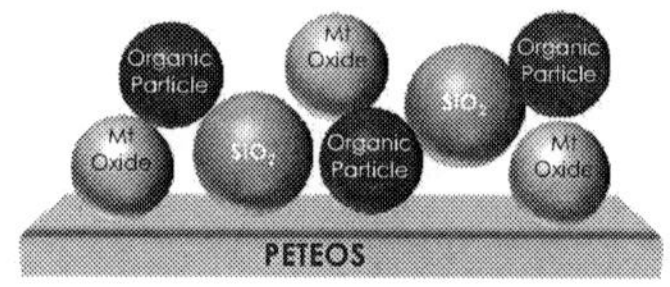


**Low pH**

- Silica brush imprints
- Good Mt removal
  ( ~$10^{10}$ atoms/cm$^2$)

**High pH**

- No Silica brush imprints
- Poor Mt removal
  (4-6 X$10^{10}$ atoms/cm$^2$)

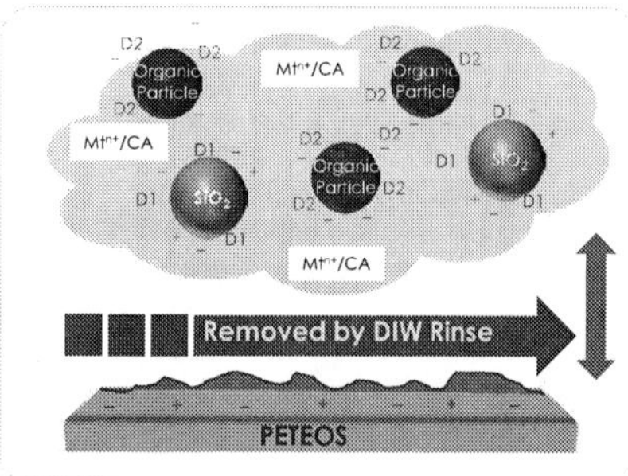


CA = Mt complexing agent
D1 = $SiO_2$ dispersant
D2 = organic residue dispersant

steric repulsion

electrostatic repulsion

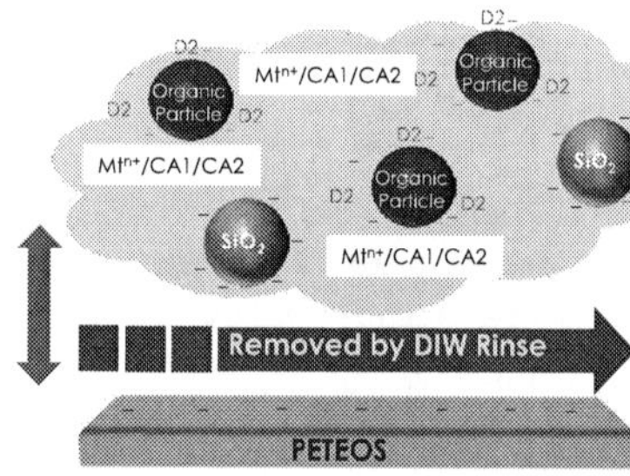


3 | CONFIDENTIAL

# Higher Tungsten Etch Rates Observed

Entegris

Increasing pH due to dissolution as Polyoxotungstate Keggin ions

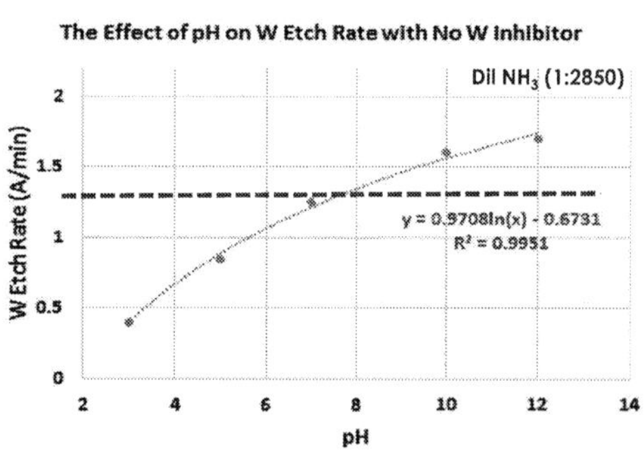


SC1: 1:50 > 6 A/min

1

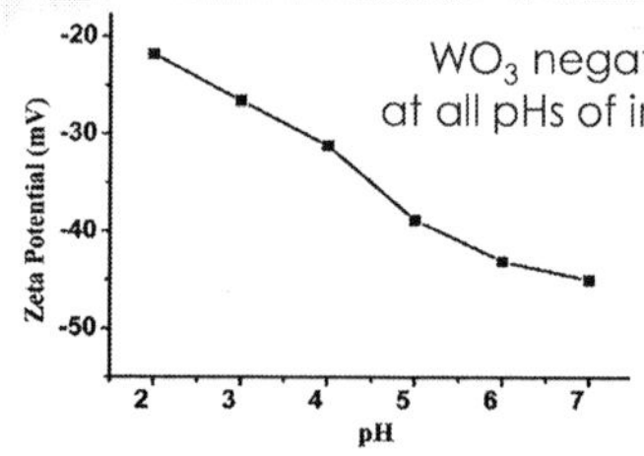


$WO_3$ negative at all pHs of interest

2

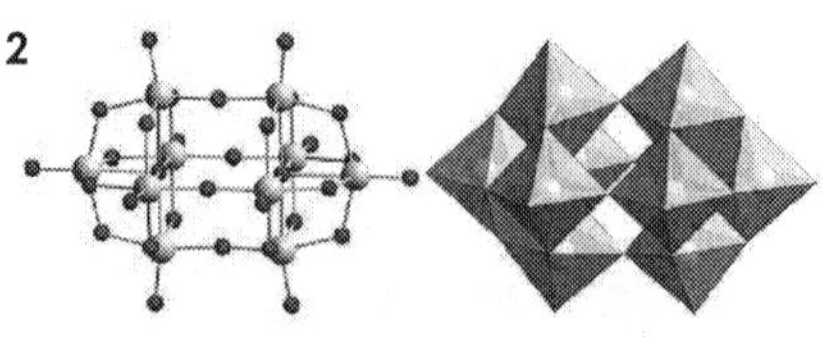

**Decatungstate ($[W_{10}O_{32}]^{4-}$)**

1 - Liu, et al. J. Mater. Chem A, Issue 6, 2014.
2 - "Hetero and lacunary polyoxovanadate chemistry: Synthesis, reactivity and structural aspects". Coord. Chem. Rev. 255 2270–2230, 2011.

4

# Surface Residue PCMP Cleaning with W Slurries[1]

Entegris

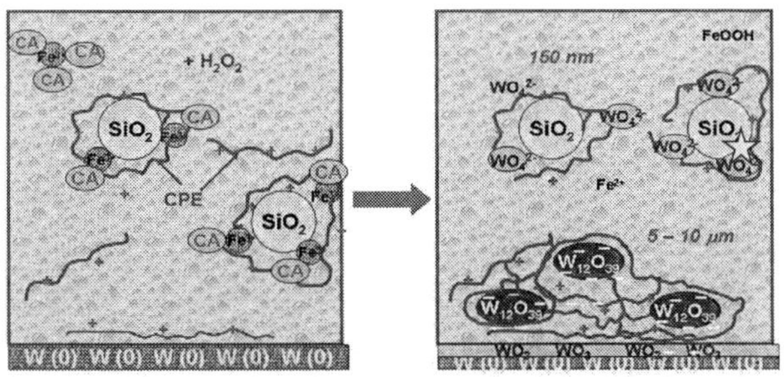


**Nanodispersion before polishing** **Dispersion during/after polishing**

- **W(0) is oxidized to W(VI)$O_3$ (pp) and W(VI)$O_4^{2-}$ (s);**
- **$WO_4^{2-}$ (s) is adsorbed on the surface of the $Fe^{3+}$/CPE coated, positively charged Si particles;**
- **$WO_4^{2-}$ precipitates @ pH < 3 as $(N^+R_4)_n(H_xW_qO_y)^{n-}$/silica aggregates that act as *additional abrasive* during polishing.**

**W CMP slurry – 15-32 ppm Fe**

- $Fe^{2+}$ /CA (complexing agent) on the silica surface and in the CMP slurry at pH = 2.5;
- End of polishing, soluble $Fe^{2+}$ complexes on the wafer surface;
- DI water rinse and high pH post-CMP W cleaners **precipitate $Fe^{3+}$ as insoluble $Fe(OH)_3$** on the PETEOS/SiN surface

↓

**FeOx Very difficult to remove**

1 - D. White et al, Materials Research Society Symposium Proceedings (2007), 991 (Advances and Challenges in Chemical Mechanical Planarization), 145-150



# Improving Organic Residue Removal

## from $Si_3N_4$ Contact Angle and FTIR

Electrostatic Repulsion during CMP

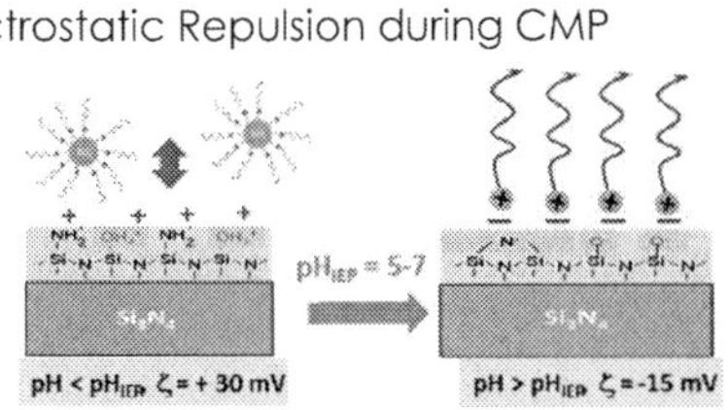


W Post-CMP Cleaner

$Si_3N_4$ surface typically highly contaminated by cationic dishing and erosion control agents

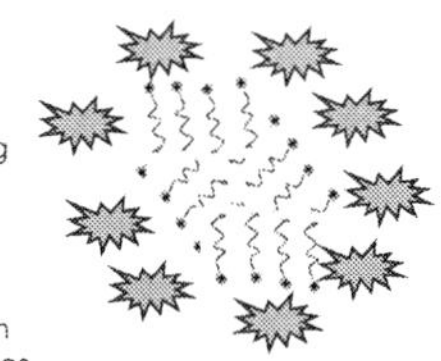

Cleaning additive removes cationic contamination from dielectric surface and disperses

Contact Angle

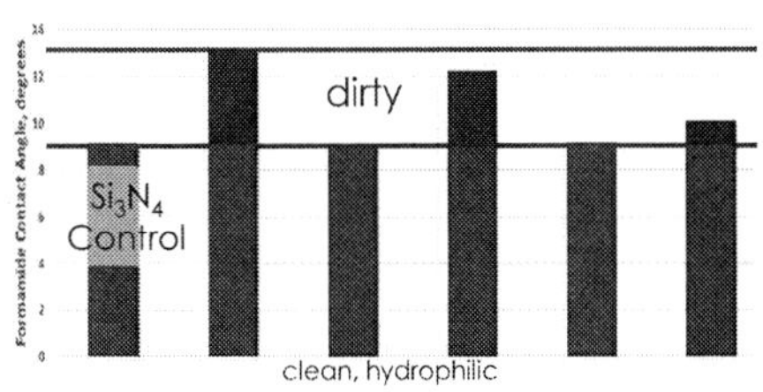


FTIR

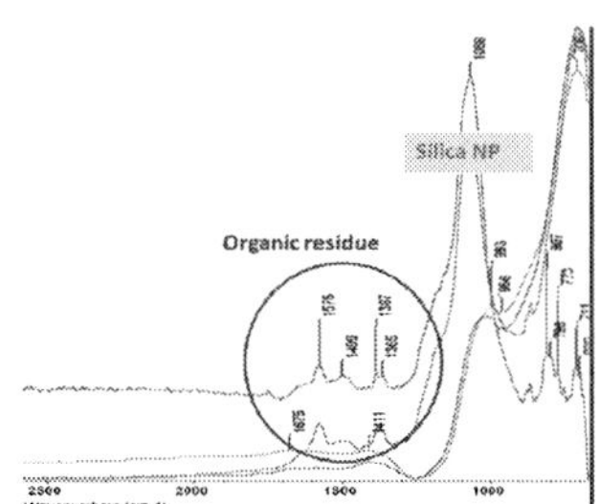


Entegris

# Defectivity Correlated to Charge Repulsion

## Between Silica Particles and Various Surfaces (W, $SiO_2$, $Si_3N_4$)

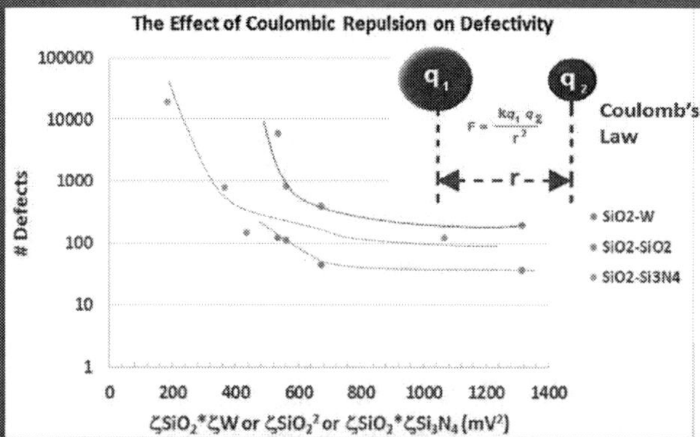


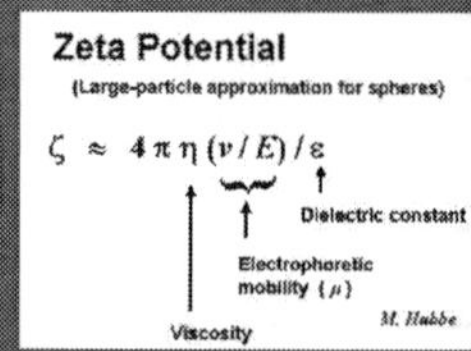


Additive increases negative charge on silica surface

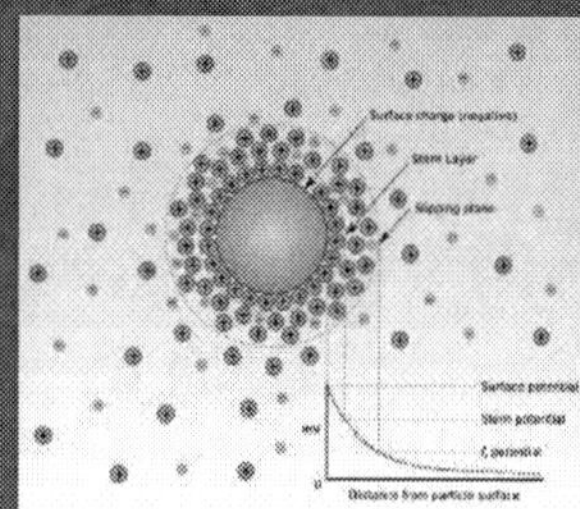


1 - White, M. L. et al, *Mater. Res. Soc. Symp. Proc.* 991, 0991-C07-02 (2007)
2 - Hedge, S. and Babu, *H. V. 2Eelectrochem. Soc. St. Lett.* V7, pp. 316-318 (2008)
3 - White , M. L. et al. *Mat. Sc. For.* 1249 E04-07 (2010).




# PlanarClean® AG-WXXX Exhibits

## No bulk corrosion and no OCP gap between W and TiN

W Control

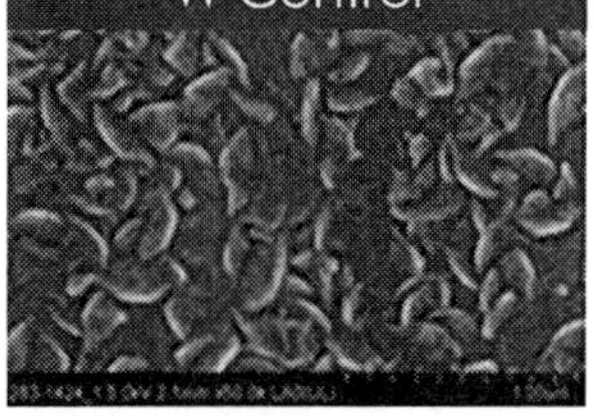

Post PC AG-WXXX

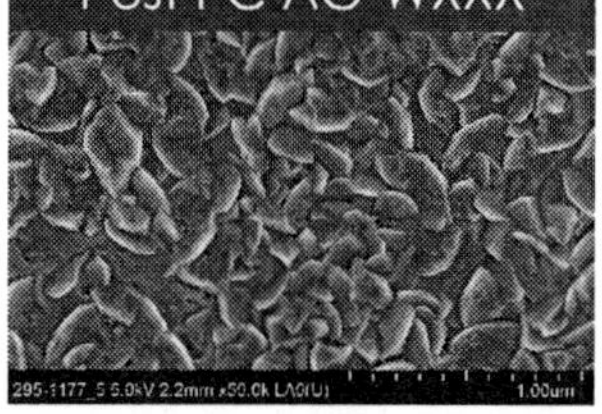

No Visible Etching of Tungsten

Etched W Surface

Post PC AG-WXXX

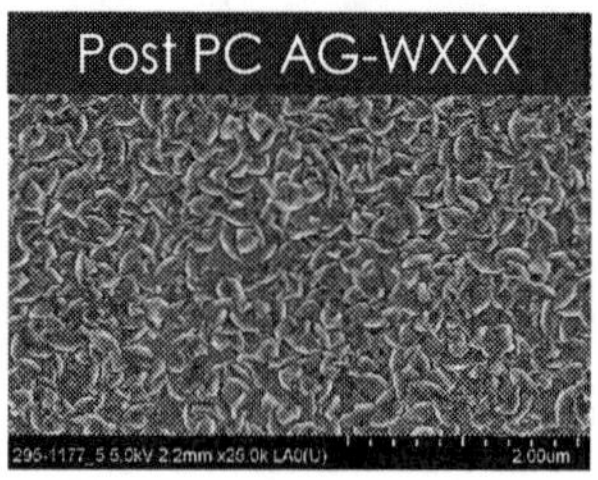

WXXX

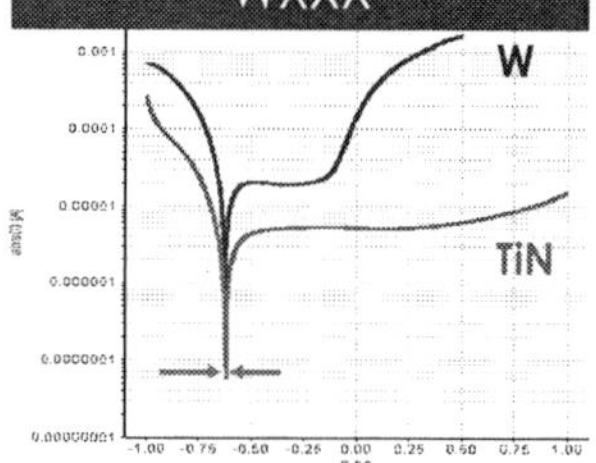


| | TiN | Galvanic Corrosion | W |
|---|---|---|---|
| | Cathode | | Anode |
| Corrosion Potential (V) | -0.622 | -0.626 | -0.625 |
| Corrosion Current (A) | 1.76 × 10-06 | 2.35 × 10-07 | 1.13 × 10-05 |
| Corrosion Rate (Å/min.) | 0.167 | 0.013 | 0.642 |
| Galvanic Corrosion (Å/min.) | 0.013 | | |

# Low pH PCMP W Cleaners

## Further reduce metal defects after CMP

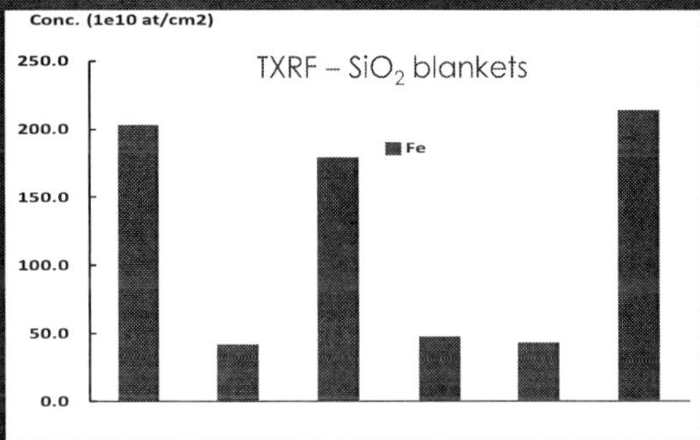


- Complexing reagents: to keep in solution soluble $Fe^{2+}$ species
- Cleaning additives: able to eliminate silica brush marks at pH = 2

Entegris



# Force of Adhesion – AFM Measurements

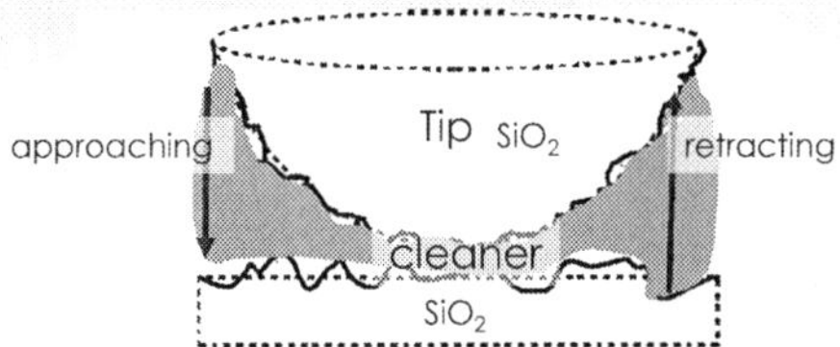


$$dF = k_c\,dZ_c + \frac{dF_{surf}}{dD}\,dD = \left(k_c - \frac{dF_{surf}}{dD}\right) dZ_c$$

**$F$ = force of adhesion**
**$k_c$ = spring constant of the cantilever**
**$Z_c$ = cantilever deflection**
**$F_{surf}$ = distance-dependent surface force**
**$D$ = separation distance**

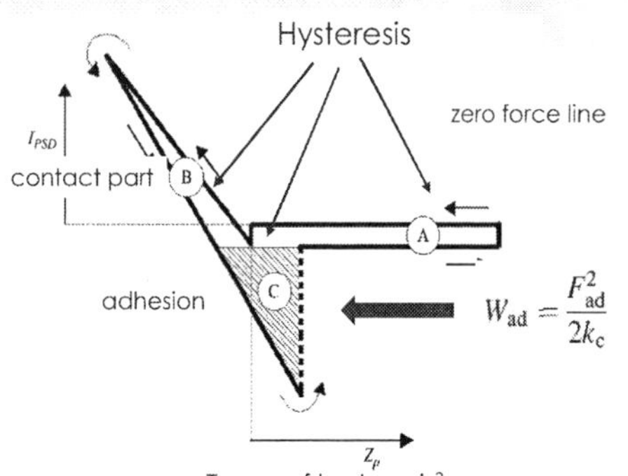


Types of hysteresis[3]

$I_{PSD}$ = detector current signal
$Z_p$ = height position of the piezoelectric translator

3. Hans-Jurgen Butt, Brunero Cappella*, Michael Kappl, Surface Science Reports 59 (2005) 1–152

Entegris

# Example of Cleaning Performance Prediction

via AFM force of adhesion measurements

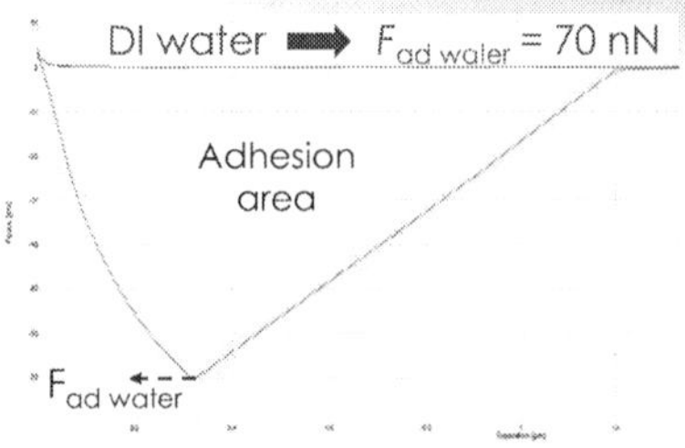


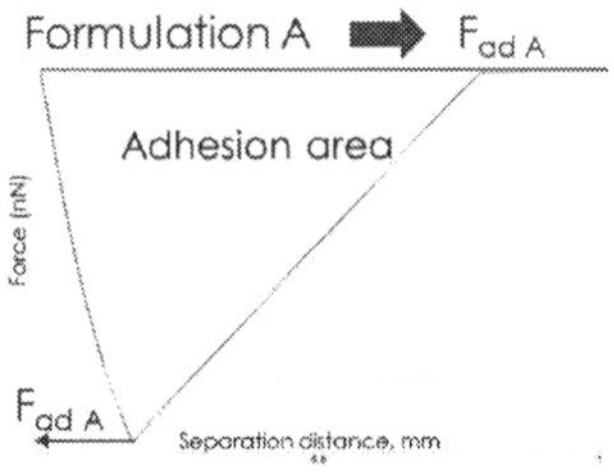


+ additive X

Formulations A & B, pH = 2

Main adhesion forces:

- Van der Waals
- Electrostatic double layer
- Hydration repulsion
- Hydrophobic attraction

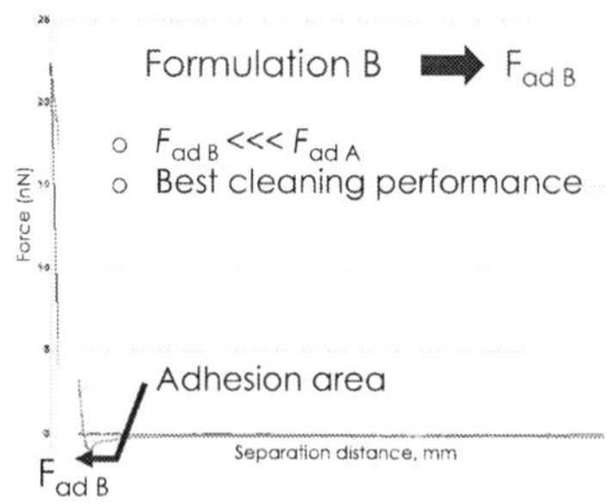


- $F_{ad\ B}$ <<< $F_{ad\ A}$
- Best cleaning performance

11 | CONFIDENTIAL

# SEM Images of PETEOS Coupons

Polished with W CMP slurry and cleaned with Formulations A and B

- Tabletop polishing
- Colloidal silica Ludox, PS = 20-30 nm
- pH = 2.3

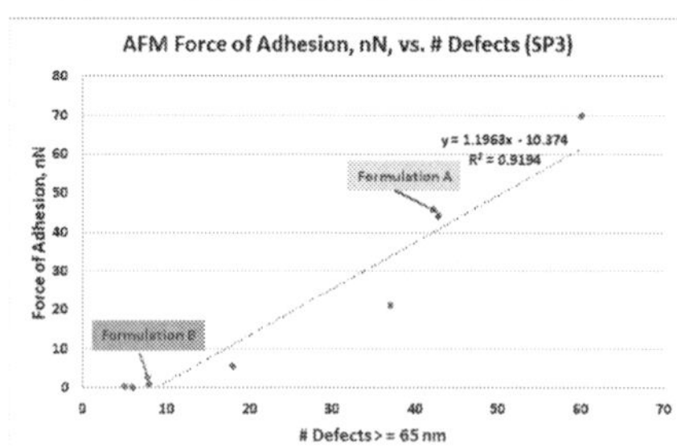


12 | CONFIDENTIAL

# Correlation SEM vs. Calculated Adhesion

Based on contact angle measurements

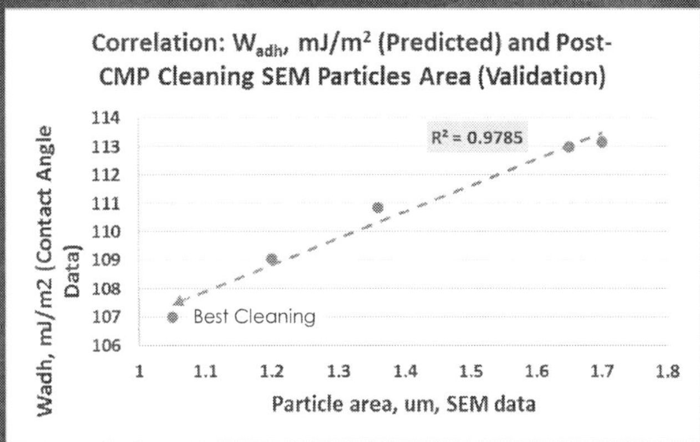


13 | CONFIDENTIAL

# PlanarClean AG-W formulations

exhibit lower defects and organic residues over traditional cleans

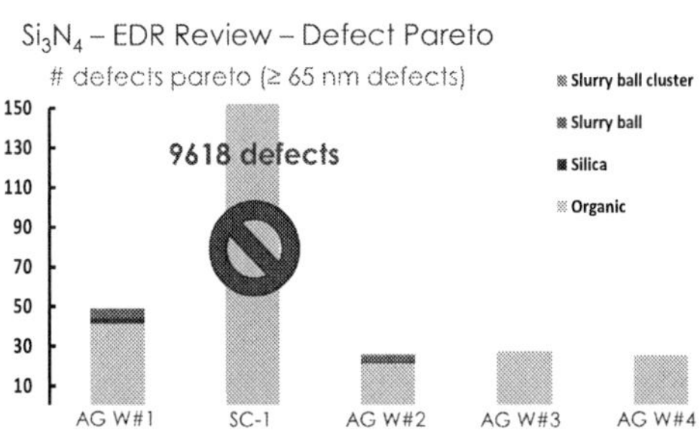


PC AG-W Series show improved performance over SC-1 on all substrates

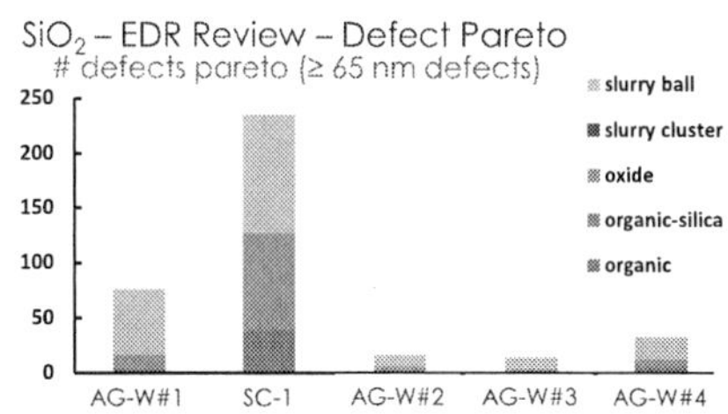


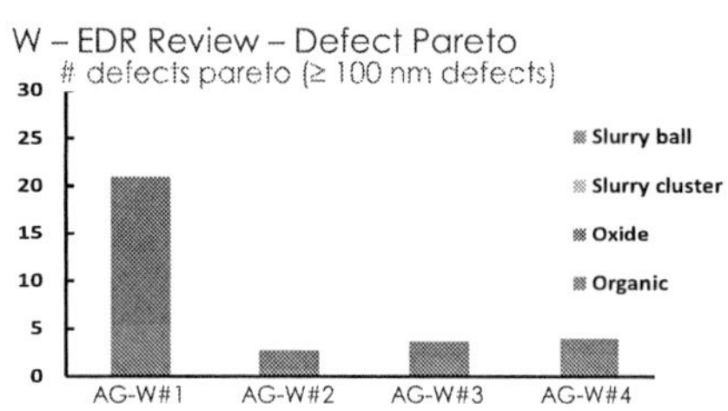


14

## Summary

Entegris

Organic contamination reduction studied by FTIR and contact angle

Work of adhesion and AFM force of adhesion show that less energy is needed to remove silica slurry particles in the presence of the cleaner

Entegris AG-W Formulations show significantly lower defects on $Si_3N_4$, TEOS and W wafers compared to commodity cleans

15 | CONFIDENTIAL

## Thank you!

Entegris

16 | CONFIDENTIAL

# AUTHOR INDEX

**Surface Preparation and Cleaning Conference (SPCC)**
c/o Mark Thirsk
Linx Consulting-PO Box 384
Mendon, MA 01756-0384
USA

ISBN 978-1-5108-8502-8

CURRAN ASSOCIATES INC.
proceedings.com